P9-CWD-945

THE OXFORD BOOK OF ROYAL ANECDOTES

THE OXFORD BOOK OF ROYAL ANECDOTES

Edited by

ELIZABETH LONGFORD

Oxford New York

OXFORD UNIVERSITY PRESS

1989

Oxford University Press, Walton Street, Oxford OX2 6DP
Oxford New York Toronto
Delhi Bombay Calcutta Madras Karachi
Petaling Jaya Singapore Hong Kong Tokyo
Nairobi Dar es Salaam Cape Town
Melbourne Auckland

and associated companies in
Berlin Ibadan

Oxford is a trade mark of Oxford University Press

British Library Cataloguing in Publication Data
The Oxford Book of royal anecodotes.
1. Great Britain. Royal families, history
I. Longford, Elizabeth, 1906–
941'.009'22
ISBN 0–19–214153–8

Library of Congress Cataloging-in-Publication Data
The Oxford book of royal anecdotes.
Includes bibliographical references and index.
1. Great Britain—Kings and rulers—Anecdotes.
2. Great Britain—Princes and princesses—Anecdotes.
3. Monarchy—Great Britain—Anecdotes. 4. Great Britain—History—Anecdotes. 5. Anecdotes—Great Britain. I. Longford, Elizabeth Harman Pakenham, Countess of, 1906–
DA28.1095 1989 941'.009'92 88–25496
ISBN 0–19–214153–8

Typeset by Wyvern Typesetting Ltd.
Printed in the United States of America

TO FRANK

Contents

CONTENTS

Introduction

My object in this introduction is to suggest the special point of royal anecdotes. At the same time, royal anecdotes have much in common with other versions of this type of entertainment. Royal anecdotes, like all others, must make some attempt at 'punch'. Ideally they should be attached to an event or happening.

King William III was said to have been too small to offer his arm to his massive wife Queen Mary. Instead he dangled from hers 'like an amulet from a bracelet'. One simile does not make an anecdote. The simile about the amulet, however, provides a good analogy for the royal anecdote and its event.

The anecdote should hang like an amulet from the arm, so to speak, of the greater event. That event in turn will be attached to an even larger and more complex body—the life of the nation of which the royal house is one part. But that is venturing too far into the realm of history. This book is not history, merely a whole series of small amulets hanging from one powerful arm.

Almost everything here will be found reduced to the concrete, the individual. No general effects; no causes of the Civil War or the Industrial Revolution, except indirectly, as in someone's diary.

Moreover, the individual anecdote is usually economical, giving it point and pith. If it is encapsulated in a neatly tailored and recognizable shape, all the better. It will often be found to begin with the words 'Once' or 'One day'. Long anecdotes are relatively rare; only occasionally the story and language genuinely need to spread themselves, as for instance in the writings of that great pen-orator, Macaulay.

These restrictions can apply equally to anecdotes of all kinds. But there is one type of freedom which other collections of anecdotes can enjoy but which royal anecdotes must forgo—and I am not referring to the freedom of indiscretion.

Royal anecdotes form part of an immense narrative history. These kings and queens are descended one from another and their story hangs together in a special sense. It is impossible to think of William and Mary without thinking of James II and Queen Anne, strung above and below on the same arm. Of course, with, say, dramatic anecdotes you could rightly argue that Marlow influenced Shakespeare and Beckett influenced Pinter. All the same, it would not be necessary to hook them too closely together.

The linking thread of chronology would be enough. Apart from that restraint, literary and other anecdotes can enjoy a certain carefree independence.

With my royal anecdotes I have needed not only chronology but also captions. Many of the latter take the form of quite long editorial notes, the penalty of trying to make my method approximate as closely as possible to history without actually assuming history's mantle. That brings me to the contents of royal anecdotes.

Ideally again, every anecdote should have a touch of something that is either 'funny/ha, ha' or 'funny/peculiar'. It will be found that the earlier kings and queens provoke more laughter of the 'funny/peculiar' kind. This sometimes slips into astonishment or even horror. That was indeed the wish of many of the chroniclers. When we reach the stories about later kings and queens, we find that wit and humour have developed alongside everything else, so that more anecdotes belong to the 'funny/ha, ha' type.

Pleasant descriptions, however skilfully drawn but which contain nothing unexpected, must be carefully watched. A coronation that went off exactly as the Lord Chamberlain hoped would not make good anecdotage. Conversely, something out of the way will bring it into the anecdotal class: it may be an unusual viewpoint, say that of a child, or something going wrong.

Things that go wrong are always candidates for royal anecdotes, whether they are mildly disturbing, utterly ridiculous, or profoundly tragic. A slide on banana skin is inherently more anecdotal than a stately procession; the foul murder of Edward II than the calm demise of Edward IV, though both can qualify. That is why royal anecdotes, together with political, legal, military, *et al.*, tend to give the impression that kings, politicians, judges, and the top brass were all potential disaster areas.

Because royal anecdotes are so closely allied to history they may also be informative. (I shall illustrate this and other points from anecdotes that do not happen to be included in the text.) We are informed by that splendid twentieth-century diarist, Sir Henry ('Chips') Channon that in 1935 Edward Prince of Wales was 'more American than ever', because of his liaison with Mrs Simpson.[1] Chips added that it did not matter 'since all the Royal Family except the Duke of Kent have German voices'. Elsewhere Chips told the story of Edward VII commenting on his niece Princess Marie-Louise's return home after divorcing her husband, the appalling Prince of Anhalt: 'Poor Marie-Louise. She came back just as she VENT'. Those two incidents tell us something, perhaps controversial, about the royal voices of a certain period.

[1] *Chips: The Diary of Sir Henry Channon*, ed. Robert Rhodes James (1967).

We may notice the ways in which entertainment of royal visitors has changed. It is conceivable that some of our modern visiting royalties would have preferred the conditions of a medieval court, where there would have been no bother with time limits to the huge feasts and stupendous days at the chase. Today a Buckingham Palace spokesman is more likely to name 'the usual things' on offer to royal guests as 'laying a wreath on the Tomb of the Unknown Warrior and having tea with the Queen Mother at Clarence House'. A nineteenth-century Shah of Persia, when he was taken to visit a famous English lunatic asylum, paid a more glowing tribute to his hosts there than he did to Queen Victoria herself, who disapproved of his having a boxing-match staged in her palace garden. His staff dined with the Queen off gold plate while the Shah himself insisted on eating with his fingers on his bedroom floor. He would have been altogether more at home with Henry VIII than at the Victorian court.

Fantastic anecdotes have their place in the royal collection. Here, however, the serious problem of 'Truth v. Myth' arises. How much legend is allowable? Lord Macaulay gives a list of the absurd rumours surrounding Charles II's death, and then goes on to justify his printing it:

> His Majesty's tongue has swelled to the size of a neat's tongue. A cake of dileterious powder had been found in his brain. There were blue spots on his breast. There were black spots on his shoulder. Something had been put into his snuff box. Something had been put into his broth. Something had been put into his favourite dish of eggs and ambergris. The Duchess of Portsmouth had poisoned him in a cup of dried pears. Such tales ought to be preserved; for they furnish us with a measure of the intelligence and virtue of the generation which eagerly devoured them.

For the same good reason, many incredible tales are included in this collection.

William of Malmesbury, one of the twelfth-century chroniclers, had a different reason for telling an unlikely tale. Sometimes he felt it his duty to do so provided he warned his readers about its dubious provenance. Thus he described the death by blinding and starvation of the Anglo-Saxon Alfred, eldest son of Ethelred the Unready, at the hands of Earl Godwin, ending with the words: 'I have mentioned these circumstances, because such is the report; but as the chronicles [the sources then available to him, such as the Anglo-Saxon Chronicle] are silent, I do not assert them for fact.' Elsewhere William of Malmesbury is more specific. While writing of King Athelstan (d. 939) he had access to a strictly contemporary account which he had recently discovered but which his own sources had evidently not seen. Appreciating his advantage he added: 'Thus far relating to the king I have written from authentic testimony: that which follows I have learnt more from old ballads, popular through succeeding times, than from books written expressly for the information of posterity. I have subjoined them, not to defend their veracity, but to put my readers in possession of all I know.'

Incidentally, the anecdotes related by the monks who were also chroniclers did not always win the respect they do today. In 1741, the first Baron Lyttelton, author of *Henry the Second* (1767), wrote to his friend the poet Alexander Pope about William Warburton, bishop and scholar, who happened to be researching in Cambridge:

> If when he is at Cambridge he should find anything in the library there relating to Henry the Second or Becket, that may be of use to me, I will take the liberty to desire him to communicate it. It will be two or three years before my book is finished . . . [Twenty-six years before it was published.] Nothing remains to me but to endeavour to draw something like history out of the rubbish of monkish annals—a disagreeable task—but yet if I can execute it well, there are materials enough to make it a work of some instruction and pleasure to my countrymen . . . Certain I am, that such an architect as you, or Mr Warburton, could not of these Gossip ruins, rude as they are, raise a new edifice that would be fit to enshrine the greatest of our English Kings, and last to eternity.[1]

It is a fine clear statement from the Age of Reason though personally, while revelling in that remarkable passage, I should change the words 'rubbish' to 'delights' and 'disagreeable' to 'absorbing'; and I should be quite happy if this particular book of kings and queens lasted for something just short of eternity.

The monks were in fact no greater fabricators than those who came long after. They would have been astounded by most of the anecdotes emerging from the late twentieth-century tabloid press. Princess Margaret once told her biographer Christopher Warwick that her father King George VI proposed to keep two notebooks: one entitled 'Things My Daughters Never Did' and another on the subject of the things his daughters never said.

Most people are interested in knowing whether a contemporary anecdote is a case of genuine reporting or mere invention. Efforts have therefore been made, where it seemed appropriate, to indicate the authenticity of stories and the reliability of the sources used. Even Byron, who had plenty of imagination himself, made the requisite enquiries when his friend the poet Tom Moore sent him a funny story about his contemporary George IV: 'I delight in your "fact, historical,"' Byron wrote back, '—*is* it a fact?'

[1] *Correspondence of Alexander Pope* (1886).

Much can be learnt about the British monarchy from anecdotes, even if they would not all get into the history books. Certain positive and negative factors, for instance, recur. One can begin to recognize them and make judgements about them. The sovereigns who had favourites were generally 'baddies'—bad for themselves and bad for their favourites. 'When royalty comes in, friendship flies out of the window', wrote Chips Channon, quoting the hostess at a party given for royalty in 1936. Our Elizabeth II can be called a 'goodie' in the sense that she does not have favourites, or if she does, they are few and unflamboyant. We can group together in our minds the sovereigns who had favourites, or loved learning, or understood what it was to be a warrior nation. Elizabeth I, high-spirited as she was, encompassed all three.

In the end we want to know what it was and is like to be royal. This must include what it was like to live in a court. Though the eighteenth-century Fanny Burney (Mme d'Arblay) is among the best of writers about royalty, she did not take to her royal duties like a duck to water. Yet her bonds with the court were as tight as matrimony. 'I am *married*, my dearest Susan,' she wrote to her sister. 'I look upon it in that light—I was averse to forming the union, and I endeavoured to escape it; but my friends interfered—they prevailed—and the knot is tied. What then remains but to make the best wife in my power?' Sometimes Fanny feels more like a nun than a wife and speaks of the court as 'this monastery'.

The Duke of Wellington managed to remain detached yet devoted. 'Without personal attachment to any of the Monarchs whom he served', wrote Greville, 'and fully understanding . . . their individual merits and demerits, he alike reverenced their great office in the persons of each of them, and would at any time have sacrificed his ease, his fortune, or his life, to serve the Sovereign.'

Very few people could have moved about courts like that, always more affected by the sovereign's office than his person. Most people seemed to feel a kind of constraint in the presence of royalty, as described by Channon. 'There can be no doubt royalty casts a strange atmosphere. It makes many people self-conscious and either wish to thrust themselves forward or else become too self-effacing; both forms of behaviour are equally tiresome.'

Tiresome for whom? Chiefly for the monarch. The writers of royal anecdotes are often sensitive to the ambivalent feelings of kings. Lord Esher, for example, remembered the feeling of destitution experienced by both George V and Queen Mary on the night that Edward VII died. Queen Mary, of all people, 'rather clung, like a child for a moment, and said, "This is going to be a terrible time for us, full of difficulties. I hope you will help us." I could not speak. They seemed to me not a King and Queen but two poor storm-battered children.'[1] George IV had felt it even more. He told Lord Eldon 'that he wished he had continued to be only Prince of Wales, and had not ascended the throne . . .' Lord Eldon, after jotting down these words in his anecdote book, added: 'I remember Dr Bailey, who had seen a great deal of his Majesty George III, said it was a dreadful bad thing to be a Physician—but to be a King was a thousand times worse.'

Courtiers as well as kings were puzzled or choked off by etiquette which seemed trivial in its variety. Yet they had to conform in order to get on. In mid eighteenth-century England, for instance, you had to bow to the King, but it was disrespectful to bow to the King of France. While it was correct to curtsy to the Emperor, you must prostrate yourself before the Sultan.

Royalties in this book are for the most part born in the purple but strive to become human. Not protocol but humanity turns out to be the winner. The point could not have been put better than it once was by Mrs Delany, an old lady who knew George III's court, lived in Windsor, and befriended Fanny Burney. She is advising Fanny how to talk to Queen Charlotte:

> I do beg of you when the King or Queen speak to you, not to answer with mere monosyllables. The queen often complains to me of the difficulty with which she can get any conversation, as she not only always has to start the subjects, but, commonly, entirely to support them: and she says there is nothing she so much loves as conversation, and nothing she finds so hard to get. She is always best pleased to have the answers that are made to lead her on to further discourse. Now, as I know she wishes to be acquainted with you, and converse with you, I do really entreat you not to draw back from her, nor to stop conversation with only answering Yes, or No.[2]

Fanny was a novelist and therefore capable of conversation, but she was too nervous to take this advice at first.

Anecdotes are nevertheless frequently shaped and founded in conversation. Not the least interesting anecdotes in this book have emerged from people who broke the royal conversation-barrier.

Finally, there is all the difference between writing about kings and queens, however sympathetically, and being one of them. Alfred the Great, freely translating the *Consolation of Philosophy* by Boethius, a Roman philosopher and senator of the sixth century AD, wrote that in order to rule efficiently a king needed to have his land fully manned: he must have praying men, fighting men and working men'. He does not seem to have wanted men of words. Or perhaps he relied for words on the praying men.

[1] James Lees-Milne, *The Enigmatic Edwardian* (1986), quoting Esher's Journal, 7 May 1910.
[2] *Diaries and Letters of Madame d'Arblay* [Fanny Burney], II, ed. by her Niece (7 vols, 1842–6).

Many years earlier, during the first century BC, the Roman philosopher Seneca was advising Lucullus, the noted epicure: 'If you want to be agreeable to great princes, do them many services, and speak to them few words.' If either Alfred's or Seneca's views had prevailed, this book might not have been written. Or at any rate there would have been fewer anecdotes to choose from.

Fortunately, however, the practice of Sir James Melville, the seventeenth-century Scottish historian, has more frequently been adopted. He not only 'advised and criticized' great princes but also wrote down 'the most remarkable Affairs of State . . . not mentioned by other Historians'.

This book has tried to cover both the royal affairs not mentioned by other contemporary historians and also those frequently mentioned and therefore well known—Alfred and the Cakes as well as Alfred and the Clocks.

A few guide-lines on details. The spelling of proper names may sometimes vary even on the same page, since I have used the familiar spelling for titles and comments but naturally preserved the author's often individualistic spelling in quotations.

'Royal Sayings' are short sections where no sources appear. This is either because the source is anonymous or the 'saying' has been quoted so often that it is attributable to no single authority. Some 'sayings' could in fact have been attributed to a definite source but because of their nature it has seemed better to include them in these sections. The place of publication of all cited sources is London, unless stated otherwise.

In the choice of anecdotes I have been guided by their intrinsic interest and not by any theoretical need to 'fill out' particular monarchs. Thus the amount of space allotted to each monarch will not necessarily reflect the length of his or her reign.

The structure of the book is as far as possible chronological, both as regards the royal dynasties, the groups of monarchs within each royal house, and the anecdotes attached to each individual monarch. Sources vary from contemporary to modern, from a centuries-old tale in metrical verse to the latest contribution from a historical journal. Details of the modern translations of contemporary accounts used can be found in the Acknowledgements. While contemporary anecdotes stand out by their authenticity and freshness, the accounts by modern historians have the great advantage of accuracy, lucidity, economy, and availability. I am fortunate in collecting my material during a period when many important chronicles and diaries are at last being re-translated and re-edited, thus outdating the rather flat-footed Victorian or earlier versions that had hitherto held the field.

I must first thank Her Majesty the Queen for gracious permission to use material from the Royal Archives.

I would also like to thank the following historians for their advice and suggestions, though they are of course in no way responsible for the finished product: Dr James Campbell, Professor John Cannon, John Gillingham, Professor Ralph Griffiths, Dr George Holmes, Dr Maurice Keen.

I am greatly indebted to the assistance of Tricia Robertson and Dr Rowena E. Archer. Dr Archer's diligent work is evident most noticeably in the medieval section of the book, and she read through the entire typescript before it went to press.

I am also deeply grateful to the staff of the London Library and of the Oxford University Press, especially the Oxford Publisher Robin Denniston and to the energy and excellent judgement of my editor, Judith Luna.

Yet again I thank Agnes Fenner for her typing.

I would like particularly to thank my friends Joseph Bryan III, Dr Brigid Boardman, Elizabeth Jenkins, Laurence Kelly, Lord Ponsonby, Jasper Ridley, A. L. Rowse, and Andrew Wilson for putting me on to new anecdotes, as also Lt.-Col. Sir Julian Paget, and Sir John Frettiwell for the story of the Conqueror's bones.

My daughter-in-law Clare introduced me to the medieval studies by her great-aunt Helen Cam; my sister-in-law Mary and her book on Edward IV were most useful; my grandson Orlando found material for me on the medieval kings; my daughter Antonia's new book on Warrior Queens has been of great help to me, as has my husband's history of the House of Lords: my thanks to them all.

Chelsea and East Sussex

GENEALOGICAL TABLES

SAXONS AND DANES
871 – 1066

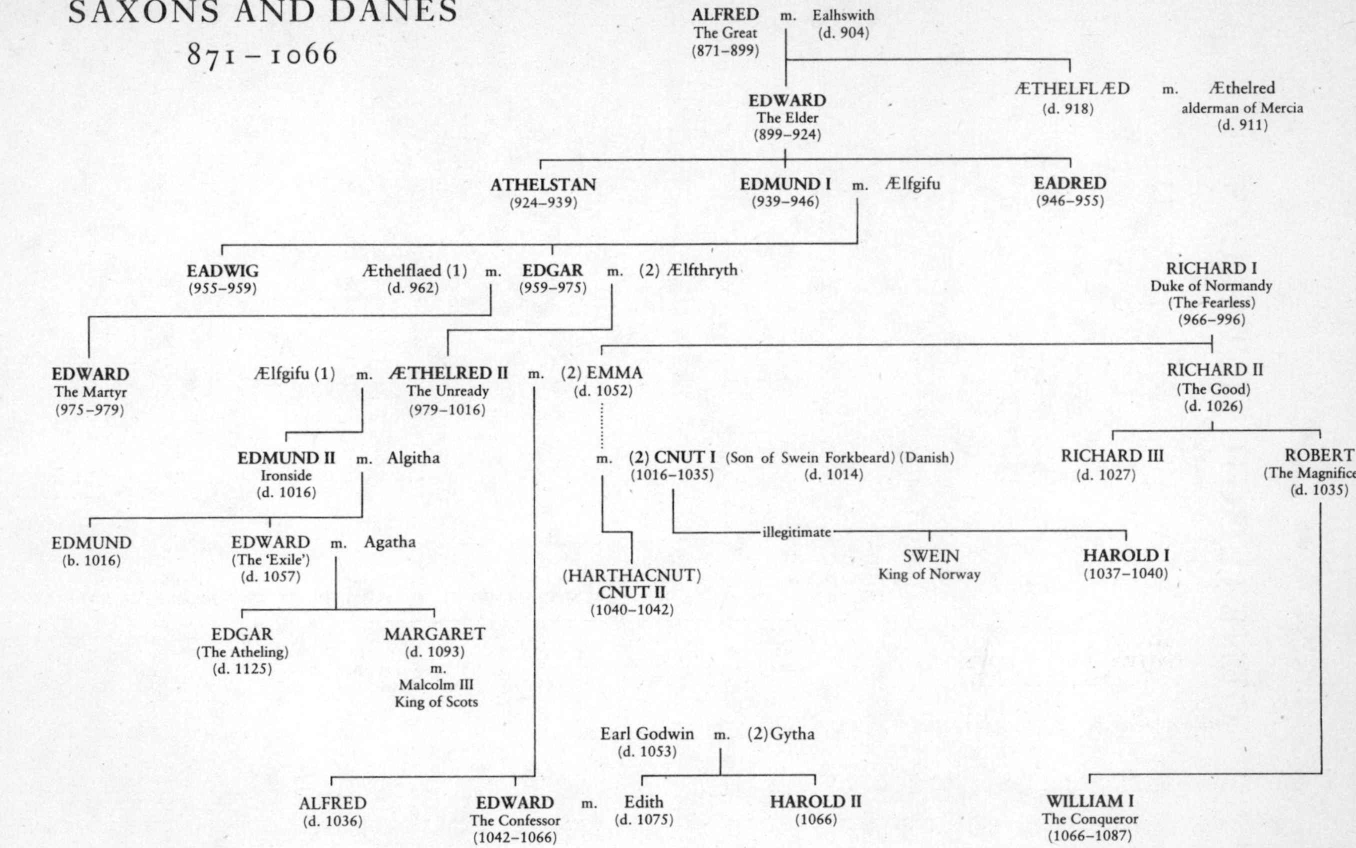

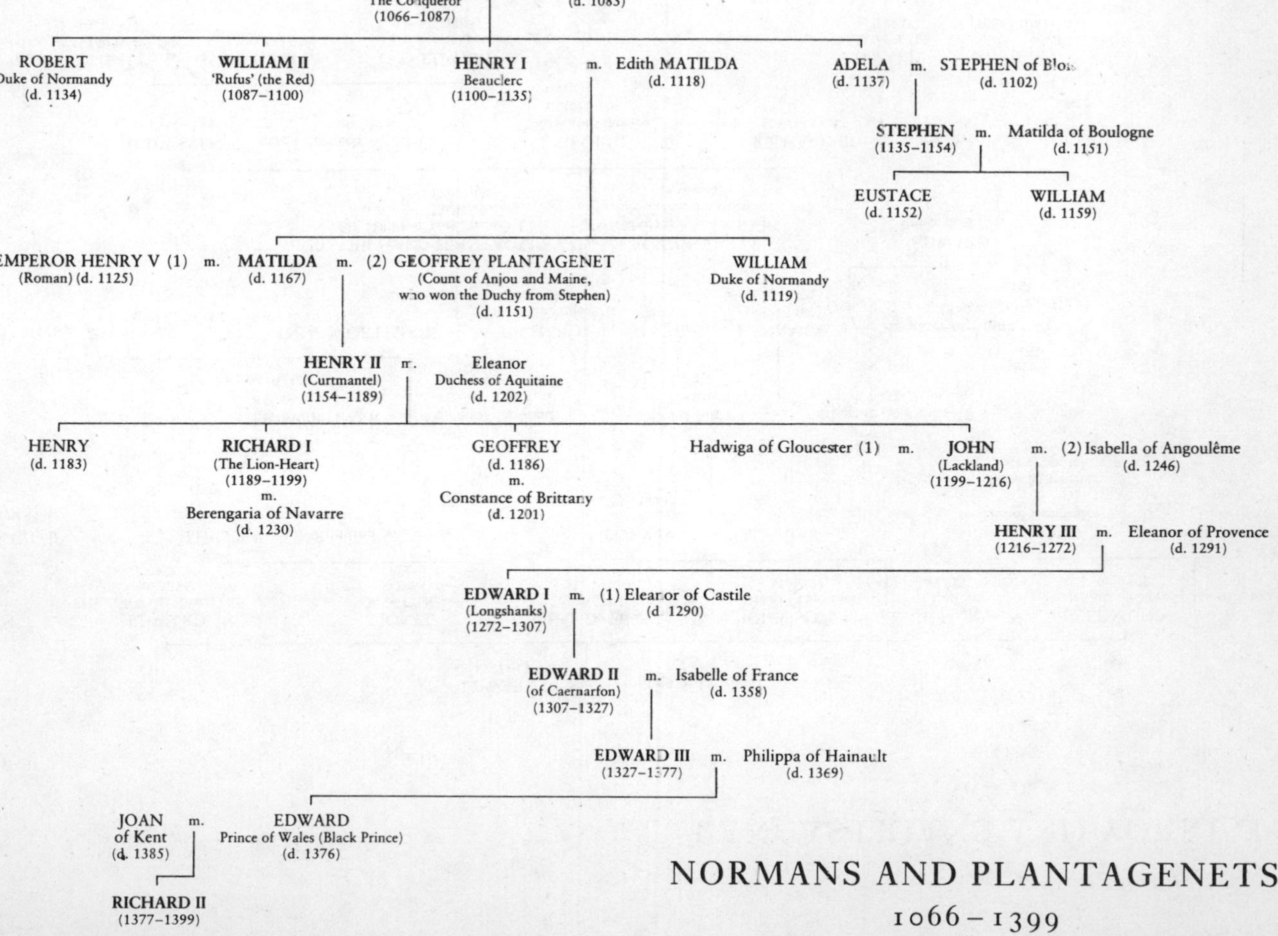

WILLIAM I
The Conqueror
(1066–1087)
m.
MATILDA of Flanders
(d. 1083)
ROBERT
Duke of Normandy
(d. 1134)
WILLIAM II
'Rufus' (the Red)
(1087–1100)
HENRY I
Beauclerc
(1100–1135)
m.
Edith MATILDA
(d. 1118)
ADELA
(d. 1137)
m.
STEPHEN of Blois
(d. 1102)
STEPHEN
(1135–1154)
m.
Matilda of Boulogne
(d. 1151)
EUSTACE
(d. 1152)
WILLIAM
(d. 1159)
EMPEROR HENRY V (1)
(Roman) (d. 1125)
m.
MATILDA
(d. 1167)
m.
(2) GEOFFREY PLANTAGENET
(Count of Anjou and Maine,
who won the Duchy from Stephen)
(d. 1151)
WILLIAM
Duke of Normandy
(d. 1119)
HENRY II
(Curtmantel)
(1154–1189)
m.
Eleanor
Duchess of Aquitaine
(d. 1202)
HENRY
(d. 1183)
RICHARD I
(The Lion-Heart)
(1189–1199)
m.
Berengaria of Navarre
(d. 1230)
GEOFFREY
(d. 1186)
m.
Constance of Brittany
(d. 1201)
Hadwiga of Gloucester (1)
m.
JOHN
(Lackland)
(1199–1216)
m.
(2) Isabella of Angoulême
(d. 1246)
HENRY III
(1216–1272)
m.
Eleanor of Provence
(d. 1291)
EDWARD I
(Longshanks)
(1272–1307)
m.
(1) Eleanor of Castile
(d. 1290)
EDWARD II
(of Caernarfon)
(1307–1327)
m.
Isabelle of France
(d. 1358)
EDWARD III
(1327–1377)
m.
Philippa of Hainault
(d. 1369)
JOAN
of Kent
(d. 1385)
m.
EDWARD
Prince of Wales (Black Prince)
(d. 1376)
RICHARD II
(1377–1399)
NORMANS AND PLANTAGENETS
1066–1399

LANCASTRIANS AND YORKISTS

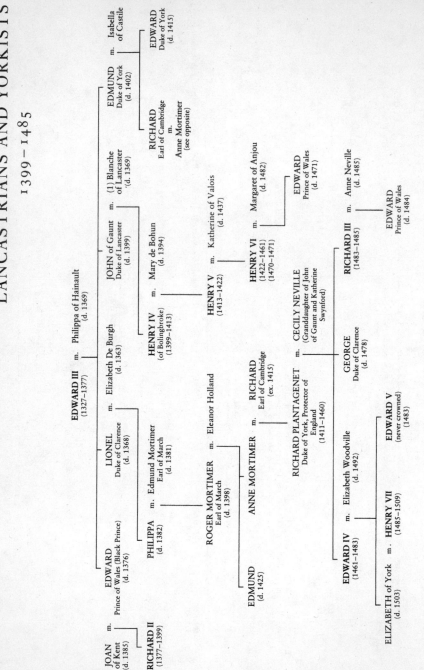

1399–1485

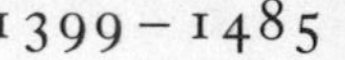

EDWARD III (1327–1377) m. Philippa of Hainault (d. 1369)

JOAN of Kent (d. 1385) m. EDWARD Prince of Wales (Black Prince) (d. 1376)

LIONEL Duke of Clarence (d. 1368) m. Elizabeth De Burgh (d. 1363)

JOHN of Gaunt Duke of Lancaster (d. 1399) m. (1) Blanche of Lancaster (d. 1369)

EDMUND Duke of York (d. 1402) m. Isabella of Castile

RICHARD II (1377–1399)

PHILIPPA (d. 1382) m. Edmund Mortimer Earl of March (d. 1381)

HENRY IV (of Bolingbroke) (1399–1413) m. Mary de Bohun (d. 1394)

RICHARD Earl of Cambridge m. Anne Mortimer (see opposite)

EDWARD Duke of York (d. 1415)

ROGER MORTIMER Earl of March (d. 1398) m. Eleanor Holland

HENRY V (1413–1422) m. Katherine of Valois (d. 1437)

EDMUND (d. 1425)

ANNE MORTIMER m. RICHARD Earl of Cambridge (ex. 1415)

HENRY VI (1422–1461) (1470–1471) m. Margaret of Anjou (d. 1482)

RICHARD PLANTAGENET Duke of York, Protector of England (1411–1460) m. CECILY NEVILLE (Granddaughter of John of Gaunt and Katherine Swynford)

EDWARD Prince of Wales (d. 1471)

EDWARD IV (1461–1483) m. Elizabeth Woodville (d. 1492)

GEORGE Duke of Clarence (d. 1478)

RICHARD III (1483–1485) m. Anne Neville (d. 1485)

ELIZABETH of York (d. 1503) m. **HENRY VII** (1485–1509)

EDWARD V (never crowned) (1483)

EDWARD Prince of Wales (d. 1484)

THE TUDORS

1485–1603

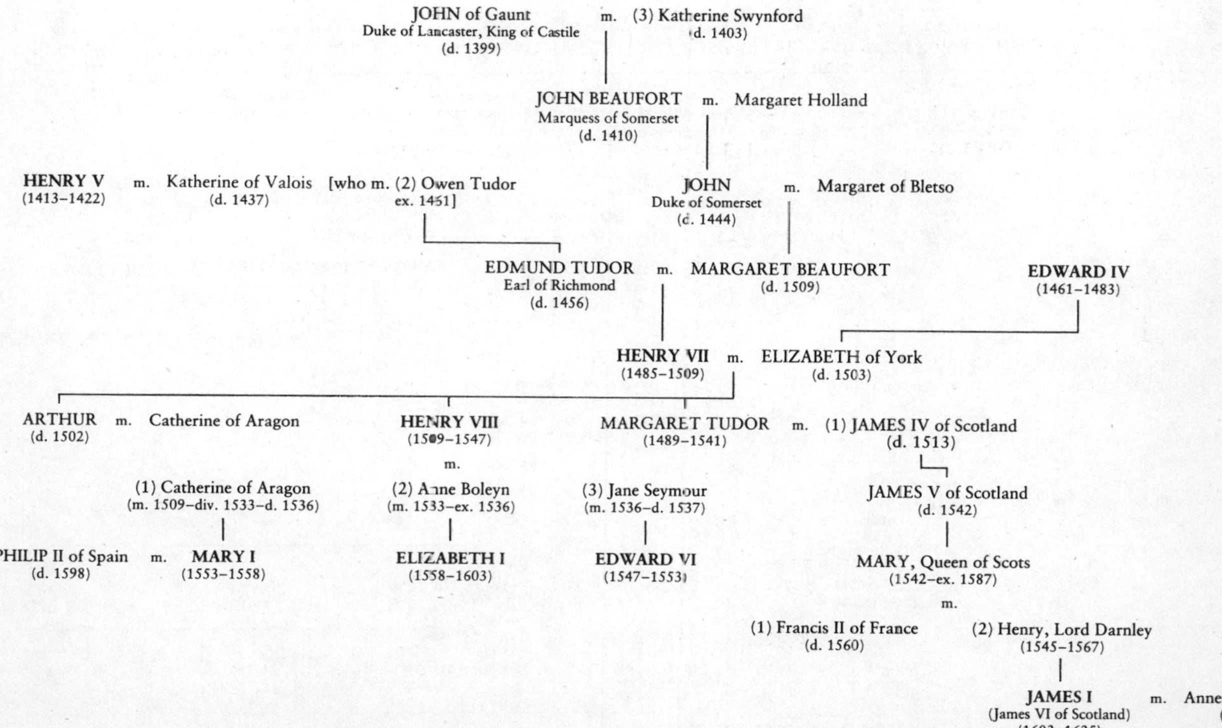

JOHN of Gaunt
Duke of Lancaster, King of Castile
(d. 1399)
m.
(3) Katherine Swynford
(d. 1403)
JOHN BEAUFORT
Marquess of Somerset
(d. 1410)
m.
Margaret Holland
HENRY V
(1413–1422)
m.
Katherine of Valois
(d. 1437)
[who m. (2) Owen Tudor
ex. 1461]
JOHN
Duke of Somerset
(d. 1444)
m.
Margaret of Bletso
EDMUND TUDOR
Earl of Richmond
(d. 1456)
m.
MARGARET BEAUFORT
(d. 1509)
EDWARD IV
(1461–1483)
HENRY VII
(1485–1509)
m.
ELIZABETH of York
(d. 1503)
ARTHUR
(d. 1502)
m.
Catherine of Aragon
HENRY VIII
(1509–1547)
m.
MARGARET TUDOR
(1489–1541)
m.
(1) JAMES IV of Scotland
(d. 1513)
(1) Catherine of Aragon
(m. 1509–div. 1533–d. 1536)
(2) Anne Boleyn
(m. 1533–ex. 1536)
(3) Jane Seymour
(m. 1536–d. 1537)
JAMES V of Scotland
(d. 1542)
PHILIP II of Spain
(d. 1598)
m.
MARY I
(1553–1558)
ELIZABETH I
(1558–1603)
EDWARD VI
(1547–1553)
MARY, Queen of Scots
(1542–ex. 1587)
m.
(1) Francis II of France
(d. 1560)
(2) Henry, Lord Darnley
(1545–1567)
JAMES I
(James VI of Scotland)
(1603–1625)
m.
Anne of Denmark
(d. 1619)

STUARTS AND HANOVERIANS
1603–1837

JAMES I (James VI of Scotland) (1603–1625) m. Anne of Denmark (d. 1619)

- HENRY, Prince of Wales (d. 1612)
- ELIZABETH (d. 1662) m. Frederick V, Elector Palatine of the Rhine (d. 1632)
  - CHARLES LOUIS (d. 1680)
  - RUPERT (d. 1682)
  - SOPHIA (d. 1714) m. Ernest Augustus, Elector of Hanover (d. 1698)
    - **GEORGE I** (1714–1727) m. Sophia Dorothea of Brunswick & Zelle (d. 1726)
      - **GEORGE II** (1727–1760) m. Caroline of Brandenburg-Anspach (d. 1737)
        - FREDERICK, Prince of Wales (d. 1751) m. Augusta of Saxe-Gotha-Altenburg (d. 1772)
          - **GEORGE III** (1760–1820) m. Sophia Charlotte of Mecklenburg-Strelitz (d. 1818)
            - **GEORGE IV** (Regent from 1811, King 1820–1830) m. Caroline of Brunswick-Wölfenbuttel (d. 1821)
              - CHARLOTTE (d. 1817)
            - FREDERICK, Duke of York (d. 1827)
            - **WILLIAM IV**, Duke of Clarence (1830–1837) m. Adelaide of Saxe-Meiningen (d. 1849)
            - EDWARD, Duke of Kent (d. 1820) m. Victoria of Saxe-Coburg
              - **VICTORIA** (1837–1901)
            - ERNEST AUGUSTUS, King of Hanover (d. 1851)
            - ADOLPHUS, Duke of Cambridge (d. 1850)
- **CHARLES I** (1625–1649) m. HENRIETTA MARIA, dr. of Henry IV of France (d. 1669)
  - **CHARLES II** (1660–1685) m. Katherine of Braganza (d. 1705)
  - **JAMES II** (1685–1688) (d. 1701) m. (1) Anne Hyde (d. 1671); m. (2) Mary of Modena (d. 1718)
    - (by Anne Hyde) **MARY II** (1689–1694) m. **WILLIAM III** (son of Mary and William of Orange) (1689–1702) (r. alone from 1694)
    - (by Anne Hyde) **ANNE** (1702–1714) m. George of Denmark (d. 1708)
    - (by Mary of Modena) JAMES Francis Edward Stuart (Old Pretender) (d. 1766)
      - CHARLES EDWARD (Young Pretender) (d. 1788)
  - MARY (d. 1660) m. William of Orange

VICTORIA AND HER DESCENDANTS

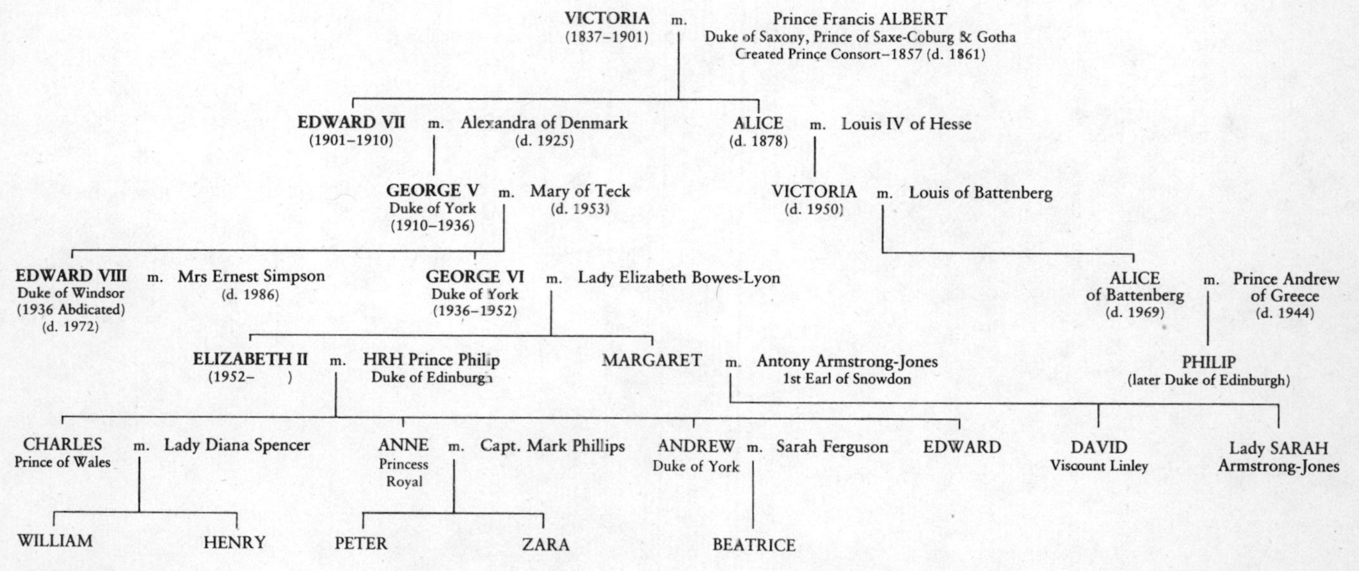

VICTORIA m. Prince Francis ALBERT
(1837–1901)
Duke of Saxony, Prince of Saxe-Coburg & Gotha
Created Prince Consort–1857 (d. 1861)
EDWARD VII m. Alexandra of Denmark
(1901–1910)
(d. 1925)
ALICE m. Louis IV of Hesse
(d. 1878)
GEORGE V m. Mary of Teck
Duke of York
(1910–1936)
(d. 1953)
VICTORIA m. Louis of Battenberg
(d. 1950)
EDWARD VIII m. Mrs Ernest Simpson
Duke of Windsor
(1936 Abdicated)
(d. 1972)
(d. 1986)
GEORGE VI m. Lady Elizabeth Bowes-Lyon
Duke of York
(1936–1952)
ALICE m. Prince Andrew
of Battenberg
(d. 1969)
of Greece
(d. 1944)
ELIZABETH II m. HRH Prince Philip
(1952–)
Duke of Edinburgh
MARGARET m. Antony Armstrong-Jones
1st Earl of Snowdon
PHILIP
(later Duke of Edinburgh)
CHARLES m. Lady Diana Spencer
Prince of Wales
ANNE m. Capt. Mark Phillips
Princess
Royal
ANDREW m. Sarah Ferguson
Duke of York
EDWARD
DAVID
Viscount Linley
Lady SARAH
Armstrong-Jones
WILLIAM
HENRY
PETER
ZARA
BEATRICE

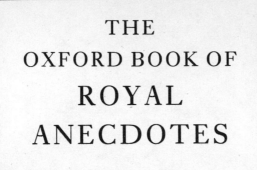

THE OXFORD BOOK OF ROYAL ANECDOTES

THE CELTS AND THE BRITONS

THE MINGLING OF THE RACES IN BRITAIN

THE era of Celt, Saxon, and Dane is like Macbeth's battle on the blasted heath. Prophecy hovers around. Horns are heard blowing in the mist, and a confused uproar of savage tumult and outrage. We catch glimpses of giant figures—mostly warriors at strife.

G. M. Trevelyan, *History of England* (1926)

Boudicca

d. AD 61

The 'Warrior Queen' has one advantage over other early rivals for fame: she was unquestionably real. As queen of the Iceni tribe she rebelled against Rome in AD *60 and two Roman historians described her challenge to the might of Rome. Cassius Dio* (c.*150–235*) *gave her a formidable mane of bright red hair, a rough voice, huge frame and the notoriety of having put men to shame:*

All this ruin was brought upon the Romans by a woman.

The Roman Governor of Britain, Suetonius, was faced with a sudden rebellion in the Province:

Prasutagus, king of the Iceni, after a life of long and renowned prosperity, had made the emperor [of Rome] co-heir with his own two daughters. Prasutagus hoped by this submissiveness to preserve his kingdom and household from attack. But it turned out otherwise. Kingdom and household were plundered like prizes of war, the one by Roman officers, the other by Roman slaves. As a beginning, his widow Boudicca was flogged and her daughters raped. The Iceni chiefs were deprived of their hereditary estates as if the Romans had been given the whole country. The king's own relatives were treated like slaves.

And the humiliated Iceni feared still worse, now that they had been reduced to provincial status. So they rebelled. With them rose the Trinobantes and others. Servitude had not broken them, and they had secretly plotted together to become free again.

The Britons besieged and captured the hated Roman settlement of Camulodunum (Colchester), spurred on by apparent supernatural help:

The statue of Victory at Camulodunum fell down—with its back turned as if it were fleeing the enemy. Delirious women chanted of destruction at hand. They cried that in the local senate-house outlandish yells had been heard; the theatre had echoed with shrieks; at the mouth of the Thames a phantom settlement had been seen in ruins. A blood-red colour in the sea, too, and shapes like human corpses left by the ebb tide, were interpreted hopefully by the Britons—and with terror by the settlers.

A Roman division was routed, the imperial agent fled to Gaul, Londinium and Verulamium (St Albans) were evacuated and sacked. Finally Suetonius decided to challenge the British tribes led by Boudicca with 10,000 Roman regulars and auxiliaries.

Boudicca drove round all the tribes in a chariot with her daughters in front of her. 'We British are used to women commanders in war,' she cried. 'I am descended from mighty men! But I am not fighting for my kingdom and wealth now. I am fighting as an ordinary person for my lost freedom, my bruised body and my outraged daughters . . . But the gods will grant us the vengeance we deserve! . . . Consider how many of you are fighting—and why; then you will win this battle, or perish. That is what I, a woman, plan to do!—let the men live in slavery if they will.'

But Suetonius confidently gave the signal for battle and the experienced infantry and cavalry annihilated the Britons, sparing not even women and baggage animals.

It was a glorious victory, comparable with bygone triumphs. According to one report almost eighty thousand Britons fell. Our own casualties were about four hundred dead and a slightly larger number of wounded. Boudicca poisoned herself.

Tacitus, The Annals of Imperial Rome, ed. Michael Grant (1963)

Boudicca was never forgotten, though her rather lumpish British name was changed by later generations to the more romantic Boadicea. They also made sure that the British Warrior Queen had her ultimate revenge, at least over the Roman Empire. A bronze statue by Thomas Thorneycroft was placed in 1902 by the London County Council at the foot of Westminster Bridge. A couplet from William Cowper's eighteenth century poem 'Boadicea' was inscribed on the right side of the plinth:

Regions Caesar never knew
Thy Posterity shall sway.

BOUDICCA

On the front of the plinth, facing the Houses of Parliament, the inscription runs:

> Boadicea (Boudicca) Queen of the Iceni
> Who died AD 61 after leading her people
> Against the Roman invader.

THE LAST WORD

Boadicea's grave 'at King's Cross'. The grave of Boadicea, the chariot-borne warrior queen who fought the Romans nearly 2,000 years ago, has been located by archæologists—they believe that it is under platform eight at King's Cross railway station.

'We have just refurbished platform eight and anyone wanting to dig it up had better come up with a strong case,' said a British Rail spokesman.

Daily Telegraph, 22 February 1988

Arthur

c. AD 500

Arthur was known as the Dux Britanniae, leader of the British or Britain. The evidence for his existence is contained in the Historia Britonum, *wrongly attributed either to the sixth-century scribe Gildas or to the eighth-century Nennius, a disciple of the bishop of Bangor. According to the author of the* Historia Britonum *Arthur was the commander in twelve victorious battles against the Saxons:*

The twelfth battle was on Mons Badonis, where in one day nine hundred and sixty men were killed in one attack of Arthur's, and no-one but himself laid them low.

Richard Barber, *King Arthur: Hero and Legend* (Woodbridge, 1986), quoting the earliest version of the *Historia Britonum*

A version produced a century later made the point that it was Arthur's valour not his birth that made him 'Dux':

Then the warrior Arthur, with the soldiers and kings of Britain, used to fight against them [the Saxons]. And though there were many of more noble birth than he, he was twelve times leader in the war and victor of the battles. Ibid.

A miraculous element entered in at the end of the later version:

In the district which is called Buelt there is another marvel. There is a pile of stones there, and one stone with the footprint of a dog on it placed on top of the heap. When he hunted the boar Trwyd, Cafall—who was the dog of the warrior Arthur—imprinted the mark of his foot on it; and Arthur afterwards assembled a heap of stones under the stone on which was the footprint of his dog, and it is called Carn Cafall. And people come and carry away the stone in their hands for a period of a day and a night, and on the following day it is found on top of the heap. Ibid.

According to the Annals of Wales, Arthur and his treacherous nephew Medraut (Modred) died at the battle of Camlann.

The Medieval Arthur

The twelfth-century bishop and chronicler Geoffrey of Monmouth bears much responsibility for the flood of Arthurian legends. But these legends of the twelve Knights of the Round Table and the second coming of the king rendered the twelve battles suspicious, particularly as Charlemagne also had twelve paladins and the prospect of a return to earth after death; not to mention Christ's twelve apostles and second coming. Arthur's grave in medieval Glastonbury was said to have borne the inscription:

Hic jacet Arthurus,
Rex quondam, rexque futurus.

(Here lies Arthur,
Former king, and future king.)

In the fifteenth century, Sir Thomas Malory's Le Morte Darthur *launched the dying King Arthur on the most celebrated version of his journey to the Vale of Avalon, said to be the Celtic Paradise. He was picked up by*

A little barge with many fair ladies in it, and among them a queen.

Sir Thomas Malory, *Le Morte Darthur*

Arthurian legend was highly valued by medieval exponents of chivalry, becoming a libretto for the grand opera of war, as we shall see in the reigns of Edward I and III. Even Henry VII 'revamped' the medieval Round Table in the Great Hall at Winchester and

had it painted with the Tudor rose as part of his own homage to Arthur, which included naming his son, born in 1486, after him.

B. Stone, 'Models of kingship: Arthur in Medieval Romance', *History Today* (November 1987)

William Caxton, the first English printer, defended the historicity of King Arthur in the Prologue to his edition of Malory, 1485:

First, in the Abbey of Westminster at St Edward's shrine, remains the print of his [Arthur's] seal in red wax closed in beryl, in which is written 'Patricius Arthurus, Britanniae Galliae Germaniae Daciae Imperator' ['Emperor of Britain, Gaul, Germany, Dacia']. Also, in the Castle of Dover you may see Gawain's skull and Caradoc's mantle; at Winchester the round table; in other places Lancelot's sword and many other things. Then all these things considered, no man can reasonably deny that there was a king of this land named Arthur.

English Historical Documents, IV, 1327–1485, ed. A. R. Myers (1969)

The Victorian Arthur

The Victorians generally dealt with Arthur much as they dealt with Boudicca, enriching and prettifying his story and even giving him, like the Prince Consort, 'the white flower of a blameless life' to wear in his buttonhole. Whereas Malory had sent 'a' queen and 'many' ladies for him in a 'little' barge, the Victorian Poet Laureate Lord Tennyson dispatched 'three' queens and a three-decker barge densely packed with mourning women.

The dying king and his faithful knight Sir Bedivere are waiting beside a lake:

Then saw they how there hove a dusky barge,
Dark as a funeral scarf from stem to stern,
Beneath them; and descending they were ware
That all the decks were dense with stately forms,
Black-stoled, black-hooded, like a dream—by these
Three Queens with crowns of gold . . .

Arthur, having been received on board by the 'tallest' queen, delivers to Bedivere a truly Victorian homily on the theme:

The old order changeth, yielding place to new . . .

Then Arthur bids him farewell:

I am going a long way . . .
To the island-valley of Avilion; . . .
Where I will heal me of my grievous wound.

Alfred Lord Tennyson, *Idylls of the King*

William Morris in his Defence of Guenevere (*1858*) *brought Arthur down from his pedestal by making Queen Guenevere speak of*

Arthur's great name, and his little love . . .

In his 'King Arthur's Tomb', Morris has Launcelot returning to Glastonbury's 'gilded towers' in search of his lover, the queen; he has a vision of the past and swoons upon a stone that he does not realize is King Arthur's gravestone:

I stretched my hands towards her and fell down,
How long I lay in swoon I cannot tell:
My head and hands were bleeding from the stone,
When I rose up, also I heard a bell

—the bell of the convent in which Queen Guenevere was said to have ended her life.

The Modern Arthur

Modern scholarship makes a bow, albeit a hesitant one, to the legendary British king.

Unfortunately he has only the most shadowy claims to historical reality . . . We can only say that there seem to have been memories of a British war-leader called Arthur, who was associated with the battle of Mons Badonicus and subsequent campaigns.

John Blair, 'The Anglo-Saxon Period', in *The Oxford Illustrated History of Britain*, ed. Kenneth O. Morgan (Oxford, 1984)

Arthurian legends apart, it seems possible to accept the existence of a British warrior-king living some four hundred years after Boudicca, whose enemy was the Saxon rather than the Roman invader. Three hundred years later a Saxon warrior-king would in turn be fighting off new enemies, the Danes.

THE SAXONS

In the fifth and sixth centuries the Britons and Celts were invaded from the East by mainly German tribes from the lands between the Rhine and the Elbe.

Ethelbert

KING OF KENT

560–616

A major step forward in English history was the mission of St Augustine which reached Kent in 597. The support he received from the king of Kent was to prove invaluable once Ethelbert had overcome his suspicions. From the Isle of Thanet Augustine sought an interview. It is described by the Venerable Bede in his history, completed in 731, which remains an invaluable account of England in the Dark Ages:

After some days, the king came to the island, and sitting down in the open air, summoned Augustine and his companions to an audience. But he took precautions that they should not approach him in a house, for he held an ancient superstition that if they were practisers of magical arts, they might have opportunity to deceive and master him.

Bede, A History of the English Church and People, ed. L. Sherley-Price (Harmondsworth, 1965)

Edwin

KING OF NORTHUMBRIA

616–633

Edwin and his people were converted to Christianity in 627 by Paulinus, who was sent by Pope Gregory to assist Augustine in his mission. The story of the pope seeing the golden-haired English boys in the slave market at Rome and asking who they were, dates from this period. On being told they were Angles, the pope said, 'Not Angles but angels.'

In those parts of Britain under King Edwin's jurisdiction, the proverb still runs that a woman could carry her new-born babe across the island from sea to sea without any fear of harm. And such was the king's concern for the welfare of his people, that in a number of places where he had noticed clear springs adjacent to the highway, he ordered posts to be erected with brass bowls hanging from them, so that travellers could drink and refresh themselves. And so great was the people's affection for him, and so great the awe in which he was held, that no one presumed to use these bowls for any other purpose. The king's dignity was highly respected throughout his realm, and whether in battle or on a peaceful progress through city, town, and countryside in the company of his thanes, the royal standard was always borne before him.

Bede, A History of the English Church and People, ed. L. Sherley-Price

Ine

KING OF THE WEST-SAXONS
688–726

Having lived riotously, King Ine was converted to a respectable old age by his queen.

Ine's queen was Ethelburga, a woman of royal race and disposition: who, perpetually urging the necessity of bidding adieu to earthly things, at least in the close of life, and the king as constantly deferring the execution of her advice, at last endeavoured to overcome him by strategem. For, on a certain occasion, when they had been revelling at a country seat with more than usual riot and luxury, the next day, after their departure, an attendant, with the privity of the queen, defiled the palace in every possible manner, both with the excrement of cattle and with heaps of filth; and lastly he put a sow, which had recently farrowed, in the very bed where they had lain.

They had hardly proceeded a mile, when she attacked her husband with the fondest conjugal endearments, entreating that they might immediately return thither . . . Her petition being readily granted, the king was astonished at seeing a place, which yesterday might have vied with Assyrian luxury, now filthily disgusting and desolate: and silently pondering on the sight, his eyes at length turned upon the queen.

Seizing the opportunity, and pleasantly smiling, she said, 'My noble spouse, where are the revellings of yesterday? Are not all these things smoke and vapour? Have they not all passed away? Woe be to those who attach themselves to such, for they in like manner shall consume away . . .' Without saying more, by this striking example, she gained over her husband to those sentiments which she had in vain attempted for years by persuasion.

William of Malmesbury, Chronicle of the Kings of England, ed. J. A. Giles (1904). *Written in the twelfth century, William of Malmesbury's account of pre-Conquest England owed a good deal to legends and stories of little historical value, current in his own time.*

Offa

KING OF THE MERCIANS

757–796

Offa was much admired by the chroniclers for his pedigree, his victories and his piety.

For in the same year [755] Offa put to flight [the king of Mercia] and ruled over Mercia for thirty-nine years. A most noble youth, Offa was the son of Wingferd . . .

There follows a string of noble ancestors culminating in Woten.

Offa was a most vigorous king; for he defeated the men of Kent in battle, likewise those of Wessex, likewise those of Northumbria. He was a religious man; for he transferred the bones of St Alban to a monastery that he himself built and greatly enriched; and he gave to the vicar of St Peter, the Pope of Rome, a permanent fixed payment from every house in the land.

Henry of Huntingdon, Historia Anglorum, ed. T. Arnold, Rolls Series (1879). *For pre-Conquest stories Henry of Huntingdon's history owes much to legend but contains a little reliable information.*

Offa did indeed make an annual payment to the pope dating from the Synod of 787, which was probably the origin of 'Peter's Pence'. The Chronicler Matthew Paris credits Offa with the actual pence—'denarii sancti Petri'—which he said he used to endow the English school in Rome and levied on all house property in England except St Albans. Offa's payment, however, was from his own treasury, for the relief of the poor and lights in St Peter's. The chronicler Florence of Worcester attributed the raising of 'Peter's Pence' to King Æthelwulf of Wessex, c.855.

The Mercian king's name is today remembered through the remains of 'Offa's Dyke'—an earthwork built from the mouth of the Wye to the mouth of the Dee, to repel raids from Wales.

In 787 King Offa's daughter Eadburg was married to Brihtric king of Wessex:

. . . in his time, Danish pirates came to England with three ships . . . These were the first Danes who landed in England.

The Chronicle of Florence of Worcester, ed. and tr. T. Forester (1854). *Florence of Worcester's work was produced in the twelfth century but he is reliable for the eleventh century as he seems to have used contemporary sources which now no longer survive.*

Caedwalla

KING OF NORTH WALES

d. 834

Caedwalla and Edwin of Northumbria were fierce rivals, though they had spent time together as boys. At one point, Caedwalla was lying in despair by the lake of Douglas, having been told by Edwin that he wanted his kingdom of North Wales. Caedwalla was aroused by his nephew Brian:

His nephew Brian covered him with his clothes.
Brian weeps and says very often Alas!
His tears accidentally wet the king's nose.

The king asks why Brian weeps.

'Fair uncle,' said Brian, 'I am not personally offended,
But I weep for Britain, from which thou art sprung;
For the Christian people which is there all vanquished,
And the false pagans, so soon increased in power . . .
Now this Edwin is so intimate with thee,
That he aims at bearing the crown from the north to the south;
Thy loss is everlasting, thy blood is debased;
When Arthur died, Britain lost its shield.

The Chronicle of Pierre de Langtoft, ed. Thomas Wright, 2 vols.; Rolls Series (1866). *Langtoft was a canon of Bridlington (Yorkshire) in the reign of Edward I. His earliest history derived from the* Historia Britonum *of Geoffrey of Monmouth and was written in French metrical verse.*

Caedwalla, though a Christian himself, made common cause with the pagan king Penda of Mercia and at the battle of Heathfield in 833 killed Edwin and his son Osfrith, breaking up the kingdom of Northumbria. But in the following year Caedwalla was killed in battle by Edwin's nephew Oswald, who planted a cross on the field with his own hands to inspire his much smaller force—which it duly did. The chronicler celebrated Caedwalla's death:

Forty-four years was Cadwallo king of the land,
Never a single year did he live without war.
His seneschal caused an image of metal to be made,
And caused the corpse of Cadwallo to be subtily drawn into it,
And caused it to be seated on a brazen horse, in London
In memory and in sign to commemorate his deeds.

Ibid.

Alfred the Great

871–899

It was in the sixteenth century that Alfred king of Wessex was first given the honorific title that is unique in British history. At least one historian has called him 'the Greatest'. His fame rests on a dual achievement: the defence of Wessex from the Vikings; and the inspiration and supervision of a revival of learning, especially the proliferation of native Anglo-Saxon. By these means he created the conditions for a future unified country. Although his courage and wisdom attracted the attention of later chroniclers, we owe our knowledge of his life above all to the biography written by the monk Asser of St David's, Alfred's chaplain and later bishop of Sherborne.

THE ADVANTAGES AND DISADVANTAGES OF ALFRED'S YOUTH

Now, he was greatly loved, more than all his brothers, by his father and mother—indeed, by everybody—with a universal and profound love, and he was always brought up in the royal court and nowhere else. As he passed through infancy and boyhood he was seen to be more comely in appearance than his other brothers, and more pleasing in manner, speech and behaviour . . . but alas, by the shameful negligence of his parents and tutors he remained ignorant of letters until his twelfth year, or even longer [*c.*860]. However, he was a careful listener, by day and night, to English poems, most frequently hearing them recited by others, and he readily retained them in his memory . . .

One day, therefore, when his mother was showing him and his brothers a book of English poetry which she held in her hand, she said: 'I shall give this book to whichever one of you can learn it the fastest.' Spurred on by these words, or rather by divine inspiration, and attracted by the beauty of the initial letter in the book, Alfred spoke as follows in reply to his mother, forestalling his brothers (ahead in years, though not in ability): 'Will you really give this book to the one of us who can understand it the soonest and recite it to you?' Whereupon, smiling with pleasure she reassured him, saying: 'Yes, I will.' He immediately took the book from her hand, went to his teacher and learnt it. When it was learnt, he took it back to his mother and recited it.

Alfred the Great: Asser's Life of King Alfred and other contemporary sources, tr. and ed. Simon Keynes and Michael Lapidge (Harmondsworth, 1983), quoting Asser's *Life of King Alfred.*

THE MENACE OF THE VIKINGS

Before Alfred could go into action on behalf of religion and learning, he had to deal with the third invasion of Wessex by the Vikings, who took him by surprise in January 878, swooping down from Mercia (the Midlands). He went into hiding at Athelney in Somerset, his fortunes at a low ebb but himself the focus for a spate of significant legends. Two of the stories were as follows:

St Cuthbert of Lindisfarne came to Alfred disguised as a pilgrim, and asked for food; the king set aside all that he had, but when an attendant took him the food, the 'pilgrim' had mysteriously disappeared; moved by the king's generosity, Cuthbert then worked a miracle on the king's behalf and appeared to him in a vision offering advice on how to beat the Vikings and indeed promising him victory and future prosperity. [Another story] relates how the king, accompanied by a single attendant, entered the Viking camp disguised as a minstrel, and surreptitiously gathered information about the enemy's plans; after several days he returned to Athelney, told his followers what he had learnt, and then led them to victory.

Ibid.

ALFRED AND THE CAKES

This story, by far the most famous of the 'Athelney' saga, does not appear in Asser's Life of Alfred *and is therefore suspect. Possibly the good monk, rightly devoted to expounding his hero's triumphant nobility, could not bear to think of his making a mistake even with a cake. The earliest version of the story occurs in an anonymous Life of St Neot, probably some hundred years after the actual date when Alfred was in hiding.*

There is a place in the remote parts of English Britain far to the west, which in English is called Athelney and which we refer to as 'Atheling's Isle'; it is surrounded on all sides by vast salt marshes and sustained by some level ground in the middle. King Alfred happened unexpectedly to come there as a lone traveller. Noticing the cottage of a certain unknown swineherd (as he later learned), he directed his path towards it and sought there a peaceful retreat; he was given refuge, and he stayed there for a number of days, impoverished, subdued and content with the bare necessities. Reflecting patiently that these things had befallen him through God's just judgement, he remained there awaiting God's mercy through the intercession of His servant Neot; for he had conceived from Neot the hope that he nourished in his heart . . .

Now it happened by chance one day, when the swineherd was leading his flock to their usual pastures, that the king remained alone at home with the swineherd's wife. The wife, concerned for her husband's return, had entrusted some kneaded flour to [the oven]. As is the custom among countrywomen, she was intent on other domestic occupations, until, when she sought the bread from [the oven], she saw it burning from the other side of the room. She immediately grew angry and said to the king (unknown to her as such): 'Look here, man,

> You hesitate to turn the loaves which you see to be burning,
> Yet you're quite happy to eat them when they come warm from the oven!

But the king, reproached by these disparaging insults, ascribed them to the divine lot; somewhat shaken, and submitting to the woman's scolding, he not only turned the bread but even attended to it as she brought out the loaves when they were ready. Ibid., quoting the appendix, 'Alfred and the cakes'

The whimsical idea of turning the woman's hexameters into Somersetshire dialect occurred to J. A. Giles, the Victorian translator of the legend. His couplet, which he boldly inserted in his version of Asser's Life, ran as follows:

> Cas'sn thee mind the ke-aks, man, an' doosen see 'em burn?
> I'm boun thee's eat 'em vast enough, az zoon az 'tiz the turn.

Six Old English Chronicles, ed. J. A. Giles (1896)

THE VICTORY OF EDINGTON AND THE PEACE, MAY 878

After emerging from Athelney, where he had built a fortress, Alfred gathered the fighting men of Somerset, Wiltshire, and Hampshire to seek out the Danes under their king Guthrum. He camped one night at Iley (modern Eastleigh Wood) in Sutton Veny.

When the next morning dawned he moved his forces and came to a place called Edington [in Wiltshire], and fighting fiercely with a compact shield-wall against the entire Viking army, he persevered resolutely for a long time, at length he gained the victory through God's will. He destroyed the Vikings with great slaughter, and pursued those who fled as far as their stronghold . . .

When he had been there [besieging them] for fourteen days the Vikings, thoroughly terrified by hunger, cold and fear, and in the end by despair, sought peace on this condition: the king should take as many chosen hostages as he wanted from them and give none to them; never before, indeed, had they made peace with anyone on such terms. When he had heard their embassy, the king (as is his wont) was moved to compassion and took as many chosen hostages from them as he wanted . . . Guthrum their king promised to accept Christianity and to receive baptism at Alfred's hand. *Alfred the Great*, quoting Asser's *Life of King Alfred*

ALFREDIAN DEFENCE

Alfred's military achievements went far beyond actual encounters with the Vikings. In the periods between campaigns he took important steps to ensure the kingdom's future security. Two of these were the reorganization of the fyrd (army) and the establishment of a complex system of defence based on burghs (fortified towns). A third has been responsible, with some exaggeration, for Alfred's reputation as the 'Father of the English Navy'.

This same year [896] the hosts in East Anglia and Northumbria greatly harrassed Wessex along the south coast with predatory bands, most of all with the warships they had built many years before. Then king Alfred ordered warships to be built to meet the Danish ships: they were almost twice as long as the others, some had sixty oars, some more; they were both swifter, steadier and with more freeboard than the others; they were built neither after the Frisian design nor after the Danish but as it seemed to himself that they could be most serviceable.

The Anglo-Saxon Chronicle tr. G. N. Garmonsway (1953). *The single most important source for the period to the Conquest, long associated with the inspiration of Alfred himself and ending in 1154.*

ALFREDIAN LAW

Rather like his shipbuilding, Alfred's lawgiving combined the tradition of others with his own preference.

Then I, King Alfred, gathered them together and ordered to be written many of the ones that our forefathers observed—those that pleased me; and many of the ones that did not please me I rejected with the advice of my councillors . . . *Alfred the Great*, quoting Alfred's Book of Laws

But there was also some hesitation:

For I dared not presume to set down in writing at all many of my own, since it was unknown to me what would please those who should come after us. But those which I found either in the days of Ine, my kinsman or of Offa, king of the Mercians, or of Ethelbert (who first among the English people received baptism), and which seemed to me most just, I collected herein, and omitted the others. Ibid.

ALFREDIAN LEARNING

Alfred attached an extraordinary importance to learning which he claimed had

declined so thoroughly in England that there were very few men on this side of the Humber who could understand their divine services in English, or even translate a single letter from Latin into English; and I suppose that there were not many beyond the Humber either. There were so few of them that I cannot recollect even a single one south of the Thames when I succeeded to the kingdom.

Alfred the Great, quoting Alfred's own preface to his translation of Pope Gregory's *Pastoral Care*

ALFRED'S SANCTIONS AGAINST ILLITERACY

The king's military successes gave him the prestige to promote learning, both by example and by threats. He would sit in at judicial hearings and then admonish his nobles if their judgements had seemed unjust to the common people. He once addressed them:

'I am astonished at this arrogance of yours, since . . . you have enjoyed the office and status of wise men, yet you have neglected the study and application of wisdom. For that reason, I command you either to relinquish immediately the offices of worldly power that you possess, or else to apply yourselves much more attentively to the pursuit of wisdom.'

Ibid., quoting Asser's *Life of King Alfred*

The officers of state were duly terrified:

As a result nearly all the ealdormen and reeves and thegns (who were illiterate from childhood) applied themselves in an amazing way to learning how to read, preferring rather to learn this unfamiliar discipline (no matter how laboriously) than to relinquish their offices of power.

Ibid.

And if anyone tried but was too old or stupid to learn to read, the king would order his son, or other relative, or literate freeman or even literate slave

'to read out books in English to him by day and night . . .' Ibid.

INVENTION OF A CLOCK TO MARK RELIGIOUS DUTIES

Alfred had vowed to dedicate to the service of God a half of every day and night in his life, but as frequent rain and mist obscured the sun, he needed some accurate way of measuring the hours. He therefore ordered his chaplains to make six wax candles [*each*] *twelve inches long and marked with the inches. But the English weather again proved an obstacle:*

Because of the extreme violence of the wind, which sometimes blew day and night without stopping through the doors of the churches or through the numerous cracks in the windows, walls, wall-panels and partitions, and likewise through the thin material of the tents, the candles on occasion could not continue burning through an entire day and night. Ibid

Alfred therefore 'ingeniously and cleverly' devised a plan to defeat the draughts:

He ordered a lantern to be constructed attractively out of wood and ox-horn—for white ox-horn, when shaved down finely with a blade, becomes as translucent as a glass vessel. Once this lantern had been marvellously constructed . . . it could not be disturbed by any gust of wind, for he had asked for the door of the lantern to be made of horn as well. Ibid.

And so the candles burnt uninterruptedly, one after the other, through the days and nights, Alfred having mastered time.

THE DEATH OF ALFRED, 899

In this year died Alfred, son of Ethelwulf, six nights before All Hallow's Day [1 November]. He was king over all England except that part which was under Danish domination, and he ruled the kingdom twenty-eight and a half years. *The Anglo-Saxon Chronicle*

Edward the Elder

899–924

The reconquest of the area settled by the Vikings—the Danelaw—was begun by Alfred's son and heir. In this he was ably assisted by his sister Æthelflæd who, upon the death of her husband, the ealdorman of Mercia, became known as the 'Lady of the Mercians'.

And here indeed Æthelflæd . . . ought not to be forgotten, as she was a powerful accession to his [Edward's] party, the delight of his subjects, the dread of his enemies, a woman of an enlarged soul, who, from the difficulty experienced in her first labour, ever after refused the embraces of her husband; protesting that it was unbecoming the daughter of a king to give way to a delight which, after a time, produced such painful consequences. This spirited heroine assisted her brother greatly with her advice, was of equal service in building cities, nor could you easily discern whether it was more owing to fortune or her own exertions, that a woman should be able to protect men at home, and to intimidate them abroad.

William of Malmesbury

The claim to be the first king of all England remains a matter of some dispute. The chronicles reported how Edward succeeded to his sister's mantle in Mercia and later enjoyed even wider recognition.

All the people of Mercia who had been under allegiance to Æthelflæd turned in submission to him. The kings of Wales, Hywel, Clydog and Idwal and all the people of Wales, gave him their allegiance . . . [And] then the king of Scots and the whole Scottish nation accepted him as 'father and lord': so also did . . . all the inhabitants of Northumbria, both English and Danish, Norwegians and others; together with the king of the Strathclyde Welsh and all his subjects. Ibid.

Athelstan

924–939

In reality the reconquest was subject to fluctuations and so some would prefer Edward's son Athelstan, as the first to be truly king of all England. Educated in the household of his redoutable aunt, he was readily accepted as king by the Mercians.

William of Malmesbury, writing in the twelfth century, has preserved a valuable portrait of Athelstan, whom he described as slender and flaxen-haired.

That he was versed in literature, I discovered a few days since, in a certain old volume . . .

Of royal race a noble stem
Hath chased our darkness like a gem.
Great Athelstan, his country's pride,
Whose virtue never turns aside;
Sent by his father to the schools,
Patient, he bore their rigid rules,
And drinking deep of science mild,
Passed his first years unlike a child.
Next clothed in youth's bewitching charms,
Studied the harsher lore of arms,
Which soon confessed his knowledge keen,
As after in the sovereign seen.
Soon as his father, good and great,
Yielded, though ever famed, to fate,
The youth was called the realm to guide,
And, like his parent, well preside.
The nobles meet, the crown present,
On rebels, prelates curses vent;
The people light the festive fires,
And show by turns their kind desires.
Their deeds their loyalty declare,
Though hopes and fears their bosoms share.
With festive treat the court abounds;
Foams the brisk wine, the hall resounds:
The pages run, the servants haste,
And food and verse regale the taste.

The minstrels sing, the guests commend,
Whilst all in praise to Christ contend.
The king with pleasure all things sees,
And all his kind attentions please. *William of Malmesbury*

William also recorded an early story of the king telling how:

his grandfather Alfred seeing and embracing him affectionately when he was a boy of astonishing beauty and graceful manners had most devoutly prayed that his government might be prosperous; indeed he had made him a knight unusually early, giving him a scarlet cloak, a belt studded with diamonds, and a Saxon sword with a golden scabbard. Ibid.

Known to posterity as 'the Glorious', Athelstan allied with Sihtric king of Northumbria and in 933 invaded and harried Scotland. His name is forever linked with a great victory at an unknown site which caused the chronicler to write in verse:

937 In this year king Athelstan, lord of warriors,
Ring giver of men, with his brother prince Edmund,
Won undying glory with the edges of swords,
In warfare around *Brunanburgh*
With their hammered blades, the sons of Edward
Clove the shield wall and hacked the linden bucklers,
As was instinctive in them, from their ancestry
To defend their land, their treasures and their homes,
In frequent battle against each enemy. *The Anglo-Saxon Chronicle*

Recording the furious and bloody battle with five kings and seven earls among the dead he concluded emphatically:

Never before in this island, as the books
Of ancient historians tell us, was an army
Put to greater slaughter by the sword
Since the time when Angles and Saxons landed. Ibid.

Edmund

939–946

In spite of Athelstan's victory, there was clearly much still to be done. The task was taken up by his brother. The chronicler was again moved to verse:

942 In this year Edmund, lord of the English,
Guardian of kinsmen, loved doer of deeds, conquered Mercia
As far as Dore and Whitwell Gap the boundary form
And Humber river, that broad ocean-stream   Ibid.

In 944 it was recorded that he brought Northumbria under his sway and the following year he

ravaged all Strathclyde and ceded it to Malcolm, king of Scots, on the condition that he would be his fellow worker both by sea and land.

Ibid.

A VOW FULFILLED

Edmund's treatment of Dunstan, the future Saint, is preserved in a hunting anecdote. When he was a young monk prepared to go into exile, a reprieve came from an odd quarter.

King Edmund was hunting a stag, which darted up through the woods to the top of Cheddar gorge. Seeing no way of escape it leapt over the cliff, followed by the baying hounds. The King saw his danger, but his horse was beyond his utmost power of control. The wrong done to Dunstan flashed through his mind and he vowed to make amends if his life were spared. On the very edge the horse stopped short and turned aside. When the king got home he sent for Dunstan and made him ride with him to Glastonbury. There he sat him in the abbot's seat and bade him rule the house he loved. J. Armitage Robinson, *The Times of Saint Dunstan* (Oxford, 1923)

A ROYAL MURDER

Edmund met an untimely end at the hand of an assassin on 26 May 946. The cryptic note of it by the Anglo-Saxon chronicler was embellished later.

A certain robber named Leofa, whom he had banished for his crimes, returning after six years' absence, totally unexpected, was sitting, on the feast of St Augustine . . . among the guests at Puckle church in Gloucestershire . . . This, while the others were eagerly carousing, was perceived by the king alone; when, hurried with indignation and impelled by fate, he leapt from the table, caught the robber by the hair, and dragged him to the floor; but he, secretly drawing a dagger from its sheath, plunged it with all its force into the breast of the king as he lay upon him. Dying of the wound, he gave rise over the whole kingdom to many fictions concerning his decease.

The robber was shortly torn limb from limb by the attendants who rushed in . . . St Dunstan, at the time abbot of Glastonbury, had foreseen his ignoble end, being fully persuaded of it from the gesticulations and insolent mockery of a devil dancing before him. *William of Malmesbury*

Eadred

946–955

Edmund's sons being then too young to rule, the last of Edward the Elder's children was chosen as king. For his success in Northumbria against the incursion of Eric Bloodaxe of Norway Eadred had a claim for recognition as king of all England. He grandly styled himself

Eadred, king, emperor of the Anglo-Saxons and the Northumbrians, governor of the pagans and protector of the Britons.

Cartularium Saxonicum ed. W. de Gray Birch, 3 vols. (1885–93), III

THE DEATH OF EADRED, 955

Being then anxious about his life by reason of his long sickness, he sent on all sides to collect his goods, to distribute them, while he could, to his followers with a willing and free disposition in his lifetime. The man of God, Dunstan, went for this purpose, just as did the other keepers of the royal treasures, to bring back to the king what he had in his custody. When some days later he was returning with this treasure in packs, the way he had come, a voice was heard coming from heaven which said to him: 'Behold, now King Eadred has departed in peace.' The horse which the man of God was riding was suddenly struck dead at this voice, because it could not endure the presence of the angel's sublimity. When he arrived, he discovered that the king had ended his life at the very time when the angel announced it to him on that journey.

English Historical Documents, I, *c.*500–1042, ed. D. Whitelock (1955), quoting the *Life of St Dunstan*

Eadwig

955–959

Eadred being childless, the throne then passed to the sons of Edmund. Though his reign was short Eadwig's behaviour was responsible for one of the best known and most colourful anecdotes concerning any Anglo-Saxon king.

After him succeeded Eadwig, the son of King Edmund, a youth indeed in age and endowed with little wisdom in government, though, when elected, he ruled in due succession and with royal title over both peoples. A certain woman, foolish, though she was of noble birth, with her daughter, a girl of ripe age, attached herself to him, pursuing him and wickedly enticing him to intimacy, obviously in order to join and ally herself or else her daughter to him in lawful marriage. Shameful to relate, people say that in his turn he acted wantonly with them, with disgraceful caresses, without any decency on the part of either. And when at the time appointed by all the leading men of the English he was anointed and consecrated king by popular election, on that day after the kingly anointing at the holy ceremony, the lustful man suddenly jumped up and left the happy banquet and the fitting company of his nobles, for the aforesaid caresses of loose women.

English Historical Documents, I, ed. Whitelock (1955)

The horrified archbishop ordered some of his suffragans to bring the king back, but fearing the royal anger they appointed Dunstan and the bishop of Lichfield to go.

When in accordance with their superiors' orders they had entered, they found the royal crown, which was bound with wondrous metal, gold and silver and gems, and shone with many-coloured lustre, carelessly thrown down on the floor, far from his head, and he himself repeatedly wallowing between the two of them in evil fashion, as if in a vile sty. They said: 'Our nobles sent us to you to ask you to come as quickly as possible to your proper seat, and not to scorn to be present at the joyful banquet of your chief men.' But when he did not wish to rise, Dunstan, after first rebuking the folly of the women, drew him by his hand from his licentious reclining by the women, replaced the crown, and brought him with him to the royal assembly, though dragged from the women by force. Ibid.

The penalty for his interference was Dunstan's exile.

Edgar

959–975

The great grandson of Alfred, Edgar became known as 'the Peaceable' largely because the victories and campaigns of his forebears had finally brought a measure of stability and freedom from outside attack. The time was ripe for a reformation of the Church which was largely the work of Dunstan, whom Edgar recalled from exile. In the stakes for recognition as the first king of England Edgar also has some claim.

His reign was prosperous and God granted him
To live his days in peace; he did his duty,
And laboured zealously in its performance.
Far and wide he exalted God's praise
And delighted in His law, improving the security
Of his people more than all the kings
Who were before him within the memory of man.

The Anglo-Saxon Chronicle

A LATE CORONATION

It was only after fourteen years on the throne that Edgar was eventually crowned in a ceremony of great significance using a new order of service which was the work of Dunstan and which long remained in use.

In this year, Edgar, ruler of the English
Was consecrated King by a great assembly,
In the ancient city of Acemannesceastee,
Also called Bath by the inhabitants
Of this island. On that blessed day,
Called and named Whit Sunday by the children of Men,
There was great rejoicing by all. As I have heard,
There was a great congregation of priests and a goodly company of monks,
And wise men gathered together.

Ibid.

Submissions to the king continued and in the same year:

The king led all his fleet to Chester and there six kings came to him to make their submission and pledged themselves to be his fellow workers by sea and land.

Ibid.

The chronicler Holinshed has a more colourful account of the same event (where the number of kings has increased to eight), in which Edgar symbolically acts as steersman while the kings are his oarsmen: [Edgar] called them to enter into a barge upon the water of the Dee, and placing himself in the forepart of the barge at the helm, he called those eight high princes to row the barge up and down the water, showing thereby his princely prerogative and royal magnificence, in that he might use the service of so many kings that were his subjects. And thereupon he said (as hath been reported) that then might his successors account themselves kings of England, when they enjoyed such prerogative of high and supreme honour.

Holinshed's Chronicles, I, 1587

Edgar's reputation was still high when he died in 975.

In this year Edgar passed away
Ruler of the English,
Friend of the West Saxons
And protector of the Mercians.
That was known far and wide
Throughout many nations,
    Kings honoured the son of Edmund
Far and wide over the gannet's bath,
And submitted to the sovereign,
As was his birth right.
    No fleet however proud
No host however strong,
Was able to win booty for itself
In England, while that noble king
Occupied the royal throne.

The Anglo-Saxon Chronicle

Edward the Martyr

975–979

The short reign of Edgar's eldest son Edward, who came to the throne aged fifteen, opened with evil portents, wrote the chronicler. (The name 'Martyr' meant 'unpopular'.)

In the beginning of the reign there appeared a comet which, doubtless, foretold the great famine which followed in the year ensuing. For at that time a certain dissolute noble, Elfhere by name, with the consent and the help of a powerful faction [of Mercia], destroyed some of the abbeys which King Edgar and Bishop Ethelwold had founded. Wherefore the Lord was moved to anger, and, as of old, brought evil on the land.

In the fourth year of the reign of St Edward, all the great men of the English nation fell from a loft at Calne, except St Dunstan, who supported himself by taking hold of a beam. Some of them were much hurt, and some were killed. It was a sign from the Most High of the impending forfeiture of his favour by the assassination of the King, and of the evils it would bring on them from various nations. *Henry of Huntingdon*

THE MURDER OF EDWARD, 978

A reaction against the monastic reform of Edgar's reign was recorded and the young king was cruelly murdered:

treasonably slain by his own family at Corfe-gate, at even-tide; and, carrying to the grave their malice towards him in life, he was buried at Wareham without royal honours, that his name might perish also . . . It is reported that his stepmother, that is the mother of King Ethelred, stabbed him with a dagger while she was in the act of offering him a cup to drink.

Ibid.

The Anglo-Saxon chronicler gave the date of the murder, which he said took place in Corfe 'passage', as 18 March. He also added a lament in verse:

No worse deed was ever done among the English
 Than this was,
Since first they sought the land of Britain.
 Men murdered him but God exalted him;

EDWARD THE MARTYR

In life he was an earthly king,
But after death he is now a heavenly saint.

The Anglo-Saxon Chronicle

THE BONES OF EDWARD THE MARTYR

From Wareham the bones of the murdered king were eventually moved to Shaftesbury Abbey, Dorset, where they became the object of regular pilgrimages. At the dissolution of the monasteries in Henry VIII's reign they disappeared, only to be rediscovered in 1931. For the year before, a family called Claridge had acquired the site on which the Abbey's ruins stood, and one day the gardener, digging in the remains of the north transept, unearthed a Tudor casket, two feet long and made of lead, containing part of a skull and some bones.

But a dispute arose between the two Claridge brothers after their mother died twenty years later. Both over eighty, one of the brothers wished the bones to return to Shaftesbury Abbey, while the other wanted them to go to a small sect of the Russian Orthodox Church in exile, who promised to buy some disused mortuary buildings at Brookwood cemetery, Woking, and there set them up in a new shrine. When last heard of the royal bones were still waiting—in a cutlery box—in the strongroom of a Woking bank for a decision on their fate.

Ethelred the Unready

979–1016

Although modern historians have called for a more balanced view, Ethelred has had a consistently bad press, starting with his collusion in the murder of his predecessor and stepbrother, Edward, and ending with his ultimate defeat through mishandling renewed Danish invasions. The surviving accounts of his reign were all written after his defeat, even The Anglo-Saxon Chronicle, *which was compiled after his death. Such failure lent itself to uncomplimentary stories.*

His damning epithet derived from a play on his name, Ethelræd, meaning 'noble counsel' to which was added un-ræd, 'no counsel', the latter being subsequently mistranslated as unready.

INCONTINENT AT BAPTISM

Ethelred, son of King Edgar, and brother of Edward, was consecrated king before all the nobles of England at Kingston. An evil omen, as St Dunstan interpreted it, had happened to him in his infancy. For at his baptism he made water in the font; whence the man of God predicted the slaughter of the English people that would take place in his time.

Henry of Huntingdon

DANES AND DANEGELD

The chief feature of Ethelred's reign was the renewed incursions by Danish raiders. A solution was adopted for controlling the Danes by paying them tribute, and though Ethelred was not the first to propose such a policy the collection of Danegeld is indelibly associated with his name. The payments have been interpreted as a sign of weakness and military incapacity but most were made on the advice of councillors and though not a permanent solution because of the numbers of separate raiding parties, they did achieve intermittent respites.

The first payment followed a raid in 994.

Then the king and his councillors determined to send to them and promise them tribute and provisions, on condition that they should cease that harrying. And they then accepted that and the whole army came then to Southampton and took winter quarters there; and they were provisioned throughout all the West Saxon kingdom and they were paid 16,000 pounds in money.

The Anglo-Saxon Chronicle

Further sums were paid in 1001, 1004, 1007 (amounting to £36,000), 1009, 1011, 1012 (£48,000), 1013, and 1014. The long-term effect was summed up sourly in the twelfth century:

And this infliction has continued to this present day, and, unless God's mercy interposes, will still continue, for we now pay to our kings, from custom, the tax which was levied by the Danes from intolerable fear.

Henry of Huntingdon

ST BRICE'S DAY MASSACRE, 1002

Ethelred's reputation is further underlined by reprisals which he took against the Danes on 13 November 1002.

And in that year the king ordered to be slain all the Danish men who were in England—this was done on St Brice's day—because the king had been informed that they would treacherously deprive him, and then all his councillors, of life, and possess this kingdom afterwards.

The Anglo-Saxon Chronicle

Ethelred himself described his action later:

A decree was sent out by me, with the counsel of my leading men and magnates, to the effect that all the Danes who had sprung up in this island, sprouting like cockle amongst the wheat, were to be destroyed by a most just extermination. *English Historical Documents*, I, quoting a charter of Ethelred

However much he might try to defend his actions, Ethelred's reward for this was the wrath of King Sweyn Forkbeard of Denmark whose sister had been a victim of the massacre.

DISGRACE AND DEATH

By 1013, when Sweyn came to England with his son Cnut, Ethelred's fortunes had reached a low ebb. His wife, Emma of Normandy, retreated with her two sons Edward and Alfred to France and Ethelred followed soon after. Sweyn's death in 1014 brought a reprieve.

Then all the councillors who were in England . . . determined to send for King Ethelred and they said that no lord was dearer to them than their natural lord, if he would govern them more justly than he did before. Then the king . . . said that he would be a gracious lord to them and reform all the things which they hated . . . [And] during the spring King Ethelred came home to his own people and he was gladly received by them all.

The Anglo-Saxon Chronicle

Ethelred's reinstatement was brief. Cnut went home only to return in the summer of 1015. He had made himself master of the old Danelaw by the time Ethelred died in London on 23 April 1016.

Edmund Ironside

1016

The Anglo-Saxon dynasty was greatly weakened in 1016. Two of Ethelred's sons perished before him and it was his third son who was elected king by the Witan (council) and crowned in London. Known as 'Ironside' for his courage, he devoted his short reign to defending his inheritance against the ravages of Cnut. In this he was severely hampered by the ignoble behaviour of one of his father's favourites—Edric Streona (Grasper). On one battlefield Edric mounted a hill and held up a severed head, saying it was Edmund's. The king removed his helmet to show himself alive. He then violently hurled his spear at Edric which, glancing off Edric's shield, pierced two soldiers standing beside him.

Defeat at the battle of Ashington (Essex) forced Edmund to make terms with Cnut and they agreed to divide the kingdom between them, Cnut taking the north and Edmund the south.

EDMUND'S MURDER

The story that Edmund was murdered is late. Contemporary sources do not record the event, though his sudden death clearly lent itself to speculation that it was not merely coincidental.

King Edmund was treasonably slain a few days afterwards. Thus it happened: one night, this great and good king having occasion to retire to the house for relieving the calls of nature, the son of the ealdorman Edric, by his father's contrivance, concealed himself in the pit, and stabbed the king twice from beneath with a sharp dagger, and, leaving the weapon fixed in his bowels, made his escape. Edric then presented himself to Canute, and saluted him, saying, 'Hail! though who art sole king of England!' Having explained what had taken place, Canute replied, 'For this deed I will exalt you, as it merits, higher than all the nobles of England.' He then commanded that Edric should be decapitated and his head placed upon a pole on the highest battlement of the Tower of London. Thus perished King Edmund Ironside, after a short reign of one year, and he was buried at Glastonbury, near his grandfather Edgar.

Henry of Huntingdon

THE DANES

Cnut

1016–1035

Although England had long suffered at the hands of successive Danish raiders, it was only in the early eleventh century that a young man of about twenty-two acceded as the first Danish king of the realm. In spite of his barbaric origins Cnut proved a beneficent ruler, giving England a code of law and, though not born a Christian, he was generous to the Church which he later joined.

TWO VERSES FROM AN ELEVEN-VERSE POEM IN PRAISE OF CNUT

The Jutes followed you out, they who were loath to flee. You arrayed the host of the men of Skane, free-handed adorner of Van's reindeer of the sail [a sea-god's sailing-ship]. The wind filled the canvas, Prince, above your head. You turned all your prows westward out to sea. Where you went, you made your name renowned.

You carried the shield of war, and so dealt mightily, chief. I do not think, O Prince, that you cared much to sit at ease. Lord of the Jutes, you smote the race of Edgar in that raid. King's son, you dealt them a cruel blow. You are given the name of stubborn.

English Historical Documents, I, quoting the 'Knutsdrapa', a poem by Ottar the Black, who was in Cnut's service

CNUT AND THE WAVES

On three occasions Cnut was said to have displayed his 'nobleness and greatness of mind'. The first was when he married his daughter to the emperor 'with an immense dowry'; the second, when he journeyed to Rome and arranged for pilgrims on the French roads to pay half the usual tolls—at his expense. The third occasion was the most celebrated:

Thirdly, when at the summit of his power, he ordered a seat to be placed for him on the sea-shore when the tide was coming in; thus seated, he shouted to the flowing sea, 'Thou, too, art subject to my command, as the land on which I am seated is mine; and no one has ever resisted my commands with impunity. I command you, then, not to flow over my land, nor presume to wet the feet and the robe of your lord.' The tide, however, continuing to rise as usual, dashed over his feet and legs without respect to his royal person. Then the King leapt backwards, saying: 'Let all men know how empty and worthless is the power of kings, for there is none worthy of the name, but He whom heaven, earth, and sea obey by eternal laws.' From thenceforth King Canute never wore his crown of gold, but placed it for a lasting memorial on the image of Our Lord affixed to a cross, to the honour of God the almighty King: through whose mercy may the soul of Canute, the King, enjoy everlasting rest. *Henry of Huntingdon*

CNUT'S GIFTS TO CANTERBURY, 1031

In this year Cnut returned to England. As soon as he arrived in England, he gave to Christchurch [Canterbury] the port at Sandwich, together with all the dues that there accrue from both sides of the harbour, so that whenever the tide is at its highest and at the full, and a ship is afloat in closest proximity to the shore, and a man is standing on that ship with a small axe in his hand, the monastery shall receive the dues from as far inland as can be reached by a small axe thrown from the ship.

The Anglo-Saxon Chronicle

Cnut was hard but just, and also delighted in song, both sacred and secular. An enactment on hunting illustrates the first point:

And I will that every man be entitled to his hunting in wood and in fields on his own possessions; and let everyone forego my hunting. Beware where I will have it untrespassed under penalty of full wite [fine].

Another story tells how a sudden burst of anger caused him to kill one of his house-carls, a crime which he expiated by paying nine times the man's worth, though the tribunal which he insisted on facing refused to convict him. A visit by the king to Ely on a feast day was remembered by his composing a song for the monks to sing as he and his knights rowed past:

Merrily sung the monks in Ely,
When Cnut the king rowed thereby;
Rowed knights near the land,
And hear we these monks sing.

Cnut died at forty, and his mighty dream of a permanent northern empire founded on a united Scandinavia and England died with him.

Harold Harefoot

1035–1040

The personal life of Cnut had led to complications in the succession. He had married Ethelred's widow, Emma of Normandy, both because she was beautiful and to establish continuity. By her he had a son, Harthacnut, who was in Denmark at the time of his father's death. Cnut's son Harold Harefoot, by his mistress Ælfgifu of Northampton, was in England and after much controversy was elected to the English throne. Harold Harefoot had been supported by the Mercians and the sea-traders of London, of which transaction William of Malmesbury wrote:

He was elected by the Danes and the citizens of London, who, from long intercourse with these barbarians, had almost entirely adopted their customs.

William of Malmesbury

A BENEFACTOR OF CROYLAND ABBEY

Though not generally lauded as a great Christian like his father, Harold received a generous paragraph from the Croyland chronicler:

He presented to our monastery [Croyland] the mantle used at his coronation, made of silk, and embroidered with flowers of gold, which the sacristan afterwards changed into a cope. And still more kindnesses would he have shown us . . . had not a speedy death prematurely carried him off, while still pausing upon the very threshold of his reign. Four years being completed, and the rule of the kingdom being but tasted of, as it were, he departed this life, and was entombed at Westminster.

Ingulph's Chronicle of the Abbey of Croyland, ed. H. T. Riley (1854). *This Chronicle was a forgery made in the later Middle Ages and attributed by the monks to their abbot Ingulf* (c.*1085–1109*) *in an effort to support its claims to privilege.*

CRUELTY TOWARDS ETHELRED'S SON

In 1036 King Ethelred's eldest surviving son crossed from Normandy to see his mother Emma. As he was the nearest male heir of the Anglo-Saxon royal house his presence in England was clearly something of a threat to the new Danish dynasty.

In this year Alfred, the blameless prince, son of King Ethelred came hither to this country, in order to visit his mother who was residing in Winchester; but earl Godwine would not permit him to . . .

But then Godwine prevented him, and placed him in captivity,
Dispersing his followers besides, slaying some in various ways;
Some of them were sold for money, some cruelly murdered,
Some of them were put in chains, and some of them were blinded,
Some were mutilated and some were scalped . . .
Threatened with every kind of injury, the prince still lived,
Until the decision was taken to convey him
To the city of Ely, in chains as he was.
As soon as he arrived, his eyes were put out on board ship,
And thus sightless, he was brought to the monks.
And there he remained as long as he lived.

The Anglo-Saxon Chronicle

Harthacnut

1040–1042

One of Harthacnut's first acts on returning to England as king was to desecrate his half-brother's dead body.

At the instigation of Elfric, archbishop of York, and of others whom I am loath to name, he ordered the dead body of Harold to be dug up, the head to be cut off, and thrown into the Thames, a pitiable spectacle to men! but it was dragged up again in a fisherman's net and buried in the cemetery of the Danes at London. [St Clement Danes]. *William of Malmesbury*

Harthacnut's need to reward the servants who had brought him over from Denmark also made him unpopular:

He imposed a rigid, and intolerable tribute upon England, in order that he might pay, according to his promise, twenty marks to the soldiers of each of his vessels. While this was harshly levied throughout the kingdom, two of the collectors, discharging their office rather too rigorously, were killed by the citizens of Worcester; upon which, burning and depopulating the city by means of his commanders, and plundering the property of the citizens, he cast a blemish on his fame and diminished the love of his subjects.

Ibid.

DEATH OF HARTHACNUT

Hardecanute was snatched away by a sudden death in the flower of his age at Lambeth, the Saxon Chronicle says, from excess of drinking, after a short reign of two years. He was of an ingenuous disposition, and treated his followers with the profusion of youth. Such was his liberality [others say greed] that tables were laid four times a day with royal sumptuousness for his whole court, preferring that fragments of the repast should be removed after those invited were satisfied, than that such fragments should be served up for the entertainment of those who were not invited. In our time it is the custom, whether from parsimony, or as they themselves say from fastidiousness, for princes to provide only one meal a day for their court. *Henry of Huntingdon*

Edward the Confessor

1042–1066

The election of King Ethelred's last surviving son Edward, upon Harthacnut's death, at last restored the Anglo-Saxon line, though Edward was half Norman through his mother Emma and had been an exile in Normandy from 1014 to 1040. Contemporary chronicles regarded Edward's gentle saintliness as a good substitute for strong rule, since God was his ally. Earl Godwin and his sons continued for the most part to enjoy royal favour.

King Edward, under obligation for his kingdom to the powerful Earl Godwin, married his daughter Edith, sister of Harold, who afterwards became king. But Earl Godwin and his sons Sweyn and Harold proved so dangerous that Edward banished them. *Henry of Huntingdon.*

CRISIS IN 1051

The banishment of the Godwins in 1051 resulted from a brawl at Dover which the earl refused to punish because it lay within his earldom of Wessex. Godwin was summoned to London to answer for his behaviour. When he sued for peace the king's response recalled one of Godwin's past crimes:

that he could hope for the king's peace when and only when he gave him back his brother [Alfred, murdered 1036] alive together with all his men and all their possessions intact which had been taken from them quick or dead. *The Life of King Edward*, ed. F. Barlow (1962)

As part of his attack upon the Godwins Edward put aside his wife Edith by whom he had so far had no children. He then offered the throne to his cousin William duke of Normandy who according to The Anglo-Saxon Chronicle *actually visited England.*

Earl Harold went west to Ireland and stayed there all winter under the protection of the king. And soon after this happened the king forsook the Lady who had been consecrated his queen and had her deprived of all that she owned in land, and in gold and in silver, and of everything, and had her committed to his sister at Wherwell.

Then soon came duke William from beyond the sea with a great retinue of Frenchmen and the king received him and as many of his companions as it pleased him and let him go again. *The Anglo-Saxon Chronicle*

William's visit remains a matter of debate but Godwin and his sons returned in 1052 and together with Edith were reinstated. The following year Godwin perished.

In the twelfth year of Edward's reign, when the king was at Winchester, where he often resided, and was sitting at table, with his father-in-law, Godwin, who had conspired against him sitting by his side, the Earl said to him, 'Sir king, I have been often accused of harbouring traitorous designs against you, but as God in heaven is just and true, may this morsel of bread choke me, if even in thought I have ever been false to you.' But God, who is just and true, heard the words of the traitor, for the bread stuck in his throat and choked him, so that death presently followed, the foretaste of the death which is eternal. *Henry of Huntingdon*

THE SUCCESSION IN CRISIS

In spite of Godwin's restoration the succession crisis remained. Edward and Edith had no children. Bishop Brithwin of Wilton was once concerned about the likely extinction of the royal race of the Angles (English) and while meditating on this subject, fell asleep:

. . . when behold! he was rapt on high, and saw Peter, the chief of the apostles, consecrating Edward, who at that time was an exile in Normandy, king; his chaste life too was pointed out, and the exact period of his reign, twenty-four years, determined; and, when enquiring about his posterity, it was answered; 'The kingdom of the English belongs to God; after you he will provide a king according to his pleasure.' *William of Malmesbury*

THE ORIGIN OF THE 'ROYAL TOUCH'

In the remembered golden age of Edward the Confessor,

A young woman had married a husband of her own age, but having no issue by the union, the humours collecting abundantly about her neck, she had contracted a sore disorder; the glands swelling in a dreadful manner. Admonished in a dream to have the part affected washed by the king, she entered the palace and the king himself performed this labour of love, by rubbing the woman's neck with his fingers dipped in water. Joyous health followed his healing hand: the lurid skin opened, so that worms flowed out with the purulent matter, and the tumour subsided. But as the orifice of the ulcers was large and unsightly, he commanded her to be supported at the royal expense until she should be perfectly cured. However, before a week was expired, a fair, new skin returned, and hid the scars so completely, that nothing of the original wound could be discovered: and within a year becoming the mother of twins, she increased the admiration of Edward's holiness. Those who knew him more intimately, affirm that he often cured this complaint in Normandy: whence appears how false is their notion, who in our times assert, that the cure for this disease does not proceed from personal sanctity, but from hereditary virtue in the royal line.

William of Malmesbury

A story told the year after Edward's canonization in 1161 shows that he was perhaps too saintly to be the royal fount of law and order.

Once, among the rest lying on his bed, and his private Treasurer, Hugoline by name, having unaware left open the chest of money in the chamber, a groome of the place being invited through . . . the seeming sleepe of the Prince, approaching boldly, tooke away from thence a good quantity thereof, and put it up in his pocket; and being glad of so happy a success, returned again a second time and so a third; when as the King who had feigned sleep till then, perceiving Hugoline to be coming brake silence, and with friendly voice, said to the wretch: 'Get thee gone for the treasurer comes; who if he chance to catch thee once will not leave thee a farthing of all thou hast.' At this voice, the fellow fled, and scarcely had got his feete forth from one doore, but Hugoline entered by the other, and finding so great a summe of treasure to be wanting he was even ready at first to faint for griefe, then entering into a rage with himselfe began to rent the aire with crys and sighes; when S. Edward arising from his bed and still dissembling the matter quietly demanded the occasion of so great a heavyness and having heard it: 'Hold thy peace man (said he) perhaps he who hath taken it away had more need thereof than we: much good may it do him: the rest I hope will serve for us.' With such quietness he passed over that act. Whence may be gathered how well subdued he had his passions.

Ailred of Rievaulx's Life of Saint Edward King and Confessor, ed. Maffey, H. Hawkins (1632)

On the subject of the succession, the ingenuous Edward was said to have been deceived on his deathbed by his nephew Harold Godwinsson, Earl of Wessex.

In the year of Our Lord 1066 . . . during the month of April, a star known as a comet [Halley's Comet] appeared in the north-west and remained visible for almost 15 days. Learned astrologers . . . declared that this portended the transfer of a kingdom. Indeed Edward king of England, son of King Ethelred . . . had died shortly before; and Harold son of Earl Godwin had usurped the kingdom of England and had already ruled it for three months and caused much harm, stained as he was by perjury, cruelty and other vices . . . The truth was that Edward had declared his intention of transmitting the whole kingdom of England to his kinsman William duke of Normandy . . . through the same Harold, and had with the consent of the English made him heir to all his rights. Moreover, Harold himself had taken an oath of fealty to Duke William of Rouen in the presence of the Norman nobles, and after becoming his man had sworn on the most sacred relics to carry out all that was required of him . . .

He deceived King Edward who was then grievously ill and near to death; he gave an account of his crossing and arrival in Normandy and mission there, but then added falsely that William of Normandy had given him his daughter to wife [there may have been a betrothal] and granted him as his son-in-law all his rights in the English Kingdom. Though the sick monarch was amazed, nevertheless he believed the story and gave his approval to the cunning tyrant's wishes.

The Ecclesiastical History of Orderic Vitalis, ed. Marjorie Chibnall, 5 vols. (Oxford, 1969–81), II. *Orderic Vitalis (1075–c.1143) was born near Shrewsbury but became a monk in Normandy. He always loved his native land and in his account of the Norman Conquest showed a marked degree of fairness and an absence of Norman bias.*

THE DEATH OF EDWARD THE CONFESSOR

Edward's great gift to his people was the building of Westminster Abbey just outside the London of his day. It was consecrated on 28 December 1065, but he was too ill to attend the service.

The consecration of the church was on Holy Innocents' day; and he passed away on the vigil of the Epiphany, and was buried in this same abbey church as is told hereafter:

Now did king Edward, lord of the English,
Send his righteous soul to Christ,
His holy spirit into God's keeping.
Here in the world he dwelt for a time
In royal majesty, sagacious in counsel;
A gracious ruler for twenty-four years
And a half, he dispensed bounties.

The Anglo-Saxon Chronicle

Harold

1066

In the battles of 1066 for the English crown, Harold had no valid hereditary claim, though he was brother-in-law of the Confessor. His mother, Gytha, was a Dane, sister-in-law of Cnut. Two hereditary claimants who intended to fight for their 'rights' were Harold Hardrada king of Norway and William duke of Normandy. Despite these rivals, Harold was elected king by the Witan (*Council*) *in London. He was forty-four when he succeeded to the throne. Although he did not marry, his mistress, Edith Swan-neck, bore his children. His attractive personality was recognized even by his enemies.*

NOTHING WANTING BUT HONOUR

Harold's defenders urged that he had sworn loyalty to William under duress. The Normans denied it and accused him of dishonourable conduct.

This Englishman was very tall and handsome, remarkable for his physical strength, his courage and eloquence, his ready jests and acts of valour. But what were all these gifts to him without honour, which is the root of all good?

Orderic Vitalis

HAROLD'S BROKEN OATH

It was about the year 1064 that Harold visited Normandy, possibly for pleasure but more probably to confirm King Edward's promise of the throne to Duke William. The voyage, so wonderfully depicted on the Bayeux Tapestry, ended with Harold's oath to William, promising, according to the Norman chroniclers, to support the duke's candidature.

William sumptuously refreshed Harold with splendid hospitality after all the hardships of his journey. For the duke rejoiced to have so illustrious a guest in a man who had been sent him by the nearest and dearest of his friends: one, moreover, who was in England second only to the king, and who might prove a faithful mediator between him and the English. When they had come together in conference at Bonneville, Harold in that place swore fealty to the duke employing the sacred ritual recognized among Christian men. And as is testified by the most truthful and most honourable men who were there present, he took an oath of his own free will in the following terms: firstly that he would be the representative of Duke William at the court of his lord, King Edward, as long as the king lived;

secondly that he would employ all his influence and wealth to ensure that after the death of King Edward the kingdom of England should be confirmed in the possession of the duke; thirdly that he would place a garrison of the duke's knights in the castle of Dover and maintain these at his own care and cost; fourthly that in other parts of England at the pleasure of the duke he would maintain garrisons in other castles and make complete provision for their sustenance. The duke on his part, who before the oath was taken had received ceremonial homage from him, confirmed to him at his request all his lands and dignities. For Edward in his illness could not be expected to live much longer . . . After this there came the unwelcome report that the land of England had lost its king, and that Harold had been crowned in his stead. This insensate Englishman did not wait for the public choice, but breaking his oath, and with the support of a few ill-disposed partisans, he seized the throne of the best of kings on the very day of his funeral, and when all the people were bewailing their loss.

English Historical Documents, II, 1042–1189, ed. D. C. Douglas and G. W. Greenaway (1953), quoting William of Poitiers, *The Deeds of William, duke of the Normans and king of the English. William of Poitiers was William's chaplain and finished his history* c.1071.

CIVIL STRIFE

Harold's problems in 1066 were compounded by the enmity of his brother Tostig who, as a result of a rebellion, had been driven out of his earldom of Northumbria. Abroad, he threw in his lot with the Norwegian king and in the event it was this invasion which first reached England in September 1066.

Three Battles

King Harold left London for the north when he heard that his brother Tostig had joined with Harold Hardrada king of Norway in a determined attempt to seize the kingdom. Forced marches would be necessary, for Harold's brothers-in-law, the earls Edwin and Morcar, had been heavily defeated south of York on 20 September at the battle of Fulford Gate. At the same time Harold was suffering from a violent pain in his leg.

He told no one about it but spent the night praying before the holy rood at Waltham. During that night Edward the Confessor was said to have appeared in a vision to the abbot of Ramsay bidding him tell Harold that he would win the coming battle. When Harold received this message his leg was miraculously cured.

On 24 September the city of York surrendered to the enemy, who were drawn up on the far side of the river when Harold arrived at Stamford Bridge.

He saw a horseman fall from his charger and, hearing it was the king of Norway, he said:

'He is a tall man and goodly to look upon, but I think that his luck has left him.'

Harold then gave his brother Tostig a chance to avoid battle by offering him a third of the kingdom. What would Harold give the Norwegian king? asked Tostig.

'Six feet of ground or as much more as he needs, as he is taller than most men.' *Henry of Huntingdon*

THE BATTLE OF STAMFORD BRIDGE, 25 SEPTEMBER 1066

The battle was desperately fought, the armies being engaged from daybreak to noonday, when, after fierce attacks on both sides, the Norwegians were forced to give way before the superior numbers of the English, but retreated in good order. Being driven across the river [Ouse], the living trampling on the corpses of the slain, they resolutely made a fresh stand. Here a single Norwegian, whose name ought to have been preserved, took post on a bridge, and hewing down more than forty of the English with a battleaxe, his country's weapon, stayed the advance of the whole English army till the ninth hour. At last someone came under the bridge in a boat, and thrust a spear into him, through the chinks of the flooring. The English having gained a passage, King Harold (of Norway), and Tostig (his ally), were slain: and their whole army were either slaughtered, or, being taken prisoners, were burnt. Three days later William Duke of Normandy landed at Pevensey in Sussex. Ibid.

THE BATTLE OF HASTINGS (SENLAC), 14 OCTOBER 1066

While Harold was celebrating his victory in York, he heard the shattering news of William's landing. Immediately he set off again, covering the 190 miles to London in six days. But this time, as he passed Waltham and prayed for another victory, the figure of Christ on the cross bowed his head as if in sorrow—or so said the sacristan. After pausing in London only long enough to collect what fresh troops were available, Harold marched towards Hastings, to cut off the Normans' advance on London. The English footsoldiers and Norman cavalry and archers met on a ridge north of Hastings and the sea.

The courageous leaders mutually prepared for battle, each according to his national custom. The English, as we have heard, passed the night without sleep, in drinking and singing, and, in the morning, proceeded without delay towards the enemy; all were on foot, armed with battle-axes, and covering themselves in front by the junction of their shields, they formed an impenetrable body, which would have secured their safety that day, had not the Normans, by a feigned flight, induced them to open their ranks, which till that time, according to their custom, were closely compacted. The king himself on foot, stood, with his brother, near the standard; in order that, while all shared equal danger, none might think of retreating. This standard William sent, after the victory, to the pope; it was sumptuously embroidered, with gold and precious stones, in the form of a man fighting.

On the other side, the Normans passed the whole night in confessing their sins, and received the sacrament in the morning: their infantry, with bows and arrows, formed the vanguard, while their cavalry, divided into wings, were thrown back. The earl, with serene countenance, declaring aloud, that God would favour his, as being the righteous side, called for his arms; and presently, when, through the hurry of his attendants, he had put on his hauberk the hind part before, he corrected the mistake with a laugh; saying, 'My dukedom shall be turned into a kingdom.' Then beginning the song of Roland, that the warlike example of that man might stimulate the soldiers, and calling on God for assistance, the battle commenced on both sides.

William of Malmesbury

AN ARROW IN THE EYE

The best-known story of Harold is probably that of his death as a result of an arrow in the eye, an event clearly illustrated in the Bayeux Tapestry but so close to a scene of a man being felled by a sword as to suggest that it took more than an arrow to dispatch the king. Both William of Malmesbury and Henry of Huntingdon emphasized the same features of the battle: the English stood with compacted shields; the feigned retreat ordered by William trapped the English into breaking their line and pursuing an enemy they thought was in flight; and their destruction by the Norman cavalry and Harold's death from an arrow followed. Henry of Huntingdon's account has been chosen because it includes the slashing of the already mortally wounded Harold by Norman knights. One of them was named as the heir to Guy of Ponthieu, and it was at Ponthieu that Harold had landed before swearing his oath to William and allegedly breaking it.

Harold had formed his whole army in close column, making a rampart which the Normans could not penetrate. Duke William, therefore, commanded his troops to make a feigned retreat. In their flights they happened unawares on a deep trench, which was treacherously covered, into which numbers fell and perished. While the English were engaged in pursuit the main body of the Normans broke the centre of the enemy's line, which being perceived by those in pursuit over the concealed trench, when they were consequently recalled most of them fell there. Duke William also commanded his bowmen not to aim their arrows directly at the enemy, but to shoot them in the air, that their cloud might spread darkness over the enemy's ranks; this occasioned great loss to the English . . . In this attack the greater part were slain; but the remainder, hewing away with their swords, captured the standard. Meanwhile, a shower of arrows fell round King Harold, and he himself was pierced in the eye. A crowd of horsemen now burst in, and the King, already wounded, was slain.

Henry of Huntingdon

One of the horsemen was said to have hacked off Harold's leg, for which unchivalrous act William dismissed him.

THE BURIAL OF HAROLD

Meanwhile the duke had finally routed the enemy and returned to the battlefield, where he gazed on a scene of destruction so terrible that it must have moved any beholder to pity. For the mangled bodies that had been the flower of the English nobility and youth covered the ground as far as the eye could see. Harold was recognised by some tokens, not by his face, and brought to the duke's camp; the conqueror commanded William Malet to bury the body near to the sea-shore, which in life he had defended so long with his armed forces. *Orderic Vitalis*

If that story was true, Harold's body was afterwards exhumed, to be reburied at Waltham Abbey. His brothers Gyrth and Leofwine perished with him and a foreign dynasty supplanted the Anglo-Saxon line.

THE NORMANS

What remained of the old Saxon aristocracy after the two bloody battles of 1066 was stamped out. At the same time Saxon institutions were taken over and adapted to the conquerors' full-blooded feudal system. Three out of the four Norman kings were strong military rulers. The Norman's home was literally his castle. It was when the magnates ran amok under Stephen that people remembered the 'good security' as well as the harshness of life under their first Norman king. In so far as the Norman kings had trouble with the powerful Church, they tended to be treated more severely than their Saxon predecessors by the monkish chroniclers. Yet they were personally pious; and renowned for their energy and courage.

William the Conqueror

1066–1087

Born c.*1028, William was the illegitimate son of Robert II duke of Normandy by Arlette, daughter of a tanner. His life was one of prolonged and successful struggle. The wealth of England, carefully organized by him, was essential to support his armies. In the chronicles he was to be William the Great. His defence of women's rights, at least in one respect, may reflect a happy marriage with Matilda of Flanders. Powerfully built and of medium height, he grew fat and lost his front hair but never lost his impressive regality.*

WILLIAM'S LANDING AT PEVENSEY BAY IN SUSSEX

A bad omen was turned by him to good account:

As the Duke left his vessel to set foot on land
He stumbled and fell with his hands on the sand.
All those who stood near him upraised a great cry,
Struck with fear at so evil an Augury.
But the Duke he exclaimed 'By the splendour of God
I thus with both hands lay my grasp on this Sod.
Prize without challenging—no man can make
So thus of all around us due seizin I take.
We shall see who are brave.

Wace, Roman de Rou et des Ducs de Normandie, ed. A. Malet (1860). *This twelfth-century poem by Canon Robert Wace is often unreliable.*

THE CONQUEROR AT HASTINGS

On the site where William had his first victory, he marked his triumph by founding the monastery, now a school, of Battle Abbey. Accounts of prowess in the field lost nothing in the telling.

Duke William had not concluded his harangue, when all the squadrons, inflamed with rage, rushed on the enemy with indescribable impetuosity, and left the Duke speaking to himself! Before the armies closed for the fight, one Taillefer [a juggler], sportively brandishing swords before the English troops, while they were lost in amazement at his gambols, slew one of their standard-bearers. A second time one of the enemy fell. The third time he was slain himself. Then the ranks met; a cloud of arrows carried death among them; the clang of sword-strokes followed; helmets gleamed, and weapons clashed. *Henry of Huntingdon*

THE CORONATION OF WILLIAM IN LONDON ON CHRISTMAS DAY 1066

This is the only recorded case of the Conqueror showing fear. The coronation service was conducted by two prelates, Archbishop Ealdred of York in English and Bishop Geoffrey of Coutences in French.

But at the prompting of the devil, who hates everything good, a sudden disaster and portent of future catastrophes occurred. For when Archbishop Ealdred asked the English, and Geoffrey bishop of Coutences asked the Normans, if they would accept William as their king, all of them gladly shouted out with one voice if not in one language that they would. The armed [Norman] guard outside, hearing the tumult of the joyful crowd in the church and the harsh accents of a foreign tongue, imagined that some treachery was on foot, and rashly set fire to some of the buildings. The fire spread rapidly from house to house; the crowd who had been rejoicing in the church took fright and throngs of men and women of every rank and condition rushed out of the church in frantic haste. Only the bishops and a few clergy and monks remained, terrified, in the sanctuary, and with difficulty completed the consecration of the king who was trembling from head to foot. Almost all the rest made for the scene of conflagration: some to fight the flames, and many others hoping to find loot for themselves in the general confusion. The English, after hearing of the perpetration of such misdeeds, never again trusted the Normans who seemed to have betrayed them, but nursed their anger and bided their time for revenge. *Orderic Vitalis*

THE DOMESDAY BOOK, 1085–6

The chroniclers took William's thorough survey of his English acquisition to be a sign of his avarice rather than efficiency. Posterity has regarded it as one of the great documents of history.

After this the king had important deliberations and exhaustive discussions with his council about this land and how it was peopled, and with what sort of men. Then he sent his men all over England into every shire to ascertain how many hundreds of 'hides' of land there were in each shire, and how much land and live-stock thc king himself owned in the country, and what annual dues were lawfully his from each shire. He also had it recorded how much land his archbishops had, and his diocesan bishops, his abbots and his earls, and . . . what or how much each man who was a land holder here in England had in land or in live-stock, and how much money it was worth. So very thoroughly did he have the enquiry carried out that there was not a single 'hide', not one virgate of land, not even—it is shameful to record it, but it did not seem shameful for him to do—not even one ox, nor one cow, nor one pig which escaped notice in his survey. And all surveys were subsequently brought to him. *The Anglo-Saxon Chronicle*

THE MAKING OF THE NEW FOREST

If Domesday embarrassed the chroniclers, the New Forest outraged them. Their accounts of William's destruction are much exaggerated. William of Malmesbury, for instance, lamented over 'the dreadful spectacle' of thirty miles of desolation dedicated to animals 'not subjected to the general service of mankind'. Henry of Huntingdon was equally bitter.

He wrung thousands of gold and silver from his most powerful vassals, and harassed his subjects with the toil of building castles for himself. If any one killed a stag or a wild boar, his eyes were put out and no one presumed to complain. But beasts of chase he cherished as if they were his children; so as to form the hunting ground of the New Forest he caused churches and villages to be destroyed, and, driving out the people, made it an habitation for deer. *Henry of Huntingdon*

WILLIAM'S SUCCESS IN KEEPING ORDER

Among other things we must not forget the good order he kept in the land, so that a man of any substance could travel unmolested throughout the country with his bosom full of gold. No man dared to slay another, no matter what evil the other might have done him. If a man lay with a woman against her will, he was forthwith condemned to forfeit those members with which he had disported himself. *The Anglo-Saxon Chronicle*

THE DEATH AND BURIAL OF WILLIAM

Despite the difficulty of discovering any certain utterances of William, his biographer believes that his dying words, as reported by Orderic Vitalis, have an authentic ring.

The king passed the night of 8 September in tranquillity, and awoke at dawn to the sound of the great bell of Rouen Cathedral.

On his asking what it signified, his attendants replied: 'My lord the bell is ringing for Prime in the church of Saint Mary.' Then the king raised his eyes and lifted his hands and said: 'I commend myself to Mary the holy Mother of God . . . that I may be reconciled to her Son our Lord Jesus Christ.' And having said this he died.

Immediate confusion followed his passing, and some of the attendants behaved as if they had lost their wits.

Nevertheless the wealthiest of them mounted their horses and departed in haste to secure their property. Whilst the inferior attendants, observing that their masters had disappeared, laid hands on the arms, the plate, the linen, and the royal furniture, and hastened away, leaving the corpse almost naked.

D. C. Douglas, *William the Conqueror* (1964), quoting Orderic Vitalis

The corpse was to fare no better in years to come. Having been broken in half while being forced into the stone coffin, it was reinterred in 1522 but vandalized by the Calvinists in 1562, only a thigh bone remaining, which, having been reburied in 1642, was thought to have been finally destroyed in 1793 by the French revolutionaries.

Today there is a further addition to the story of William I's burial. A thigh bone, said by the French authorities to be genuine, was discovered in the old tomb and reburied under a new tombstone on 9 September 1987. Queen Elizabeth II had visited the old tomb on 6 June 1984 but did not know then that one of her ancestor's bones was still inside.

William Rufus

1087–1100

Though he reigned for no more than thirteen years and died at only forty, William II or the Younger, Rufus, or the Red, received the kind of verbal larruping usually reserved for evildoers on the grand scale. Irreligion, avarice, and perversion were imputed to him by the Church with which he quarrelled. But was he really worse than a typical bully of a soldier?

A RECENT PORTRAIT OF WILLIAM RUFUS

Here he is built up from the very bricks thrown at him by his monkish detractors into a not altogether unattractive personality.

William Rufus had great physical presence. Although a small man, he was dangerous, like an animal. Archbishop Anselm once likened him to a wild bull. The only portrait is from the pen of William of Malmesbury, who is unlikely to have seen him in the flesh; but even if it is second hand . . . it represents how this vivid actor was remembered.

William of Malmesbury . . . describes the king as thick-set and muscular with a protruding belly; a dandy dressed in the height of fashion, however outrageous, he wore his blond hair long, parted in the centre and off the face so that his forehead was bare, and in his red, choleric face were lively eyes of changeable colour, speckled with flecks of light. In private with his boon-companions he was easy-going. He cracked jokes as he dealt with the business of the day, and was never so facetious as when he was doing wrong, for he hoped that the witticism would dispel the stigma. In public, however, his lack of eloquence was noticeable, and when angry he was reduced to stuttering incoherence. He was apt to assume a ferocious manner, and with his haughty, inflamed face, threatening eyes, and loud, hectoring voice sought to intimidate the company. We may infer that the young king, whose high sense of importance and of the reverence due to the crown was not matched by natural dignity or suitable powers of expression, was driven to this boorish, bullying behaviour as a substitute. But although he easily took offence and was not slow to pay back an imagined insult with interest, he appreciated boldness in others and was easily pacified. Magnanimity was one of his virtues. Orderic Vitalis viewed him in much the same way: he was every inch a soldier, at times violent and swollen with anger, but in his dealings with soldiers of noble birth always courteous, jovial and bountiful.

Despite his reluctance to enrich his followers at the expense of the royal demesne he had inherited from his father, he was renowned for his generosity, a truly royal virtue. F. Barlow, *William Rufus* (1986)

WILLIAM AND THE JEWS OF ROUEN

These stories skilfully manage to combine William's greed with his alleged contempt for Christianity.

It was reported by such travellers that about this time, when King William was staying at Rouen, some Jews who lived in that city came to him and complained that some of their co-religionists had then recently abandoned Judaism and become Christians. They asked that for a price paid to him he should compel them to throw over Christianity and to return to Judaism. He agreed and taking the price of apostasy ordered the Jews in question to be brought before him. To cut the story short, he made most of them, broken by threats and intimidation, deny Christ and return to their former error. *Eadmer's History of Recent Events in England,* ed. G. Bosanquet (1964). *This was written by Archbishop Anselm's own chaplain between 1110 and 1143.*

The second story concerns a steadfast young Jew who is converted to Christianity. His distracted father persuades the king to try, for a large sum of money, to reconvert him. William fails ignominiously, but still claims the reward.

The man replied, 'My son is now still more firm in his confession of Christ and has become still more hostile to me than he was before, and do you say "I have done what you asked me to do, pay up what you promised?" First carry out what you have undertaken to do and then talk about promises. That was the agreement between us.' 'I have done what I could,' said the King. 'Although I have not succeeded, I am certainly not going to have worked for nothing.' In the result the Jew, hard put to it, with difficulty obtained the concession that on his giving half the promised sum of money he should be allowed to keep the other half. Ibid.

WILLIAM'S OSTENTATION, COVETOUSNESS, AND IMMORALITY

The author of this indictment, William of Malmesbury, knew the contemporary writings of Eadmer and took his cue from him.

He [William] was anxious that the cost of his clothes should be extravagant, and angry if they were purchased at a low price. One morning, indeed, while putting on his new boots, he asked his chamberlain what they cost; and when he replied, 'Three shillings,' indignantly and in a rage he cried out, 'You son of a whore, how long has the king worn boots of so paltry a price? Go, and bring me a pair worth a mark of silver.' He went, and bringing him a much cheaper pair, told him, falsely, that they cost as much as he had ordered: 'Aye,' said the king, 'these are suitable to royal majesty.'

... The rapacity of his disposition was seconded by Ralph [Ranulf Flambard], the inciter of his covetousness; a clergyman of the lowest origin, but raised to eminence by his wit and subtlety.... At this person's suggestion, the sacred honours of the church, as the pastor died, were exposed to sale: for whenever the death of any bishop or abbot was announced, directly one of the king's clerks was admitted, who made an inventory of everything, and carried all future rents into the royal exchequer.

... Then was there flowing hair and extravagant dress; and then was invented the fashion of shoes with curved points; then the model for young men was to rival women in delicacy of person, to mince their gait, to walk with loose gesture, and half naked. Enervated and effeminate, they unwillingly remained what nature had made them; the assailers of others' chastity, prodigal of their own. Troops of pathics, and droves of harlots followed the court; so that it was said, with justice, by a wise man, that England would be fortunate if Henry I [William's brother] could reign.

William of Malmesbury

A ROYAL OATH

Rufus has also left his particular stamp on history through a well-attested use of oaths, often tinged with an element of blasphemy. At the siege of Mont St Michel in Normandy he declared approval of one of his company thus:

By the Holy face of Lucca, henceforth you shall be mine and included in my roll of honour and will receive the rewards of knighthood. Ibid.

THE QUARREL WITH THE CHURCH

William's reputation suffered badly in the hands of monastic writers because of his quarrel with Archbishop Anselm, whom he regarded with suspicion and a certain amount of personal animosity. A biased account of the dispute survives in the work of Anselm's chaplain and devotee, Eadmer. As his hostility to Anselm mounted, so did William's ill-temper, which at its worst provoked a famous outburst:

William bade them tell the Archbishop that he hated him much yesterday, that he hated him much today, and that he would hate him more and more to-morrow and every other day.

E. A. Freeman, *The Reign of William Rufus*, 2 vols. (Oxford, 1882), I

THE BUILDING OF WESTMINSTER HALL

Posterity must be grateful to William for his rebuilding of Westminster Hall, though the chronicler considered his action discreditable.

William the younger came over to England [from Nomandy] in the twelfth year of his reign, and kept court for the first time in the new palace at Westminster. Upon his entering the hall to inspect it, some of his attendants observed that it was large enough, others that it was much larger than was necessary; to which the king replied, that it was not half large enough: a speech fitting a great king, though it was little to his credit. Soon afterwards, news was brought to him, while hunting in the New Forest, that his family were besieged in Maine. He instantly rode to the coast, and took ship, whereupon the sailors said to him, 'Wherefore, great king, will you have us put to sea in this violent storm? Have you no fear of perishing in the waves?' to which the King replied, 'I never yet heard of a king who was drowned.' *Henry of Huntingdon*

In fact William's nephew and the heir to Henry I was to be drowned in the White Ship *in 1120.*

DEATH IN THE NEW FOREST

Was William shot accidentally by Walter Tirel, or shot by the secret command of Henry I, or not shot by Tirel at all? All these theories were advanced but none was proven. His dramatic death and the fate of his corpse, however, were agreed to be fitting punishments for his evil life.

The next morning King William sat at meat with his intimates, making preparations to go hunting in the New Forest after dinner. As he was laughing and joking with his attendants and pulling on his boots, a smith arrived and offered him six arrows. He took them eagerly, praised the maker for his work and, ignorant of what was in store, kept four for himself and handed two to Walter Tirel [knight of Poix]. 'It is only right', said the king, 'that the sharpest should be given to the man who knows how to shoot the deadliest shots.'

A warning letter was sent to the king by the abbot of Gloucester, which William rejected disdainfully.

'Does he think I act after the fashion of the English, who put off their journeys and business on account of the snores and dreams of little old women?' Saying this he galloped into the wood.

The king and Walter of Poix were stationed in the wood with a few companions; as they stood on the alert waiting for their prey with their weapons ready a beast suddenly ran between them. The king drew back from his place, and Walter let fly an arrow. He fell to the ground and, dreadful to relate, died at once. When this one mortal perished many were thrown into great confusion, and terrible shouts that the king was dead rang through the wood.

Orderic Vitalis

According to one chronicler, John of Salisbury, Walter Tirel always denied having fired the shot or having been in that part of the forest or indeed in the forest at all. His burial was said to have been ignominious; as another writer suggested:

A few countrymen conveyed the body, placed on a cart, to the cathedral at Winchester; the blood dripping from it all the way. Here it was committed to the ground within the tower, attended by many of the nobility, though lamented by few. Next year [in fact in 1107] the tower fell; though I forbear to mention the different opinions on this subject, lest I should seem to assent too readily to unsupported trifles, more especially as the building might have fallen, through imperfect construction, even though he had never been buried there.

William of Malmesbury

WILLIAM'S OBITUARY

The verdict of the Anglo-Saxon chronicler was harsh, influenced in large part by Rufus's heavy hand on the Church.

In his days, therefore righteousness declined and evil of every kind towards God and man put up its head. He oppressed the Church of God . . . I may be delaying too long over all these matters, but everything that was hateful to God and to righteous men was the daily practice in this land during his reign. Therefore he was hated by almost all his people and abhorrent to God. This his end testified, for he died in the midst of his sins without repentance or any atonement for his evil deeds.

The Anglo-Saxon Chronicle

Henry I

1100–1135

The youngest son of the Conqueror, Henry earned the title 'Lion of Justice', for his unremitting labours in the field of administration; and later the nickname 'Beauclerc', which reflected his good education. His marriage to the Scottish princess Matilda linked him to the old royal line of Wessex and he made a claim to be porphyrogenitus because he was born after his father had become king of England. He holds the record of more acknowledged bastard offspring than any other English king but failed to leave a legitimate male heir to succeed. He was muscular, had thick black hair and was renowned for his energy and cruelty and a nagging fear of assassination.

THE SHAVING AT CARENTON IN NORMANDY, EASTER 1105

Before setting about the pacification of Normandy against his eldest brother Duke Robert Curthose, Henry made a typical gesture of reconciliation with the Church. Bishop Erlo of Sées had welcomed him on landing and exhorted him not to fight for increased power but for the defence of his people.

The king was encouraged by the bishop's words; . . . he said, 'I will rise up to work for peace in the name of the Lord . . . and the tranquillity of the Church of God.'

The bishop was also encouraged for he resumed his 'holy discourse':

'All of you wear your hair in woman's fashion, which is not seemly for you who are made in the image of God and ought to use your strength like men . . . It is not for beauty or pleasure that penitents are instructed not to shave or cut their hair but so that those who, in the sight of God, are bristling with sins . . . may walk outwardly bristling and unshorn before men.'

The bishop added that he supposed they would not clip lest the stumps should prick their mistresses' faces.

Long beards give them the look of he-goats, whose filthy viciousness is shamefully imitated by the degradation of fornicators and sodomites, and they are rightly abominated by decent men for the foulness of their vile lusts.'

When he had finished speaking the king consented in a mood of elation, as did all his magnates, and the bishop, ready for action, immediately drew scissors from his cloak-bag and proceeded to cut the hair, first of the king, and then of . . . most of the magnates with his own hands. The king's whole household and all who flocked to follow their example were close-shorn; dreading a royal decree, they anticipated it by cutting off the tresses they had hitherto treasured, and trod their once-cherished locks under foot as contemptible refuse. *Orderic Vitalis*

They then received the sacrament, clean-shaven, and were ready for business.

HENRY AND HIS CHAMBERLAIN

His Chamberlain Payne FitzJohn used customarily to draw every night a sexterce [liquid measure] of wine to allay the royal thirst; and it would be asked for once or twice in the year, or not at all. So Payne and the pages had no scruple about drinking it all up, and often did so early in the night. It happened that the King in the small hours called for wine, and there was none. Payne got up, called the pages, and found nothing. The King discovered them hunting for wine and not finding it. So he summoned Payne, all trembling and afeared, and said: 'What is the meaning of this? Do you not always have wine with you?' He timidly answered: 'Yes, lord, we draw a sexterce every night, and by reason of your leaving off to be thirsty, or to call for it, we often drink it either in the evening or after you have gone to sleep: and now we have confessed the truth, we beg forgiveness of your mercy.' *The King*: 'Did you draw no more than one for the night?' *Payne*: 'No.' 'That was very little for the two of us: in future draw two every night from the butlers, the first for yourself, the second for me.' Thus his true confession rid Payne of his reasonable fear, and soothed the King's displeasure; and it was characteristic of the royal courtesy and liberality to recompense him with gladness and gain in place of scolding and anger.

Walter Map, Courtiers' Trifles, tr. M. R. James, revised C. N. L. Brooke and R. A. B. Mynors (Oxford, 1983). *Master Walter Map, c.1130–c.1209, was a secular clerk born on the English/Welsh border who was a friend of Gerald of Wales, studied in Paris, and served Henry II, rising to be archdeacon of Oxford. His* Courtiers' Trifles *have been called 'a marvellous guide to a fascinating lumber-room', containing a jumble of twelfth-century mental furniture.*

THE WHITE SHIP DISASTER

Late on 25 November 1120 Henry and his family set out from Normandy. Having the swiftest ship in the fleet, Henry's children delayed departure. An evening's carousing ensued and shortly after embarkation the drunken helmsman ran the ship aground. Only one man survived to tell the tale. The heir, William, his brother, Richard, and a sister perished with devastating consequences for the succession.

Apart from the king's treasure and the casks of wine, Thomas's boat carried only passengers, and they commanded him to try to overtake the king's fleet, which was already sailing in the open sea. As his judgment was impaired by drink, he trusted in his skill and that of his crew, and recklessly vowed to leave behind all those who had started first . . . As the drunken oarsmen were rowing with all their might, and the luckless helmsman paid scant attention to steering the ship through the sea, the port side of the *White Ship* struck violently against a huge rock, which was uncovered each day as the tide ebbed and covered once more at high tide. Two planks were shattered and, terrible to relate, the ship capsized without warning. Everyone cried out at once in their great peril, but the water pouring into the boat soon drowned their cries and all alike perished.

Only two men grabbed hold of a spar from which the sail hung and, clinging to it for the greater part of the night, waited for help to come from any quarter. One was a butcher of Rouen named Berold, and the other a noble lad called Geoffrey, the son of Gilbert of Laigle . . . The night was frosty, so that the young man . . . finally lost his grip and . . . fell back to perish in the sea and was never seen again. But Berold, who was the poorest of all and was dressed in a pelisse made of rams' skins, was the only one of the great company to see the day. In the morning he was taken aboard a light vessel by three fishermen and reached dry land alone. Later, when he was somewhat revived, he told the whole sad tale to those who wished to learn . . .

The sad news spread swiftly from mouth to mouth through the crowds along the sea coast, and came to the ears of Count Theobald . . . but that day no one dared announce it to the anxious king, who earnestly asked for news . . . However, on the following day, by a wise plan of Count Theobald's, a boy threw himself, weeping, at the king's feet, and the king learned from him that the cause of his grief was the wreck of the *White Ship*. Immediately Henry fell to the ground overcome with anguish, and after being helped to his feet by friends and led into a private room, gave way to bitter laments.

Orderic Vitalis

That after the wreck of the White Ship *Henry I 'never smiled again' seems to be a respectable Victorian legend. In 1840 Agnes Strickland, in her* Lives of the Queens of England *(I, 164), wrote that 'all the Chroniclers' agreed to this. They did all agree that Henry's grief was great, but it was apparently Agnes Strickland who first crystallized it in an unforgettable phrase. Within twenty years the phrase was established as history, little Princess Louise, aged 10, writing a French essay for her parents, Victoria and Albert, on the* White Ship *and Henry I which concluded that though he lived on for fifteen years 'he was never seen to smile!'*

HENRY AND THE DEATH OF THE BISHOP OF LINCOLN, 1123

On Wednesday, 10 January it happened that the king was riding in his deer park, with the bishop of Salisbury on one side of him and Robert Bloet, bishop of Lincoln, on the other; and they were talking as they rode. Then the bishop of Lincoln sank down in the saddle, and said to the king: 'Lord king, I am dying.' The king sprang down from his horse, and caught him in his arms, and had him carried home to his lodging, but he died immediately.

The Anglo-Saxon Chronicle

A FAVOURABLE ASSESSMENT OF HENRY'S MORAL QUALITIES

If he could he conquered without bloodshed; if it was unavoidable, with as little as possible. He was free, during his whole life, from impure desires; for, as we have learnt from those who were well informed, he was led by female blandishments, not for the gratification of inconstancy, but for the sake of issue; nor condescended to casual intercourse, unless where it might produce that effect; in this respect the master of his natural inclinations, not the passive slave of lust. He was plain in his diet, rather satisfying the calls of hunger, than surfeiting himself by variety of delicacies. He never drank but to allay thirst; execrating the least departure from temperance, both in himself and in those about him. He was heavy to sleep, which was interrupted by frequent snoring. His eloquence was rather unpremeditated than laboured; not rapid but deliberate. He was inferior in wisdom to no king in modern times; and, as I may almost say, he clearly surpassed all his predecessors in England.

William of Malmesbury

DEATH FROM A SURFEIT OF LAMPREYS, 1135

Henry had married his widowed daughter, Empress Matilda, to Geoffrey, count of Anjou:

The year following [1134] King Henry remained in Normandy, by reason of his great delight in his grandchildren, born of his daughter by the count of Anjou.

Henry intended to return to England, writes the chronicler, but never did so.

His daughter detained him on account of sundry disagreements, which had their origin in various causes, between the King and the count of

Anjou, and which were fomented by the arts of his daughter. These disputes irritated the King, and roused an ill feeling, which some have said resulted in a natural torpor, which was the cause of his death. For, returning from hunting in the 'Wood of Lions', he partook of some lampreys, of which he was fond, though they always disagreed with him; and though his physician recommended him to abstain, the King would not submit to his salutary advice . . . This repast bringing on ill humours, and violently exciting similar symptoms, caused a sudden and extreme disturbance, under which his aged frame sank into a deathly torpor; in the reaction against which, nature, in her struggles produced an acute fever, while endeavouring to throw off the oppressive load. But when all power of resistance failed, this great king died on the first day of December 1135.

Henry of Huntingdon

Stephen

1135–1154

Following the death of his son, Henry I had nominated his daughter Matilda, widow of the German emperor, as his heir, and married her off to Geoffrey Plantagenet, count of Anjou. In 1135, however, Stephen of Blois claimed that his royal uncle had changed his mind on his deathbed and recognized his nephew instead. Stephen moved swiftly to get himself crowned but had not the ruthless temperament to control the ensuing turmoil and civil war that his dispute with Empress Matilda provoked.

A CHIVALRIC KING

Stephen was reputed the handsomest man in England, brave and so good-natured that he would gladly eat with simple folk. Chivalrous and generous, Stephen was unable to command his barons.

When the traitors saw that Stephen was a good-humoured kindly and easy going man who inflicted no punishment, then they committed all manner of horrible crimes. They had done him homage and sworn oaths of fealty to him but not one of their oaths was kept. *The Anglo-Saxon Chronicle*

Other writers were in agreement:

He was adept at the martial arts but in other respects little more than a simpleton. *Walter Map*

and:

If he had legitimately acquired the kingdom and had administered it without lending trusting ears to the whispers of malevolent men, he would have lacked little which adorns the royal character. *William of Malmesbury*

THE EMPRESS MATILDA

Whatever the defects of Stephen's character, he was a more popular choice than Matilda. She was viewed by most as a foreigner; handicapped by the mere fact of being a woman; married to the hated Angevin enemy; and eventually proved to be proud and overbearing.

What was a sign of extreme haughtiness and insolence, when the King of Scotland and the Bishop of Winchester and her brother, the Earl of

Gloucester, the chief men of the whole kingdom, whom she was then taking round with her as a permanent retinue, came before her with bended knee to make some request, she did not rise respectfully, as she should have, when they bowed before her, or agree to what they asked, but repeatedly sent them away with contumely, rebuffing them by an arrogant answer and refusing to hearken to their words; and by this time she no longer relied on their advice, as she should have, and had promised them, but arranged everything as she herself thought fit and according to her own arbitrary will. The Bishop of Winchester, seeing these things done without his approval, and a good many others without his advice, was sufficiently vexed and irritated, yet he disguised all his feelings with caution and craft, and watched silently to see what end such a beginning would have.

Gesta Stephani, ed. K. R. Potter (Oxford, 1976). *This near-contemporary account of the reign is by an unknown chronicler recently identified as Robert of Lewes, bishop of Bath.*

TWO BAD OMENS BEFORE THE BATTLE OF LINCOLN, 1141

Stephen's fortunes reached a low point in 1141 when he faced the rebellious earl of Chester and Matilda's loyal half-brother, Robert, earl of Gloucester. That débâcle, at Lincoln, ended in Stephen's capture.

Meanwhile King Stephen, in much tribulation of mind, heard Mass celebrated with great devotion; but as he placed in the hands of Bishop Alexander the taper of wax, the usual royal offering, it broke . . . The pix also, which contained Christ's body, snapped its fastening, and fell on the altar, while the bishop was celebrating; a sign of the king's fall from power.

Henry of Huntingdon

STEPHEN'S BRAVERY AT THE BATTLE OF LINCOLN

Before the battle began, Stephen was unable to address the army in the usual way as he 'lacked an agreeable voice', but had to get the noble knight Baldwin FitzGilbert to speak for him. Once in the fight, Stephen took the lead.

Then the king's power really shone as with a great battle axe he felled some and scattered others. A new shout went up: 'Everyone onto him! Him against everyone!' At length the king's axe was shattered by the repeated blows. Then Stephen drew his sword, worthy of the royal arm, and wrought wonders until it too was broken. Seeing this, William of Cahagnes, a valiant knight, charged the king and seizing him by the helmet shouted, 'Here everyone, here, I've got the king!' They all rushed in and the king was captured . . .

And so God's judgement was passed on King Stephen: he was led before the Empress Matilda, and imprisoned in Bristol castle. Ibid.

THE ESCAPING EMPRESS

Ultimately Matilda failed to get herself crowned. This was owing more to her own haughtiness than to any lack of pluck. She became associated with two colourful and successful escapes. In 1141 she gained her freedom from Devizes disguised as a corpse, dressed in grave clothes and bound to a bier with ropes. She was carried thus to the safety of Gloucester. Better known is her flight from Oxford Castle in 1142. The chroniclers agreed that Matilda's escape unscathed was a miracle. Details supplied by other chroniclers included the fact that she was let down from the castle tower by a rope and wore a white robe to camouflage her in the snow.

For, when food and every means of sustaining life were almost exhausted in the castle and the king was toiling with spirit to reduce it by force and siege-engines, very hard pressed as she was and altogether hopeless that help would come she left the castle by night, with three knights of ripe judgement to accompany her, and went about six miles on foot, by very great exertions on the part of herself and her companions, through the snow and ice (for all the ground was white with an extremely heavy fall of snow and there was a very thick crust of ice on the water). What was the evident sign of a miracle, she crossed dry-footed, without wetting her clothes at all, the very waters that had risen above the heads of the king and his men when they were going over to storm the town, and through the king's pickets, which everywhere were breaking the silence of the night with the blaring of trumpeters or the cries of men shouting loudly, without anyone at all knowing except her companions and just one on the king's side who revealed her departure, went away from the castle unhindered, as has been said, and unharmed, and by very great effort reached the town of Wallingford during the night. *Gesta Stephani*

THE GALLANT KING AND HIS CHEEKY NEPHEW

As hope of success for Matilda faded her son Henry entered the political arena. His visit to England in 1147 ended in ignominy.

While these things were proceeding thus Henry, son of the Count of Anjou, the lawful heir and claimant to the kingdom of England, came to England from overseas with a fine company of knights. At his arrival the kingdom was straightway shaken and set in a turmoil, because the report of his arrival, to spread more widely in its accustomed way, stated falsely that he was at the head of many thousand troops, soon to be very many thousand, and had brought with him a countless quantity of treasure, sometimes that one district had been plundered, sometimes that another had been ravaged with fire. So his adherents joyfully pricked up their ears, and it seemed to them that a new light had dawned, whereas the king's supporters, as though cowering beneath a dreadful thunderclap, were for some space of time disheartened. *Gesta Stephani*

But when Henry's force, in reality a small body, was twice defeated by Stephen's men, Henry's followers deserted him.

Overwhelmed, and with good cause, by the affliction of this disaster he appealed to his mother, but she herself was in want of money and powerless to relieve his great need. He also appealed to his uncle, the Earl of Gloucester, but he, brooding like a miser over his moneybags, preferred to meet his own requirements only. As all in whom he trusted were failing him in this critical moment he finally, it was reported, sent envoys in secret to the king, as to a kinsman, and begged him in friendly and imploring terms to regard with pity the poverty that weighed upon him and hearken compassionately to one who was bound to him by close ties of relationship and well-disposed to him as far as it depended on himself. On receiving this message the king, who was ever full of pity and compassion, hearkened to the young man . . . And though the king was blamed by some for acting not only unwisely, but even childishly, in giving money and so much support to one to whom he should have been implacably hostile, I think that what he did was more profound and more prudent, because the more kindly and humanely a man behaves to an enemy the feebler he makes him and the more he weakens him. Ibid.

In the contest between Stephen and Matilda a story of 1152 has come down, via Henry of Huntingdon, emphasizing Stephen's humanity in dealing with even the most turbulent barons. While besieging John Marshal in Newbury Castle he granted Marshal a truce on condition that he gave his young son William as a hostage, to guarantee that he would use the truce to negotiate, not to provision the castle. William's father promptly used the truce to send in reinforcements.

Stephen's entourage urged him to hang William at once, but the king was unwilling to execute the child without giving his father a chance to save him by surrendering Newbury. But John Marshal, having four sons and a fruitful wife, considered the youngest of his sons of far less value than a strong castle. He cheerfully told the king's messenger that he cared little if William were hanged, for he had the anvils and hammers with which to forge still better sons. When he received this brutal reply, Stephen ordered his men to lead William to a convenient tree. Fearing that John planned a rescue, the king himself escorted the executioners with a strong force.

William, who was only five or six years old, had no idea what this solemn parade portended. When he saw William, earl of Arundel, twirling a most enticing javelin, he asked him for the weapon. This reminder of William's youth and innocence was too much for King Stephen's resolution, and taking the boy in his arms, he carried him back to the camp. A little later some of the royalists had the ingenious idea of throwing William over the castle walls from a siege engine, but Stephen vetoed that scheme as well. He had decided to spare his young prisoner.

For some two months William was the guest of King Stephen while the royal army lay before Newbury. One day as the king sat in a tent strewn with varicolored flowers William wandered about picking plantains. When the boy had gathered a fair number he asked the king to play 'knights' with him. Each of them would take a 'knight' or plantain, and strike it against the one held by the other. The victory would go to the player who with his knight struck off the clump of leaves that represented the head of his opponent's champion. When Stephen readily agreed to play, William gave him a bunch of plantains and asked him to decide who should strike first. The amiable king gave William the first blow with the result that the royal champion lost his head.

S. Painter, *William Marshal* (Baltimore, 1933)

The child William Marshal lived to become Regent of England, and to defeat another generation of rebellious magnates, thus preserving the child-king Henry III. One good turn deserved another. Meanwhile in 1153 Stephen agreed the Treaty of Westminster with Matilda's son Henry of Anjou: Stephen should remain king for life (in the event less than one more year), and then Henry should succeed him.

GLOOMY SUMMARY OF THE REIGN

Men said openly that Christ and His saints slept.

The Anglo-Saxon Chronicle

THE ANGEVINS OR PLANTAGENETS

Geoffrey, count of Anjou, father of Henry II by the Empress Matilda, gave his name to the new dynasty—Angevin being the adjective derived from Anjou. As for Plantagenet, this was a nickname given to Geoffrey because of his habit of wearing a sprig of the broom plant (genêt) *in his helmet. Its first known use was by Richard duke of York, in the fifteenth century. Geoffrey's descendants were fine physical specimens, energetic, able, hot-tempered, and at the same time charismatic. The violent temper was popularly attributed to their descent from Melusine, daughter of Satan, who had long ago married a count of Anjou and it prompted St Bernard of Clairvaux's observation, 'From the Devil they came and to the Devil they will return.'*

Henry II

1154–1189

A man of action and learning and a great king, Henry II had a freckled 'lion-like' face, a strong, stocky frame and unbounded energy. His hands were said never to be empty but always held a bow or a book. Sometimes known as Henry FitzEmpress, after his mother, he was already count of Anjou, duke of Normandy and, by his marriage to Eleanor the cast-off wife of Louis VII of France, duke of Aquitaine, when he ascended the throne at the age of twenty-one. His realm, which stretched from Ireland to the Pyrenees, has been dubbed the 'Angevin Empire' and his success in governing it stemmed from his own genius. The disaster of his reign—Becket's murder—stemmed partly from his ungovernable temper.

GOOD MANNERS

The Lord King, Henry The Second, of late was riding as usual at the head of all the great concourse of his knights and clerks, and talking with Dom Reric, a distinguished monk and an honourable man. There was a high wind; and lo! a white monk was making his way on foot along the street and looked round, and made haste to get out of the way. He dashed his foot against a stone and was not being borne up by angels at the moment, and fell in front of the feet of the king's horse, and the wind blew his habit right over his neck, so that the poor man was candidly exposed to the unwilling eyes of the lord king and Reric. The king, that treasure-house of all politeness, feigned to see nothing, looked away and kept silence; but Reric said, *sottovoce*, 'A curse on this bare-bottom piety.' *Walter Map*

GOOD TEMPER

There does not seem anyone beside him possessed of such good temper and affability. Whatever way he goes out he is seized upon by the crowds and pulled hither and thither, pushed wither he would not, and, surprising to say, listens to each man with patience, and though assaulted by all with shouts and pullings and rough pushings, does not challenge anyone for it, nor show any appearance of anger, and when he is hustled beyond bearing silently retreats to some place of quiet. Ibid.

GOOD DEEDS

Some time ago I crossed the Channel with him with twenty-five ships which had the obligation of carrying him over without payment. But a storm scattered them all and drove them upon rocks and shores unmeet for ships, except his own, which by God's grace was conveyed into harbour. So in the morning he sent, and to each sailor restored the estimated amount of his loss, though he was not obliged to do so; and the whole sum came to a large amount and perhaps there have been kings who have not paid even their just debts. Ibid.

THE KING IN FURY

Many stories have been preserved about Henry's bouts of rage and fury, none more graphic than the account of his anger at a servant who spoke well of one of his enemies.

I heard that when the king was at Caen and was vigorously debating the matter of the king of Scotland, he broke out in abusive language against Richard du Hommet for seeming to speak somewhat in the king of Scotland's favour, calling him a manifest traitor. And the king, flying into his usual temper, flung his cap from his head, pulled off his belt, threw off his cloak and clothes, grabbed the silken coverlet off the couch, and sitting as it might be on a dung heap started chewing pieces of straw.

W. L. Warren, *Henry II* (1973), quoting a contemporary letter

THE UNPREDICTABLE, RESTLESS KING

One of Henry's chaplains Peter of Blois, a Breton who sometimes acted as secretary to Henry and his wife Eleanor of Aquitaine, described travelling in the royal retinue:

If the king has said he will remain in a place for a day—and particularly if he has announced his intention publicly by the mouth of a herald—he is sure to upset all the arrangements by departing early in the morning. And you then see men dashing around as if they were mad, beating packhorses, running carts into one another—in short giving a lively imitation of hell. If on the other hand, the king orders an early start, he is sure to change his mind, and you can take it for granted that he will sleep until midday. Then you will see the packhorses loaded & waiting, the carts prepared, the courtiers dozing, traders fretting & everyone grumbling . . . I hardly dare say it, but I believe that in truth he took a delight in seeing what a fix he put us in.

Ibid., quoting a letter of Peter of Blois

A SCENE BETWEEN HENRY AND HUGH OF LINCOLN, BISHOP AND SAINT

Bishop Hugh had been summoned to Woodstock to explain his offences in excommunicating a royal forester and refusing a clerical appointment to one of the king's secular friends.

As he approached the royal hunting-lodge . . . the king, who was extremely angry with him, rode off into the forest with his barons, and finding a pleasant spot sat himself on the ground, with the members of the court dispersed in a circle around him. The bishop followed them, but Henry bade everyone ignore his presence. No one rose to greet the bishop or said a word to him, but Bishop Hugh, undaunted, eased an earl out of his place beside the king and sat himself down too. There was a long, brooding silence, broken finally by Henry who, unable to do nothing, called for needle and thread and began to stitch up a leather bandage on an injured finger. Again there was a heavy silence until Bishop Hugh, contemplating the king at his stitching, casually remarked, 'How like your cousins of Falaise you look.' At this the king's anger fled from him, and he burst into laughter which sent him rolling on the ground. Many were amazed at the bishop's temerity, others puzzled at the point of the remark, until the king, recovering his composure, explained the gibe to them: William the Conqueror was a bastard, and his mother was reputedly the daughter of one of the leather-workers for which the Norman town of Falaise was famous.

Ibid., from the chronicler Adam of Eynsham. *Adam was biographer of Hugh of Lincoln.*

HENRY'S COMPLEX PERSONALITY

Henry's character baffled his friends and later his biographers. In 1170, shortly before the murder of Becket, Henry had a set-to with Bishop Roger of Worcester, son of his kinsman Earl Robert of Gloucester. Henry denounced Roger as a 'traitor' for not turning up at the coronation of his son and heir Henry the Younger, unaware that it was not Roger's fault. The bishop retaliated by accusing the king of ill-treating his allegedly beloved Gloucester kinsmen:

'That's how you treat your kinsmen and friends. That's what people receive for serving you. Take my revenues if you will; although I should have thought you might have been satisfied with those of the archbishop [Becket] and the vacant bishoprics and abbeys—surely that's enough on your conscience.'

A member of the party, who thought to please the king, bitterly reviled the bishop; but the king turned on him angrily and abused him soundly, saying, 'You miserable lout; do you think that just because I say what I like to my kinsman and bishop, it gives you any right to dishonour him with your tongue? I don't want to hear a word against him.' So they came to their lodgings, and after dinner the king and the bishop had a friendly talk about what might be done about the archbishop.

Warren, *Henry II*, quoting William FitzStephen. *FitzStephen (d. 1190) was a close friend and biographer of Thomas Becket, writing* c.*1170.*

Henry and Thomas Becket

Undoubtedly the best-known story of Henry's reign is that of his quarrel with his erstwhile friend and chancellor Thomas Becket. Though it occupied scarcely eight years out of the thirty-four that Henry was king it took on a disproportionate significance partly because of what it revealed of the two characters. The chroniclers recorded anecdotes bearing on their early intimacy and later enmity when they clashed over clerical privilege.

THE POOR MAN AND THE CLOAK

The period of their friendship sometimes gave Henry the opportunity to tease and to mock:

One day they were riding together through the streets of London. It was a hard winter and the king noticed an old man coming towards them, poor and clad in a thin ragged coat. 'Do you see that man?' said the king. 'Yes, I see him', replied the chancellor. 'How poor he is, how frail, and how scantily clad!' said the king.' Would it not be an act of charity to give him a thick warm cloak?' 'It would indeed; and right that you should attend to it, my king.'

Drawing near the poor man, Henry offered him a cloak.

Said the king to the chancellor, 'You shall have the credit for this act of charity,' and laying hands on the chancellor's hood tried to pull off his cape, a new and very good one of scarlet and grey, which he was loth to part with. A great din and commotion then arose and the knights and great men of their retinue hurried up wondering what was the cause of this sudden strife. But it was a mystery; both of them had their hands fully occupied and more than once seemed likely to fall off their horses. At last the chancellor reluctantly allowed the king to overcome him, and suffered him to pull the cape from his shoulder and give it to the poor man. Ibid.

Becket became Henry's archbishop in May 1162. By the end of the following year a series of disputes had put them at loggerheads. In an effort at reconciliation the two met in a field at Northampton. Henry addressed his archbishop:

'Have I not raised you from the poor and humble to the pinnacle of honour and rank? It hardly seemed enough for me unless I not only made you father of the kingdom but also put you even before myself. How can it be that so many favours, so many proofs of my affection for you, which everyone knows about, have so soon passed from your mind, that you are now not only ungrateful but oppose me at every turn?'

'Far be it from me, my lord,' said the archbishop. 'I am not unmindful of the favours which, not simply you, but God the bestower of all things has deigned to confer on me through you, so far be it from me to show myself ungrateful or to act against your wishes in anything, so long as it is agreeable to the will of God . . . You are indeed my liege lord, but He is lord of both of us, and to ignore His will in order to obey yours would benefit neither you nor me . . . Submission should be made to temporal lords, but not against God, for as St Peter says, "We ought to obey God rather than men." '

To this the king replied 'I don't want a sermon from you: are you not the son of one of my villeins?'

'It is true', said the archbishop, 'that I am not of royal lineage; but then, neither was St Peter . . .'

Ibid., quoting the chronicler Roger of Pontigny, biographer of Thomas Becket

Contemporaries commented that Becket had experienced a religious conversion on being appointed archbishop of Canterbury:

However, as he put on those robes reserved, at God's command, to the highest of his clergy, he changed not only his apparel but his cast of mind.

The Plantagenet Chronicles ed. E. Hallam (1986), quoting Ralph of Diceto, *Images of History*. *Ralph was a meticulous archivist who became dean of St Paul's and died* c.*1201*.

The change in Becket was more vividly described by one of his biographers:

Clad in a hair shirt of the roughest kind, which reached to his knees and swarmed with vermin, he mortified his flesh with the sparest diet, and his accustomed drink was water used for the cooking of hay.

Warren, *Henry II*, quoting William FitzStephen

THE KING OFFENDED BY BECKET, 1170

Becket's self-imposed exile in France (1164–70), spent partly at the abbey of Pontigny, was punctuated by a series of conferences aimed at bringing about a peace.

After the coronation of his son the king crossed the channel. A conference was held at Montmirail between him and Archbishop Thomas, where the king of France was present. But after much else, when it came to the embrace, because the archbishop said, 'I kiss you in honour of God,' the king refused the kiss as made only conditionally.

The Plantagenet Chronicles, quoting Ralph of Diceto

MURDER IN THE CATHEDRAL, 29 DECEMBER 1170

'Will no one rid me of this turbulent priest?' Did the king say it—or only something like it? The chroniclers and the biographers differ in their accounts and the authenticity of the words must remain uncertain.

At some point, probably on Christmas Day itself, Henry, maudlin with anger at Thomas's ingratitude and railing at the cowardice of his vassals, uttered the fatal words reported by Edward Grim [Thomas's biographer who was wounded during the murder], 'What miserable drones and traitors have I nourished and promoted in my household, who let their lord be treated with such shameful contempt by a low-born clerk!'

F. Barlow, *Thomas Becket* (1986)

Four of Henry's knights, probably not very intelligent, heard some such outburst and left Bures in Normandy for Canterbury, the spokesman for the four being Reginald FitzUrse. The murder scene in Canterbury Cathedral, where Becket had gone on this dark winter evening to hear vespers, had begun with one of the four striking the archbishop on the shoulder with the flat of his sword. William FitzStephen says there was a warning cry of 'Fly, you are a dead man', perhaps suggesting that they did not at first intend to kill him. But when Becket stood firm and resisted their attempts to drag him outside, the four butchered him. Far from being pleased at the handiwork of his faithful knights, Henry seems to have been horrified.

HENRY'S REMORSE

Friends of the king like Arnulf of Lisieux feared that his extreme penitence, beginning with sackcloth, ashes, and three days' solitary starvation, would lose them their king as well as their archbishop:

The king burst into loud lamentations and exchanged his royal robes for sackcloth and ashes, behaving more like a friend than the sovereign of the dead man. At times he fell into a stupor, after which he would again utter groans and cries louder and more bitter than before. For three whole days he remained shut up in his chamber and would neither take food nor admit anyone to comfort him, until it seemed from the excess of his grief that he had determined to contrive his own death. The state of affairs was lamentable and the reason for our grief and anxiety was now changed. First we had to bewail the death of the archbishop, now, in consequence, we began to despair of the life of the king, and so by the death of the one we feared in our misery that we might lose both.

Warren, *Henry II*, quoting a letter of Arnulf, bishop of Lisieux

THE MURDERERS EXPOSED BY THE ARCHBISHOP'S TABLE

On the second or third day after this horrible crime, the murderers stopped for a night's lodging at the archbishop's manor-house of Malling [in Kent], as if their act had been well done. Suddenly, while they were gathered round the fire after their meal, the main table at which the archbishops were accustomed to dine began to shake. So violent was the shaking that everything lying on it, including the knights' pack-saddles, was hurled to the ground with a terrific crash. The servants rushed in immediately with lights and were absolutely amazed that such a strong, solid board should shake like that. Less than an hour passed before the same table again threw down all the things that lay on it with a crash that was even louder, longer and more frightening than before. Then the knights as well as the servants made a thorough search to see whether there was anything under the table to make it totter. When nothing was found, one of the four knights said: 'Take away those pack-saddles of ours which the table itself seems to consider unworthy of lying upon it. And from this we can judge what kind of deed we have perpetrated.'

Gerald of Wales: Giraldus Cambrensis, Opera, ed. J. S. Brewer, J. F. Dimock, and *G. F. Warner, 8 vols.; Rolls Series (1861–91), VIII. Gerald of Wales* (c.*1147*–c.*1216*), *was a royal chaplain and archdeacon and a prolific writer. Resentment over the king's refusal to admit him to the bishopric of St David's led to his personal animosity to Henry II and his sons. Despite his bias he is informative and very valuable for the period.*

HENRY'S PENITENCE

Eventually, Henry did public penance for what had happened to Becket.

When he came near Canterbury, he dismounted from his horse, and laying aside all the emblems of royalty, with naked feet, and in the form of a penitent and supplicating pilgrim, arrived at the cathedral on Friday the thirteenth of June . . . where, prostrate on the floor, and with his hands stretched to heaven, he continued long in prayer . . . Meanwhile the bishop of London was commanded by the king to declare, in a sermon addressed to the people, that he had neither commanded, nor wished, nor by any device contrived the death of the martyr, which had been perpetrated in consequence of his murderers having misinterpreted the words which the king had hastily pronounced: wherefore he requested absolution from the bishops present, and baring his back, received from three to five lashes from every one of the numerous body of ecclesiastics who were assembled.

Roger of Wendover, Flowers of History, ed. J. A. Giles (1849). *Roger, a monk at St Albans was the first of that house to write a series of chronicles which covered the period from the creation to 1440. Wendover's personal contribution dates from about 1215, his account of earlier years drawing upon such contemporary writers as Roger of Howden and Ralph of Diceto. Contemporary accounts dried up about 1201 and so for the reign of John (1199–1215), Wendover, though a primary authority, is unsafe.*

HENRY COULD FLAY THE ALMIGHTY HIMSELF

In 1189 the city of Le Mans was accidentally burnt and Henry forced to flee. His rage burst out against the author of the crime.

Lord, today you have shabbily removed, to my great confusion and dishonour, the city I loved most, in which I was born and bred, and where my father was buried and the body of Saint Julian likewise lies hid. Therefore, Lord, I shall retaliate as best I can by damaging that part of me in which you take most delight [his soul]. *Gerald of Wales* VIII

HENRY AND HIS FAMILY

The king was no more successful in controlling his family than he was in controlling the Church. Believing his own security demanded the imprisonment of his wife from 1173 onwards, he seems to have anticipated the disloyalty of all four of his sons. Gerald of Wales tells the story of how one space in the painted chamber of Winchester Castle was ordered by Henry to be left blank. He was to have it filled in in later years with a poignant design:

There was an eagle painted, and four young ones of the eagle perched upon it, one on each wing and a third upon its back, tearing at the parent with talons and beaks, and the fourth, no smaller than the others, sitting upon its neck and awaiting the moment to peck out its parent's eyes. When some of the king's close friends asked him the meaning of the picture, he said, 'The four young ones of the eagle are my four sons, who will not cease persecuting me even unto death. And the youngest, whom I now embrace with such tender affection, will someday afflict me more grievously and perilously than all the others.' Ibid.

The love of Henry's life was Rosamund Clifford, the 'Fair Rosamund' of legend who died in 1176, poisoned, it was said, by the queen; though Eleanor could have done it only by bribing a servant, since she herself was in prison. Eleanor, disguised as a man, had been caught by Henry trying to join her three rebellious sons at the French court. The sons, Young Henry, Richard, and John, were attempting to increase their continental power in collaboration with the French king, at Henry's expense. His illegitimate son, Geoffrey, remained loyal, Henry saying of him:

'Baseborn indeed have my other children shown themselves; this alone is my true son!'

In 1183 Henry's heir, Young Henry, died and by Henry's last year, 1189, Richard was again deep in the struggle with his father for his rights as heir. Henry, ill and abandoned by all but his natural son Geoffrey, now Archbishop of York, was forced to give Richard the kiss of peace, while cursing him under his breath. When he found that the name of his favourite son John was on the list of Richard's supporters, he turned his face to the wall, according to the chroniclers, and said:

'Enough; now let things go as they may; I care no more for myself or for the world . . . Shame, shame on a conquered king.'

Two days later he died, but reconciled, thanks to the ministrations of his son Geoffrey.

HENRY II'S LYING-IN-STATE, 7 JULY 1189

The bad impression made by the quarrels of Henry's sons with their father is reflected in a story narrated by the chronicler Benedict of Peterborough. The king had died on 6 July.

On the morrow of his death, however, he was carried out for burial adorned with regal pomp: a golden crown on his head, gloves on his hands, a gold ring on his finger, holding a sceptre, wearing shoes of gold fabric with spurs on his feet and girded with a sword. He lay with his face exposed.

When this was announced to his son Richard by an attendant, Richard hastened to meet him. And as he arrived on the scene, immediately blood flowed from the nostrils of the dead king, as if his spirit were angered at Richard's approach. Then, the attendant weeping and wailing, Richard accompanied the body of his father to Fontevraud, where he had him buried.

Benedict of Peterborough, Chronicle of the Reigns of Henry II and Richard, ed. W. Stubbs, 2 vols.; Rolls Series (1867), II. *The chronicle is now attributed to Roger of Howden* (*d. 1201*).

Richard I

1189–1199

The Lionheart or Coeur de Lion used the vast resources of the Angevin Empire which he had inherited to make war. He is known also as a patron and artist, having composed a famous ballad in captivity. Though he possessed fewer of his father's skills as a ruler, his long absences from England, either fighting or in prison, did nothing to diminish the popularity of this six-foot five crusader paladin, with his golden hair and blue eyes. In a formalized medieval 'portrait', the king's crown seems to fit him much better than did Stephen's or John's for whom it looked much too big; his curls are short, his mouth firm instead of smiling (as with them) and he carries a broadsword in his right hand.

RICHARD AND THE KING OF FRANCE

The legend of Richard's homosexuality found clear expression in the following story. Modern historians do not accept it. In 1187 Richard, in the midst of a quarrel with his father, made off to Paris, to consort, defiantly, with his father's great enemy, King Philip Augustus.

Philip so honoured him that every day they ate at the same table, shared the same dish and at night the bed did not separate them. Between the two of them there grew up so great an affection that King Henry was much alarmed and, afraid of what the future might hold in store, he decided to postpone his return to England until he knew what lay behind this great friendship.

Roger of Howden, Gesta Regis Henrici Secundi, ed. W. Stubbs, 2 vols.; Rolls Series (1867). *Of Howden (Yorks.), Roger began writing c.1170 and died in 1201. He was a clerk to Henry II and his work is of great value for this brief period.*

RICHARD'S CORONATION

The solemn occasion, which was followed by the coronation banquet, was marred by a terrible massacre of the Jews.

After they had banqueted, however, the leaders of the Jews arrived against the express decree of the king. And since the previous day the king had forbidden by public notice that any Jew or Jewess could come to his coronation, the courtiers laid hands on the Jews and stripped them and flogged them and having inflicted blows, threw them out of the king's court. Some they killed, others they let go half dead. One of those Jews was so badly injured with slashes and wounds that he despaired of his life, and so terrified was he by the fear of death that he accepted baptism of William, prior of the church of St Mary at York, and was christened William. And in this way he avoided the danger of death, and the hands of his persecutors.

The people of London, following the courtier's example, began killing, robbing and burning the Jews.

Yet a few of the Jews escaped that massacre, shutting themselves up in the Tower of London or hiding in the houses of their friends.

The Plantagenet Chronicles, quoting Ralph of Diceto

RICHARD'S GENERAL LIFE-STYLE

The Church, through a noted preacher of the time, Fulk of Neuilly, once accused Richard of begetting three shameless daughters: Pride, Avarice, and Sensuality. The king's witty reply was a shrewd dig at the Church's own self-indulgence.

'I give my daughter Pride to the Knights Templar, my daughter Avarice to the Cistercians, and my daughter Sensuality to the Princes of the Church.'

LONDON FOR SALE!

Richard made enormous financial demands upon his kingdom to finance his crusade, for his ransom, and for his war against France. His chronicler Richard of Devizes, a monk of St Swithin's, Winchester, is accurate and original, and gives the best contemporary account of the reign, though unfortunately it stops in 1192. He has caught an echo of Richard's desperate efforts to raise money for his crusade:

The king most obligingly unburdened all those whose money was a burden to them, and he gave to whomever he pleased whatever powers and possessions they chose. Joking one day with his companions who were standing by, he made this jest: 'If I could have found a buyer I would have sold London itself.'

The Chronicle of Richard of Devizes of the Time of King Richard the First, ed. J. T. Appleby (1963)

Richard and the Third Crusade 1189–92

Philip Augustus king of France and Leopold duke of Austria were Richard's allies in the struggle to wrest Jerusalem from Saladin, the Saracens' leader. But though the crusade was followed by disaster for Richard, his passion for success and prowess on the field placed him on the pinnacle of chivalrous fame.

RICHARD'S INTERVIEW WITH ABBOT JOACHIM, 1190

After capturing Messina in Sicily, Richard sumptuously entertained King Philip of France on Christmas Day and afterwards talked with the aged and sainted Joachim, abbot of Corrazo. Richard was informed by Joachim that Saladin was the sixth head of the Dragon of the Apocalypse, while Antichrist was the seventh and last head: Richard would be victorious over Saladin and slay him. This was good news; but Richard wanted to know more about Antichrist:

While all those who heard Joachim's words were lost in wonder, King Richard said to the abbot, 'Where is the birthplace of Antichrist? And where will he reign?' Joachim replied . . . that Antichrist was believed to be born already in Rome and would occupy the apostolic throne there . . . Then the king said to him, 'If Antichrist is born in Rome and will there possess the papal throne, I think he is that very Clement who is now Pope.' The king said this because he disliked the Pope. *Benedict of Peterborough*

LIMITED SUCCESS

The siege and capture of Acre in 1191 was the only success of the crusade, though even this was marred by quarrels with his allies and the massacre of his enemies.

The king of the English, who would brook no delay, on the third day of his coming to the siege had his wooden castle that had been built in Sicily and was called 'The Griffon-Killer' built and set up. Before dawn on the fourth day the machine stood against the walls of Acre, and because of its great height it overlooked the city beneath it. As soon as the sun rose, the bowmen upon it kept up an unceasing rain of arrows on the Turks and Thracians. The stone-throwers, skilfully placed, broke down the walls by repeated shots. Even more effective than these were the miners, who opened a way for themselves underground and dug under the foundations of the walls. Ladders also were placed against the walls, and the troops on the ramparts kept watch for an entry. The king himself ran about through the ranks, ordering, exhorting, and inspiring, and he was thus everywhere beside each man, so that to him alone might be ascribed what each man did. *Richard of Devizes*

THE ANGEVIN TEMPER IN RICHARD

The king, having recovered from sickness at Jaffa, planned to storm Jerusalem, not realizing that a truce had been made between his supporters and the brother of Saladin. Richard was enraged at being apparently deserted.

No-one, indeed, upon his unexpected recovery dared to tell him what, unknown to him, they had taken upon themselves to do through fear of his death. Hubert Walter, bishop of Salisbury, however, having conferred with Count Henry concerning the truce, won an easy agreement concerning the best thing to do. When they were discussing by what artifice they could, without danger to themselves, avoid the perilous battle, they hit upon the one artifice out of a thousand: that the people should be dissuaded from the fight. And the astonishing thing came to pass. The courage of the men who were going to fight so failed them, even without dissuasion, that on the appointed day, when the king, as was his custom, went out in front to lead the army, of all the knights and shield-bearers only nine hundred were found. At the defection the king grew extremely angry and raged with wrath. He chewed up the pine staff that he carried in his hand into small bits. He gave vent to his indignation in these words: 'O God,' he said, ' "O God, O God, why hast Thou forsaken me?" For whose sake did we foolish Christians, for whose sake did we Englishmen come here from the farthest part of the world to bear arms? Was it not for the God of the Christians? How good indeed Thou art to Thy servants, who are now for Thy name's sake to be delivered over to the sword and to be the prey of wolves! Oh, how unwillingly would I desert Thee in such a grave hour of need, if I were to Thee what Thou art to me, my lord and advocate! Henceforth my banners will lie prostrate, not for me, indeed, but for Thee. Not because of the cowardice of my army, indeed, art Thou, my king and my God, and not Thy wretched little king, this Richard, defeated this day.'

Richard of Devizes

RICHARD GLIMPSES JERUSALEM

Despite old Joachim's prophecy, Richard never seemed able to capture Jerusalem and slay the Dragon's sixth head, Saladin. It filled him with sorrow.

One day as he was riding out over the hills above Emmaus, King Richard suddenly saw a distant view of the walls and towers of Jerusalem. Hastily he covered his face with his shield, that he might not fully behold the City which God had not allowed him to deliver.

S. Runciman, *A History of the Crusades*, III (Cambridge, 1954)

RICHARD'S IDENTITY BETRAYED

King Richard, having thus landed in Austria, he sent his boy to the town of Gynatia to market, to buy food for his hungry attendants. The boy, on going to the market, made a show of several bezants [gold coins], and behaved in a haughty and pompous manner, on which he was seized by the citizens, who asked who he was, to which he replied that he was the servant of a rich merchant, who had arrived at that town after a three days' journey:

they on this let him go, and he went stealthily to the secret dwelling of the king, and advised him to fly at once, telling him what had happened to him.

The king, however, wished, after his harassing voyage, to rest for a few days and . . . this same boy often went to the public market: and on one occasion . . . he happened incautiously to carry his master the king's gloves under his belt. The magistrates . . . had him again apprehended, and after inflicting various tortures on him, and beating him, threatened to put out his tongue and cut it off, if he did not at once confess the truth. The boy at length was compelled by these tortures to tell them how the matter stood.

Roger of Wendover

PRISONER OF DUKE LEOPOLD, 1192

After the truce at Jaffa Richard's only feasible route home was overland. Inadequately posing as a pilgrim, he fell into the hands of Leopold of Austria.

A German chronicler says that when Richard was captured he was found in a kitchen, roasting meat on a spit, hoping that by doing this servile work he would escape recognition. Unfortunately the kitchen hand was wearing a magnificent ring, worth many years' wages. The details of this story are probably false but in common with the accounts in other chronicles it suggests that the travellers—despite their elaborate pilgrim's attire, long hair and flowing beards—did not take enough trouble to conceal their wealth. . . . So, shortly before Christmas 1192, less than fifty miles from the safety of the Moravian border, Richard fell into the hands of Leopold of Austria. After this time no one in England had a kind word for the Austrians: 'they are savages who live more like wild beasts than men,' wrote Ralph of Diss.

Leopold sent Richard to a strong castle built high on a rocky slope overlooking the Danube: the castle of Dürnstein. The castle is in ruins today, but a legend still clings to its broken walls, the legend of Blondel, the faithful minstrel who travelled the length and breadth of Germany in search of his missing lord. He visited castle after castle and outside each one sang the first lines of a song which he and Richard had composed together. At last, at Dürnstein, he heard the refrain. In its earliest known form, the legend was told by a Rheims minstrel in the second half of the thirteenth century. There is not a shred of evidence to indicate that there is any truth in the story—but it was good publicity for minstrels.

John Gillingham, *Richard the Lionheart* (1978)

On payment of a huge ransom Richard was released by the German emperor on 4 February 1194. The message passed by Philip of France to Richard's brother John ran: 'Look to yourself; the devil is loose.'

RAISING THE RANSOM MONEY

The greater churches came up with treasures hoarded from the distant past, and the parishes with their silver chalices. It was decided that the archbishops, bishops, abbots, priors, earls and barons should contribute a quarter of their annual income; the Cistercian monks . . . their whole year's wool crop, and clerics living on tithes one-tenth of their income.

The Plantagenet Chronicles, quoting Ralph of Diceto

RICHARD'S DEATH AT CHÂLUS, APRIL 1199

In March Richard arrived to subdue the rebel castle of Châlus near Limoges. The crossbowman who shot him on 26 April was reduced to using a frying-pan as a shield. Richard carried a shield but no body armour for his evening ride round the walls.

On the same day, when the king of England and Marchadès [the famous soldier of fortune, Mercadier] were reconnoitring the castle on all sides . . . a certain arbalister, Bertram de Gurdun by name, aimed an arrow from the castle, and struck the king on the arm, inflicting an incurable wound. The king, on being wounded . . . rode to his quarters, and issued orders to Marchadès and the whole of the army to make assaults on the castle without intermission, until it should be taken . . . After its capture, the king ordered all the people to be hanged, him alone excepted who had wounded him, whom, as we may reasonably suppose, he would have condemned to a most shocking death if he had recovered.

After this, the king gave himself into the hands of Marchadès, who, after attempting to extract the iron head, extracted the wood only, while the iron remained in the flesh; but after this butcher had carelessly mangled the king's arm in every part, he at last extracted the arrow. When the king was . . . in despair of surviving he . . . ordered Bertram de Gurdun, who had wounded him, to come into his presence, and said to him, 'What harm have I done to you, that you have killed me?' On which he made answer, 'You slew my father and my two brothers with your own hand and you had intended now to kill me; therefore, take any revenge on me that you may think fit, for I will readily endure the greatest torments you can devise, so long as you have met with your end, after having inflicted evils so many and so great upon the world.' On this the king ordered him to be released and said, 'I forgive you my death . . . To the conquered faction now let there be bright hopes, and the example of myself.' And then, after being released from his chains, he was allowed to depart, and the king ordered one hundred shillings of English money to be given him. Marchadès, however, the king not knowing of it, seized him, and after the king's death, first flaying him alive, had him hanged. *Roger of Howden*

SAYING OF RICHARD

To the Holy Roman Emperor:

'I am born of a rank which recognizes no superior but God.'

John

1199–1216

Henry II dubbed his fourth son Lackland because of the difficulties of finding adequate provision for John. As king, John's failure to hold on to his father's 'empire' seemed to confirm the appropriateness of the early epithet but the unrelieved blackness of John's reputation is as exaggerated as his brother Richard's preternatural lustre. John's bad press owed much to the work of a generation of monastic chroniclers who never knew him but recorded in detail scurrilous stories with a largely didactic purpose in mind.

'KING JOHN WAS NOT A GOOD MAN'!

So many damaging tales were told about John that it is hard to be selective. On his first expedition to Ireland (1185), he and his companions laughed at the beards of the Irish chieftains when they came to pay their homage. Richard I scorned his efforts when faced with his rebellion in 1194, saying he was not a man to win land if there was the merest show of resistance and pardoned his 27-year-old brother because he was only a child! When in 1199 he was invested as duke of Normandy he was giggling and gossiping so much that he dropped the ceremonial lance. Soon after Richard's death he went in the company of the saintly bishop Hugh of Lincoln to visit the tombs of his father and brother at Fontevrault.

On the holy feast of the Resurrection of our Lord he approached the altar to make his offering, as was the custom, to the bishop who was assisting there. His chamberlain placed in his palm the twelve gold pieces which are the customary oblation of kings. Surrounded by a large crowd of nobles, he stood in front of the bishop, gazing on these coins and playing with them, and delayed making his offering for so long a period that everyone gaped at him in amazement. At last, the bishop, annoyed at such behaviour at this particular time and place, said, 'Why do you look at them so intently?' He answered, 'I am looking at these gold pieces and thinking that if I had had them a few days ago I would not have delivered them to you, but have pocketed them; but now you can take them.' The holy and generous soul of the man of God was outraged, and, blushing with shame on his behalf, he drew back his arm, and refused to touch the gold, or let such greedy lips kiss his hand. He groaned and shook his head at him, saying 'Put down what you are clutching, and go away.' Throwing the money into the silver basin for oblations, he withdrew.

Adam of Eynsham, Magna Vita Sancti Hugonis ed. D. L. Douie and D. H. Farmer (Oxford, 1961, corr. repr. 1985)

THE DEATH OF PRINCE ARTHUR

John's first problem in 1199 was the claim of his nephew Arthur, the son of John's older brother, Geoffrey of Brittany. Philip Augustus had supported Arthur in his claim to England but had abandoned him again in 1200 when peace was made, a peace which rather unjustly earned John a new epithet, 'Softsword'. When in 1202 John and Philip renewed their war, the latter once more took up Arthur's cause. John in a spectacular bid to rescue his mother Eleanor at Mirebeau captured his nephew.

After King John had captured Arthur and kept him alive in prison for some time, at length, in the castle of Rouen, after dinner on the Thursday before Easter (3 April 1203), when he was drunk and possessed by the devil he slew him with his own hand, and tying a heavy stone to the body cast it into the Seine. It was discovered by a fisherman in his net, and being dragged to the bank and recognised, was taken for secret burial, in fear of the tyrant, to the priory of Bec called Notre Dame des Prés.

W. L. Warren, *King John* (1961), quoting the Margam Annals. *Produced in Glamorganshire, these annals were contemporary and may well be accurate regarding the disappearance of Arthur, though the case is still not proven.*

THE FALL OF CHÂTEAU GAILLARD, 6 MARCH 1204

This castle, built at vast expense, by Richard I, was the key to Rouen, John's Norman capital. An ingenious scheme, typical of John, to raise the French siege of Saucy Castle by sending supply boats in on the evening tide while simultaneously attacking the enemy camp, miscarried because of the strength of the current. Later chroniclers preferred to record a different view of John's efforts.

At length messengers came to king John with the news, saying the king of the French has entered your territories as an enemy, has taken such and such castles, carries off the governors of them ignominiously bound to their horses' tails and disposes of your property at will without anyone gainsaying him. In reply to this news, king John said, 'Let him do so; whatever he now seizes on I will one day recover': and neither these messengers, nor others who brought him the like news, could obtain any other answer.

Roger of Wendover

JOHN AND THE EMIR OF NORTH AFRICA

Like many of his predecessors, John became embroiled in a dispute with the Church. Refusing to accept the Pope's nominee, Stephen Langton, as archbishop of Canterbury, John endured, first an interdict (1208) and second a ban of excommunication (1209). Here was fertile ground for the later chronicler Matthew Paris, who told the improbable story of John offering England as a tributary of the emir and himself embracing Islam. John's envoys, however, found the emir reading St Paul's epistles in Greek. The emir decided there was nothing wrong with Paul except his apostacy from the Jewish faith. The emir was curious about John's personality and Paris claimed to have heard the story from the very envoy who gave the reply. Matthew Paris (d. 1259), was a monk of St Albans who worked with Roger of Wendover. He was no more reliable than Roger in describing John, but was a more lively writer. He wrote a considerable time after John's death and had a strong animus against him.

The emir wanted to know if King John were a man of sound morals, if he were virile and bore lusty sons, adding that if Robert did not reply truthfully he would never believe a Christian again, certainly not a tonsured one. So Robert, promising on his word as a Christian to tell the truth, was obliged to admit that John was a tyrant not a king, a destroyer instead of a governor, crushing his own people and favouring aliens, a lion to his subjects but a lamb to foreigners and rebels. He had lost the duchy of Normandy and many other territories through sloth, and was actually keen to lose his kingdom of England or to ruin it. He was an insatiable extorter of money: he invaded and destroyed his subjects' property; and he had bred no worthy children but only such as took after their father. He detested his wife and she him. She was an incestuous and depraved woman, so notoriously guilty of adultery that the king had given orders that her lovers were to be seized and throttled on her bed. He himself was envious of many of his barons and kinsfolk, and seduced their more attractive daughters and sisters. As for Christianity he was unstable and unfaithful. When the emir heard this he no longer merely despised John: he loathed him.

Here, for the first time, emergent in full colour, is the traditional impression of King John.

Warren, *King John*, summarizing the account of Matthew Paris

JOHN'S BLASPHEMY

A fondness for oaths was not unique to John, as has been seen with William Rufus. John's particular choices were, 'By God's teeth' and 'By God's feet', according to Roger of Wendover and Matthew Paris. The latter had a further tale of the king's blasphemy:

During this same period the king was leading such a dissipated life that he ceased to believe in the resurrection of the dead and other similar articles of the Christian faith. He made blasphemous and ribald remarks, one of which I recall. One day out hunting a particularly fine fat stag was brought down and skinned in the presence of the king. 'Oh what a good life that beast has led,' laughed the king mockingly, 'and yet it has never heard holy Mass!' *Matthew Paris, Chronica Majora*, ed. H. R. Luard, Rolls Series (1872–4)

JOHN'S LECHERY

Though the story of John's lechery as recounted to the emir may owe a great deal to the pen of Matthew Paris, the fact is that some seven or eight bastards were born to the king. William of Newburgh (d. 1198), told how John had lusted after the wife of the baron Eustace de Vesci who contrived to smuggle a prostitute into the king's bed in her place. When John next day coarsely told him how good his wife had been in bed, de Vesci confessed—and fled. John's first wife, Isabella of Gloucester, bore him no children and he divorced her in 1199 after ten years to marry the twelve-year-old Isabella of Angoulême. Far from hating her, John was said to have been infatuated with her. He failed to return to France in 1205 because:

he was enjoying all the pleasures of life with his queen, in whose company he believed that he possessed everything he wanted . . .

Roger of Wendover

JOHN'S GOVERNMENT

The improvement in John's reputation owes most to the surviving records of his government, which show his great concern and interest in administration. Contemporary comments on this, though few, were fairly objective. Ralph of Coggeshall (d. 1216), recorded that he ruled 'satis laboriosae' (energetically enough). He travelled widely in England, as few of his predecessors had done, often dealing with mundane financial and legal matters. Some credit was readily given. An anonymous chronicler provides a valuable and informative account of John's last years:

There was now [1212] no one in Ireland, Scotland or Wales who did not bow to his nod, a situation which, as is well known, none of his predecessors had achieved. He would thus have seemed successful and overflowing with promise for his successors except that he had been despoiled of his foreign lands and subjected to the sentence of excommunication. *The Plantagenet Chronicles*

The same commentator gave his conclusion thus:

John was indeed a great prince but scarcely a happy one and, like Marius, he experienced the ups and downs of fortune. He was munificent and liberal to outsiders but a plunderer of his own people, trusting strangers rather than his subjects, wherefore he was eventually deserted by his own men and, in the end, little mourned. Ibid.

MAGNA CARTA—LIBERTY WITHIN THE LAW

In sealing the Great Charter on 15 June 1215 in the meadows of Runnymede, John stood as always for the Crown's authority. The rebellious barons insisted that the Crown's authority should not be synonymous with arbitrary power.

King George VI, while returning to Windsor after carrying out an engagement during World War II, caught sight of Runnymede and said:

'That's where it all began . . .'

Roger of Wendover described the scene in 1215, beginning with the barons in London receiving a conciliatory message from John:

They in their great joy appointed the fifteenth of June for the king to meet them, at a field lying between Staines and Windsor. Accordingly, at the time and place pre-agreed on, the king and nobles came to the appointed conference, and when each party had stationed themselves apart from the other they began a long discussion about terms of peace and the aforesaid liberties . . . At length, after various points on both sides had been discussed, King John, seeing that he was inferior in strength to the barons, without raising any difficulty, granted the underwritten laws and liberties.

Roger of Wendover

THE ACCIDENT TO JOHN'S BAGGAGE, OCTOBER 1216

This much-chronicled disaster towards the end of John's life gave rise to the traditional 'laundry' joke, that 'King John lost his crown in the Wash'. The following account is from the notebook of Ralph, abbot of a Cistercian house at Coggeshall in Essex, which is a valuable contemporary source for the reign:

Moreover the greatest distress troubled him, because on that journey from Lynn he had lost his chapel with his relics, and some of his packhorses with divers household effects at the Wellstream [which flowed into the Wash], and many members of his household were submerged in the waters of the sea, and sucked into the quicksand there, because they had set out incautiously and hastily before the tide had receded.

Radulphi de Coggeshall Chronicon Anglicanum, ed. J. Stevenson, Rolls Series (1875)

Roger of Wendover added that John had lost not only relics and household effects but also

the treasures, precious vessels, and the other things which he cherished with special care . . .

Roger of Wendover

Did John's special treasures include the regalia, with the Empress Matilda's imperial crown, bequeathed to John by his grandmother? Professor Warren writes:

It is possible then that they were indeed swallowed up by the sands of the Wellstream estuary. But anyone who hopes to find them there should note the further possibility that John was robbed of them on his deathbed: a priest who went to Newark to say a mass for the dead king's soul subsequently told the abbot of Coggeshall that he had seen men leaving the city laden with loot. Warren, *King John*

Writing in the sixteenth century John Stow found some good to say of bad King John: in 1213 he repented of his illegal acts and was forgiven by Stephen Langton, archbishop of Canterbury.

The 17 of August, Stephen Langton . . . and all the others that were banished, arrived at Dover, and went to Winchester to the king, who meeting them in the way, fell flat upon the earth before their feet, and with tears beseeched them to take pity on him, and on the realm of England. The Archbishop and Bishops likewise, with tears took him up from the ground, and brought him unto the doors of the Cathedral church and with the psalm of *Miserere* absolved him; then the King took an oath to call in all wicked laws, and to put in place the laws of King Edward [the Confessor]. Divine service being ended, the King, Archbishop, Bishops, and Nobles, dined all at one table. *Annals of John Stow* (1631)

The reconciliation did not last and a year or two later the barons put on the screw of Magna Carta.

Another story of King John's death, attributed by Stow to 'a nameless author'.

King John was poisoned by a monk of Swinsted Abbey in Lincolnshire, for saying that if he might live half a year, he would make a half penny loaf of bread worth twenty shillings. Ibid.

Henry III

1216–1272

Borrowing from the chronicle of Nicholas Trevet, the writer William Rishanger portrayed Henry III as being of medium height and thickset, with a narrow forehead and a drooping eyelid. It is an unimpressive portrait. Succeeding his father at the age of nine—the first royal minor in English history—he lived to be sixty-five. For such a long reign, it was undistinguished except in two important respects, one of them negative as far as Henry was concerned: the famous parliament of 1265 (Simon de Montfort's parliament), summoned by Henry under duress; and the king's contribution to culture and religion through superb church building. The rebuilding of the Confessor's abbey at Westminster, partly inspired by the construction of the Sainte Chapelle by his brother-in-law, Louis IX of France, was probably Henry's most spectacular architectural achievement.

THE MINORITY 1217–23

Apart from the threat to stability implicit in the new king's youth, the civil war of John's reign was still being waged with the Dauphin, Louis of France occupying London. A hasty coronation was organized at Gloucester by a few trusty barons and the elderly William Marshal, earl of Pembroke, though over seventy, was prevailed upon to act as regent.

'By God's sword,' said William Marshal, 'this advice is true and good; it goes so straight to my heart, that if everyone else abandoned the king, do you know what I would do? I would carry him on my shoulders, step by step, from island to island, from country to country, and I would not fail him, not even if it meant begging my bread.' 'You cannot say more and God will be with you,' replied his friends.

Chronicles of the Age of Chivalry, ed. Elizabeth Hallam (1987), quoting the Barnwell Annalist

Magna Carta was reissued in 1217 as a sign of good will to rebellious barons (thereby guaranteeing its future importance), and with courage and resolve the French were driven out, so that by the time William Marshal died in 1219 the risk of Plantagenet collapse had receded. His final exchange with his king (as given in an Anglo-Norman epic poem written before 1231, which is an excellent source for the period) was typical of the man:

'Sir, I pray God, if ever I have done anything that has pleased Him, to grant you the grace to be a good man (*homme de bien*). And if it so happen that you follow the example of any criminal ancestor (*ancêtre felon*), I pray God not to grant you a long life.' 'Amen', replied the king.

Histoire de Guillaume le Marechal, ed. P. Meyer, 3 vols.; Société de l'histoire de France (1891–1901), III

Henry felt William's death was yet another punishment for the murder of Thomas Becket.

When the King, who loved William devotedly, heard the news of his death and saw his dead body covered with a cloth, he heaved a deep sigh and said: 'Alas, woe is me, is the blood of blessed Thomas the martyr not even yet avenged?' *Matthew Paris*

Henry's Piety and Devotion

Henry's extreme piety gave rise to many stories. It even developed a competitive element in his relationship with his royal brother-in-law. Henry's insistence on hearing mass every time he met a priest while travelling to meet Louis IX so delayed his arrival that on a future occasion the French king had all priests banned from Henry's route!

On one occasion, St Louis, the French king, when in conversation with him . . . said that 'the attention ought not always to be devoted to the hearing of masses, but that we ought to hear sermons as often as possible'; to which the King of England with witty urbanity replied, 'that he would rather often see a friend, than hear speak of him, although he should hear good spoken of him'. *Matthew Paris*

HENRY'S LIFELONG AWE OF THE CHURCH

Whilst we are speaking of the virtues of the noble king, we ought not to omit to mention, that as soon as he was crowned, he always afforded strict justice to everyone, and never allowed it to be subverted by bribery . . . he held all ordained prelates and especially religious men in such respect . . . that once on a time when all the prelates of the kingdom were assembled before the king by order of the pope, to make a grant of the twentieth part of all moveable property for the assistance of the Holy Land, and were sitting apart discussing the matter, the king said in a low voice to Geoffrey Fitz-Peter and William Briwere, who sat at his feet, 'Do you see those prelates who are sitting there?' They answered, 'We do, my lord.' The king then said to them, 'If they knew how much I, in my reverence of God, am afraid of them and how unwilling I should be to offend them, they would trample on me as on an old and worn-out shoe.' *Roger of Wendover*

Henry's Reputed Avarice

Royal expenses were swelled by calls for wars in France, Ireland, Wales, and Scotland, as well as the Holy Land and for the Church in general. Though he was frequently lavish, Henry could be tight-fisted and he gained a reputation for greed, as three anecdotes reveal.

AT THE BIRTH OF THE HEIR

On the night of the 16th of June, a son was born at Westminster to the King by his wife Eleanor. At this event all the nobles of the kingdom offered their congratulations, and especially the citizens of London, because the child was born at London; and they assembled bands of dancers, with drums and tambourines, and at night illuminated the streets with large lanterns . . . A great many messengers were sent to make known this event, who returned loaded with costly presents. And now the King deeply clouded his magnificence as a king, for, as the messengers returned, the King enquired of each what he had received, and those who had received least, although they brought valuable presents with them, he ordered to send them back with contempt; nor was his anger appeased till each person had given satisfactory presents at the will of the messengers. Of this a certain Norman wittily remarked, – 'God gave us this child, but the king sells him to us.' *Matthew Paris*

WRUNG FROM THE JEWS

About the same time, the King extorted from the most unfortunate Jews a heavy ransom in gold and silver. To say nothing of the rest, he defrauded one Jew, viz. Aaron of York, of four marks of gold and four thousand marks of silver. The King received from each Jew, whether man or woman, the gold into his own hand, becoming, from a king, a new kind of tax-gatherer.

Ibid.

WRUNG FROM THE LONDONERS

When they were all assembled [in Westminster Hall] the King humbly, and as if with rising tears, entreated that each and all of the citizens would with mouth and heart forgive him for his anger, malevolence, and rancour towards them; for he confessed openly, that he often, and his agents oftener, had done them manifold injury, by unjustly taking from them and retaining possession of their property, and by often violating their liberties, for which he now begged them to grant him their pardon. The citizens, therefore, seeing that it was not expedient for them to act otherwise, assented to his request; but no restitution of the property taken from them was made to them. Ibid.

HENRY'S COMPASSION FOR SINNERS

A certain knight of Norfolk, named Godfrey de Millers, of noble birth and distinguished in knightly deeds, being shamefully led astray, secretly entered the lodgings of John the Briton, a knight, for the purpose of lying with his daughter, but was seized by some persons concealed, with the connivance of the harlot herself, who was afraid of being thought a married man's mistress, violently thrown to the ground, and severely beaten and wounded. After this he was suspended to a beam, with his legs stretched apart, and, when thus exposed to the will of his enemies, he was disgracefully mutilated to such a degree that he would have preferred decapitation, and, thus wounded and mutilated, was ejected, half-dead, from the house. A complaint of this proceeding having reached the King, the authors of this great cruelty were seized, and John the Briton being found guilty of it, he was disinherited and banished for ever . . . About the same time, too, a certain handsome clerk, the rector of a ridged church, who surpassed all the knights living round him in giving repeated entertainments and acts of hospitality, was involved in a similar misfortune. However, the King, touched with compassion and deeply grieved, ordered it to be proclaimed as a law by herald, that no one should presume to mutilate another for adultery except in the case of his wife. Ibid.

THE BATTLE OF EVESHAM

Simon de Montfort had captured King Henry and his son Prince Edward at the battle of Lewes, but Edward escaped and raised an army against the rebels. It was said that Simon de Montfort foresaw the death of himself and one of his sons in the final battle against his enemies, including Lord Edward (Edward I). Simon held Henry as a hostage.

Some persons . . . stated that on one occasion the bishop [St Robert Grossetête of Lincoln] placed his hand on the earl's eldest son, and said to him, 'My well-beloved child, both thou and thy father shall die on one day, and by one kind of death; but it will be in the cause of justice and truth.' Report goes, that Simon, after his death, was distinguished by the working of many miracles, which, however, were not made publicly known, for fear of kings . . . Before the above battle, as some say, Simon having gone out of the town of Evesham, and seen with what prudence and skill the ranks of his adversaries were drawn up, said to his companions, 'By St James's arm' (such was his usual oath), 'they are approaching with wisdom, and they have learnt this method from me, not of themselves. Let us, therefore, commend our souls to God, for our bodies are theirs.' *Matthew Paris*

A ROYAL MENAGERIE

As premier zoo-keeper of Western Europe, Henry kept in his royal menagerie at the Tower a camel, buffaloes, the first elephant in England, a bear from the King of Norway, three leopards from the Emperor Frederick II, and a lion from Louis IX.

The King to the Sheriffs of London, greeting.

We bid you to find necessaries for our lion and his keeper while they are in the Tower of London, and this shall be reckoned to you at the Exchequer.

The King to the same, greeting.

We bid you to cause William, the keeper of our lion, to have 14*s*. which he spent on buying chains and other things for the use of the said lion, and this shall be reckoned to you at the Exchequer.

M. A. Hennings, *England Under Henry III* (1924)

The arrival of the royal elephant had been remarkable enough for one of the chroniclers to note it, under the heading, 'Of an elephant in England'.

About this time, too, an elephant was sent to England by the French king as a present to the king of England. We believe that this was the only elephant ever seen in England, or even in the countries on this side of the Alps; wherefore the people flocked together to see the novel sight.

Matthew Paris

THE ROYAL FISH DISH

The death of the first Henry from too many lampreys clearly did not dampen the royal preference for these eel-like fish. With so many fast days fish consumption was considerable. Sometimes it was difficult to bear:

As to sending lampreys to the king. Order is given to the sheriff of Gloucester that since after lampreys all fish seem insipid to both the king and the queen, the sheriff shall procure by purchase or otherwise as many lampreys as possible in his bailiwick, place them in bread and jelly, and send them to the king while he is at a distance from those parts by John of Sandon, the king's cook, who is being sent to him. When the king comes nearer, he shall send them to him fresh. And the king will make good any expense to which the sheriff may be put in this connection when he comes to those parts. Witness the king at Canterbury on the fourth day of March [1237]. Hennings, *England Under Henry III*

Edward I

1272–1307

Born in 1239, Edward had a long and formative political apprenticeship during which he had revealed disturbing traits of disloyalty, impetuosity, and recklessness. He learned, however, from his mistakes and has left behind a reputation as a strong king, as his many epithets reveal: 'the English Justinian', for his contribution to law; 'the Hammer of the Scots', for his persistent, though ultimately unsuccessful, war against his northern neighbour; and 'the Flower of Chivalry' and 'the best lance in the world', for his prowess.

'LONG SHANKS'

On 2 May 1774, Edward's tomb at Westminster Abbey was opened and the king's body was found to be intact. Tall and wiry and known as Long Shanks from the way he gripped his saddle, Edward is described plainly, if a little too conventionally.

In boyhood, his hair was silvery blond in colour; in his youth, it began to turn from fair to dark, while in old age it became a magnificent swan-like white. His forehead was broad, and the rest of his features were regular, except that his left eyelid drooped, which gave him a resemblance to his father.

Despite a lisp, he was not lacking in eloquence when persuasive arguments were needed. His long arms were in proportion to his supple body; no man was ever endowed with greater muscular strength for wielding a sword. His breast swelled above his stomach, and the length of his legs ensured that he was never dislodged from his seat by the galloping and jumping of horses.

Chronicles of the Age of Chivalry, quoting William Rishanger. *A monk of St Albans, Rishanger wrote a continuation of the chronicles associated with the Hertfordshire monastery.*

THE BOY EDWARD

He was famed as one who enjoyed the enduring protection of the Lord of Heaven. For example, as a boy, he was once in the middle of a game of chess with one of his knights in a vaulted room, when suddenly, for no apparent reason, he got up and walked away. Seconds later, a massive stone, which would have crushed completely anyone who happened to be underneath it, fell from the roof on to the very spot where he had been sitting. *William Rishanger*

PLANTAGENET TEMPER

Bad temper was a characteristic common to many of our kings. Edward was no exception and surviving records tell of a coronet in need of repair because he had thrown it into the fire and of a page to whom he paid 20 marks (about £13), in compensation because he had wounded the boy badly when he struck him in anger at his daughter's wedding. Fury, however, could evaporate as rapidly as it erupted. He once had occasion to deal with a courtier who failed to take care of a royal falcon after it brought down a wild duck in the reeds of a river.

Edward rebuked him, and when the man seemed to be ignoring his words, he added threats, to which the other, seeing that there was neither bridge nor ford in the vicinity, replied: 'Oh, I'm safe enough with this river between us!' Infuriated, the prince plunged on horseback into the river (whose depth he did not know) and his horse swam across to the opposite bank, which was hollowed out by the downward flow of the water. After emerging with some difficulty, Edward drew his sword and set off in pursuit of the courtier, who had taken to flight.

The courtier was soon overtaken and baring his head, knelt humbly to the prince, who immediately forgave him. They went together to the assistance of the falcon. Ibid.

THE KING'S EVIL

He is the first King for whom there is solid documentary evidence for the practice of touching for the king's evil. This is not to argue that the practice was an innovation under Edward: it probably developed under Henry III, in emulation of St Louis of France. The numbers of sufferers from scrofula who came to be blessed by Edward were very considerable. Over 600 appeared in 1276–7, the first year for which records survive, 1,700 in 1289–90, and just over 2,000 in 1305–6 . . . and the scale of the exercise of Edward I's curative touch is indicative of his great prestige. Not surprisingly, few Welshmen came forward, but many Scots seem to have been touched: almost 1,000 seem to have received the royal blessing when Edward was in the north in 1303–4. It is not possible to determine whether the royal touch had any beneficial effects, but there were no complaints that it failed, and at least, unlike many forms of medical treatment, it could do no harm.

M. Prestwich, 'The Art of Kingship: Edward I 1272–1307', *History Today* (May 1985)

THE SIEGE OF ACRE, 1272

Edward and his first wife, Eleanor of Castile, were deeply devoted, as a story from Edward's crusading demonstrates.

In that month [June 1272] a member of the secret society of the Assassins, fanatics whose name became synonymous with murderers, employed by one of the Emirs in negotiation with Edward, obtained a private interview with him under pretence of important secret business and suddenly attacked him with a dagger, wounding him in the arm. Edward repelled him with a vigorous kick and seizing a stool knocked him down and snatched the dagger from him, but in so doing wounded himself in the forehead. The dagger being poisoned, Edward's wounds gave cause for great anxiety; he made his will, appointing executors and guardians for his children, and its very brevity is significant. Popular legend loves to depict the faithful and devoted Eleanor sucking the poison from her husband's arm, but it must regretfully be admitted that in the fullest account by any English contemporary the only reference to Eleanor is less romantic, as the first step taken by the surgeon to whose skill the prince's recovery was attributed was to order the removal of the weeping wife, saying that it was better that she should shed tears than that all England should mourn.

L. F. Salzman, *Edward I* (1968)

EDWARD'S REFORM OF THE CURRENCY, 1279

Edward had proclaimed an edict throughout England to the effect that clipped money should no longer be circulated, nor should anyone be forced to accept it. In addition he designated a small number of places in the kingdom, in towns and cities, where the money could be exchanged. For each pound of the non-current coin, an extra sixteen pennies were to be paid for the exchange, and people would receive one pound of unclipped coin . . . Within a short time [after the edict] no one would consider accepting it.

Chronicles of the Age of Chivalry, quoting Thomas Wykes. *An Augustinian canon of Osney Abbey, near Oxford, Wykes is very informative on some matters.*

A PROBLEM OF POLLUTION, 1306

In 1306 Edward, through his government, dealt with a nuisance of heating in London. The description is taken from the sixteenth century from one who lamented that by his time, the reign of James I, Edward's command was no longer being followed.

Upon complaints of the clergy and nobility resorting to the City of

London, touching the great annoyance and danger of contagion growing by reason of the stench of burning sea-coal, which divers fire makers in Southwark, Wapping and East Smithfield now used to make their common fires because of the cheapness thereof . . . The King expressly commanded the mayor and sheriffs of London forthwith to make proclamation that all those fire makers should cease their burning sea-coal, and make their fires of such fuel of wood and coal as had been formerly used.

Annals of John Stow

Edward and Wales

Great success attended Edward's campaigns against Llywelyn and the Welsh princes. A lasting memorial to that victory survives in the massive coastal castles which he built as part of his conquest. The capture of Anglesey in 1277 drew a triumphant comment from the king:

'Our friend Llywelyn has lost the first feather in his tail.'

Edward Jenks, *Edward Plantagenet* (1902)

VICTORY, 1284

It was not until 1284, after three campaigns, that Wales succumbed. To celebrate, Edward held a 'Round Table', near Caernarvon, with knightly tournaments and dancing. Six years earlier he had attended the opening of the tombs of King Arthur and Queen Guenevere at Glastonbury. He was doubtless delighted with his trophies in 1284.

A large piece of the Cross of Our Lord, called 'Croizneth', in the Welsh language, was among many famous relics handed over to King Edward. The crown of the renowned Arthur, once king of Britain, was also given to King Edward, along with other jewels.

Chronicles of the Age of Chivalry, quoting the Worcester Annalist. *This, and other monastic annals, are important contemporary sources.*

A PRINCE OF WALES, 1284 OR 1301?

The story that Edward presented to the native Welsh a son born in the Principality at the time of his victory who 'coulde speake never a word of English', is a sixteenth-century invention. Edward's eldest son, Alfonso, was still alive and it was not until 1301 that this second child, named after his father, was invested as Prince of Wales and earl of Chester. The contemporary account of his birth is far from glamorous.

On 24 April, at Caernarvon, Queen Eleanor of Castile gave birth to a son who was named Edward after his father. Ibid.

KNIGHTHOOD, 1306

The knighting of the young prince proved to be a more spectacular affair. The 'Feast of Swans' at which some 300 were knighted, combined Edward's passion for Arthur and vows to pursue the Scottish cause with greater vigour.

There was such a crush of people before the high altar that two knights died, several fainted and at least three had to be taken away and attended to.

Then two swans were brought before King Edward I in pomp and splendour, adorned with golden nets and gilded reeds, the most astounding sight to the onlookers. Having seen them, the king swore by the God of Heaven and by the swans that he wished to set out for Scotland and, whether he lived or died, to avenge the wrong done.

Ibid., quoting *The Flowers of History. Writing at Westminster Abbey, the anonymous author recorded events at first hand.*

DEATH OF QUEEN ELEANOR

The Queen died near Lincoln on 10 December 1290 while the king was on the way to Scotland. Grief-stricken, he accompanied her body back to London.

The king gave orders that in every place where her bier had rested, a cross of the finest workmanship should be erected in her memory, so that passers-by might pray for her soul. He arranged that the Queen's portrait should be painted on to each cross. William Rishanger

The cross associated with Charing Cross station in London is probably the best known of them all, though the present cross is a copy. The original was destroyed by Cromwell's soldiers. Original crosses remain at Waltham, Geddington and Hardingstone, just outside Northampton, though Edward had had twelve erected.

Edward and Scotland

In March 1286 Alexander III of Scotland was killed by a fall from his horse. His heir was his granddaughter, Margaret the 'Maid of Norway', then aged three. A match between the Maid and Edward's heir was thwarted by Margaret's death. In the ensuing dispute—the Great Cause—Edward supported John Balliol as candidate for the Scottish throne in preference to Robert Bruce but Edward's subsequent high-handedness led to war which lasted beyond the king's death.

THE STONE OF SCONE

Early victory made Edward over-confident and in 1296 he removed the 'Stone of Destiny' on which the inauguration of Scottish kings had long been performed and which has been used at English coronations ever since.

Edward I returned to Berwick upon Tweed by way of Scone, and gave orders that the stone, on which the kings of Scotland used to be enthroned as their equivalent of coronation, should be removed and transported to London, as a sign that the kingdom had been renounced and conquered.

Chronicles of the Age of Chivalry, quoting Walter of Guisborough. *An Augustinian canon, Walter wrote an account most valuable for events in the north.*

BATTLES AND SIEGES

Throughout the conflict Scottish resilience was marked by a seemingly endless supply of leaders. In 1298 Edward met William Wallace at the battle of Falkirk. Walter of Guisburgh reports that the king slept on the ground the night before, his shield for a pillow, his horse beside him. The horse, however, stepped on his royal master as he lay asleep. In the confusion of darkness the alarm spread that he was wounded. Only slightly hurt, Edward went into battle in the morning. His victory that day was never followed up.

THE SIEGE AND SURRENDER OF STIRLING CASTLE, 1304

Between April and July Edward sat before the great stone fortress of Stirling. His accounts for these months refer to payments for the construction of vantage points in the royal house from which Edward's new queen and her ladies could watch his great siege engines in action.

One day during the siege, something not merely unusual, but even miraculous happened to the king of England. While he was riding repeatedly round the castle, unarmed, someone hurled a javelin, called in English a springald, from a sling in a tower. It struck between Edward I's legs, neither touching his flesh nor even wounding his feet. Immediately, this was held to be a miracle. For not once, but a hundred times, weapons directed at him fell to his right and to his left, never harming him but frequently wounding those around him.

Chronicles of the Age of Chivalry, quoting *The Flowers of History*

When the same thing happened to Wellington at the battle of Waterloo—officers cut down beside him while he remained unscathed—he called it the Finger of Providence.

ENGINES OF WAR

They had an engine, and brought it out to cast;
The rod broke, afterwards it was of no use.
The engines without are put to work,
And cause the stones to pass walls and towers;
They overthrow the battlements around,
And throw down to the ground the houses inside.

In the midst of these doings the king causes to be built of timber
A terrible engine, and to be called Ludgar;
And this at its stroke broke down the entire wall.
Three months and eight days, reckoning by days,
Lasted the storm; the endurance was hard
To wretches within who had nothing to eat.

Chronicle of Pierre de Langtoft

Edward and His Barons

For the most part Edward maintained a good relationship with his barons but two stories of note have been preserved for posterity.

THE RUSTY SWORD

In 1278 the royal justices were instructed to investigate alleged usurpation of the king's rights by his magnates through a legal action known as Quo Warranto (*By what warrant*). *The stalwart John de Warenne reputedly drew an old rusty sword in the court:*

'Behold, my lords, behold my warrant!' said John. 'For my ancestors came over with William the Bastard and conquered their lands by the sword; and I shall defend them by the sword against any usurpers. That king did not conquer and subject the land by himself, but our forebears were his partners and helpers'

The Chronicle of Walter of Guisborough, ed. H. Rothwell, Camden Society, lxxxix (1957)

THE EARL MARSHAL'S REPLY

In 1297 Edward faced a major threat from magnates unwilling to serve in Gascony. Roger Bigod, earl of Norfolk refused to go unless the king accompanied the army.

'I will gladly go with you, O king, in front of you in the first line of battle, as is my hereditary right.' The king replied, 'You will go without me, too, with the others.' But he answered, 'I am not bound, nor is it my will, O king, to go without you.' In anger the king, it is said, retorted with these words, 'By God, O earl, you will either go or hang.' But he replied, 'By the same oath, O king, I will neither go, nor hang.' And he departed without taking his leave.

Ibid.

EDWARD'S LUCK

In Palestine he survived the murderous attack of the assassin by almost a miracle; in Paris the lightning passed over his shoulder and slew two of his attendants; at Winchelsea when his horse leapt the town wall he was uninjured; at the siege of Stirling a bolt from a crossbow struck his saddle as he rode unarmed and a stone from a mangonel brought his horse to the ground. Even illness seemed to pass him by and his last years found him as vigorous and upright as a palm tree with eyes and brain undimmed and the teeth still firm in his jaws, able to bite hard literally as well as figuratively, at the table as in the field.

Salzman, *Edward I*

AN EPITAPH

While still campaigning against the Scots, Edward died at Burgh by Sands on 7 July 1307. Even at death, stories about him grew, one report stating that his last request was for his body to be boiled so that his bones could accompany every army against the Scots until they were defeated. What is certain is that his reputation was not impaired by his inability to bring the Scots to heel.

Speak of king Edward and of his memory
As of the most renowned combatant on steed.
Since the time of Adam never was any time
That prince for nobility, or baron for splendour,
Or merchant for wealth, or clerk for learning,
By art or by genius could escape death.
Of chivalry, after king Arthur,
Was king Edward the flower of Christendom.
He was so handsome and great, so powerful in arms,
That of him may one speak as long as the world lasts.
For he had no equal as a knight in armour
For vigour and valour, neither present nor future.

Chronicle of Pierre de Langtoft

Edward II

1307–1327

The genes seemed to have gone wrong, except for the imposing Plantagenet physique of Edward of Caernarvon. Edward I's son was a failure in war (Bannockburn), had unregal tastes (bricklaying) and alienated wife, son, barons, and Church through his 'favourites'.

A QUARREL WITH HIS OVERPOWERING FATHER

Young Prince Edward, terrified of his father's furies, had sent Bishop Walter Langton to request and negotiate the gift of Pontieu to Piers Gaveston, the Prince's catamite.

The king was mightily enraged. 'Who are you that dares to ask such a thing? As the Lord lives, you shall not escape my hands unless you can prove that you undertook this negotiation against your will, through fear of the prince. Now, however, you shall not leave until you see what he who sent you has to say.' Having called for his son, the king said, 'What negotiation have you promoted through this man?' His son replied, 'That I might, with your acquiescence, give Ponthieu to my lord Piers de Gaveston.' 'You baseborn whoreson,' shouted the king, 'do you want to give away lands now, you who never gained any? As the Lord lives, if it were not for fear of breaking up the kingdom you should never enjoy your inheritance.' And seizing a tuft of the prince's hair in each hand, he tore out as much as he could, until he was exhausted, when he threw him out.

Walter of Guisborough

A HOPEFUL START

The legacy of unfinished war and debt from Edward I was not a happy one. 'Edward II sat down to the game of kingship with a remarkably poor hand and he played it very badly' (Denholm-Young). Many, however, were optimistic and enthusiastic about the 23-year-old heir.

For God had endowed him with every gift, and had made him equal to or indeed more excellent than other kings. If anyone cared to describe those qualities which ennoble our king, he would not find his like in the land. His ancestry, reaching back to the tenth generation, shows his nobility. At the beginning of his reign he was rich, with a populous land and the goodwill of his people. He became the son-in-law of the King of France, and first cousin of the King of Spain. If he had followed the advice of the barons he would have humiliated the Scots with ease. If he had habituated himself to the use of arms, he would have exceeded the prowess of King Richard. Physically this would have been inevitable, for he was tall and strong, a fine figure of a handsome man. But why linger over this description? If only he had given to arms the labour that he expended on rustic pursuits, he would have raised England aloft; his name would have resounded through the land. What hopes he raised as Prince of Wales!

Vita Edwardi Secundi, ed. N. Denholm-Young (1957). *This anonymous life of the king was the work of a well-informed and perceptive clerk and is a key source until it stops abruptly in 1325.*

EDWARD'S TASTES

There are many references to Edward's unkingly and rustic pursuits. An impostor from Oxford, claiming to be of royal blood, cited these as proof that the prince was not the king's son.

Some people did actually believe the man because the Lord Edward did not resemble his father in any of his virtues.

For it was commonly reported that he [Edward] had devoted himself privately from his youth to the arts of rowing and driving chariots, digging pits and roofing houses; also that he wrought as a craftsman with his boon companions by night, and at other mechanical arts, besides other vanities and frivolities wherein it doth not become a king's son to busy himself.

When news of this impostor reached the king he demanded the man be brought to him and he greeted him ironically with the words 'Welcome my brother'.

The Chronicle of Lanercost, tr. H. E. Maxwell (Glasgow, 1913). *Probably written at the Franciscan house in Carlisle, this source is chiefly of value for events in the north.*

Edward's Favourite

The first crisis of the reign developed over Edward's recall and subsequent patronage of Gaveston, favouritism so extreme that it has given rise to speculation about Edward's homosexuality. A mere Gascon knight, Piers was created earl of Cornwall and in 1308 acted as regent. His arrogance was intolerable to the barons, who were also offended by his nicknames. Led by Guy of Warwick, whom Gaveston called 'the black dog of Arden', the barons eventually isolated and cornered their prey.

THE DEATH OF PIERS GAVESTON, 1312

[Warwick] took a strong force, raised the whole countryside and secretly approached the place where he knew Piers to be. Coming to the village very early one Saturday he entered the gate of the courtyard and surrounded the chamber. Then the earl called out in a loud voice: 'Arise traitor, thou art taken.' When Piers heard this, seeing that the earl was there with a superior force and that his own guard did not resist, he dressed himself and came down. In this fashion Piers was taken and led forth not as an earl but as a thief; and he who used to ride on a palfrey is now forced to go on foot.

When they had left the village a little behind, the earl ordered Piers to be given a nag that they might proceed more quickly. Blaring trumpets followed Piers and the horrid cry of the populace. They had taken off his belt of knighthood, and as a thief and a traitor he was taken to Warwick, and coming there was cast into prison. *Vita Edwardi*

In spite of a pledge by the earl of Pembroke that Gaveston would be protected, the other earls, led by Warwick, seized their opportunity.

About the third hour Piers was led forth from prison; and the Earl of Warwick handed him over bound to the Earl of Lancaster, and Piers, when he saw the earl, cast himself on the ground and besought him, saying, 'Noble earl, have mercy on me.' And the earl said 'Lift him up, Lift him up. In God's name let him be taken away.' The onlookers could not restrain their tears. For who could contain himself on seeing Piers, lately in his martial glory, now seeking mercy in such lamentable straits. Piers was led out from the castle and hastened to the place where he was to suffer the last penalty; and the other earls followed at a distance to see his end, except Count Guy who remained in his castle.

When they had come to a place called Blacklow, which belonged to the Earl of Lancaster, an envoy from the earl ordered that Piers should remain there; and immediately by the earl's command he was handed over to two Welshmen, one of whom ran him through the body and the other cut off his head. Ibid.

THE KING'S REACTION

When the king was notified that Piers was dead, he was saddened and grieved very much and after a little said to the bystanders: 'By God's soul, he acted as a fool. If he had taken my advice he would never have fallen into the hands of the earls . . . What was he doing with the Earl of Warwick who was known never to have liked him?' . . . When this light utterance of the king became public it moved many to derision. But I am certain that the king grieved for Piers as a father grieves for his son. Ibid.

THE BURIAL OF PIERS AT WINDSOR

The king made his way to Windsor where he held his birthday. And on the day after the octave of St John the Evangelist the body of the oft-mentioned Piers was embalmed with aromatic herbs on the king's orders and buried in the church of the preaching brothers in a great ceremony by the venerable father of Canterbury, four other bishops, together with abbots and inumerable other churchmen. Few, however, of the princes of this world were willing to attend this particular interment with the king.

Johannes de Trokelowe, Chronica et Annales, ed. H. T. Riley, Rolls Series (1866). *Possibly the work of William Rishanger, this chronicle represents the continuation of the work being done at St Albans and covers the first quarter of the fourteenth century.*

Bannockburn

The unfinished Scottish war reached a crisis in 1314. Robert Bruce's advances had come to a halt before Stirling Castle at Midsummer 1313, when it was agreed between the Scots and the English garrison that unless an English relieving army came to Stirling before Midsummer 1314 the garrison would surrender. At the eleventh hour Edward's army advanced into Scotland where, on 24 June, on the ground around the Bannock stream the Scots in 'schiltrom' formation (a hedge of infantry armed with spears) inflicted a humiliating defeat on the English cavalry. Edward himself, who had 'fought like a lion' and had one horse killed under him, was whisked to safety.

O day of vengeance and disaster, day of utter loss and shame, evil and accursed day, not to be reckoned in our calendar; that blemished the reputation of the English, despoiled them and enriched the Scots, in which our costly belongings were ravished to the value of £200,000! So many fine noblemen and valiant youth, so many noble horses, so much military equipment, costly garments and gold plate—all lost in one unfortunate day, one fleeting hour. *Vita Edwardi*

THE STONE OF DESTINY, 1324

Robert Bruce demanded the return of the Stone as part of a permanent truce, but Edward, in rejecting Bruce's other terms, rejected this one also.

The Scots also demanded that the royal stone should be restored to them, which Edward I had long ago taken from Scotland and placed at Westminster by the tomb of St Edward. This stone was of famous memory amongst the Scots, because upon it the kings of Scotland used to receive the symbols of authority and the sceptre. Scota, daughter of Pharaoh, brought this stone with her from the borders of Egypt when she landed in Scotland and subdued the land. For Moses had prophesied that whoever bore that stone with him should bring broad lands under the yoke of his lordship. Whence from Scota the land is called Scotland which was formerly called Albany from Albanactus. Ibid.

The She-Wolf of France

Isabella of France is probably the most vilified of English queens. Married to Edward in 1308 at the age of twelve, she was the only medieval queen known to have been an adulteress and she it was who led the revolution which ended in her husband's deposition. Reputedly a great beauty, her early years as queen were difficult, given Edward's attachment to Gaveston, but in 1313 she produced a son, the first of her four children. It was the elevation of new favourites—the elder and younger Despenser—that brought her final alienation from her husband. Edward's attitude to her was hardly loving, according to one source. In 1334 at his trial the bishop of Winchester was indicted because:

In his preaching at Walyngeforde he said that the king carried a knife in his hose to kill queen Isabella, and had said that if he had no other weapon he would crush her with his teeth.

Chartulary of Winchester Cathedral, ed. A. W. Goodman (Winchester, 1927)

THE FINAL BREACH WITH QUEEN ISABELLA

In 1325 Isabella went to France to the court of her brother Charles IV. There she took as her lover the English baron Roger Mortimer, recently escaped from prison in London. In the company of her son she plotted Edward's downfall in spite of the latter's commands for her return.

Amongst other things, when the king sent his son to France, he ordered his wife to return to England without delay. When this command had been laid by the messengers before the King of France and the queen herself, she replied, 'I feel that marriage is a joining together of man and woman, maintaining the undivided habit of life, and that someone has come between my husband and myself trying to break this bond; I protest that I will not return until this intruder is removed, but, discarding my marriage garment, shall assume the robes of widowhood and mourning until I am avenged of this Pharisee.' The King of France not wishing to seem to detain her said, 'The queen has come of her own will, and may freely return if she so wishes. But if she prefers to remain in these parts, she is my sister, and I refuse to expel her.' The messengers returned and reported all this to the king. *Vita Edwardi*

THE FIRST DEPOSITION, 1327

On 24 September 1326, Isabella reached England and was enthusiastically welcomed. The king was rejected in favour of his eldest son. Isabella moved to Bristol, where she took her revenge on the Despensers and captured her husband.

When the King and Sir Hugh [Despenser] were brought to Bristol, the King was sent, by the advice of all the barons and knights, to the strong castle at Berkeley, on the Severn, and was handed over to the lord of Berkeley Castle, to be closely guarded. He was ordered to serve the King and look after him well and honourably, with proper people in attendance on him, but on no account to let him leave the castle . . . And Sir Hugh was handed over to Sir Thomas Wake, the marshal of the army.

The Queen, with all the army, set out for London, which is the principal city of England. Sir Thomas Wake had Sir Hugh tied on to the meanest and poorest horse he could find: and he had him dressed in a tabard over his clothes, embroidered with the coat of arms that he bore, and so conducted him along the road in the Queen's procession as a public laughing-stock. And in all the towns they passed through, he was announced by trumpets and cymbals, by way of greater mockery, till they reached the good city of Hereford. There the Queen and all her company . . . kept the feast of All Saints . . .

When the feast was over, Sir Hugh . . . was brought before the Queen and all the barons and knights in full assembly. A list of all his misdeeds was read out to him, to which he made no reply . . . Then he was tied on a tall ladder in full view of all the people both high and low, and a large fire was lit. Then his private parts were cut off, because he was held to be a heretic, and guilty of unnatural practices, even with the King, whose affections he had alienated from the Queen . . . so as to bring shame and disgrace on the country, and had caused the greatest barons in England to be beheaded, by whom the kingdom should have been supported and defended; and he had encouraged the King not to see his wife or son, who was to be their future king; indeed, both of them had had to leave the country to save their lives. When the other parts of his body had been disposed of, Sir Hugh's head was cut off and sent to London. His body was then divided into quarters, which were sent to the four next largest cities in England.

Polychronicon Ranulphi Higden monachi Cestrensis with the English translations of John Trevisa, ed. J. R. Lumby, Rolls Series (1965), VIII. *Higden was a Benedictine monk of St Werburgh's, Chester whose work stops in 1340. His account is scrappy but lively.*

Edward's own punishment, barbaric as it was, set out to leave no trace when the body was later produced in public—the perfect murder. The treacherous bishop of Hereford, in collusion with Queen Isabella, plotted Edward's death by a trick, sending his gaolers in Berkeley Castle a letter which seemed innocent but was in fact a coded order to kill.

Then began the most extreme part of Edward's persecution which was to continue until his death.

Firstly he was shut up in a secure chamber, where he was tortured for many days until he was almost suffocated by the stench of corpses buried in a cellar hollowed out beneath. Carpenters, who one day were working near the window of his chamber, heard him, God's servant, as he lamented that this was the most extreme suffering that had ever befallen him.

But when his tyrannous warders perceived that the stench alone was not sufficient to kill him, they seized him on the night of 22 September as he lay sleeping in his room.

There with cushions heavier than fifteen strong men could carry, they held him down suffocating him.

Then they thrust a plumber's soldering iron, heated red hot, guided by a tube inserted into his bowels, and thus they burnt his innards and vital organs. They feared lest, if he were to receive a wound in those parts of the body where men generally are wounded, it might be discovered by some man who honoured justice, and his torturers might be found guilty of manifest treason.

. . . As this brave knight was overcome, he shouted aloud so that many heard his cry both within and without the castle and knew it for a man who suffered a violent death. Many in both the town and castle of Berkeley were moved to pity for Edward, and to watch and pray for his spirit as it departed this world.

Chronicles of the Age of Chivalry, quoting the *Chronicle* of Geoffrey le Baker. *Writing c.1341, Geoffrey le Baker is not in general reliable but he does have some good material for the end of Edward's reign.*

A HEARTLESS QUEEN

The story that Isabella had Edward's heart removed from his body and put in a silver case that was later buried with her has not helped redeem her reputation. The epithet, though late, has stuck:

She-wolf of France, with unrelenting fangs,
That tear'st the bowels of thy mangled mate.

T. Gray, 'The Bard' (1757), II, 1

WHERE IS EDWARD II?

A single uncorroborated story of contemporary origin casts some doubt on Edward's murder. Writing to Edward III Manuele de Fieschi, a papal notary, told a remarkable story.

In the name of the Lord, Amen. Those things that I have heard from the confession of your father I have written with my own hand and afterwards I have taken care to be made known to Your Highness.

He informed the young king of the last days of the reign until:

Finally they sent him to the castle of Berkeley. Afterwards the servant who was keeping him, after some little time, said to your father: Lord, Lord Thomas Gourney and Lord Simon Barford, knights, have come with the purpose of killing you. If it pleases, I shall give you my clothes, that you may better be able to escape. Then with the said clothes, at twilight, he went out of the prison; and when he had reached the last door without resistance, because he was not recognized, he found the porter sleeping, whom he quickly killed; and having got the keys of the door, he opened the door and went out, and his keeper who was keeping him. The said knights who had come to kill him, seeing that he had thus fled, fearing the indignation of the queen, even the danger to their persons, thought to put that aforesaid porter, his heart having been extracted, in a box, and maliciously presented to the queen the heart and body of the aforesaid porter as the body of your father, and as the body of the king the said porter was buried in Gloucester.

G. P. Cuttino and Thomas W. Lyman, 'Where is Edward II?', *Speculum*, liii (1978)

The letter goes on to record his departure to Ireland and subsequent travels in Europe, including a visit to the pope before settling down at a hermitage.

He was in this last hermitage for two years or thereabout, always the recluse, doing penance, and praying God for you and other sinners. In testimony of which I have caused my seal to be affixed for the consideration of Your Highness.

Your Manuele de Fieschi, notary of the lord pope, your devoted servant.

Ibid.

A POET'S VIEW OF EDWARD'S DEPOSITION

Christopher Marlowe, the Elizabethan dramatist, represented Edward as relinquishing the crown while in prison in Kenilworth Castle. He vividly portrayed a medieval king's ambivalent feelings about the physical crown: a passionate feeling that he is not a person at all unless he wears the crown, contrasted with a strong apprehension of its meaningless pomp. To emphasize this duality, Marlowe makes Edward first wear the crown, then take it off and hand it over, then put it on again, and finally give it up.

Here, take my crown; the life of Edward too;
[*Taking off the crown.*]
Two kings in England cannot reign at once.
But stay awhile, let me be king till night,
That I may gaze upon this glittering crown;
So shall my eyes receive their last content . . .
[*He puts on the crown.*]
. . . . .
And in this torment comfort find I none,
But that I feel the crown upon my head,
And therefore let me wear it yet awhile.
. . . . .
Here, here! Now sweet God of heaven,
[*He gives them the crown.*]
Make me despise this transitory pomp,
And sit for ever enthronized in heaven!
Come, death, and with thy fingers close my eyes,
Or if I live, let me forget myself.

Christopher Marlowe, *The Tragedy of Edward the Second*

Verdicts on Edward II

The condemnation of Edward has been almost universal. In an age when success depended so much on the personality of the monarch, it remains difficult to excuse his failures.

HIS BIOGRAPHER BY 1313

For our King Edward has now reigned six full years and has till now achieved nothing praiseworthy or memorable, except that by a royal marriage he has raised up for himself a handsome son and heir to the throne.

Vita Edwardi

HIS COUNTRY IN 1327

Item, he has stripped his realm, and done all that he could to ruin his realm and his people, and what is worse, by his cruelty and lack of character he has shown himself incorrigible without hope of amendment, which things are so notorious that they cannot be denied.

Articles of Deposition in *Select Documents of English Constitutional History, 1307–1485*, ed. S. B. Chrimes and A. L. Brown (1961)

A SEVENTEENTH-CENTURY BIOGRAPHER

But you may object. He fell by Infidelity and Treason, as have many others that went before and followed him. 'Tis true; but yet withal observe, here was no second Pretendents, but those of his own, a Wife and a Son, which were the greatest traitors: had he not indeed been a Traitor to himself, they could not all have wronged him.

Henry Cary, Viscount Falkland (d. 1633), quoted in H. F. Hutchison. *Edward II: The Pliant King* (1971), *'The Pliant King' is a phrase used in Marlowe's* Edward the Second.

HIS MODERN BIOGRAPHER

And in a story which has to be overfull of failures, jealousies and tragedies, it is pleasant to conclude with two items in lighter mood. It is recorded of Edward of Caernarvon's sense of humour that Jack of St Albans, the royal painter, was given fifty shillings by the king's own hand for having danced on a table before the king 'and made him laugh beyond measure'. And perhaps the kindest reference to Edward of Caernarvon in all the medieval chronicles is in Walsingham. He writes that 'when Scotland would openly rebel against him and all England would rid herself of him, then the Welsh in a wonderful manner cherished and esteemed him, and, as far as they were able, stood by him grieving over his adversities both in life and in his death, and composing mournful songs about him in the language of their country . . .' The first English Prince of Wales found his only mourners in the land of his birth. Ibid.

Edward III

1327–1377

It has been said of Edward III that he 'inherited a kingdom but bequeathed a nation'. A boy-king of only fifteen at his accession, Edward learnt from his father's mistakes and though he became prematurely senile, his reign, in a century that saw two depositions, was a period of political stability. Freeing himself from the tutelage of his mother and her lover Mortimer, he ruled in harmony with his magnates and with the outbreak of the Hundred Years War in 1337 carried the invincible longbowmen deep into France. Over six feet tall, Edward was lithe and active, gallant and courtly, and was said to have 'a face like a god'.

EDUCATION OF A PRINCE

Edward was educated by Richard of Bury, the first great English bibliophile who wrote, perhaps for his pupil's edification, the following homily on books:

You may happen to see some headstrong youth lazily lounging over his studies and when the winter's frost is sharp, his nose running from the nipping cold drips down, nor does he think of wiping it with his pocket-handkerchief until he has bedewed the book before him with the ugly moisture. Would that he had before him no book, but a cobbler's apron! His nails are stuffed with fetid filth as black as jet, with which he marks any passage that pleases him. He distributes a multitude of straws, which he inserts to stick out in different places so that the halm may remind him of what his memory cannot retain. These straws, because the book has no stomach to digest them, and no one takes them out, first distend the book from its wonted closing, and at length, being carelessly abandoned to oblivion, go to decay.

He does not fear to eat fruit or cheese over an open book, or carelessly to carry a cup to and from his mouth; and because he has no wallet at hand he drops into books the fragments that are left.

The Philobiblon of Richard de Bury, ed. E. C. Thomas with a new Foreword by Michael Maclagan (Oxford, 1960)

THE COUP OF 1330

At the age of eighteen Edward made a pre-emptive strike against the ambitious Roger Mortimer. Rumours were rife that he sought the throne for himself. Helped by friends among the nobility, he made his bid at Nottingham castle, where a scout, Robert Eland, used his knowledge of its secret passages to guide them in.

The scout, by the light of torches, took the king along a secret underground passage which began at some distance from the castle and led to the kitchen and to the hall of the main tower, in which the queen had her lodging.

Then the king's friends leapt from the passageway that ran under the earth and, with their swords drawn, made for the queen's bed-chamber, which, through the grace of God, they found open. King Edward III, his weapon at the ready, kept watch for them outside the chamber of his enemies, lest his mother should see him.

Then they found the queen, who was preparing herself for her night's rest, and the object of their search, the earl of March. They seized him and took him to the hall, while the queen cried out, 'My son, my son, have pity on gentle Mortimer!' For although she had not seen her son, she suspected his presence.

Chronicles of the Age of Chivalry, quoting Geoffrey le Baker

No mercy was shown. He was found guilty of all manner of treasons and felonies.

In London, he was thrown into the Tower as once before, and on 29 November, while the parliament of England was sitting at Westminster, he was drawn and hanged.

Ibid.

The Hundred Years War

Edward's reputation rests mainly on his war with France, which gave rise to many noble deeds and a great explosion of chivalric fervour. Tributes to Edward even came from the French.

When this noble king Edward first gained England in his youth, nobody thought anything of the English and nobody spoke at all of their valour or their hardiness . . . but now they have learnt to bear arms in the time of this noble king Edward, who has often put them to work, and they are the most noble and the most daring soldiers known.

Chronique de Jean le Bel, ed. J. Viard and E. Deprez, Société de l'histoire de France, 2 vols. (Paris, 1904–5), I. *Jean le Bel's admiration of Edward stemmed from a personal delight in things chivalric.*

NAVAL VICTORY AT SLUYS, 1340

A period of 'phoney war' at the beginning of the great conflict between England and France was broken in June 1340. Edward was in personal command of an

army of soldiers and archers on board ship. The destruction of the French was such that one reporter suggested that if fish could speak they could have learnt French from the drowned!

So, anchoring his ships off Sluys, at some distance from the shore, King Edward deliberated the best plan of action for a whole day.

However, early on the morning of 24 June, the French fleet separated into three sections and moved about a mile towards where the English fleet lay.

When he saw this, the king of England announced that they should wait no longer but arm themselves and be at the ready. In the early afternoon, with the wind and sun behind him and the current flowing so as to aid his attack, Edward III divided his fleet into three sections and delivered the longed for attack upon the enemy.

A hail of iron, of bolts from crossbows, and of arrows from longbows, brought death to men in their thousands. Those who were willing and daring enough, fought at close quarters with spears, axes and swords. Some hurled stones from the towers of the masts, thus killing many.

In short it would be no distortion to say that this great naval battle was so fearful that he would have been a fool who dared to watch it even from a distance.

Geoffrey le Baker

FATHER AND SON AT THE BATTLE OF CRÉCY, 1346

Edward's eldest son was born at Woodstock on 15 June 1330 and named after his father. Known to posterity as the Black Prince—probably from the colour of his armour—he was made earl of Chester in 1333, duke of Cornwall in 1337—the first English dukedom—and Prince of Wales in 1343. Glorious victories won him fame, though there was a darker side to his character. Edward III showed himself in command of his son, as well as of the army, at young Edward's début. Seeing the prince in the thick of battle, his knights appealed to the father:

'Sir, the earl of Warwick, the lord Stafford, the lord Reginald Cobham, and the others who are about your son, are vigorously attacked by the French; and they entreat that you would come to their assistance with your battalion, for, if their numbers should increase, they fear he will have too much to do.' The king replied, 'Is my son dead, unhorsed, or so badly wounded that he cannot support himself?' 'Nothing of the sort, thank God,' rejoined the knight; but he is in so hot an engagement that he has great need of your help.' The king answered, 'Now, sir Thomas, return back to those that sent you, and tell them from me, not to send again for me this day, or expect that I shall come, let what will happen, as long as my son has life; and say that I command them to let the boy win his spurs; for I am determined, if it please God, that all the glory and honour of this day shall be given to him, and to those into whose care I have intrusted him.'

Sir John Froissart's Chronicles of England, France, Spain and the adjoining countries, ed. T. Johnes, 2 vols. (1839). *Froissart's excellent chronicle contains the most graphic contemporary account of the Hundred Years War. Froissart was born in the Low Countries, at Hainaut; he was in England from 1360 to 1367 and made a further visit to the court of Edward III's grandson and heir, Richard II.*

THE BLACK PRINCE'S ADOPTION OF HIS EMBLEM AND MOTTO

In time of war he would bear the arms of England; at tournaments, in time of peace, the ostrich feathers and 'Ich Dien'.

It is still widely held that the Black Prince adopted the emblem of three ostrich feathers and the motto '*Ich Dien*' on the death of the King of Bohemia at Crécy. The usual account holds that Edward III took three ostrich feathers from the dead King's helmet and offered them to the Black Prince by right of conquest. It is such a good story, in keeping with the best traditions of medieval history—the young Prince on the threshold of a brilliant military career, inspired by the chivalric death of the blind old man—that it is a pity it is unlikely to be true. John of Bohemia's personal crest was the entire wings of a vulture 'besprinkled with linden leaves of gold', and this is the crest carved on the Croix de Bohême at Crécy. However an ostrich feather seems to have been the badge of his family, and he may occasionally have used it.

As regards the motto '*Ich Dien*', 'I serve', it might have been used by John of Bohemia, but there is no evidence available to prove it . . . We find in the illustrated manuscripts of Froissart that the Black Prince is always distinguished by three feathers in his helmet—though one must also note that several of his brothers used an ostrich feather on their seals. '*Ich Dien*' was certainly one of his mottoes, his second one being 'Houmont', 'High spirits'. The latter was his 'badge of war', '*Ich Dien*' being used in peace time. A 'signature' of the Black Prince using both '*Ich Dien*' and 'Houmont' can be seen in the Museum of the Public Record Office.

B. Emerson, *The Black Prince* (1976)

QUEEN PHILIPPA AND THE BURGHERS OF CALAIS, 1347

Philippa of Hainault enjoys an unequalled reputation among medieval queens. She produced twelve children for her husband and often used her influence upon him to good effect. Froissart enjoyed her patronage and tended to portray her in a romantic role. Following victory at Crécy, Edward successfully besieged Calais and planned his revenge on its six chief citizens. Philippa's fellow countryman, the chivalric knight Sir Walter Mauny, pleaded in vain that their deaths would harm the king's reputation.

The queen of England, who at that time was very big with child, fell on her knees, and with tears said, 'Ah, gentle sir, since I have crossed the sea with great danger to see you, I have never asked you one favour: now, I most humbly ask as a gift, for the sake of the Son of the blessed Mary, and for your love to me, that you will be merciful to these six men.' The king looked at her for some time in silence, and then said, 'Ah, lady, I wish you had been anywhere else than here: you have entreated in such a manner that I cannot refuse you; I therefore give them to you, to do as you please with them.' The queen conducted the six citizens to her apartments, and had the halters taken from round their necks, after which she new clothed them and served them with a plentiful dinner: she then presented each with six nobles, and had them escorted out of the camp in safety.

Froissart

EDWARD III'S PROJECT FOR AN ARTHURIAN ROUND TABLE, 1344

The new Age of Chivalry which seemed to emerge from the exploits in France drew inspiration from the Arthurian legend.

In this place [Windsor chapel] the lord king and all the others stood up together, and having been offered the book, the lord king, after touching the Gospels, took a corporal oath that he himself, at a certain time appointed for this, provided the power should remain in him, would begin a Round Table, in the same manner and estate as the lord Arthur, formerly King of England, maintained it, namely to the number of 300 knights, a number always to be maintained, and he would support and cherish it according to his power . . . When this was done, the trumpets and drums sounded together, and the guests hastened to a feast; which feast was complete with richness of fare, variety of dishes, and overflowing abundance of drinks.

English Historical Documents, IV, 1327–1485, ed. A. R. Myers (1969), quoting the *Chronicle* of Adam Murimuth. *Adam Murimuth was a contemporary canon of St Paul's, London.*

THE FOUNDING OF THE MOST NOBLE ORDER OF THE GARTER

Although the first formal Garter ceremony was probably held by Edward III at Windsor on 23 April (St George's Day) 1348, the precise details of the foundation of this chivalric order remain a mystery. Edward may have used the famous Winchester round table and certainly the notions of equality and meritorious knightly service carried Arthurian overtones. Froissart confused the event with the Round Table ceremony of 1344; and the earliest account dated the first ceremony 1350.

On St George's day, 23 April, King Edward III caused a great feast to be held at Windsor Castle, where he established a chantry of twelve priests and set up a hostel for impoverished knights who could not afford to support themselves. There, their daily needs taken care of, they would live in the service of the Lord, supported by the perpetual charity of the founders of the college.

Others besides the king promised their support for this foundation. Among them were the king's eldest son, Prince Edward, the earl of Northampton, the earl of Warwick, the earl of Suffolk, the earl of Salisbury, and other barons. Some knights too were of their number, including Walter Mauny. All were true gentlemen blessed with great riches.

Geoffrey le Baker

Including the sovereign, there were twenty-six founder knights.

All these men, together with the king, were dressed in robes of russet and wore garters of blue on their right legs. The robes of the order were completed by a blue mantle, embroidered with the arms of St George . . .

Afterwards, they attended a feast where they ate together at a common table in honour of the blessed martyr to whom this noble brotherhood was particularly dedicated, for it was called the order of St George of the Garter.

Ibid.

HONI SOIT QUI MAL Y PENSE

The reason for the foundation of the order is equally obscure. It derives from an apocryphal story of a lost garter and was first given by Polydore Vergil, an Italian who visited England in the sixteenth century and lived for a time at court.

But the reason for founding the order is utterly uncertain; popular tradition nowadays declares that Edward at some time picked up from the ground a garter from the stocking of his queen or mistress, which had become unloosed by some chance, and had fallen. As some of the knights began to laugh and jeer on seeing this, he is reputed to have said that in a very little while the same garter would be held by them in the highest honour. And not long after, he is said to have founded this order . . . Such is popular tradition.

Popular tradition added that as the king tied the blue garter round his own leg he enunciated the Order's motto: 'Honi soit qui mal y pense'—*'Shame on him who thinks ill of it'. In the sixteenth century, William Camden invented the fact that the king's 'mistress' and owner of the garter was Joan Countess of Salisbury, who later married the Black Prince; there is no evidence for the Continental story that the king had raped her. And in the seventeenth century, Elias Ashmole suggested that the motto, 'Shame on him who thinks ill of it', really referred to Edward III's claim to rule France. Blue and gold, the Garter colours, were also the French colours.*

A FIFTEENTH-CENTURY NOVEL ABOUT THE GARTER

The Valencian knight Joanot Martorell (d. 1468) described the garter foundation in his novel Tirant Lo Blanc *which was published in 1490. A story of chivalry, courtship, and war, it names the hero Tirant among the first of the knights, though these are only numbered twenty-three, of the Order. The account begins with the loss of a garter by a court lady called Honeysuckle.*

One of the knights near the king said: 'Honeysuckle, you have lost your leg armour. You must have a bad page who failed to fasten it well.'

She blushed slightly and stooped to pick it up, but another knight rushed over and grabbed it. The king then summoned the knight and said: 'Fasten it to my left stocking below the knee.'

Joanot Martorell, *Tirant Lo Blanc*, tr. D. H. Rosenthal (1984)

Edward wore the garter for more than four months until a favourite servant told him that it was causing a scandal.

The king replied: 'So the queen is disgruntled and my guests are displeased!', and he said in French: '*Puni soit qui mal y pense*. Now I swear before God that I shall found a new knightly order upon this incident: a fraternity that shall be remembered as long as the world endures.' Ibid.

THE BLACK DEATH, AUGUST 1348

The arrival of bubonic plague, brought from the east by fleas carried on black rats, was one of the greatest natural disasters of history. It undermined the English economy and created a huge shortage of manpower and labour. The following account records what was done on royal authority, though not necessarily by order of Edward himself.

Meanwhile the king sent into each shire a message that reapers and other labourers should not take more than they had been wont to do under threat of penalties defined in the statute, and for this purpose he introduced a statute. The workmen were, however, so arrogant and obstinate that they did not heed the king's mandate, but if anyone wanted to have them he had to give them what they asked; so he either had to satisfy the arrogant and greedy wishes of the workers or lose his fruit and crops. When the king was told that they were not observing his order, and had given higher wages to their workmen, he levied heavy fines on abbots, priors, knights of greater and lesser consequence and others, both great and small, throughout the countryside, taking 100*s*. from some, 40*s*. or 20*s*. from others, according to their ability to pay . . . Then the king caused many labourers to be arrested, and sent them to prison; and many of them escaped and fled to the forests and woods for a time, and those who were captured were severely punished. And most of such labourers swore that they would not take daily wages in excess of those allowed by ancient custom, and so they were set free from prison . . .

After the pestilence many buildings both great and small in all cities, towns, and boroughs fell into total ruin for lack of inhabitants; similarly many small villages and hamlets became desolate and no houses were left in them, for all those who had dwelt in them were dead, and it seemed likely that many such little villages would never again be inhabited.

English Historical Documents IV, quoting Henry Knighton. *Henry Knighton was an Augustinian canon of Leicester, most valuable for the reign of Richard II.*

THE BATTLE OF POITIERS, 1356

In spite of the effect of the plague the war against France continued and the Black Prince proved himself once more on the field. The cardinal of Périgord who allowed his men to fight though he was supposed to be negotiating a truce was deemed to be unchivalric and the prince had his revenge.

As the prince was thus advancing upon his enemies, followed by his division, and upon the point of charging them, he perceived the lord Robert de Duras (nephew of the cardinal) lying dead near a small bush on his right hand, with his banner beside him and ten or twelve of his people; upon which he ordered two of his squires and three archers to place the body upon a shield, carry it to Poitiers, and present it from him to the cardinal of Perigord, and say that 'I salute him by that token'. *Froissart*

The greatest prize of the battle was the king of France himself. The prince sent two lords to find him and they eventually spied him surrounded by men. They were squabbling with his captor:

The king of France was in the midst of them, and in great danger; for the English and Gascons had taken him from sir Denys de Morbeque, and were disputing who should have him, the stoutest bawling out, 'It is I that have got him': 'No, no,' replied the others, 'we have him.' The king, to escape from this peril, said, 'Gentlemen, gentlemen, I pray you conduct me and my son in a courteous manner to my cousin the prince; and do not make such a riot about my capture, for I am so great a lord that I can make all sufficiently rich.' Ibid.

Proper treatment was accorded to the king of France, as chivalry demanded:

The prince himself served the king's table, as well as the others, with every mark of humility, and would not sit down at it, in spite of all his entreaties for him to do so, saying, that 'he was not worthy of such an honour, nor did it appertain to him to seat himself at the table of so great a king, or of so valiant a man as he had shown himself by his actions that day'. Ibid.

THE PRINCE AND THE CONSTABLE

The capture of John II of France led to a treaty of peace. When war broke out again—in Spain—in 1367, leadership devolved almost exclusively upon the Black Prince. An early prize in battle was the great constable of France, Bertrand du Guesclin. Against advice the Prince struck a bargain with his prisoner which was met within one month.

Now it happened (as I have been informed) that one day when the prince was in great good humour, he called sir Bertrand du Guesclin, and asked him how he was. 'My lord', replied sir Bertrand, 'I was never better: I cannot otherwise but be well, for I am, though in prison, the most honoured knight in the world.' 'How so?' rejoined the prince. 'They say in France,' answered sir Bertrand, 'as well as in other countries, that you are so much afraid of me, and have such a dread of my gaining my liberty, that you dare not set me free; and this is my reason for thinking myself so much valued and honoured.' The prince, on hearing these words, thought sir Bertrand had spoken them with much good sense . . . 'What sir Bertrand do you imagine that we keep you a prisoner for fear of your prowess? By St George, it is not so; for, my good sir, if you will pay one hundred thousand francs you shall be free.' Ibid.

THE DUKE OF AQUITAINE

As the health of his father declined the Black Prince became more prominent, especially in the English lands in France. Wherever he travelled he attracted attention:

He rode night and morning until he reached Plymouth and abode there until his great array was ready. And it befell right speedily afterward that he had all his vessels loaded with victuals and jewels, hauberks, helmets, lances, shields, bows, arrows, and yet more; he let ship all his horses and anon embarked, and all the noble knights. There one might see the flower of chivalry and of right noble bachelery.

Chandos Herald, Life of the Black Prince, ed. M. K. Pope and E. C. Lodge (Oxford, 1910). *The herald who served Sir John Chandos wrote a metrical life in French based in part on eyewitness accounts of events.*

His marriage to Joan, the fair Maid of Kent, earned a certain notoriety for she had already been twice widowed but it was evidently a love match and she was reputed to be a woman of great beauty. Their fondness for each other was cause for comment, as when they were parted in 1366:

Very sweetly did they embrace and take farewell with kisses. Then might you see ladies weep and damsels lament; one bewailing her lover and another her husband. The Princess sorrowed so much that, being then big with child, she through grief delivered and brought forth a very fair son, the which was called Richard. Great rejoicings did all make, and the Prince also was right glad at heart, and all say with one accord: 'Behold a right fair beginning.'

and reunited in 1367, when they went on an informal walkabout:

The Princess came to meet him, bringing with her her first born son [Edward, who was to die in childhood] . . . very sweetly they embraced when they met together. The gentle prince kissed his wife and son. They went to their lodging on foot, holding each other by the hand. Ibid.

THE SACK OF LIMOGES, 1370

The Age of Chivalry was not always an uncomplicated tale of knights in shining armour and the Black Prince carries a prominent blot on his escutchion for his punishment of the citizens of Limoges who rebelled against Edward's heavy financial exactions:

Soon the prince besieged the city and battered it with deadly assaults and attacks, giving no respite to those within its walls. Then the walls were undermined; they collapsed, and the city was taken. The conquered city of Limoges was destroyed almost down to the ground. Those found there were killed, very few being taken prisoner and spared their lives.

Chronicles of the Age of Chivalry, quoting Thomas Walsingham. *Thomas was a monk of St Albans, continuing the chronicle tradition of that house.*

DEATH OF A PRINCE, 1376

Worn out by his campaigning in France, the Black Prince died in June 1376 amidst great mourning.

On 8 June, during the parliament at Westminster, Edward, prince of Wales, King Edward's eldest son, died. That day was the feast of the Holy

Trinity, which the prince had always been accustomed to celebrate every year, wherever he was, with greatest solemnity.

For as long as he lived and flourished, his good fortune in battle, like that of a second Hector, was feared by all races, Christian and pagan alike.

When Prince Edward died, all the hope of his people died too: for while he lived, the English dreaded no enemy who might invade. Ibid.

DEATH OF A KING, 1377

The last years of Edward III's reign were marred by the presence of his mistress Alice Perrers, a former lady-in-waiting of Queen Philippa. She bore him three children, but it was political influence over him, unmatched by any other medieval royal mistress, that earned her universal opprobrium. Her very presence seemed to mar his last hours when he died at Sheen on 21 June.

Shameful to relate, during the whole time that he was bed-ridden, King Edward had been attended by that infamous whore Alice Perrers, who always reminded him of things of the flesh. She never discussed nor permitted any discussion about the safety of his soul, but continually promised him a healthy body, until she saw a sure sign of his death with the failing of his voice.

When she realized that he had lost the power of speech and that his eyes had dulled, and that the natural warmth had left his body, quickly that shameless doxy dragged the rings from his fingers and left. Ibid.

Though the chronicler concluded that Edward's reputation was by then tarnished, he could recall some good:

Indeed this king among all other kings and princes of the world had been glorious, gracious, merciful and magnificent, and was called *par excellence* 'Most Gracious' for his pre-eminent and outstanding grace. His face was more like an angel's than a man's, for there was such a miraculous light of grace in it, that anyone who looked openly into it or dreamed of it at night, might hope that comforting delights would come to him that very day.

Ibid.

The effigy of Edward III in Westminster Abbey, made soon after his death, shows a long face, level brows, and forked beard. There is a look of nobility and also exhaustion conveyed by the hollow cheeks and drooping mouth. He had been ill for many weeks and died of a stroke.

Richard II

1377–1399

Known as Richard of Bordeaux from the town of his birth, Richard was only ten when he succeeded his grandfather Edward III. His brief life seems to be polarized between pity and terror, thus offering Shakespeare a truly tragic subject. He was handsome and pleasure-loving but abrupt in speech and had a stammer. After his death the legend grew up that he had been born without a skin and nourished in the skin of goats; but there was art and luxury in his palaces and when he stood alone he had one 'finest hour'.

A BAD OMEN

An imaginative chronicler, with the benefit of hindsight, recorded Richard's coronation.

At the coronation of this lord three ensigns of royalty foreshadowed for him three misfortunes. First, in the procession he lost one of the coronation shoes; whence, in the first place, the commons who rose up against him hated him ever after all his life long: secondly, one of the golden spurs fell off; whence, in the second place, the soldiery opposed him in rebellion: thirdly, at the banquet a sudden gust of wind carried away the crown from his head; whence, in the third and last place, he was set aside from his kingdom and supplanted by king Henry.

Chronicon Adae de Usk, 1377–1421, ed. E. M. Thompson (1904). *Adam of Usk, an Oxford-trained lawyer from Monmouth, wrote an account based on his own experience, but he is inclined to be credulous.*

Wat Tyler and The Peasants' Revolt, 1381

Opposition to an oppressive poll-tax first broke out in Essex and then spread to Kent, which found an articulate leader in Wat Tyler. Accompanied by the priest John Ball, Tyler led his rebels to London seeking an interview with Richard. The young king at 14 seems to have played a leading role.

And when the king heard of their doings he sent his messengers to them on the Tuesday next after Trinity Sunday, to ask them why they were behaving in this fashion and why they were raising a rebellion in his land. And they returned answer by the messengers that they were rising to deliver him and to destroy the traitors to him and to his kingdom. The king sent again to them to bid them cease their doings, in reverence for him, until he could speak with them, and he would make reasonable amends, according to their will, of all that was amiss; and the commons begged him, by the said messengers, that he would be pleased to come and talk with them at Blackheath. And the king sent again the third time to say that he would willingly come next day at the hour of prime to hear their purpose; and then the king, who was at Windsor, removed with all speed to London.

English Historical Documents, IV, quoting *the Anonimalle Chronicle. Written by an eyewitness this is the best account of the revolt.*

The commons expressed great loyalty:

And the commons had among themselves a watch word in English, 'With whom hold you?' and the response was, 'With king Richard and with the true commons', and those who could not or would not so answer were beheaded and put to death. Ibid.

However, the ensuing destruction in London, including the Savoy home of Richard's most senior uncle, John of Gaunt, was scarcely reassuring. Nor could Richard get much sense from his advisers:

At this time the king was in a turret of the great Tower of London, from which he could see the manor of the Savoy and the hospital of Clerkenwell, and the buildings of Simon Hosteler near Newgate, and John Butterwick's house, all on fire at once. And he called all his lords about him to his chamber and asked their advice as to what he should do in such an emergency; and none of them could or would give him any counsel. Wherefore the young king said that he would send to the mayor of the city to bid him order the sheriffs and aldermen to have it cried round their wards that all men between the ages of 15 and 60, on pain of life and members, should go on the morrow, Friday, to Mile End and meet him at seven in the morning. He did this so that all the commons who were surrounding the Tower would raise the siege and go to Mile End to see and hear him, and all those who were in the Tower could go away safely whither they would and save themselves. Ibid.

Richard came to Mile End on 14 June, with only a handful of supporters:

And when he was come the commons all knelt down to him, saying: 'Welcome, our lord, King Richard, if it pleases you, and we will have no other king but you.' And Wat Tyghler, their leader and chief, prayed to him in the name of the commons that he would suffer them to take and hold all the traitors who were against him and the law; and the king granted that they should take at their wish those who were traitors and could be proved traitors by the law. And Wat and the commons were carrying two banners and many pennons and pennoncelles, while they made their petition to the king. And they required that no man should be a serf, nor do homage or any manner of service to any lord, but should give fourpence rent for an acre of land, and that no one should serve any man but at his own will, and on terms of regular covenant. And at this time the king caused the commons to arrange themselves in two lines, and caused a proclamation to be made before them that he would confirm and grant them their freedom and all their wishes generally, and that they should go through the realm of England and catch all traitors and bring them to him in safety and that he would deal with them as the law required. Under colour of this grant Wat Tyghler and the commons took their way to the Tower, to seize the archbishop and the others, the king being at Mile End.

The Anonimalle Chronicle

The ensuing murders by the rebels—of the archbishop of Canterbury, the treasurer and others—did not end the affair. A further meeting with them took place at Smithfield.

And when he was called by the mayor, Wat Tyghler by name, of Maidstone, he came to the king in a haughty fashion, mounted on a little horse so that he could be seen by the commons. And he dismounted, carrying in his hand a dagger which he had taken from another man, and when he had dismounted he half bent his knee, and took the king by the hand, and shook his arm forcibly and roughly, saying to him, 'Brother, be of good comfort and joyful, for you shall have within the next fortnight 40,000 more of the commons than you have now and we shall be good companions.' And the king said to Wat; 'Why will you not go back to your own country?' And the other replied with a great oath that neither he nor his fellows would depart until they had their charter such as they wished to have, and such points rehearsed in their charter as they chose to demand.

Ibid.

Wat made his demands and Richard assented, but the rebel leader grew bolder:

And to this the king gave an easy answer, and said that he should have all that could fairly be granted saving to himself the regality of the crown. And then he commanded him to go back to his home without further delay. And all this time that the king was speaking no lord nor any other of his council dared nor wished to give any answer to the commons in any place except the king himself.

Presently Wat Tyghler, in the king's presence, called for a flagon of water to rinse his mouth because he was in such a heat, and when it was brought he rinsed his mouth in a very rude and disgusting fashion before the king; and then he made them bring him a flagon of ale of which he drank a great deal, and in the king's presence mounted his horse. Ibid.

Scuffles broke out and Wat was mortally wounded, leaving Richard exposed to great personal danger:

And in this scuffle a yeoman of the king's household drew his sword and ran Wat two or three times through the body, mortally wounding him. And the said Wat spurred his horse, crying to the commons to avenge him, and the horse carried him some four score paces, and there he fell to the ground half dead. And when the commons saw him fall, and did not know for certain how it was, they began to bend their bows and to shoot; wherefore the king himself spurred his horse and rode out to them, commanding them that they should all come to him at the field of St John of Clerkenwell. Ibid.

The loyal mayor intervened to help the king's cause and Richard presided in person over the dispersal of the rebels.

And the mayor went thither and found him and had him carried to Smithfield in the presence of his fellows and there he was beheaded. And so ended his wretched life. And the mayor caused his head to be set upon a pole and carried before him to the king who still abode in the fields. And when the king saw the head he had it brought near him to abash the commons and thanked the mayor warmly for what he had done. And when the commons saw that their leader, Wat Tyghler, was dead in such a manner, they fell to the ground among the wheat like men discomforted, crying to the king for mercy for their misdeeds. And the king benevolently granted them mercy and many of them took to flight; and the king ordered two knights to lead the rest of the Kentishmen through London and over London Bridge without doing them any harm, so that each of them could go in peace to his own home. Then the king ordered the mayor, William Walworth, to put on his head a helmet in anticipation of what was going to happen; the mayor asked why he was to do so and the king replied that he was greatly obliged to him and therefore was going to confer on him the order of knighthood. Ibid.

Richard and his Queen

Tall and fair-haired, Richard was married to Anne of Bohemia in 1382, in spite of some reservations about a huge loan made to her father, Wenceslas, and her own substantial dower.

Worthy to enjoy manna,
To Englishmen is given the noble Anna;

but to those with an eye for the facts it seemed that she represented a purchase rather than a gift, since the English king laid out no small sum to secure this tiny scrap of humanity.

The Westminster Chronicle, 1381–1394, ed. L. C. Hector and B. F. Harvey (Oxford, 1982). *The author of this work was well placed to receive news and wrote within a few years of the events he describes.*

A TRAGIC END

Richard's deep affection for Anne found violent expression in his grief at her early death in 1394. The earl of Arundel miscalculated on his lord's grief. Having absented himself from her funeral procession, he came late to the service and made the mistake of asking to be excused from attendance. Richard, seizing the nearest baton from a servant, struck him to the floor. Arundel's bleeding necessitated a further service to free the church from pollution of blood and the earl was sent to the Tower. The king expressed his grief further.

In the year of our Lord 1394, on Whitsun-day (7th June), died that most gracious lady Ann, queen of England, at the manor of Shene, which lies on Thames near to Brentford. Which manor, though a royal one and very fair, did king Richard, by reason that that lady's death happened therein, command and cause to be utterly destroyed. After the ceremony of her funeral, which was carried out with becoming honours on the morrow of Saint Peter ad Vincula (2nd August), the king, clad, with his train, in weeds of mourning, straightway passed over into Ireland with a great power, to subdue the rebellion of the Irish. *Adam of Usk*

A PRODIGAL KING

Richard's distribution of patronage quickly earned strong disapproval and led to political crisis.

In his early years this king of ours was so open-handed that to make any legitimate request of him was to have it immediately granted; indeed at times he anticipated the wishes of petitioners and he used often to give more than had been asked for. So lavish was his bounty, however, that all the property attaching to the Crown, in common with the revenues belonging to the royal exchequer, was virtually dealt out piecemeal to various people who presented demands for this or that. Having thus handed out his own substance to others, he had perforce to come down on the commons, with the result that the poor are loud in their complaints and declare that they cannot go on supporting the burden. If only the king would arrange matters so as to give them some relief! He would, I think, reap no small benefit by doing so. *The Westminster Chronicle*

A BAD-TEMPERED KING

Richard compounded his prodigality with a violent temper. Several appalling outbursts are recorded. In 1384 the earl of Arundel had had his first experience of royal anger, when he dared to complain of bad government.

White with the passion which, at these words, pervaded his whole being, the king scowled at the earl. 'If it is to my charge', he said, 'that you would lay this, and it is supposed to be my fault that there is misgovernment in the kingdom, you lie in your teeth. You can go to the Devil!' A complete hush followed as these words were heard, and there was nobody among the company who dared to speak. Then the duke of Lancaster broke the silence and delivered a speech in which he skilfully glossed the earl's remarks, so that the king's anger was assuaged. Ibid.

In 1385 Richard had to be physically restrained from using his sword on the archbishop of Canterbury when he complained about the king's courtiers. The following year demands were made for the removal of the king's favourites and Richard replied:

. . . that he would not, for them, remove the meanest scullion in his kitchen from his office.

Chronicon Henrici Knighton, ed. J. R. Lumby, 2 vols.; Rolls Series (1895), II

ROYAL FAVOURITES

The crisis that overtook Richard in 1388 was largely the result of his fondness for Robert de Vere, for whom he had created the first English marquisate—Dublin—and for Michael de la Pole. Richard came to be suspected of Edward II's vice. The favourites escaped to exile, where de Vere died in 1392.

This year [1395], in November, the king of England brought from Louvain the body of his former friend Robert de Vere whom he had created Duke of Ireland.

Thomas Walsingham, Historia Anglicana, ed. H. T. Riley, 2 vols.; Rolls Series (1864), II

Richard ordered a magnificent funeral at Colne, in Essex.

And he accompanied the balsam coffin in which the embalmed body lay and ordered it to be opened that he might look upon the face and clasp the fingers. Ibid.

Richard's Interests

The royal court became a cultural centre during the reign. To Richard's credit is the rebuilding of Westminster Hall with its splendid roof; the fine tomb for himself and his queen, Anne; and two surviving portraits.

BUILDING

From 1387 onwards he gave £100 a year towards the completion of the nave of Westminster abbey, a work directed by his own architect Henry Yeveley. He contributed also to the new choir of York Minster and the west front of Canterbury cathedral; the privileges he granted to Winchester College amounted to a most substantial subsidy to Wykeham's foundation. For himself Richard refitted his father's manor-house at Kennington, improved King's Langley, and in the period around 1385 had a great deal done to modernize both Eltham and Sheen. Eltham had a new royal bathroom fitted and a dancing chamber built.

John Harvey, *The Black Prince and His Age* (1976)

COOKERY

The nature of the court feasts was determined by his interest in fine cooking and the zest for new combinations of contrasting flavours. His court cookery book *The Forme of Cury* has been preserved: the manuscript came into the possession of the Staffords, was presented by Edmund Stafford to Queen Elizabeth and later was part of the Harleian Collection; it was printed for the Society of Antiquaries in the reign of George III. It is stated in its prologue that Richard is accounted 'the best and ryallest [most royal] vyander of all Christian Kings' and that the book is compiled by his master cook with the 'assent and advisement of maisters of phisik and of phielosophy' that dwelt in his court. It consists of 196 recipes, and throughout there is an emphasis on the exotic; the recipe for cooking oysters in Greek wine seems characteristic. A considerable luxury trade is presupposed: spices are in common use; there is much pepper, sometimes whole, sometimes powdered, and much ginger; there are frequent references to cinammon, cardamom, nutmeg and saffron and in one case to spikenard; sugar of Cyprus seems specially prized, but there is also white sugar and sugar clarified with wine.

Gervase Mathew, *The Court of Richard II* (1968)

DIVINATION AND ASTROLOGY

A Book of Divinations, *Libellus Geomancei* . . . is stated to have been prepared for the solace of King Richard by the least of his servants in

March 1391, and probably reflects the deepest private interests of the King. After sections on Kingship, Physiognomy and Dreams the book culminates on . . . a list of questions which when asked under an appropriate conjunction of planets, and with the use of diagrams would enable the King to receive answers on such problems as the strength of chastity, the welfare of the King and kingdom and the fidelity of friends. The volume is pervaded by trust in the power of the planets: a study of planetary conjunctions might provide a clue to several of Richard's actions in the political crises of his reign. Ibid.

ANIMALS

It would seem that a pelican made its way to court in 1393.

The king kept Christmas at Eltham, where he was waited upon in great state about Epiphany [6 January] by the Londoners and presented by them with a dromedary ridden by a boy: to the queen they gave a large and remarkable bird with an enormously wide gullet. *The Westminster Chronicle*

CULPABLE LUXURY OF RICHARD'S COURT

By the sixteenth century the stories of Richard's court had reached legendary proportion.

He . . . maintained the most plentiful house that ever any king in England did either before his time or since. For there reported daily to his court above ten thousand persons that had meat and drink there allowed them. In the kitchen there were three hundred servitors, and every other office was furnished after the like rate. Of ladies, chamberers, and launderers, there were above three hundred at the least. And in gorgeous and costly apparel they exceeded all measure, not one of them that kept within the bounds of his degree. Yeomen and grooms were clothed in silks, with cloth of grain and scarlet, over-sumptuous, you may be sure for their estates. And this vanity was not only used in the court in those days, with embroideries, rich furs, and goldsmiths' work, and every day there was devising of new fashions, to the great hindrance and decay of the commonwealth. *Holinshed's Chronicles* (Westport, Conn., 1976), I

A ROYAL VISIT TO IRELAND, 1394

Richard's ventures westwards, rather than into France—the first king to set foot there since John in 1210—were not unsuccessful. He made his first excursion just after Anne of Bohemia died and a second in the final year of his reign. One of his squires, Henry Cristall, informed Froissart of Richard's dealings with four Irish kings.

The king of England intended these four kings should adopt the manners, appearance and dress of the English, for he wanted to create them knights. He gave them first a very handsome house in the city of Dublin for themselves and attendants, where I was ordered to reside with them, and never to leave the house without an absolute necessity. I lived with them for three or four days without any way interfering that we might become accustomed to each other, and I allowed them to act just as they pleased. I observed that as they sat at table they made grimaces, that did not seem to me graceful or becoming, and I resolved in my own mind to make them drop that custom . . .

They had another habit I knew to be common in the country which was the not wearing breeches. I had, in consequence, plenty of breeches made of linen and cloth, which I gave to the king's and their attendants, and accustomed them to wear them . . . In riding, they neither used saddles nor stirrups, and I had some trouble to make them conform in this respect to the English manners. *Froissart*

Gradually the kings were trained and prepared for knighthood.

The four kings were very richly dressed, suitable to their rank, and that day dined at the table of king Richard, where they were much stared at by the lords and those present: not indeed without reason; for they were strange figures, and differently countenanced to the English or other nations. We are naturally inclined to gaze at anything strange, and it was certainly, sir John, at that time, a great novelty to see four Irish kings. Ibid.

The Deposition of Richard: Two Views

Richard met his fate at the hands of his cousin Henry Bolingbroke, the son of John of Gaunt, to whom he had unwisely denied his inheritance. Bolingbroke eventually confronted the king in Wales. Richard surrendered and was imprisoned in the Tower of London. The Official Lancastrian account made Richard's end seem like voluntary abdication.

And after the king had spoken apart with the duke and the archbishop, looking from one to the other with a cheerful countenance, as it seemed to the bystanders, at last the king, calling all those present to him, declared publicly in their presence, that he was ready to make the renunciation and resignation according to his promise. And although, to save the labour of such a lengthy reading, he might, as he was told, have had the resignation and renunciation, which was drawn up in a certain parchment schedule, read by a deputy, the king willingly, as it seemed, and with a cheerful countenance, holding the same schedule in his hand, said that he would read it himself, and he did read it distinctly. And he absolved his lieges and made renunciation and cession, and swore this . . . and he signed it with his own hand, as is more fully contained in the schedule, of which the tenor follows in these words . . .

English Historical Documents, IV, quoting the Rolls of Parliament

The truth of the matter was somewhat different:

The Duke of Lancaster went on the morrow to the Tower, with the Duke of York and the Earl of Arundel in his company; and, when there, he desired the Earl of Arundel to send King Richard to him. The Earl went to deliver his message; and when the king heard it, he replied, 'Arundel, go tell Henry of Lancaster from me that I will do no such thing, and that, if he wishes to speak with me, he must come to me; otherwise I will not speak to him.' The earl reported his answer to the duke, upon which he and the other lords went to the king . . . and . . . the king asked the Duke of Lancaster, 'Why do you keep me so closely guarded by your men-at-arms? I wish to know if you acknowledge me as your lord and king, or what you mean to do with me?' The duke replied, 'It is true you are my lord and king, but the council of the realm have ordered that you should be kept in confinement until the day of the meeting of parliament.' The king again swore, and desired that the queen his wife might come to speak to him. 'Excuse me, my lord!' replied the duke, 'it is forbidden by the council.' Then was the king in great wrath, but he could not help himself, and said to the duke, that he did great wrong both to him and to the crown. The duke replied, 'My lord, we cannot do otherwise till the parliament meets.' The king was so enraged by this speech that he could scarcely speak, and paced twenty-three steps down the room without uttering a word; and presently he broke out thus: '. . . you have acknowledged me as your king these twenty-two years, how dare you use me so cruelly? I say that you behave to me like false men, and like false traitors to their lord; and this I will prove, and fight four of the best of you, and this is my pledge'; saying which the king threw down his bonnet. The Duke of Lancaster fell down on his knees, and besought him to be quiet till the meeting of parliament, and there everyone would bring forward his reason. 'At least, fair sirs, for God's sake let me be brought to trial, that I may give an account of my conduct, and that I may answer to all they would say against me.' Then said the Duke of Lancaster, 'My lord, be not afraid, nothing unreasonable shall be done to you.' And he took leave of the king, and not a lord who was there durst utter a word.

After this began the parliament; and when Henry of Lancaster entered, he found there already seated all the prelates of the kingdom, to wit, eighteen bishops and thirty-two mitred abbots, besides the other prelates. [The duke seated himself on the throne and claimed the kingdom on the ground of the misdeeds of Richard II.]

English Historical Documents, IV, quoting the *Chronique de la traison et mort de Richard Deux, roy D'engleterre*

THE CROWN AS SUPREME SYMBOL OF KINGSHIP

Shakespeare's kings reveal their profoundest feelings about their situation, good or bad, through their thoughts of the physical crown. King John offered his obedience to the pope by handing his crown—'the circle of my glory'—*to the papal legate, who registered his acceptance by handing it back.*

Richard II, when his fortunes begin to fail, sees his crown as a round tower occupied by Death.

For God's sake, let us sit upon the ground,
And tell sad stories of the death of kings—
How some have been deposed, some slain in war,
Some haunted by the ghosts they have deposed,
Some poisoned by their wives, some sleeping killed,
All murdered. For within the hollow crown
That rounds the mortal temples of a king
Keeps Death his court; and there the antic sits,
Scoffing his state and grinning at his pomp,
Allowing him a breath, a little scene,
To monarchize, be feared, and kill with looks,
Infusing him with self and vain conceit,
As if this flesh which walls about our life
Were brass impregnable; and humoured thus,
Comes at the last, and with a little pin
Bores through his castle wall; and farewell, king.

Shakespeare, *Richard II*, III. ii

SHAKESPEARE ON RICHARD'S ABDICATION

Unlike the chronicler, Shakespeare represents Richard's feelings as tragic, and again he explores them through the presence of the physical crown.

KING RICHARD Give me the crown. Here, cousin, seize the crown.
Here, cousin, On this side my hand, and on that side thine.
Now is this golden crown like a deep well
That owes two buckets filling one another,

The emptier ever dancing in the air,
The other down, unseen, and full of water.
That bucket down and full of tears am I,
Drinking my griefs, whilst you mount up on high.

BOLINGBROKE [HENRY IV] I thought you had been willing to resign.

KING RICHARD My crown I am, but still my griefs are mine.
You may my glories and my state depose,
But not my griefs; still am I king of those.

Ibid. IV. i

REPORTED SAYING OF RICHARD II

He said expressly, with harsh and determined looks, that the laws were in his own mouth, sometimes he said that they were in his breast.

English Historical Documents, IV, quoting the official Articles of Deposition from the Rolls of Parliament

A DISCERNING DOG!

Eventually Richard was deserted even by his dog.

Then, too, I saw with king Henry a greyhound of wonderful nature, which, on the death of his master the earl of Kent, found its way by its own instinct to king Richard, whom it had never before seen and who was then in distant parts; and whithersoever the king went, and wheresoever he stood or lay down, it was ever by his side, with grim and lion-like face, until the same king, as is before told, fled at midnight by stealth and in craven fear from his army; and then, deserting him, and again led by instinct and by itself and with no guide, it came straight from Caermarthen to Shrewsbury to the duke of Lancaster, now king, who lay at that time in the monastery with his army, and, as I looked on, it crouched before him, whom it had never before seen, with a submissive but bright and pleased aspect. And when the duke had heard of its qualities, believing that thereby his good fortune was foretold, he welcomed the hound right willingly and with joy, and he let it sleep upon his bed. And after the setting aside of king Richard, when it was brought to him, it cared not to regard him at all other than as a private man whom it knew not; which the deposed king took sorely to heart.

Adam of Usk

RICHARD'S DEATH, 1400

Following his deposition, Richard was forcibly disguised as a forester and imprisoned first at Leeds Castle and finally at Pontefract. His death remains a matter of controversy and Shakespeare's adoption of a murder story was certainly not confirmed by the examination of the skeleton made in 1871. Adam of Usk associated his demise with disappointment at a rebellion in Richard's favour in the winter of 1400 which Henry IV crushed.

And now those in whom Richard, late king, did put his trust for help were fallen. And when he heard thereof, he grieved more sorely and mourned even to death, which came to him most miserably on the last day of February, as he lay in chains in the castle of Pontefract, tormented by sir [Thomas] Swinford with starving fare. *Adam of Usk*

Public exposition of the body followed:

The body of lord Richard, late king of England, was brought to the church of Saint Paul in London, the face not covered but shown openly to all; and the rites being there celebrated on that night and a mass on the morrow, he was buried at Langley among the Dominican friars. My God!, how many thousand marks he spent on burial-places of vainglory, for himself and his wives, among the kings at Westminster! But Fortune ordered it otherwise. Ibid.

EPILOGUE

As with Edward II, there were stories of Richard's escape, the most popular that he had got away to Scotland and died in Stirling Castle. Henry IV never entirely quelled the rumours that he was alive and would return. After eight years:

bills were set up in divers places of London, and on the door of Paul's church in which was contained that king Richard being alive and in health, would come shortly with great magnificence and power to recover again his kingdom: but the contriver of this device was quickly found out, apprehended, and punished according to his demerits. *Holinshed*

THE LANCASTRIANS

Henry IV

1399–1413

A revolutionary with a guilty conscience—this was what Henry Bolingbroke duke of Lancaster, eldest son of 'time-honoured' John of Gaunt, became. A medieval monarch, he repented his own success in breaking down the hedge around Richard's divinity. His face in effigy is square with level brows, slightly sagging cheeks and lower lip: a conventionally good face but more like an official than a king. He inherited Richard's troubles as well as his throne and bequeathed to the country he longed to serve two new ills: the idea of religious persecution and a disputed succession.

Two Views of the Coronation

PESSIMISTIC

Looking back over Henry's reign, as he had done with that of Richard II, Adam of Usk recalled bad omens at the coronation:

Henry the fourth, after that he had reigned with power for fourteen years, crushing those who rebelled against him, fell sick, having been poisoned; from which cause he had been tormented for five years by a rotting of the flesh, by a drying up of the eyes, and by a rupture of the intestines; and at Westminster, in the abbot's chamber, within the sanctuary, thereby fulfilling his horoscope that he should die in the Holy Land, in the year of our Lord 1412–13, and on the twentieth day of the month of March, he brought his days to a close. And he was carried away by water, and was buried at Canterbury. That same rotting did the anointing at his coronation portend; for there ensued such a growth of lice, especially on his head, that he neither grew hair, nor could he have his head uncovered for many months. One of the nobles, at the time of his making the offering in the coronation-mass, fell from his hand to the ground; which then I with others standing by sought for diligently, and when found, it was offered by him.

Adam of Usk

OPTIMISTIC

The office of 'king's champion' is peculiar to England and was first exercised by Sir John Dymoke at the coronation of Richard II. The caparisoned knight challenged to single combat any who disputed the king's right to reign. The same family performed the office until it lapsed after the coronation of George IV. For a usurper like Henry IV, the office had particular relevance.

When dinner was half over, a knight of the name of Dymock entered the hall completely armed and mounted on a handsome steed, richly barbed with crimson housings. The knight was armed for wager of battle, and was preceded by another knight bearing his lance: he himself had his drawn sword in one hand, and his naked dagger by his side. The knight presented the king with a written paper, the contents of which were, that if any knight or gentleman should dare to maintain that king Henry was not a lawful sovereign, he was ready to offer him combat in the presence of the king, when and where he should be pleased to appoint. The king ordered this challenge to be proclaimed by heralds in six different parts of the town and the hall, to which no answer was made. *Froissart*

HOLY OIL

Henry's legitimacy was endorsed further by the first use of a phial of heavenly oil.

On the day of the translation of St Edward, King and Confessor, Henry IV was crowned King at Westminster by the hands of the Lord Thomas, Archbishop of Canterbury . . . and as an augury of richer grace in the future, as it was believed, he was anointed with that heavenly oil which the Blessed Mary, Mother of God, once entrusted to the keeping of the Blessed Thomas, martyr, Archbishop of Canterbury, while he was in exile, prophesying to him that the Kings of England who should be anointed with this oil would be defenders and friends of the Church. This oil, kept in a golden eagle and stone phial, was hidden for a long time, but at last was miraculously revealed when the lord Henry, first Duke of Lancaster, was fighting the king's battles across the seas. The eagle was handed over by a certain holy man, who found it by divine revelation. He gave it to the most noble prince Edward, first-born of the illustrious King of England, so that he might be anointed with this oil as king after his father's death. Prince Edward deposited this oil in the Tower of London, enclosing it in a chest secured with many locks; and there it lay hidden, either through forgetfulness or neglect, until the time of King Richard, son of the prince.

In 1399 King Richard, carefully investigating the treasures left to him by his predecessors, unexpectedly found the eagle and phial and the writing or prophecy of the Blessed Thomas the martyr with them. And when he had learnt what power this oil possessed, he asked the lord Thomas, Archbishop of Canterbury, to anoint him again with this oil. The latter quite refused, however, saying that it was sufficient for him to have received holy unction at his hands in his first coronation, and he ought not to have a repetition of such unction. King Richard carried this eagle, with the phial, with him when he set out for Ireland, and brought it back with him to this land. The Archbishop asked him for it at Chester, and he handed it over, saying that now he clearly perceived how it was not the divine will for him to be anointed with this oil, and how another ought to receive such a noble sacrament. The archbishop, preserving such treasures under reverent custody, kept them until the time of the coronation of the present king, who was the first king of England to be anointed with such a precious liquid.

English Historical Documents, IV, quoting Thomas Walsingham

HENRY ENTERTAINED THE GREEK EMPEROR, 1400

It was costly to king it on the grand scale but Henry needed royal prestige, and Manuel II was appealing to a former crusader for help against the Turks,

. . . abiding with him at very great cost for two months, being also comforted at his departure with very great gifts . . . This emperor always walked with his men, dressed alike and in one colour, namely white, in long robes cut like tabards; he finding fault with the many fashions and distinctions in dress of the English, wherein he said that fickleness and changeable temper was betokened. *Adam of Usk*

A LEPER KING

The early years of the reign were troubled by rebellions, some in Richard's favour, some led by the elusive Welsh hero Owen Glendower and others, because of disillusionment with the new régime. Henry was sorely pressed and in 1405 meted out severe sentences in the north. The archbishop of Canterbury pleaded in vain for the life of his fellow metropolitan.

Then were the Archbishop of York and the lord Mowbray condemned to death, and Sir William Plympton with them, and were beheaded outside the city of York.

And when the archbishop should die, he said: 'Lo! I shall die for the laws and good rule of England.' And then he said unto those who were to die with him: 'Let us suffer death meekly, for we shall this night, by God's grace, be in Paradise.' Then said the archbishop to the man who had to cut off his head, 'For His love that suffered five wounds for all mankind, give me five strokes, and I forgive you my death.' And so he did and thus they died.

And immediately afterwards, as it was said, the king was smitten with a leprosy; for the archbishop Almighty God soon afterwards wrought many great miracles.

When the pope heard of the death of the Archbishop of York, he cursed all those that slew him, and all those who assented to his death or advised it, and commanded the archbishop of Canterbury that he would denounce all those who were cursed; but the Archbishop would not do it alone.

An English Chronicle, 1377–1461, ed. J. S. Davies, Camden Society (1856). *This anonymous chronicle was compiled before 1471.*

DE HAERETICO COMBURENDO

The Church, of which Henry was a devout son, was threatened at this time by the Great Schism in the papacy with the spectacle of two and, at one stage, three popes; and by heretical movements, such as the Lollards in England. Henry's efforts to suppress the latter were supported by Archbishop Arundel of Canterbury and led to the Statute for Burning Heretics in 1401. In 1410 the king's son was involved in one such case, the burning of John Badby.

A certain layman, a smith [faber] by trade, obstinately defended his heresy, that it is not the Body of Christ which is sacramentally carried in the church, but an inanimate object, worse than a toad or a spider, which are animate animals, and he would not forsake this opinion; so he was handed over to the secular arm. And when he was condemned, and enclosed in a barrel in Smithfield, the lord Prince Henry, the king's first-born, who was present, came up to him, and counselled and warned him to repent; but the lost worthless wretch paid no heed to the warnings of the prince, and chose rather to be burnt than to give reverence to the living sacrament. Wherefore he was enclosed in the barrel, and was struck by the devouring flame, and groaned miserably amidst the burning. The lord prince was moved by his pitiful cries, and ordered the materials of the fire to be taken away, and the man to be removed from the flames. He comforted the half-dead man and promised him even now life and pardon, and a grant of threepence a day from the royal treasury, if he would repent; but the unhappy man, whose spirit had revived, spurned the offer of so much honour, undoubtedly possessed by an evil spirit. Wherefore the lord prince ordered him to be enclosed once more in the barrel, with no possibility of pardon thenceforward; and so it followed that there the foolish man was burnt to ashes, and died miserably in his sins.

English Historical Documents, IV, quoting Thomas Walsingham

THE KING, THE PRINCE, AND THE LAW

This story was first told by Sir Thomas Elyot in The Governor, *1531. Prince Hal's servant had been arrested for some misdeed. The Prince, in a furious rage and brandishing a weapon, ordered Chief Justice Gascoigne to release him forthwith. The judge bravely accused the young Prince of not setting a good example of obedience to the law.*

'And now, for your contempt and disobedience, go you to the prison of the King's Bench, whereunto I commit you, and remain ye there prisoner until the pleasure of the King your father be further known.' With which words being abashed, and wondering also at the marvellous gravity of that worshipful Justice, the noble Prince laying his weapon apart . . . went to the King's Bench as he was commanded. C. L. Kingsford, *Henry V* (1901)

When King Henry IV heard the news he said:

'O merciful God, how much am I bound to your infinite goodness, specially for that you have given me a judge, who feareth not to minister Justice, and also a son who can suffer semblably and obey Justice.'

Ibid.

The Death of Henry IV

Henry's guilt remained with him until the end and his entire reign was made harsh by his disease. He requested burial at Canterbury, rather than the traditional place of Westminster and made no demands regarding a suitable tomb. Legends quickly grew up about his last illness and death. One told of sudden sickness as he prepared to depart for the Holy Land. Henry was rushed to the abbot's lodgings.

At length when he was come to himself, not knowing where he was, he inquired of such of them as were about him, what place that was. They showed him that it belonged to the Abbot of Westminster; and because he felt himself so sick, he commanded them to ask if that chamber had any special name; whereunto it was answered that it was named Jerusalem. The said the king: 'Praise be to the Father of Heaven, for now I know I shall die in this chamber, according to the prophecy of me beforesaid that I should die in Jerusalem.' And so after that he made himself ready and died shortly afterwards, upon the day of St Cuthbert, or the 20 March, when he had reigned 13 years 5 months 21 days.

R. Fabyan, The New Chronicles of England and France, ed. H. Ellis (1811)

Another wrote of his last confession. His confessor, John Tille, was urged by the lords standing by to instruct the king to repent three sins—the death of Richard, the execution of Archbishop Scrope and the usurpation. Henry replied:

For the first two pointes I wrote unto the pope the very truth of my conscience. And he sent me a bull with absolution and penance assigned which I have fulfilled. And as for the third pointe, it is hard to set remedy, for my children will not suffer that the regalie go oute of our lineage.

John Capgrave, The Chronicle of England, ed. F. C. Hingeston, Rolls Series (1858). *Capgrave enjoyed the patronage of Henry's youngest son, Humphrey duke of Gloucester. He probably met Henry once and he wrote his biography of the king from memory and from information from those who knew him better.*

Even Henry's final hours augured ill:

He was so sorely pressed at the latter end of his sickness that those who attended him, not perceiving him breathe, concluded he was dead and covered his face with a cloth. It was the custom in that country, whenever the king was ill, to place the royal crown on a cushion beside his bed, and for his successor to take it on his death. The prince of Wales, being informed by the attendants that his father was dead, had carried away the crown; but shortly after the king uttered a groan, and his face was uncovered—when, on looking for the crown, he asked what was become of it? His attendants replied, that 'my lord the prince had taken it away'. He bade them send for the prince; and on his entrance, the king asked him why he had carried away the crown? 'My lord,' answered the prince, 'your attendants, here present, affirmed to me that you were dead; and as your crown and kingdom belong to me as your eldest son, after your decease, I had it taken away.' The king gave a deep sigh, and said, 'My fair son, what right have you to it? for you well know I had none.' 'My lord,' replied the prince, 'as you have held it by right of your sword, it is my intent to hold and defend it the same during my life.' The king answered, 'Well, act as you see best: I leave all things to God, and pray that he would have mercy on me!' Shortly after, without uttering another word, he departed this life.

The Chronicles of Enguerrand de Monstrelet, ed. T. Johnes. *This French chronicler takes up where Froissart leaves off and though colourful is not always reliable.*

HENRY'S VISION OF THE UNQUIET CROWN

How many thousand of my poorest subjects
Are at this hour asleep? O sleep, O gentle sleep,
Nature's soft nurse, how have I frighted thee,
That thou no more wilt weigh my eyelids down
And steep my senses in forgetfulness? . . .
Canst thou, O partial sleep, give thy repose

To the wet sea-boy in an hour so rude,
And in the calmest and most stillest night,
Deny it to a king? Then happy low, lie down.
Uneasy lies the head that wears a crown.

Shakespeare, *2 Henry IV*, III. i

HENRY'S BURIAL

In accordance with Henry's own wish his coffin was placed in Becket's chapel, near the tomb of the Black Prince, in Canterbury Cathedral. In 1691 Henry Wharton published the purported story of an esquire who accompanied the body by sea from London to Canterbury and who, during a storm, cast it into the sea. On 21 August 1832 the truth was established by opening the tomb at Canterbury. Parts of the rough wooden coffin were cut away to reveal a leaden coffin. A section of the latter being removed, wrappers provided the final obstacle.

These wrappers were cut through and lifted off, to the astonishment of all present, the face of the deceased king was seen in complete preservation. The nose was elevated, the cartilage even remaining, though on the admission of the air it sunk rapidly away, and had entirely disappeared before the examination was finished. The skin of the chin was entire, of the consistency and thickness of the upper leather of a shoe, brown and moist; the beard thick and matted, and of a deep russet colour.

The jaws were perfect and all the teeth in them, except one fore-tooth, which had probably been lost during the king's life.

Dr Spry, 'A Brief Account of the Examination of the Tomb of King Henry IV in the Cathedral of Canterbury', *Archaeologia Cantiana*, viii (1872)

Henry V

1413–1422

The greatest man that ever ruled England.

K. B. McFarlane of Magdalen College, Oxford

A mere condottiere. A. J. P. Taylor of Magdalen College, Oxford

Until quite recently only the first of these judgements seemed possible. Today there is a feeling that Henry of Monmouth's obsession with his right to France demanded too much in human lives. Edward III's costly campaigns for his 'legal' rights has evoked a similar response. In Henry V's own day, two lines of verse seemed to suggest that commerce was better for England than conquest:

Cherish merchandise, keep the admiralty,
That we be masters of the narrow sea.

Nevertheless, Henry was a national hero who reached out to posterity by inspiring Shakespeare. His lean, hairless face and curiously pouting or pursed lips give the impression of an aesthete rather than a fighter, and his distant descendant, Queen Elizabeth II, once asked whether the portrait at Windsor was really like him.

Prince Hal

We know definitely that Henry's youth was not one of extreme piety but there our certain knowledge ends. He quarrelled with his father and was reconciled in 1412. Details of his reckless actions survive in a Tudor translation of an early biography extended by stories told by the earl of Ormonde and at this distance it is impossible to pronounce judgement on their authenticity. One such story told of veritable robbery.

[Henry] accompanied with some of his young lords and gentlemen would await in disguised array for his own receivers, and distress them of their money. And some time at such enterprises both he and his company were surely beaten; and when his receivers made to him their complaints, how they were distressed and robbed in their coming unto him, he would give them discharges of so much money as they had lost, and besides that they should not depart from him without great rewards for their trouble and vexations. And he that best and most manly had resisted him and his company in their enterprize, and of whom he had received the greatest and most strokes, should be sure to receive of him the greatest and most bounteous rewards.

The First English Life of Henry V, ed. C. L. Kingsford (Oxford, 1911). *This sixteenth-century translation of the biography of Henry V by Titus Livius, an Italian humanist, commissioned by Humphrey, duke of Gloucester, was amplified by stories from Monstrelet, English chronicles, and Ormonde. Bar the last, it is the most authentic biography.*

A CONTEMPORARY ACCOUNT OF PRINCE HAL'S CONVERSION

The stage was thereby set for a supposed reform when the old king died. The chronicler described how before he was king, the Prince of Wales

intended greatly to riot, and drew to wild company; & divers Gentlemen and Gentlewomen followed his will & his desire at his commandment; and likewise all his men of his household were attending & pleased with his governance, except three men of his household, which were fully heavy and sorry of his governance; and they counselled him ever contrary, and fain would have had him to do well, & forsake riot. And therefore he hated them three most of all men in his house, unto the time that his father was dead.

And then he began to reign for King, & he remembered the great charge and worship that he should take upon him; And anon he commanded all his people that were attendant to his misgovernance afore time, & all his household, to come before him. And when they heard that, they were full glad, for they supposed that he would have promoted them into great offices, & that they should have stood in great favour & trust with him, & nearest of counsel, as they were afore time. And trusting hereupon, they were the homelier & bolder unto him, & nothing dread him; insomuch, that when they were come before him, some of them winked on him, & some smiled, & thus they made nice semblance unto him, many a one of them. But for all that, the Prince kept his countenance full sadly unto them, and said to them: 'Sirs, ye are the people that I have cherished & maintained in Riot & wild governance; and here I give you all in commandment, & charge you, that from this day forward you forsake all misgovernance, & live after the laws of Almighty God, & after the laws of our land. And who that doeth contrary, I make faithful promise to God, that he shall be truly punished according to the law, without any favour or grace.' And charged them, on pain of death, that they should never give him comfort nor counsel to fall to riot no more; for he had taken a charge on him, that all his wits & power were too little, without the help of God & good governance. And so he rewarded them richly with gold & silver, & other jewels, and charged them all to avoid his household, & live as good men, & never more to come in his presence, because he would have no occasion nor remembrance whereby he should fall to riot again. And thus he voided all his household, saving those three persons that he hated most . . . & them he loved afterward best, for their good counsel and good governance, & made them afterward great lords: and thus was left in his household none but those three men.

The Brut or The Chronicles of England, II, ed. F. W. D. Brie, Early English Text Society (1908, spelling modernized). *These chronicles began as the French* Brut d'Engleterre *and there were English translations and continuations.*

THE CASTING OFF OF FALSTAFF

Part of the legend of Henry's misspent youth is devoted to Sir John Falstaff—an invention of the Elizabethan dramatists. It served the Bard's purpose of illustrating Henry's reform but the character cannot be identified either with Sir John Oldcastle, erstwhile friend and Lollard, executed by Henry in 1417, or the veteran soldier Sir John Fastolf (d. 1429). Shakespeare has Henry dismiss his boon companion with stinging words:

I know thee not, old man. Fall to thy prayers.
How ill white hairs become a fool and jester!
I have long dreamt of such a kind of man,
So surfeit-swelled, so old, and so profane;
But being awake, I do despise my dream.

Shakespeare, *2 Henry IV*, v. v

THE KING EMPHASIZED THE RULE OF JUSTICE AND LAW

The chronicler illustrated Henry V's success in the story of two knights' quarrel and how the king dealt with it.

For, in the first year of his reign, there were two knights at great debate: the one was of Lancashire, & the other of Yorkshire; and they made them as strong of people as they could, & skirmished together; and men were slain & hurt on both parties.

And when the King heard thereof, he sent for them: & they came to the King to Windsor, as he was going to his dinner. And when the King understood that they were come, he commanded them to come before him; and then he asked them, 'whose men they were'. And they said, his liege men. 'And whose men be those that ye have raised up to fight for your quarrel?' And they said, 'his men'. '& what authority or commandment had ye, to raise up my men or my people, to fight & slay each other for your quarrel? In this ye are worthy to die.' And they could not excuse themselves, but besought the King of his grace. And then the King said, 'by the faith that he owed to God & to Saint George, unless they agreed & accorded, by the time that he had eaten his oysters, they should both be hanged before he had supped.' And then they went apart, & agreed by themselves, and came in again when the King had eaten his oysters. And then the King said: 'Sirs, how standeth it with you?' And then they kneeled down, and said: 'If it please your good grace, we be agreed & accorded.'

The Brut

HENRY V AND THE TENNIS BALLS

Henry's reign is dominated by his fight for his heritage in Normandy and Gascony and his victories against the French. Herein lies the origin of the story of the tennis balls sent by the king of France with the message that gaming was more suited to his youthfulness than arms.

The French, in the blindness of harmful pride having no foresight, with words of gall answered foolishly to the ambassadors of the King of England, that because King Henry was young they would send him little balls to play with, and soft cushions to rest on, until what time he should grow to a man's strength. At which news the King was much troubled in spirit, yet with short, wise, and seemly words, he thus addressed those who stood about him: 'If God so wills and my life lasts, I will within a few months play such a game of ball in the Frenchmen's streets, that they shall lose their jest and gain but grief for their game. If they sleep too long upon their cushions in their chamber, perchance before they wish it I will rouse them from their slumbers by hammering on their doors at dawn.'

The First English Life, quoting the manuscript of John Strecche. *John Strecche wrote early in the reign of Henry VI.*

War with France

If the story of the studied insult is true the king of France must have quickly regretted the provocation. Henry crossed to Harfleur in August 1415; successfully besieged that town which he entered 'both barefooted and bare legged'; and, in spite of heavy losses in his army due to dysentery, decided to march to Calais. His decisiveness is illustrated from his habit of only giving one of two answers to a question: 'It shall be done', or 'It is impossible'. Discipline was a key element of this monarch's success and he was willing to sacrifice precious manpower to maintain it on the road to Calais.

And there was brought to the king in that field a certain robber, an Englishman who, in God's despite and contrary to the royal decree, had stolen and carried off from a church (perhaps thinking it was made of gold) a pyx of copper-gilt in which the Host was reserved, that pyx having been found in his sleeve. And in the next hamlet where we spent the night, by command of the king, who was punishing in the creature the wrong done to the Creator (as Phinehas did with Zimri), and after sentence had been passed, he met his death by hanging.

Gesta Henrici Quinti, ed. F. Taylor and J. S. Roskell. *This reliable contemporary account of Henry's deeds was written by the king's chaplain.*

AGINCOURT, 25 OCTOBER 1415

On the long march to Calais, Henry was opposed by the French before the village of Agincourt. The English were vastly outnumbered:

And amongst other things which I noted as said at that time, a certain knight, Sir Walter Hungerford, expressed a desire to the king's face that he might have had, added to the little company he already had with him, ten thousand of the best archers in England who would have been only too glad to be there. 'That is a foolish way to talk', the king said to him, 'because, by the God in Heaven upon Whose grace I have relied and in Whom is my firm hope of victory, I would not, even if I could, have a single man more than I do. For these I have here with me are God's people, whom He deigns to let me have at this time. Do you not believe', he asked, 'that the Almighty, with these His humble few, is able to overcome the opposing arrogance of the French who boast of their great number and their own strength?' as if to say, He can if He wishes. And, as I myself believe, it was not possible, because of the true righteousness of God, for misfortune to befall a son of His with so sublime a faith, any more than it befell Judas Maccabeus until he lapsed into lack of faith and so, deservedly, met with disaster. Ibid.

SHAKESPEARE'S VERSION

[WARWICK] O that we now had here
But one ten thousand of those men in England
That do no work today.
HENRY What's he that wishes so?
My cousin Warwick? No, my fair cousin.
If we are marked to die, we are enough
To do our country loss; and if to live,
The fewer men, the greater share of honour.
God's will, I pray thee wish not one man more.

. . . .

We few, we happy few, we band of brothers.
For he today that sheds his blood with me
Shall be my brother; be he ne'er so vile,

This day shall gentle his condition.
And gentlemen in England now abed
Shall think themselves accursed they were not here,
And hold their manhoods cheap whiles any speaks
That fought with us upon Saint Crispin's day. Shakespeare, *Henry V*, IV. iii

AN UNCHIVALRIC ACT

The victory at Agincourt was marred by a royal order following upon rumour of the arrival of fresh troops. This resulted in an act of considerable barbarity according to the various sources, and one which cannot easily be excused.

During the heat of the conflict, when the English had gained the upper hand and made several prisoners, news was brought to king Henry that the French were attacking his rear, and had already captured the greater part of his baggage and sumpter-horses . . . This distressed the king very much, for he saw that though the French army had been routed they were collecting on different parts of the plain in large bodies, and he was afraid they would renew the battle. He therefore caused instant proclamation to be made by sound of trumpet, that every one should put his prisoners to death, to prevent them from aiding the enemy, should the combat be renewed. This caused an instantaneous and general massacre of the French prisoners. *Monstrelet*

'CRY—GOD FOR HARRY! ENGLAND! AND SAINT GEORGE!'

Henry's reputation owes much to Agincourt though it was in reality little more than a morale booster. The heavily armed mounted French cavalry columns were bogged down in mud and crammed in too deep to manoeuvre. The English bowmen, spread in a long line with their arrows of ash, barbed with iron heads and winged with goose feathers, took deadly aim from behind palisades. Henry's losses were in the hundreds, the French in the thousands.

When the King of England saw that he was master of the field and had got the better of his enemies he humbly thanked the Giver of victory, and he had good cause, for of his people there died on the spot only about sixteen hundred men of all ranks, among whom was the Duke of York, his great uncle, about whom he was very sorry.

When evening came the King of England, being informed that there was so much baggage accumulated at the lodging places, caused it to be proclaimed everywhere with sound of trumpet that no one should load himself with more armour than was necessary for his own body, because they were not yet wholly out of danger from the King of France. And this night the corpses of the two English princes, that is to say, the Duke of

York and the Earl of Oxford, were boiled, in order to separate the bones and carry them to England.

Chronicles of John de Wavrin, ed. W. and E. L. C. P. Hardy, Rolls Series (1864–91), II

THE HERO-KING

Henry himself was the simple, one-piece national hero throughout and he returned to England to a rapturous welcome.

Nor do our older men remember any prince ever having commanded his people on the march with more effort, bravery, or consideration, or having, with his own hand, performed greater feats of strength in the field. Nor, indeed, is evidence to be found in the chronicles or annals of kings of which our long history makes mention, that any king of England ever achieved so much in so short a time and returned home with so great and so glorious a triumph. To God alone be the honour and the glory, for ever and ever. Amen.

Gesta Henrici Quinti

THE TREATY OF TROYES AND BETROTHAL TO CATHERINE OF VALOIS

In 1417 Henry returned to France with a larger army and by a series of sieges reduced the duchy of Normandy. The climax of his wars came in 1420 when two aspects of his great French fantasy were celebrated on the same day, 21 May 1420: at Troyes the political treaty made him regent and heir to the mad king of France, Charles VI; and in Troyes Cathedral he was solemnly betrothed to Charles's daughter Catherine. When in June Henry captured the cathedral city of Sens, whose ousted archbishop had performed the marriage ceremony earlier that month, he paid the dignitary a graceful compliment:

'You have given me a wife, now I restore you your own—your church'

C. L. Kingsford, *Henry V* (1911)

SARCASTIC EXCHANGES

Henry's negotiations for the hand of Catherine of France had earlier been abortive.

King Henry was very desirous to marry her, and not without cause, for she was very handsome, of high birth, and of the most engaging manners . . . When the conference was broken off, the enclosure was destroyed, the tents and pavilions pulled down—and the two parties returned to Pontoise and Mantes.

The King of England was much displeased at the breaking off of the conference, as it prevented him from gaining his ends, and was very indignant against the duke of Burgundy [the chief French negotiator], whom he considered as the cause of it, he being the principal leader of the government. The last day they were together, seeing that his demands would not be complied with as to his marriage with the lady Catherine, he said to the duke of Burgundy, 'Fair cousin, we wish you to know that we will have the daughter of your king, and all that we have asked, or we will drive him and you out of his kingdom.' The duke replied, 'Sire, you are pleased to say so; but before you can drive my lord and me out of his kingdom I make no doubt but that you will be heartily tired.'

Monstrelet

ROYAL HAUTEUR, 1420

The Treaty of Troyes did not end the war, for its implicit disinheritance of Charles VI's son committed Henry V to interminable war against the Dauphin. The sieges went on. He was successfully conducting that of Melun while his ally the lord de l'Isle Adam was sent to garrison Joigny.

When he had remained there some time, and had properly posted his men, he returned to the siege of Melun. He had caused to be made a surcoat of light grey, in which he waited on the king of England relative to some affairs touching his office. When he had made the proper salutations, and had said a few words respecting his business, king Henry, by way of joke, said, 'What, l'Isle-Adam! is this a dress for a marshal of France?' to which he replied, looking the king in the face, 'Sire, I have had it thus made to cross the Seine in the boats.' The king added, 'How dare you look a prince full in the face when you are speaking to him?' 'Sire,' answered l'Isle-Adam, 'such is the custom of us Frenchmen; and if anyone addresses another, whatever may be his rank, and looks on the ground, he is thought to have evil designs, and cannot be an honest man, since he dare not look in the face of him to whom he is speaking.' The king replied, 'Such is not our custom.'

Monstrelet

THE JERUSALEM ANTECHAMBER, 31 AUGUST 1422

Henry died of dysentery, a campaigner's disease. Like his pious father, he had wanted his last war to be fought for the Holy Land. To the medieval kings, Jerusalem was the great staging-post between earth and heaven. On his death-bed, when Henry's chaplain reached the words in the Psalms for the dying, 'that the walls of Jerusalem may be built', Henry interrupted:

'O good Lord, thou knowest that mine intent hath been and yet is, if I might live, to re-edify the walls of Jerusalem.'

HENRY'S BURIAL

Following his death at the Bois de Vincennes outside Paris elaborate preparations were made before his funeral cortège set out for England.

The said king Henry was overtaken by death at about the age of forty. His body was dismembered and divided into several parts which were boiled in a cauldron. These different parts were enclosed with the bones in a leaden coffin filled with all manner of spices and the water in which they were boiled was put in a cemetery. Then the body was placed on a chariot hung with black cloth and driven to the church of Saint-Denys-en-France.

Chronique du religieux de Saint Denys, ed. L. Bellaquet, 6 vols. (Paris, 1839–52), VI. *The author of this French chronicle was well informed. Like many Frenchmen he expressed a great admiration for Henry in spite of his invasion.*

SAMUEL PEPYS ON CATHERINE OF VALOIS

Believe it or not, Samuel Pepys once kissed Catherine of Valois. The beautiful French wife of Henry V died in 1437 at the age of 36. Her 'skelleton' lay beside her husband's in the Confessor's chapel at Westminster and was sometimes shown to special visitors, 'the Bones firmly united, and thinly cloth'd with Flesh, like Scrapeings of tann'd Leather . . .' *It still lies in the Abbey but now in Henry V's chantry.*

I now took them [my wife and the girls] to Westminster Abbey and there did show them all the tombs very finely, having one with us alone . . . and here we did see, by perticular favour, the body of Queen Katherine of Valois, and had her upper part of her body in my hands. And I did kiss her mouth, reflecting upon it that I did kiss a Queen, and that this was my birthday, 36 years old, that I did first kiss a Queen.

The Diary of Samuel Pepys, ed. R. C. Latham and W. Matthews, 11 vols. (1970–83), IX, 23 February 1669

Henry VI

1422–1461, 1470–1471

King at nine months old, Henry never quite grew out of his childlike gentleness and naïvety. Lancastrian piety reached its apogee in him but he could not cope with the damnosa hereditas *of his dual monarchy. The Hundred Years War was scarcely over before the Wars of the Roses began. Henry's incompetence was made worse by recurring bouts of madness, a legacy perhaps from his lunatic grandfather, Charles VI of France. Efforts to have him canonized are still in progress but his inability to rule had lost him the hearts of his subjects by 1461 and his restoration in 1470 was brief. As K. B. McFarlane remarked, 'Henry VI's head was too small for his father's crown.'*

EARLY EVIDENCE OF HENRY'S PIETY

When Henry was not more than two, his mother intended taking the young king from Staines to Westminster on a Sunday but the child demonstrated a curious respect for the Sabbath:

Upon the morrow, being then Sunday, the king was borne towards his mother's chair, and he shrieked and cried and sprang, and would not be carried further; wherefore he was borne again into the inn, and there he abode on Sunday all day; and on the Monday he was borne to the chair, and then he being glad and merry, cheered.

A Chronicle of London, ed. N. H. Nicolas and E. Tyrell (1827)

At the age of eight he may have attended the trial of Joan of Arc, who was later canonized. Perhaps there were two saints in the courtroom.

Further Virtues of King Henry VI

A SIMPLE LIFE-STYLE

Further of his humility in his bearing, in his clothes and other apparel of his body, in his speech and many other parts of his outward behaviour—it is well known that from his youth up he always wore round-toed shoes and boots like a farmer's. He also customarily wore a long gown with a rolled hood like a townsman, and a full coat reaching below his knees, with shoes, boots and foot-gear wholly black, rejecting expressly all curious fashion of clothing.

Henry the Sixth: A Reprint of John Blacman's Memoir, ed. M. R. James (Cambridge, 1919). *John Blacman was a Carthusian monk and former cantor of Eton College whose biography of Henry VI was based on personal knowledge of the king and completed shortly after Henry's death.*

GENEROSITY

At another time when the executors of his uncle, the most reverent lord cardinal the bishop of Winchester came to the King with a very great sum, namely £2,000 of gold to pay him, for his own uses, and to relieve the burdens and necessities of the realm, he utterly refused the gift . . . saying: "... Do ye with his goods as ye are bound: we will receive none of them." The executors were amazed at this his saying, and entreated the King's majesty that he would at least accept that gift at their hands for the endowment of his two colleges which he had then newly founded, at Cambridge and Eton. This petition and gift the King gladly accepted.

Henry the Sixth

NO BAD LANGUAGE

Also he would never use any other oath to confirm his own truthful speech than the uttering of these words: 'Forsothe and forsothe', to certify to those whom he spoke of what he said. So also he restrained many both gentle and simple from hard swearing either by mild admonition or harsh reproof; for a swearer was his abomination.

Ibid.

COMPASSION AND MERCY

Once when he was coming down from Saint Albans to London through Cripplegate, he saw over the gate there the quarter of a man on a tall stake, and asked what it was. And when his lords made answer that it was the quarter of a traitor of his, who had been false to the king's majesty he said: 'Take it away. I will not have any christian man so cruelly handled for my sake.' And the quarter was removed immediately. He that saw it bears witness.

Ibid.

CHASTITY

The dominant characteristic of Henry is perhaps his puritanical life-style, which is well recorded in several anecdotes.

This King Henry was chaste and pure from the beginning of his days. He eschewed all licentiousness in word or deed while he was young; until he was of marriageable age, when he espoused the most noble lady, Lady Margaret, daughter of the King of Sicily, by whom he begat but one only son, the most noble and virtuous Prince Edward; and with her and toward her he kept his marriage vow wholly and sincerely, even in the absences of the lady, which were sometimes very long: never dealing unchastely with any other woman. Neither when they lived together did he use his wife unseemly, but with all honesty and gravity.

It is an argument of his watch upon his modesty that he was wont utterly to avoid the unguarded sight of naked persons, . . . therefore this prince made a covenant with his eyes that they should never look unchastely upon any woman. Hence it happened once, that at Christmastime a certain great lord brought before him a dance or show of young ladies with bared bosoms who were to dance in that guise before the King, perhaps to prove him or to entice his youthful mind. But the King was not blind to it, nor unaware of the devilish wile, and spurned the delusion, and very angrily averted his eyes, turned his back upon them, and went out to his chamber, saying: Fy, fy, for shame, forsothe ye be to blame.

At another time, riding by Bath, where are warm baths in which they say the men of that country customarily refresh and wash themselves, the King, looking into the baths, saw in them men wholly naked with every garment cast off. At which he was displeased, and went away quickly, abhoring such nudity as a great offence . . .

Besides, he took great precautions to secure not only his own chastity but that of his servants. For before he was married, being as a youth a pupil of chastity, he would keep careful watch through hidden windows of his chamber, lest any foolish impertinence of women coming into the house should grow to a head, and cause the fall of any of his household. Ibid.

HENRY THE EDUCATOR

In a reign that was ultimately so disastrous, arguably the one ray of joy was Henry's personal achievement in founding Eton, and King's College, Cambridge. Eton was founded in two stages, in 1440 and 1446, with a plan to endow a college with places for twenty-five poor scholars in the earlier year, rising to seventy in the later year. Henry laid the foundation stone at King's in 1441 where he also specified a capacity for seventy fellows and scholars. His fatherly interest in the Eton scholars was recorded by one who had been there himself:

He sought out everywhere the best living stones, that is, boys excellently equipped with virtue and knowledge . . . And with regard to the boys or youths who were brought to him to be put to school, the king's wish was that they should be thoroughly educated and nourished up both in virtue and in the sciences. So it was that whenever he met any of them at times in the castle of Windsor, whither they sometimes repaired to visit servants of the king who were known to them, and when he ascertained that they were of his boys, he would advise them concerning the following of the path of virtue, and, with his words, would also give them money to attract them, saying: 'Be you good boys, gentle and teachable, and servants of the Lord.'

Ibid.

CRISIS

Henry's minority ended in 1437 in a spirit of optimism. The boy was deemed intelligent and even precocious and there is no evidence that he was abnormal in any way. Shortly after attaining his majority signs of prodigality and neglect of his duties emerged. His distribution of patronage came to resemble that of Edward II and Richard II in its lavishness and unevenness. 1450 was a year of crisis. Kent rebelled under the leadership of Jack Cade in May and a rebel manifesto complained, blaming others for Henry's failures:

Also we say our sovereign lord may understand that his false council has lost his law, his merchandise is lost, his common people is lost, the sea is lost, France is lost, the king himself is so beset that he may not pay for his meat and drink, and he owes more than ever any King of England ought, for daily his traitors about him, when anything should come to him by his laws, at once ask it from him.

Three Fifteenth Century Chronicles, ed. J. Gairdner, Camden Society (1880)

THE KING AT FAULT

In July 1450 a Sussex yeoman refused to let Henry shelter behind his 'false councillors' and accused him directly and even prophetically, saying:

that the king was a natural fool and would often hold a staff in his hands with a bird on the end, playing therewith as a fool, and that another king must be ordained to rule the land, saying that the king was no person able to rule the land. *English Historical Documents*, IV, quoting the indictment of the accused

VETTING A QUEEN

In 1445 Henry VI was married to Margaret of Anjou. His first inspection of her was apparently contrived without her knowledge.

When the queen landed in England the king dressed himself as a squire, the Duke of Suffolk doing the same, and took her a letter which he said the King of England had written. When the queen read the letter the king took stock of her, saying that a woman may be seen over well when she reads a letter, and the queen never found out that it was the king because she was so engrossed in reading the letter, and she never looked at the king in his squire's dress, who remained on his knees all the time. After the king had gone, the Duke of Suffolk said: Most serene queen, what do you think of the squire who brought the letter? The queen replied: I did not notice him, as I was occupied in reading the letter he brought. The duke remarked: Most serene queen, the person dressed as a squire was the most serene

King of England, and the queen was vexed at not having known it, because she had kept him on his knees.

Calendar of State Paper . . . of Milan, I, ed. A. B. Hinds (1912): Letter of 24 October 1458 to the duchess of Milan

THE ONSET OF MADNESS

In August 1453 Henry suffered complete mental collapse and remained in that state some seventeen months unable even to recognize individuals. During that time Margaret of Anjou, after eight years of marriage, gave birth to an heir and named him Edward but the presentation of the baby in January 1454 failed to elicit any response.

At the Prince's coming to Windsor the Duke of Buckingham took him in his arms and presented him to the King in godly wise, beseeching the King to bless him. And the King gave no manner of answer. Nevertheless the Duke abode still with the Prince by the King. And when he could no manner of answer have, the Queen came in and took the Prince in her arms and presented him in like form as the Duke had done, desiring that he should bless it. But all their labour was in vain, for they departed thence without any answer or countenance, saving only that once he looked on the Prince and cast down his eyes again, without any more.

The Paston Letters, ed. J. Gairdner, 1904. *The private correspondence of the Paston family of Norfolk is a uniquely valuable source for the history of the period.*

RECOVERY AT CHRISTMAS 1454

During Henry's illness the government of the country fell to Richard, duke of York, the king's heir presumptive until the birth of Prince Edward, a man whose counsel the king had long spurned and who deplored the state of the realm. Government improved under his management. Though it is true that, 'if Henry's insanity had been a tragedy, his recovery was a national disaster' (R. L. Storey), contemporaries rejoiced.

Blessed be God, the king is well amended and hath been since Christmas Day; and on St John's Day [27 December] commanded his almoner to ride to Canterbury with his offering, and commanded the secretary to offer at Saint Edward. And in the monday afternoon the queen came to him and brought my lord prince with her; and then he asked what the prince's name was and the queen told him Edward; and then he held up his hands and thanked God thereof. And he said he never knew him untill that time, nor knew what was said to him, nor knew where he had been while he hath been sick till now.

R. L. Storey, *The End of the House of Lancaster* (1966); *The Paston Letters*

THE QUEEN AND THE PRINCE

Henry's recovery brought York's dismissal and a return to partisan rule. York and Lancaster first came to blows at St Albans in May 1455 and the former's victory, coupled with the latter's second collapse, gave York a second protectorate of the realm. Henry recovered but by then the driving force was Queen Margaret of Anjou, motivated by fierce maternal instincts and hatred for York. The paternity of her son was attributed to Edmund Beaufort, duke of Somerset and complaints of her behaviour grew:

The queen is a great and strong laboured woman, for she spareth no pain to sue her things to an intent and conclusion to her power.

The Paston Letters

In this same time the realm of England was out of all good governance as it had been many days before, for the king was simple and led by covetous counsel, and owed more than he was worth . . .

The queen with such as were of her affinity ruled the realm as she liked, gathering riches innumerable . . . The queen was defamed and denounced, that he who was called prince was not her son but a bastard gotten in adultery; wherefore she, dreading that he should not succeed his father in the crown of England, sought the alliance of all the knights and squires of Cheshire, to have their goodwill, and held open household among them. And she made her son, called the prince, give a livery of swans to all the gentlemen of the countryside and to many others throughout the land, trusting through their strength to make her son king, and making secret approaches to some of the lords of England to stir the king that he should resign the crown to her son; but she could not bring her purpose about.

An English Chronicle, 1377–1461, ed. J. S. Davies, Camden Society (1856)

QUEEN MARGARET AND THE ROBBERS

Margaret's defence of her little son provided rich fodder for dramatic stories:

Now this forest was the haunt of merciless, murderous cut-throats and the unhappy queen met one of hideous and horrible aspect who made to seize her. Not caring whether she lived or died but sensible only of maternal devotion, she began to address him: 'O man, born in a fortunate moment if you, having committed so many evil deeds, could change yourself so as to do one good act which will be remembered through the centuries! . . . In order to convert you from your usual callous behaviour to compassion, I put myself into your hands, I, the sorrowful queen of England, worse treated by fortune than anyone ever read of in books. Save the child, your future king, your king's only son. I make you to-day the stomach of my child; I appoint you to be his breast; I make you his father and mother . . .'

The brigand, seeing the tears of her who was queen of the country, felt pity for her and was softened by the Holy Spirit. He threw himself at her feet, began to weep with her and swore to suffer a thousand deaths before he would abandon the noble child. Begging mercy of the queen for his past crimes, he vowed to God and to her never to return to his former state of wickedness. So the queen kissed her son, weeping and groaning, and left him in the hands of the brigand who nobly performed his duty to him.

G. Chastellain, Œuvres, ed. K. de Lettenhove, Académie royale de Belgique (Brussels, 1863–6). *Chastellain was official historian to the duke of Burgundy.*

OUR LADY'S DAY AS 'LOVEDAY', 1458

A final and futile effort was made to bring the houses of Lancaster and York to harmony. The only achievement was formal accord on compensation for the chief victims of the first battle of St Albans and it was celebrated with extravagant symbolism:

And the Thursday after Midlent the king came to Westminster; and on the morne there was made a general procession to pray for the peace and at afternoon the queen came to the king and the week following the lords, by the king's commandment went in treaties between the other lords so that on our Lady's even, in Lent, it was Friday, they were made accorded at Westminster before the king and each took other by the hand and so came forth together arm in arm as friends and that afternoon the king sent writing to the mayor and commanded him to proclaim through the city how the lords were accorded and on the morn that was Our Lady's day the king and the queen and all the lords went on procession at St Paul's, solemnly thanking God that the lords were accorded; and there was seen that day one of the greatest multitude of people that day that ever was seen in Paul's.

Six Town Chronicles, ed. R. Flenley (Oxford, 1911)

An accurate assessment of 1458 was made by a London writer:

And that same year all these lords departed from the parliament, but they came never all together after that time to no parliament nor council, but if it were in the field with spear and shield.

Gregory's Chronicle, in *Historical Collections of a London Citizen*, ed. J. Gairdner, Camden Society (1876)

LAUGHTER AND BATTLES

A round of bloody battles began in autumn 1459. The forced recognition of York as Henry's heir one year later only fuelled the resolve of Margaret, and York's head with a mocking paper crown was impaled, with that of his chief ally, Richard Neville, earl of Salisbury on the gates of York city, following the battle of Wakefield. Overjoyed at her victory, Margaret swept south, her troops plundering. Too late she realized the damage done thereby to her cause. London refused her admission and Salisbury's son, Warwick (shortly, not without exaggeration, to become known as the 'Kingmaker'), confronted her at St Albans. Her victory could not repair the damage done by her march but her rescue of Henry VI forced the Yorkists to make their own king. Rumour suggests that Henry was in no fit state to rule:

The king was placed under a tree a mile away, where he laughed and sang, and when the defeat of the earl of Warwick was reported, he detained upon his promise the two princes who had been left to guard him. Very soon the duke of Somerset and the conquerors arrived to salute him and he received them in friendly fashion and went with them to St Albans to the queen.

Calendar of State Papers . . . of Milan, I: Letter of the Milanese Ambassador to his master of 9 March 1461, reporting on the second battle of St Albans

THE MAKING OF THE TUDORS: A BAD BEGINNING

The Yorkists response was to elevate the heir of Richard of York as Edward IV. While Margaret was at St Albans he had been victorious against a Lancastrian army at Mortimer's Cross in Wales. There he captured Owen Tudor, an unknown Welsh squire who had earlier fallen in love with, and secretly married, Henry V's widow. The succession of his grandson was as yet inconceivable but Edward dispatched him there and then.

This Owen Tudor was father to the earl of Pembroke and had wedded Queen Katherine, mother to King Henry VI. He thought and trusted all along that he would not be beheaded until he saw the axe and block, and when he was in his doublet he trusted on pardon and grace until the collar of his red velvet doublet was ripped off. Then he said, 'That head shall lie upon the stock that was wont to lie on Queen Katherine's lap', and put his heart and mind wholly on God, and very meekly took his death.

Gregory's Chronicle

A PATHETIC RESTORATION, 1470–1

The bloodiest and bitterest battle of the civil wars—Towton—fought in a snowstorm on Palm Sunday (29 March) 1461, persuaded many of the virtue of Edward IV's claim. Henry VI, Margaret of Anjou, and Prince Edward fled to Scotland and Edward IV set about crushing Lancastrian diehards. In 1465 Henry was captured and imprisoned in the Tower and there the civil wars might have ended had it not been for the treachery of Richard Neville, earl of Warwick, whose greed the king could not satisfy. In 1469 he raised rebellion and Louis XI of France presided over the most unlikely and unholy alliance of Margaret of Anjou and the ambitious earl. Warwick enjoyed a brief success. Henry VI, taken from the Tower, was an uninspiring sight, paraded by Edward IV's disloyal brother George, duke of Clarence:

The said duke accompanied with the earls of Warwick and of Derby and of Shrewsbury and the lord Stanley with many other noble men rode unto the Tower and set thence King Henry and conveyed him so through the high streets of the city, riding in a long gown of blue velvet unto Paul's . . . and thus was the ghostly and virtuous prince . . . restored unto his right and regally, of the which he took no great rejoice.

The Great Chronicle of London, ed. A. H. Thomas and I. D. Thornley (1938). *Discovered early in this century, this chronicle is essentially a compilation of now lost contemporary London chronicles put together probably in the 1490s.*

And when Edward IV returned in 1471 the same chronicler made a damning report on a second parading of this shabby monarch:

The which was more like a play than the showing of a prince to win men's hearts, for by this means he lost many and won none or right few, and ever he was showed in a long blue gown of velvet as though he had no more to change with.

Ibid.

THE DEATH OF KING HENRY, 1471

The restoration of Henry VI was shortlived. With Burgundian backing Edward IV returned, was reconciled to his wayward brother Clarence, and defeated and killed the Kingmaker at Barnet. He then turned west to deal with Margaret of Anjou and Prince Edward. They met at Tewkesbury and there the prince was killed and Margaret taken prisoner. Henry VI conveniently died on 21 May 1471—the official and unconvincing Yorkist line:

that of pure displeasure and melancholy he died.

Historie of the Arrivall of King Edward IV, ed. J. Bruce, Camden Society (1838; spelling modernized). *This is the official Yorkist account of the return of Edward.*

SOMETHING SINISTER

The truth was not something which the Yorkists could admit:

I shall say nothing, at this time, about the discovery of King Henry's lifeless body in the Tower of London; may God have mercy upon and give time for repentance to him, whoever it might be, who dared to lay sacrilegious hands on the Lord's Anointed! And so, let the doer merit the title of tyrant and the victim that of glorious martyr.

The Croyland Chronicle Continuations 1459–1486, ed. N. Pronay and J. Cox (1986). *Probably the very best single source for the period, written by a well-informed canon lawyer.*

THE MAKING OF AN EVIL REPUTATION

Other writers were more direct. Richard, duke of Gloucester, first accused of murdering Prince Edward at Tewkesbury, was constable of England at the time of Henry's death. For later writers bent on establishing the evil reputation of the man who was soon to occupy centre stage, 1471 was a significant year.

So was he thence conveyed unto the waterside and from thence unto Chertsey and there buried, for whom shortly after God showed sundry miracles; of whose death the common fame then went that the duke of Gloucester was not all guiltless.

The Great Chronicle

THE YORKISTS

Edward IV

1461–1483

Edward IV was born at Rouen in 1442, the son of Richard duke of York and Cecily Neville, youngest and twenty-third child of Ralph Neville, earl of Westmoreland. It may have been from this 'rose of Raby', that Edward inherited good looks but he had little of her other famed characteristic—piety. When his coffin was opened in 1789 it revealed a skeleton of 6 ft. 3½ in. His appearance was frequently commented upon by chroniclers at home and abroad. His achievements as king were immense but his lack of political foresight brought about the destruction of his own dynasty.

Now is the winter of our discontent
Made glorious summer by this son of York . . .

Shakespeare, *Richard III*, I. i

A HANDSOME KING

One foreign observer commented in 1475 on his good looks:

The king of England wore a black velvet cap upon his head, with a large fleur de lys made of precious stones upon it: he was a prince of a noble and majestic presence, but a little inclining to corpulence. I had seen him before when the Earl of Warwick drove him out of his kingdom; then I thought him much handsomer, and to the best of my remembrance, my eyes had never beheld a more handsome person.

The Memoirs of Philip de Commines, ed. A. R. Scoble, 2 vols. (1896), I. *Historian of the reign of Louis XI, Commines was present at the meeting of that king and Edward in France in 1475.*

A VAIN KING

Another foreigner commented on the king's consciousness of his appearance:

He was easy of access to his friends and to others, even the least notable. Frequently he called to his side complete strangers, when he thought that they had come with the intention of addressing or beholding him more closely. He was wont to show himself to those who wished to watch him, and he seized any opportunity that the occasion offered of revealing his fine stature more protractedly and more evidently to on-lookers. He was so genial in his greeting, that if he saw a newcomer bewildered at his appearance and royal magnificence, he would give him courage to speak by laying a kindly hand upon his shoulder.

Dominic Mancini, The Usurpation of Richard III, ed. C. A. J. Armstrong (Oxford, 1969). *A highly educated Italian, Mancini lived many years in France and wrote his history for his patron Angelo Cato. He visited England in 1483, possibly his second visit, and he was extremely well-informed and careful about his sources.*

A GREEDY KING

There is universal agreement that Edward grew fat in later life.

In food and drink he was most immoderate: it was his habit, so I have learned, to take an emetic for the delight of gorging his stomach once more. For this reason and for the ease, which was especially dear to him after his recovery of the crown, he had grown fat in the loins, whereas previously he had been not only tall but rather lean and very active.

Ibid.

A LICENTIOUS KING

Only three of Edward's bastard children are known by name but there may well have been more than these.

He was licentious in the extreme: moreover it was said that he had been most insolent to numerous women after he had seduced them, for, as soon as he grew weary of dalliance, he gave up the ladies much against their will to the other courtiers. He pursued with no discrimination the married and unmarried the noble and lowly: however he took none by force. Ibid.

A TUDOR PORTRAIT

In spite of his less endearing characteristics the overall impression which he gave was positive.

He was a goodly personage and very princely to behold: of heart courageous, politic in counsel, in adversity nothing abashed, in prosperity rather joyful than proud, in peace, just and merciful, in war sharp and fierce, in the field bold and hardy and nevertheless no further than wisdom would, adventurous. Whoso well consider his wars, shall no less commend his wisdom where he withdrew than his manhood where he vanquished. He was of visage lovely; of body mighty, strong and clean made; howbeit in his latter days, with over liberal diet, somewhat corpulent and burly but nevertheless not uncomely. He was in youth greatly given to fleshly wantonness, from which health of body in great prosperity and fortune, without a special grace, hardly refrains. This fault not greatly grieved the people, for no one man's pleasure could stretch and extend to the displeasure of very many, and it was without violence; over that, in his latter days, it lessened and was well left.

In which time of his latter days this realm was in quiet and prosperous estate.

Thomas More, History of King Richard III, ed. R. S. Sylvester (1976). *Sir Thomas More's work was composed about 1513 and though it owed much to informants who had lived through the period, he was one of the chief architects of the 'Tudor Myth' about Yorkist England.*

MISTRESS SHORE

Long known incorrectly as Jane, this mistress of Edward IV, Elizabeth Shore, wife of a London merchant, is charmingly described by Sir Thomas More and indeed she was still alive when he wrote.

Proper she was and fair: nothing in her body that you would have changed, unless you would have wished her somewhat higher . . . Yet delighted not men so much in her beauty as in her pleasant behaviour. For a proper wit had she and could both read well and write, merry in company, ready and quick of answer, neither mute nor full of babble, sometimes taunting without displeasure and not without disport. The king would say that he had three concubines, who in three diverse properties diversely excelled. One, the merriest; another the wiliest; the third, the holiest harlot in his realm, as one whom no man could get out of the church lightly to any place but it were to his bed . . . But the merriest was this Shore's wife, in whom the king therefor took special pleasure. For many he had, but her he loved.

Ibid.

A SECRET MARRIAGE

Whereas earlier monarchs had indulged a fondness for ladies with, at least political, impunity, Edward eventually allowed his lust to rule his mind in the matter of his marriage. The consequences were disastrous, even if they have left to posterity an entertaining story.

One of the ways he indulged his appetites was to marry a lady of humble origin, named Elizabeth, despite the antagonism of the magnates of the kingdom, who disdained to show royal honours towards an undistinguished woman promoted to such exalted rank. She was a widow and the mother of two sons by a former husband: and when the king first fell in love with her beauty of person and charm of manner, he could not corrupt her virtue by gifts or menaces. The story runs that when Edward placed a dagger at her throat, to make her submit to his passion, she remained unperturbed and determined to die rather than live unchastely with the king. Whereupon Edward coveted her much the more, and he judged the lady worthy to be a royal spouse, who could not be overcome in her constancy even by an infatuated king. *Dominic Mancini*

MAY MORNING 1464

Exactly when Edward first met Elizabeth Woodville, widow of Sir John Grey who had died fighting for Henry VI at the second battle of St Albans in 1461, mother of two boys, is not known. En route *to confront a Lancastrian army in 1464 Edward stopped at Stony Stratford. Early on May morning he slipped away to Grafton Regis, where he was duly married:*

at which marriage was no persons present but the spouse, the spousess, the duchess of Bedford her mother, the priest, two gentlewomen, and a young man to help the priest sing. After which spousals ended, he went to bed, and so tarried there upon three or four hours, and after departed and rode again to Stony Stratford, and came in manner as though he had been out hunting, and there went to bed again.

R. Fabyan, The New Chronicles of England and France, ed. H. Ellis (1811). *Making due allowance for embellishments to a work composed in the following century, the story seems to be substantially true.*

ARCHAEOLOGICAL DISCOVERY

The ceremony may have taken place in a small building known as the Hermitage, which was only a short walk away from Grafton manor house and which, at that time, was hidden in the forest. Recent excavations have uncovered a tiled floor, and some of the tiles bear the Woodville arms and some the white rose, as though Edward was connected with the place in a special way. Edward and Elizabeth could have reached the Hermitage from different directions and without attracting attention, and afterwards they could have arrived separately at Grafton manor house.

Mary Clive, *This Sun of York* (1973)

DEFENCE OF HIS MARRIAGE

For five months Edward kept his secret—a quite extraordinary and unique state of affairs—only presenting his wife as the earl of Warwick was vigorously negotiating a French marriage agreement for him! The revelation caused outrage. Edward's mother, Duchess Cicely, can hardly have been satisfied by her son's reputed and apocryphal defence:

'That she is a widow and has already children—by God's blessed Lady,

I am a bachelor and have some too! And so each of us has a proof that neither of us is like to be barren.' *Thomas More*

THE WOODVILLE CLAN

Quite apart from her unsuitability as queen—her lowly origins, her extant children and her widowed status—Elizabeth saddled Edward with the need to provide for her numerous relatives. This was partly achieved through advantageous marriages of which the most notorious, that of Edward's aunt, the dowager duchess of Norfolk, Katherine Neville, at the age of sixty-five, to Elizabeth's sixteen-year-old brother, was to be dubbed the maritagium diabolicum. *Woodville dominance offended the Kingmaker, pushing him ever nearer to rebellion and in 1469 Edward acknowledged criticism of another brother, Anthony Woodville, Lord Rivers, made in a colourful pun. He received a jester:*

clad in a short coat cut by the points and a pair of boots upon his legs . . . and in his hand a long pike, when the king had beheld his apparel, he inquired of him what was the cause of his long boots and of his long staff; upon my faith sir, said he, I have passed through many counties of your realm, and in places that I have passed the *Rivers* have been so high that I could hardly escape through them. *The Great Chronicle*

POLITIC RULE

In his government of the realm Edward paid great attention to detail. He made himself readily accessible to all comers and no matter was too small for his consideration. Contemporaries remarked upon a particular characteristic which stood him in good stead:

Men of every rank, condition and degree of experience in the kingdom marvelled that such a gross man so addicted to conviviality, vanity, drunkenness, extravagance and passion could have such a wide memory that the names and circumstances of almost all men, scattered over the counties of the kingdom, were known to him just as if they were daily within his sight even if in the districts where they operated, they were reckoned of somewhat inferior status. *Croyland Chronicle*

Nor could he always be relied upon to be easygoing:

Edward was of a gentle nature and cheerful aspect: nevertheless should he assume an angry countenance he could appear very terrible to beholders . . . To plaintiffs and to those who complained of injustice he lent a willing ear. *Dominic Mancini*

A SHARP REBUKE

In 1469 Edward went to Norwich to deal with the quarrel between the Pastons and the duke of Norfolk over the lands of Sir John Fastolf. Meeting Sir William Brandon, councillor to the duke, Edward showed that he was very well aware of that knight's evil influence. His memory and his displeasure were much in evidence, as John Paston reported:

Thomas Wigfield told me, and swore unto me, that when Brandon moved the king, and besought him to show my lord favour in his matter against you, that the king said unto him again, 'Brandon, though thou can beguile the duke of Norfolk, and bring him about the thumb as thou list, I let thee wit thou shalt not do me so; for I understand thy false dealing well enough.' And he said unto him, moreover, that if my lord of Norfolk left not of his hold of that matter, that Brandon should repent it, every vein in his heart, for he told him that he knew well enough that he might rule my lord of Norfolk as he would; and if my lord did anything that were contrary to his laws, the king told him he knew well enough that it was by nobody's means but by his; and thus he departed from the king. *The Paston Letters*

TENNIS IN DISREPUTE

The maintenance of law and order was a major but elusive expectation of medieval kingship. In the wake of Henry VI's lamentable failures Edward at least appeared by his actions to be more likely to succeed. In 1479 Bristol town council appealed directly to the king against Thomas Norton, who had accused the mayor of treason. They, in turn, complained:

The said Thomas hath retained in form before said diverse and many idle and misgoverned persons and is a common haunter of taverns and sitteth there with them and other such, railing for the more nightly unto midnight and draweth unto his company riotous and evil disposed persons and doth not associate himself with honest nor well ruled company; lieth in his bed till it be nine or ten at the bell daily, as well the holidays as the working days not attending to divine service as belongeth to a gentleman of his degree, spending the afternoons when sermons and evensong been, seen in playing at the tennis and other such frivolous disports whereby but if due redress be set in short time in repressing of his said riotous disposition and others of his association great mischief is like thereof hastily to ensue.

The Great Red Book of Bristol, ed. E. W. W. Veale, Bristol Record Society, xviii (1953)

Edward summoned both parties into his presence.

And for as much as the said Thomas Norton could allege no special treason against the said mayor, nor thing sounding to treason nor yet any offence committed by the said mayor against the King our most dread sovereign lord's laws, therefore it pleased his highness . . . like a right wise natural sovereign lord and a verray justicer dismissed the said mayor of all accusations made by the said Thomas Norton. Ibid.

A FAIR MIRACLE AFTER EDWARD'S RETURN TO ENGLAND

Though Edward himself must bear much blame for losing his throne, he is entitled to credit for its recovery. His extraordinary complacency of 1470 was transformed to determination and drive in 1471. Nevertheless, Edward willingly acknowledged a need for support from on high. On Palm Sunday 1471 the king went in procession to the parish church of Daventry where, as he knelt, he saw the image of a saint to whom he had often prayed when in exile:

. . . a little image of Saint Anne, made of alabaster, standing fixed to the pillar, closed and clasped together with four boards, small painted . . . as such images be wont to be made for to be sold and set up in churches, chapels, crosses, and oratories, in many places. And this image was thus shut, closed, and clasped, according to the rules that . . . all images be hid from Ash Wednesday to Easterday morning . . . And even suddenly, at that season of the service, the boards compassing the image about gave a great crack, and a little opened, which the King well perceived and all the people about him. And anon, after, the boards drew and closed together again, without any man's hand, or touching, and, as though it had been a thing done with a violence, with a greater might it opened all abroad, and so the image stood, open and discovered, in sight of all the people there being. The King, this seeing, thanked and honoured God, and Saint Anne, taking it for a good sign, and token of good and prosperous adventure.

The Arrivall

'False, Fleeting, Perjur'd Clarence'

Though he was reconciled to his brother George in 1471, Edward found him persistently troublesome. In 1472 Clarence quarrelled with his youngest brother Richard of Gloucester over Warwick the Kingmaker's property. His scheming was confounded and Edward imposed a peace.

After King Henry's son (to whom the earl of Warwick's younger daughter, the lady Anne, was married) had fallen at the battle of Tewkesbury . . . Richard, duke of Gloucester, sought to make the same Anne his wife; this desire did not suit the plans of his brother, the duke of Clarence (married previously to the earl's elder daughter), who therefore had the girl hidden away so that his brother would not know where she was, since he feared a division of the inheritance . . . The duke of Gloucester, however, was so much the more astute, that having discovered the girl dressed as a kitchen-maid in London, he had her moved into sanctuary in St Martin's.

Croyland Chronicle

TREASONOUS CLARENCE

It is hard to avoid the conclusion that Clarence proved incorrigible. In 1477 Edward accused him of treason and on 18 February 1478 the duke was put to death. But how?

The mind recoils from describing what followed in the next parliament—so sad was the dispute between two brothers of such noble character. No-one argued against the duke except the king; no-one answered the king except the duke . . . Why make a long story of it? . . . within a few days the execution, whatever form it took, was carried out secretly in the Tower of London.

Ibid.

A BUTT OF MALMSEY

The obscure comment of the Croyland Chronicle *suggests that, whatever else, Clarence was not disposed of in the normal manner, that is by beheading. There is in fact no concrete evidence regarding the manner of his death but within five years a story of drowning was current.*

The mode of execution preferred in this case was, that he should die by being plunged into a jar of sweet wine.

Dominic Mancini

By the end of the century this tale had gained ground.

This year and 18 day of February George duke of Clarence and brother unto the king, that a certain time before had been holden in the Tower as a prisoner, for consideration the king moving upon the foresaid 18 day was put secretly to death within the Tower and as the fame ran drowned in a barrell of malmsey.

The Great Chronicle

Foreign writers repeated the story and certainly there was no denial of such a mode of death. The investigation of Clarence's tomb in Tewkesbury has neither confirmed nor denied the tale, not least because others had later been buried in the same vault and there could be no certainty regarding the bones. It must remain, therefore, a mystery.

EDWARD'S DYING WORDS, 9 APRIL 1483

At the age of forty-one he succumbed to a fever, perhaps typhoid brought on by his excessive activity, good and bad.

'Wherefore, in these last words that ever I look to speak with you, I exhort you and require you all, for the love that you have ever borne to me, for the love that I have ever borne to you, for the love that our Lord beareth to us all, from this time forward, all griefs forgotten, each of you love other . . .'

And therewithal, the king no longer enduring to sit up, laid him down on his right side, his face toward them, and none was there present that could refrain from weeping. But the lords consoling him with as good words as they could and answering, for the time, as they thought to stand with his pleasure, there in his presence (as their words appeared) each forgave other and joined their hands together, when (as after appeared by their deeds) their hearts were far asunder.

As soon as the King was departed, the noble prince, his son, drew toward London, which at the time of his decease, kept his household at Ludlow in Wales. *Thomas More*

Edward V

1483

Edward V was only twelve when the duty of kingship fell to him unexpectedly in April 1483. Crisis had already touched his brief life, for he was born in the sanctuary to which Elizabeth Woodville had fled when Edward IV had lost his throne in 1470. In spite of his father's death-bed desire for peace nothing could long mask the divisions at court between the unpopular Woodvilles and the Protector, Richard duke of Gloucester. The young prince was residing at Ludlow in the care of his maternal uncle, Anthony Woodville, Lord Rivers, in the calm before the storm.

A FITTING SUCCESSOR

Edward IV had carefully regulated the education of his son.

Each day, after hearing matins and mass, and taking his breakfast, the boy was to spend his mornings 'occupied in such virtuous learning as his age shall now suffice to receive'. His midday meal was accompanied by the reading aloud to him of 'such noble stories as behoveth a Prince to understand; and know that the communication at all times in his presence be of virtue, honour, cunning, wisdom, and deeds of worship, and of nothing that should move or stir him to vices'. 'In eschewing of idleness' after his meal, he was to be further occupied about his learning, and then should be shown 'such convenient disports and exercises as behoveth his estate to have experience in'. After evensong and supper, he might be allowed 'such honest disports as may be honestly devised for his recreation'.

C. D. Ross, *Edward IV* (1974), quoting that king's 1474 ordinances

LESSONS WELL LEARNED

In word and deed he gave so many proofs of his liberal education, of polite, nay rather scholarly, attainments far beyond his age . . . There is one thing I shall not omit, and that is, his special knowledge of literature, which enabled him to discourse elegantly, to understand fully, and to declaim most excellently from any work whether in verse or prose that came into his hands, unless it were from among the more abstruse authors. He had such dignity in his whole person, and in his face such charm, that however much they might gaze he never wearied the eyes of beholders.

Dominic Mancini

LONG LIVE THE KING!

Gloucester was in the north when he heard the news of his brother's death. The council in London fixed the coronation for 4 May and Lord Rivers set out from Wales with his charge.

In the meanwhile the duke of Gloucester wrote the most pleasant letters to console the queen; he promised to come and offer submission, fealty and all that was due from him to his lord and king, Edward V, the first-born son of his brother the dead king and the queen. He therefore came to York with an appropriate company, all dressed in mourning, and held a solemn funeral ceremony for the king, full of tears. He bound, by oath, all the nobility of those parts in fealty to the king's son; he himself swore first of all.

Croyland Chronicle

THE HOLD-UP ON THE KING'S HIGHWAY, APRIL 1483

Much argument still rages about Gloucester's motives and sincerity. There is much evidence that he and others feared and deplored Woodville dominance. Richard met Lord Rivers on 29 April at Northampton, where a night of good humour ended next day in disaster.

When first they arrived they were greeted with a particularly cheerful and merry face, and sitting at the duke's table for dinner, they passed the whole time in very pleasant conversation. Eventually Henry, duke of Buckingham, also arrived, and, because it was late, they went off to their various lodgings.

When morning came, and a wretched one as it afterwards appeared, after a plan had been made during the night, all the lords set out together to present themselves to the new king at Stony Stratford, a place a few miles from Northampton. Behold! When those two dukes had nearly reached the entrance to this place they arrested Earl Rivers and his nephew Richard, the king's (uterine) brother and certain others who came with them and ordered them to be taken to the North in captivity.

Ibid.

THE KING DISMAYED

Immediately after the arrests Gloucester and Buckingham informed the young king and made him due reverence as their sovereign lord. Gloucester explained that those arrested had been guilty of treason and conspiracy against the Protector.

He said that he himself, whom the king's father had approved, could better discharge all the duties of government, not only because of his experience of affairs, but also on account of his popularity. He would neglect nothing pertaining to the duty of a loyal subject and diligent protector. The youth, possessing the likeness of his father's noble spirit besides talent and remarkable learning, replied to this saying that he merely had those ministers whom his father had given him; and relying on his father's prudence, he believed that good and faithful ones had been given him. He had seen nothing evil in them and wished to keep them unless otherwise proved to be evil. As for the government of the kingdom, he had complete confidence in the peers of the realm and the queen, so that this case but little concerned his former ministers. *Dominic Mancini*

SANCTUARY AGAIN

The news of the hijack of her son alarmed the queen. Taking the king's brother Richard duke of York, and her daughters, and her son Thomas Grey, marquess of Dorset, she sought refuge.

When this news was announced in London the unexpectedness of the event horrified everyone. The queen and the marquess, who held the royal treasure, began collecting an army, to defend themselves, and to set free the young king from the clutches of the dukes. But when they had exhorted certain nobles who had come to the city, and others, to take up arms, they perceived that men's minds were not only irresolute, but altogether hostile to themselves. Some even said openly that it was more just and profitable that the youthful sovereign should be with his paternal uncle than with his maternal uncles and uterine brothers. Comprehending this, the queen and marquess withdrew to the place of refuge at Westminster Abbey standing close to the royal palace, and called by the English a sanctuary.

Ibid.

GROWING SUSPICIONS

The decision was made to postpone the coronation until 24 June, and then 22 June. During the rest of May Gloucester seemed to observe propriety, though the detention of the Woodvilles—imprisoned since their arrest at Pontefract, where Richard II had died—caused anxiety. All the sources of course were written in the knowledge of what eventually happened and none is wholly impartial.

All praised the duke of Gloucester for his dutifulness towards his nephews and for his intention to punish their enemies. Some, however, who understood his ambition and deceit, always suspected whither his enterprises would lead. Ibid.

Conspiracy

Events moved to a climax in early June. Richard was anxious to have the king's brother and mother out of sanctuary. He claimed in a letter that a conspiracy against him was in hand.

. . . aid and assist us against the queen, her blood adherents and affinity which have intended and daily doth intend to murder and utterly destroy us and our cousin, the duke of Buckingham, and the old royal blood of this realm . . . *York Civic Records*, ed. A. Raine, Yorkshire Archaeological Society (1939), I

FRIDAY, 13 JUNE 1483

High drama was enacted at a council meeting in June. Richard had found that Edward IV's loyal chamberlain, William Lord Hastings, for whatever reason, was no longer prepared to support him. Tudor accounts of the meeting are colourful, not least in the reference to Richard's withered arm, which had been caused by the sorcery of the queen, Hastings, and mistress Shore. Mancini and the Croyland Chronicler, though less dramatic, were yet disturbed.

On the previous day, with remarkable shrewdness, the protector had divided the council so that, in the morning, part met at Westminster, part in the Tower of London, where the king was. On 13 June, the sixth day of the week, when he came to the Council in the Tower, on the authority of the protector, Lord Hastings was beheaded. Two senior prelates, moreover, Thomas, archbishop of York and John, bishop of Ely, saved from capital punishment out of respect for their order, were imprisoned . . . In this way, without justice or judgment, the three strongest supporters of the new king were removed and with all the rest of his faithful men expecting something similar these two dukes thereafter did whatever they wanted. *Croyland Chronicle*

Thus fell Hastings, killed, not by those enemies he had always feared, but by a friend whom he had never doubted. But whom will insane lust for power spare, if it dares violate the ties of kin and friendship?

Dominic Mancini

TO THE TOWER

Queen Elizabeth Woodville was at last persuaded to surrender the duke of York to Thomas Bourchier, cardinal archbishop of Canterbury. For one correspondent there was much to report.

Worshipfull Sir, I commend me to you and for tidings I hold you happy that ye are out of the press, for with us is much trouble and every man doubts other. As on Friday last was the Lord Chamberlain headed soon upon noon. On Monday last was at Westminster great plenty of harnessed men. There was the deliverance of the duke of York to my lord cardinal, my lord chancellor and many other lords temporal. And with him met my Lord of Buckingham in the midst of the hall of Westminster, my Lord Protector receiving him at the Star Chamber door with many loving words and so departed with my Lord Cardinal to the tower, where he is, blessed be Jesus, merry.

The Stonor Letters and Papers, ed. C. L. Kingsford, Camden Society (1919). *Written on 21 June 1483 by Simon Stallworth to his friend Sir William Stonor of Oxfordshire. This is another invaluable collection of contemporary letters.*

BASTARD SLIPS SHALL NOT TAKE ROOT

On 22 June sermons were preached by Dr Ralph Shaa at St Paul's Cross advocating the succession of Gloucester. The claim was based on the alleged bastardy of Edward's children, though there was a little confusion about who was guilty.

Edward said they, was conceived in adultery and in every way was unlike the late duke of York, whose son he was falsely said to be, but Richard, duke of Gloucester, who altogether resembled his father, was to come to the throne as the legitimate successor. *Dominic Mancini*

The accepted story, however, is that Edward's marriage to Elizabeth Woodville was contracted in adultery. The only person who knew was Robert Stillington, bishop of Bath and Wells, who made his secret public at this most—perhaps too—convenient moment. Commines condemned him as 'ce mauvais évêque'.

It was put forward by means of a supplication contained in a certain parchment roll, that King Edward's sons were bastards, by submitting that he had been precontracted to a certain Lady Eleanor Boteler before he married Queen Elizabeth and, further, that the blood of his other brother, George, duke of Clarence, had been attainted so that, at the time, no certain and uncorrupt blood of the lineage of Richard, duke of York, was to be found except in the person of the said Richard, duke of Gloucester.

Croyland Chronicle

The Fate of the Princes in the Tower

Richard became king on 26 June 1483. No other single historical question has received such continuous attention as that concerning the ultimate fate of Edward IV's sons. Vast quantities of ink have been used in researching every conceivable source of information but the mystery remains unsolved. Many certainly believed that Richard was guilty but even at the time there were doubts.

He and his brother were withdrawn into the inner apartments of the Tower proper, and day by day, began to be seen more rarely behind the bars and windows, till at length they ceased to appear altogether. The physician Argentine, the last of his attendants whose services the king enjoyed, reported that the young king, like a victim prepared for sacrifice, sought remission of his sins by daily confession and penance, because he believed that death was facing him . . . I have seen many men burst forth into tears and lamentations when mention was made of him after his removal from men's sight; and already there was a suspicion that he had been done away with. Whether, however, he has been done away with, and by what manner of death, so far I have not at all discovered.

Dominic Mancini

Maintaining a certain obscurity, for whatever reason, the Croyland author recorded rumours and finally a poem about the three Richards of England.

The third, after exhausting the quite ample store of Edward's wealth, was not content until he suppressed his brother's progeny. *Croyland Chronicle*

Foreign reporters were soon accusing Richard:

Look, I pray you at the events which have happened in that land since the death of King Edward. Reflect how his children, already big and courageous, have been killed with impunity, and the crown has been transferred to their murderer by the favour of the people.

English Historical Documents, IV, quoting the speech of the chancellor of France on 15 January 1484

In London there was concern about their disappearance and much rumour:

The children of King Edward were seen shooting and playing in the garden of the Tower by sundry times . . . But after Easter [1484] much whispering was among the people that the king had put the children of King Edward to death. *Great Chronicle*

And in 1485:

Considering the death of King Edward's children, of whom as then men feared not openly to say that they were rid out of this world but of their death's manner was many opinions, for some said they were murdered between two feather beds; some said they were drowned in malmsey and some said that they were sticked with a venomous poison. But how so ever they were put to death certain it was before that day they were departed from this world, of which cruel deed Sir James Tyrell was reported to be the doer. But others put that weight upon an old servant of King Richard's . . . Ibid.

THE TUDOR VIEW

By the time Sir Thomas More was writing his history the finger was unswervingly pointed at Richard III. More's account is highly readable; its veracity is quite another matter.

For Sir James Tyrell devised that they should be murdered in their beds, to the execution whereof he appointed Miles Forest, one of the four that kept them, a fellow fleshed in murder before time. To him he joined one John Dighton, his own horse-keeper, a big broad, square strong knave. Then, all the other being removed from them, this Miles Forest and John Dighton about midnight (the sely [innocent] children lying in their beds) came into the chamber and suddenly lapped them up among the clothes—so bewrapped them and entangled them, keeping down by force the featherbed and pillows hard unto their mouths, that within a while, smored and stifled, their breath failing, they gave up to God their innocent souls into the joys of heaven, leaving to the tormentors their bodies dead in the bed. Which after that the wretches perceived, first by the struggling with the pains of death, and after long lying still, to be thoroughly dead, they laid their bodies naked out upon the bed and fetched Sir James to see them. Which, upon the sight of them, caused those murderers to bury them at the stair foot, meetly deep in the ground, under a great heap of stones . . . And thus as I have learned of them that knew much and little cause had to lie, were these two noble princes—these innocent, tender children, born of most royal blood, brought up in great wealth, likely long to live, to reign and rule in the realm—by traitorous tyranny taken, deprived of their estate, shortly shut up in prison, and privily slain and murdered; their bodies cast God wot where by the cruel ambition of their unnatural uncle and his dispiteous tormentors. *Thomas More*

HISTORICAL INGENUITY

Some very fertile minds have been at work ever since to exonerate Richard from blame. Firstly, by blaming others—the duke of Buckingham, John Howard duke of Norfolk, Henry VII—but little mud has stuck here. Secondly, by suggesting that one or both children survived—according to one theory based on a coded Holbein painting, concealed in the household of Sir Thomas More—which certainly inspired the Pretenders in the reign of the first Tudor. Thirdly, by suggesting that they died from multifarious illnesses, most recently sweating sickness. All the arguments have been fuelled by successive discoveries of bones in the Tower. One of Richard's earlier defenders, Sir George Buck, writing in 1619, dismissed one such find with considerable flair:

This was the carcase and bones of an ape which was kept in the Tower and that in his old age he either chose that place to die in, or else had clambered up thither, according to the light and idle manner of those wanton animals, and after, being desirous to go down, and looking downward, and seeing the way to be very steep and deep and the precipice to be very terrible to behold, he durst not adventure to descend, but for fear he stayed and starved there.

The History of King Richard the Third by Sir George Buck, Master of the Revels, ed. A. N. Kincaid (Gloucester, 1979)

His conclusion was simple:

I verily think that he died of a natural sickness and of infirmity . . . And then there is reason and natural cause that they should both die of the like diseases and natural infirmities.

Ibid.

Bones—1674, 1933, 1987

According to Thomas More the burial 'at the stair foot, meetly deep', displeased Richard who had them exhumed and buried secretly, as More concluded 'God wot where'. The discovery in 1674 of two skeletons, during demolition work on a stone staircase, evoked the first part of More's account but not the second, resulting in an automatic but quite unfounded presumption of authenticity. An anonymous eyewitness reported:

This day I, standing by the opening, saw working men dig out of a stairway in the White Tower, the bones of those two Princes who were foully murdered by Richard III . . . they were small bones, of lads in their teens and there were pieces of rag and velvet about them . . . Being fully recognised to be the bones of those two Princes they were carefully put aside in a stone coffin or coffer.

P. W. Hammond and W. J. White, 'The Sons of Edward IV: A Re-examination of the Evidence on their Deaths and on the Bones in Westminster Abbey', in *Richard III: Loyalty, Lordship and Law*, ed. P. W. Hammond (1986), quoting the report

THE REPUTATION THAT WON'T GO AWAY

The bones were duly, if inappropriately, reburied, in an urn in the Henry VII chapel at Westminster Abbey. In 1933 permission was given for their exhumation and study by L. E. Tanner and W. Wright, but the presumption of authenticity which so prevailed was such that neither even bothered to sex the skeletons. There is now much pressure for a further exhumation with a view to carbon dating. Passions still run high, as Norman Hammond doubtless discovered within days of publishing an article in The Times *on 21 May 1987. The lingering reputation of Richard III is quite clear from the author's captions—on the front page: 'Modern Science Convicts Richard III of Murder'; and by the time readers reached the back page: 'Science Convicts Richard III of Double Murder!' A fine eye-catching piece of journalese, no doubt, and no less unproven.*

New evidence suggests that the skeletons long-reputed to be those of the 'Princes in the Tower' are indeed the remains of King Edward V and his brother Richard, Duke of York and that the man blamed for their murders, King Richard III, was guilty.

If the skeletons found in the Tower *are* those of the little princes, and the dental and skeletal evidence strongly supports that identification, the date of death can also be calculated.

The most likely date would be some time in 1484 which would be compatible with the Great Chronicle of London, compiled some 20 years later, which said that after Easter 1484 there was 'much whispering among the people that the King had put the children to death'.

It would rule out two of the three most-canvassed perpetrators of the murders: the Duke of Buckingham, executed for high treason in 1483, and Henry VII, who did not win control of the tower until August 1485.

While Shakespeare's image of Richard III as a monstrous killer has been strongly challenged by historians, a new study of the bones indicates that it is based in fact.

The skeletons were found in a wooden chest buried in the Tower of London in 1674, and were immediately assumed, without further evidence, to be those of the two princes, the sons of Edward IV who died in 1483.

They were buried in Westminster Abbey in a marble urn, and the remains were examined in 1933 by Mr Lawrence Tanner and Professor William Wright.

They concluded that the skeletons were likely to be those of the princes, and that they had died in the summer of 1483, shortly after Richard III's coronation, but some of their assessments, including the presence of bloodstains due to smothering and even the age of the elder child's skeleton, were subsequently challenged.

Moreover, as the historian Paul Murray Kendall pointed out, many people had been buried in the tower over the centuries, and there was no firm evidence that the skeletons were of the fifteenth century.

The new evidence makes the case much less circumstantial: the crucial data comes from the skeleton of Lady Anne Mowbray, daughter of the Duke of Norfolk and married at the age of six to the four-year-old Richard, Duke of York, in 1478, three years before her death.

Her lead coffin was found in London in 1965 and a full analysis of the bones was carried out before she was reburied. Since she was known to have been a month short of her ninth birthday when she died, such vital factors as the rate of growth of long bones and the pattern of eruption of adult teeth can be pinned down precisely.

Dr Theya Molleson, of the Natural History Museum, has compared the characteristics of Lady Anne's skeleton to those of the princes, and shown that the relationship between dental and skeletal maturity of the latter suggests that both are pre-pubescent boys.

The presence of large extra bones within the sutures of the skulls of both, and their similar size and position at the back of the head is strongly suggestive that the boys were related to each other, she says in the current issue of the *London Archaeologist*.

More striking is the evidence which Dr Molleson has found for a blood relationship between the two boys and Anne Mowbray, who had been related to the princes through common great-grandparents and great-great-grandparents.

Both Anne Mowbray and the elder boy have a number of permanent teeth missing, a rare phenomenon known as hypodontia. The form in which the two skeletons have it is present in less than 1 per cent of a contemporary sample, and is thought to be strongly hereditary.

Other traits include distinctive features of the hands and feet. The combined evidence of bones and teeth suggests that the boys were related to Anne Mowbray.

The age of the children at death could also be established from the teeth.

BONES OF CONTENTION FROM TOWER

This, The Times*'s own caption over the letters of reply, is an appropriate one. This issue is far from settled but one final remark must be made. Even if final proof of the authenticity of these bones 'beyond reasonable doubt' is established (a most unlikely event even with the as yet young science of carbon dating), the question of who committed the murders, if indeed the princes were murdered, will remain unanswered.*

Sir: May I as co-author, with Peter Hammond, of a recent review of the subject ('The Sons of Edward IV . . . and the Skeletons in Westminster Abbey', in *Richard III: Loyalty, Lordship and the Law*, editor P. W. Hammond, 1986) comment upon the contribution to the 'Princes in the Tower' controversy made by Mr Norman Hammond (report, May 21). The latest publication in the *London Archaeologist* does not justify the view,

'Modern science convicts Richard III of murder' and indeed Theya Molleson, in her article, was careful to make no such allegation.

Your Archaeology Correspondent found Miss Molleson's argument for family relationship very persuasive, especially the presence of extra bones in the sutures of the skulls of the two skeletons said to have been found buried at the Tower of London in 1674. Sutural bones of the above type may indeed be a rarity in the modern world and suggestive of a close relationship, but even in the 17th century one third of Londoners showed these extra bones, whether related or no. (In earlier times the frequency of the trait could be even higher: 71 per cent of a sample of Romano-British skulls, Don Brothwell, *Digging Up Bones*, 1981.)

If it is the position, size and shape of the sutural bones that is of significance in the kinship claims, why does the pattern differ in the skull of Anne Mowbray, the presumptive relative in the thesis? Professor Roger Warwick, who examined Anne Mowbray's skull, informed me that it contained at least 14 small ossicles in this region. Furthermore, the skull in Tewkesbury Abbey attributed to George Duke of Clarence, the paternal uncle of the 'Princes', shows no bones in the sutures. Similarly, Mr Hammond's 'distinctive features of the hands and feet' have been challenged in the columns of the *London Archaeologist.*

The evidence of the ages at which the children died is the strongest part of the case. Despite the elegant mathematical treatment of the dental condition in the article there is recourse to more exotic arguments in order to account for the retarded development of certain bones in the spine of the elder of the two children. There is also the uncomfortable fact that Miss Molleson found that the younger child appeared to be rather tall to have been the age calculated.

The study discussed above is a welcome addition to a long-running debate. However, although there is no doubt about the age and date for the death of Anne Mowbray (these are known from the inscription on her coffin) those for the remains attributed to the 'Princes' remain unsatisfactory since they are dependent upon circular arguments.

Radiocarbon dating of the skeletons is a highly desirable step which would determine whether or not they are derived from the 15th century and hence deserve the detailed treatment currently accorded them. One could guarantee that they would receive a more respectful handling than was the fate of Dante's remains.

Yours sincerely,
William White

Sir: The report about the article by Theya Molleson shows only that it is possible to take the results of an examination in 1933 of some bones in Westminster Abbey (dating from any time before 1674 when they were found) and interpret them to show that the bones are those of two children who died aged between 8.6 to 10.7 years and 12.9 to 16 years.

I must leave criticism of the anatomical proof of consanguinity and sex in the article to those better qualified than I to point out the serious flaws and omissions it contains. I will say, though, that a proof of age which relies for support on *The Great Chronicle of London* rests on shaky foundations indeed. The chronology of the *Chronicle* at this point is demonstrably wrong. Of the main events described under 1484 other than the rumours of the death of the Princes, one, the rebellion and death of the Duke of Buckingham, took place in 1483, and the other, the death of Queen Anne Neville, took place in 1485.

Could it be that Theya Molleson is 12 months in error in her calculations? If so, what does this do to conclusions drawn from her results?

Yours faithfully,
P. W. Hammond

Richard III

1483–1485

Richard III is almost unique among Englishmen in having a society dedicated to his name and devoted to restoring some semblance of balance in determining his reputation. His defenders, if they have not proved their main case—that someone else disposed of the Princes—have established much in his favour by concentrating on more contemporary sources in preference to the Tudor writers. Richard was not a physical monster (crookbacked, born with teeth, talons, and long hair) and his brief reign—twenty-six months—bears hallmarks of competence in government. Richard had spent most of his life before 1483 in northern England and northern sources preserve an image radically different from their southern and later counterparts. In the end, Richard's power base was too narrow to enable him to keep his throne.

The Making of a Myth

No contemporary description of Richard survives—a serious lack for historians in search of objectivity. He was evidently not as tall as some of the Plantagenets had been but extant portraits which suggest his anxiety are not unflattering. The earliest slur was recorded briefly in 1491. One John Payntour was accused of having said that Richard was:

an hypocrite, a crook back and buried in a ditch like a dog.

York Civic Records, II

John Rous, the Warwickshire antiquary who first wrote during Richard's reign and then, rewriting in the next in the hope of royal patronage, changed his story, bears much blame for the Tudor myth.

Richard was born at Fotheringhay in Northamptonshire, retained within his mother's womb for two years and emerging with teeth and hair to his shoulders.

Richard III and His Early Historians, 1483–1535, ed. A. Hanham (Oxford, 1975), quoting Rous's second version

Sir Thomas More warmed to the theme:

Richard, the third son, of whom we now entreat, was in wit and courage equal with either of them, in body and prowess far under them both: little of stature, ill-featured of limbs, crook-backed, his left shoulder much higher than his right, hard favored of visage, and such as is in states called warly [warlike], in other men otherwise. He was malicious, wrathful, envious, and from afore his birth, ever froward. It is for truth reported that the duchess his mother had so much ado in her travail, that she could not be delivered of him uncut, and that he came into the world with the feet forward, as men be borne outward and (as the fame runneth), also not untoothed . . . He was close and secret, a deep dissimuler, lowly of countenance, arrogant of heart, outwardly coumpinable [friendly] where he inwardly hated, not letting to kiss whom he thought to kill.

Thomas More

SHAKESPEARE ON RICHARD III

The Bard finished the job; the damage done has not been so easily undone. To Henry VI he gave the first distortion:

The owl shrieked at thy birth—an evil sign;
The night-crow cried, aboding luckless time;
Dogs howled, and hideous tempests shook down trees;
The raven rooked her on the chimney's top;
And chatt'ring pies in dismal discords sung.
Thy mother felt more than a mother's pain,
And yet brought forth less than a mother's hope,
To wit, an indigested and deformèd lump,
Not like the fruit of such a goodly tree.
Teeth hadst thou in thy head when thou wast born,
To signify thou cam'st to bite the world . . .

Shakespeare, *3 Henry VI*, v. vi

To Richard himself, Shakespeare gave the confirmation:

But I, that am not shaped for sportive tricks
Nor made to court an amorous looking-glass,
I that am rudely stamped and want love's majesty
To strut before a wanton ambling nymph,
I that am curtailed of this fair proportion,
Cheated of feature by dissembling nature,
Deformed, unfinished, sent before my time
Into this breathing world scarce half made up—
And that so lamely and unfashionable
That dogs bark at me as I halt by them . . .

Shakespeare, *Richard III*, I. i

A MORE FAVOURABLE OPINION

The accusations against Richard came thick and fast: that he murdered Henry VI and his son: that he murdered his brother George duke of Clarence; that he had designs upon the throne from the moment Edward IV was dead; that he unlawfully ordered the execution of Lord Hastings and the Woodvilles whom he had sent to Pontefract (these last on 25 June 1483); that he violently broke the sanctuary at Westminster in order to secure the younger Prince; and finally that he deposed and murdered his nephews. He was portrayed as Antichrist. Here is a quite different view:

I trust to God soon, by Michaelmas, the king shall be at London. He contents the people wherever he goes best that ever did prince; for many a poor man that hath suffered wrong many days have been relieved and helped by him and his commands in his progress. And in many great cities and towns were great sums of money given him which he hath refused. On my trouth, I never liked the conditions of any prince so well as his; God hath sent him to us for the weal of us all.

Christ Church Letters, ed. J. B. Sheppard, Camden Society (1877). *A letter of Thomas Langton, bishop of St Davids, to the prior of Christ Church in 1483 commenting on the royal progress after the coronation.*

Buckingham's Rebellion, October 1483

In all the crises leading to his accession Richard had found a willing ally in Henry Stafford, duke of Buckingham. The rebellion in 1483 was in fact the brain-child of Woodville and Tudor factions capitalizing on rumours about the fate of the Princes.

In the meantime and while these things were happening the two sons of King Edward remained in the Tower of London with a specially appointed guard. In order to release them from such captivity the people of the South and of the West of the kingdom began to murmur greatly, to form assemblies and to organize associations to this end . . . When at last the people . . . just referred to, began considering vengeance, public proclamation having been made that Henry, duke of Buckingham, then living at Brecknock in Wales, being repentant of what had been done would be captain-in-chief in this affair, a rumour arose that King Edward's sons, by some unknown manner of violent destruction, had met their fate.

Croyland Chronicle

THE KING DISMAYED

The news of Buckingham's treason took Richard by complete surprise. The rumour that the Princes were dead had caused the rebels to invite the exiled Henry Tudor to come to England and lead them. Richard responded with speed, writing to the chancellor from Grantham to request that he be sent the Great Seal. In an emotive postscript in his own hand Richard gave vent to his true feelings:

Here loved be God is all well and truly determined and for to resist the malice of him that had best cause to be true the duke of Buckingham the most untrue creature living whom with God's grace we shall not be long till we will be in those parts and subdue his malice. We assure you was never false traitor better provided for, as bearer, Gloucester [herald], shall show you.

Richard III: The Road to Bosworth Field, ed. P. W. Hammond and A. F. Sutton (1985; spelling modernized, quoting Richard's letter)

EXECUTION

Bad organization, bad weather, and gradual desertion brought the rebels down. The duke did not long remain at liberty.

The duke, meanwhile, was staying at Weobley, the home of Walter Devereux, Lord Ferrers, together with the bishop of Ely and his other advisers. Realising that he was hemmed in and could find no safe way out he secretly changed his attire and forsook his men; he was finally discovered in the cottage of a certain poor man because the supply of provisions taken there was more abundant than usual; he was seized and taken to the city of Salisbury where the king had arrived with a great army and he suffered capital punishment in the public market place of that city on All Soul's day, notwithstanding the fact that it fell on a Sunday in that year.

Croyland Chronicle

SUDDEN GRIEF

Though 1484 began well, with parliament confirming his title and with an oath of allegiance to Richard's son as the heir, disaster struck in April.

Shortly afterwards, however, they learned how vain are the attempts of man to regulate his affairs without God. In the following April, on a day not far off King Edward's anniversary, this only son, on whom, through so many solemn oaths, all hope of the royal succession rested, died in Middleham castle after a short illness, in 1484 and in the first year of King Richard's reign. You might have seen the father and mother, after hearing the news at Nottingham where they were then staying, almost out of their minds for a long time when faced with the sudden grief. *Croyland Chronicle*

QUEEN ANNE

Within a year Richard also lost his queen and rumours that he had poisoned her and planned to marry his niece gained credence. Anne, in fact, appears to have succumbed to tuberculosis.

It should not be left unsaid that during this Christmas feast too much attention was paid to singing and dancing and to vain exchanges of clothing between Queen Anne and Lady Elizabeth, eldest daughter of the dead king, who were alike in complexion and figure . . . and it was said by many that the king was applying his mind in every way to contracting a marriage with Elizabeth either after the death of the queen, or by means of a divorce for which he believed he had sufficient grounds . . . A few days later the queen began to be seriously ill and her sickness was then believed to have got worse and worse because the king himself was completely spurning his consort's bed. Ibid.

ELIZABETH OF YORK

At the time of the October rebellion the rebels had urged Henry Tudor to take to wife Edward IV's eldest daughter, something which he eventually did to shore up his dubious claims to the crown. In 1484 Richard was forced to deny any plans to marry her himself. His northerners voiced the strongest view:

These men told the king, to his face, that if he did not deny any such purpose and did not counter it by public declaration . . . the northerners, in whom he placed the greatest trust, would all rise against him, charging him with causing the death of the queen, the daughter and one of the heirs of the earl of Warwick and through whom he had obtained his first honour, in order to complete his incestuous association with his near kinswoman, to the offence of God. Ibid.

A LOST LETTER

As with other aspects of Richard's life, the veracity of his plans regarding Elizabeth of York is not easily established. In 1619 Sir George Buck referred to a letter which indicated that Elizabeth was herself keen on the suit. The letter has never been seen since. She wrote to Richard's friend the duke of Norfolk:

First she thanked him for his many courtesies and friendly offices and then she prayed him as before to be a mediator for her in the cause of the marriage to the king, who, as she wrote, was her only joy and maker in this world, and that she was his in heart and in thoughts, in body and in all. And then she intimated that the better half of February was past, and that she feared the queen would never die. And all these be her own words, written with her own hand, and this is the sum of her letter, whereof I have seen the autograph or original draft under her own hand. *Sir George Buck*

The letter, Buck said, was kept by Norfolk's descendant, the earl of Arundel:

And he keepeth that princely letter in his rich and magnificent cabinet, among precious jewels and rare monuments. Ibid.

POETIC TREASON

In December 1484 William Collingbourne and John Turburville were tried for treasonous dealings, probably concerning Henry Tudor. Their greatest fame, however, derived from their efforts in making a not very subtle dig at Richard's supporters, Francis, Viscount Lovell, Sir Richard Radcliffe, and William Catesby.

In these days were chief rulers about the king, the lord Lovell, and ij gentlemen being named Mr Ratcliff & Mr Catysby, of the which persons was made a sedicious ryme & fastened upon the Cross in Chepe & other places of the City whereof the sentence was as followeth,

The Catt the Ratt, and Lovell our dog
Rullen all England, under an hog. *Great Chronicle*

Bosworth Field, 22 August 1485

Henry Tudor landed at Milford Haven, in his native Wales, on 7 August 1485. Support was slow in coming. Richard swiftly summoned his men, who in turn summoned theirs. Tudor writers, preparing their readers for the eventual outcome, cast the king in a poor light.

But Richard in the mean time, being then at Nottingham was certified that Henry and the other exiles who took his part were come into Wales, and that he was utterly unfurnished and feeble in all things, contrary wise that his men whom he had disposed for defense of that province were ready in all respects. That rumour so puffed him up in mind that first he esteemed the matter not much to be regarded, supposing that Henry, having proceeded rashly, considering his small company, should surely have an evil end.

Three Books of Polydore Vergil's English History, ed. H. Ellis, Camden Society (1844). *Completed in 1513 and polished up for publication in 1534, Polydore's history was according to the author commissioned by Henry VII. Though recognizable as a piece of history, in the modern sense, its purpose was nevertheless a panegyric on the house of Tudor.*

Far more revealing is the letter of Richard's ally John duke of Norfolk to John Paston, written in obvious haste:

Well-beloved friend, I commend me unto you, letting you to understand that the king's enemies be a land, and that the king would have set forth as upon Monday but only for Our Lady's Day; but for certain he goeth forth as upon Tuesday, for a servant of mine brought to me the certainty.

Wherefore, I pray you that ye meet with me at Bury, for, by the grace of God, I purpose to lie at Bury as upon Tuesday night and that ye bring with you such company of tall men as ye may goodly make at my cost and charge, beside that ye have promised the King; and I pray you ordain them jackets of my livery, and I shall content you at your meeting with me.

The Paston Letters

THE KING WORE HIS CROWN

By 20 August 1485, with many of his troops not yet assembled, notably the northerners, the king moved to Leicester, taking up residence, according to local tradition, at the inn named for his personal badge, the White Boar. Norfolk was already there, soon to be joined by others until a goodly company was formed.

On the king's side there was a greater number of fighting men than there had ever been seen before, on one side, in England. On Sunday before the feast of Bartholomew the Apostle [24 August], the king left Leicester with great pomp, wearing his diadem on his head, and accompanied by John Howard, duke of Norfolk and Henry Percy, earl of Northumberland and other great lords, knights and esquires and a countless multitude of commoners. *Croyland Chronicle*

NIGHTMARES

Richard passed an uneasy night, tormented by dreams or apparitions.

At dawn on Monday morning the chaplains were not ready to celebrate mass for King Richard nor was any breakfast ready with which to revive the king's flagging spirit. The king, so it was reported, had seen that night, in a terrible dream, a multitude of demons apparently surrounding him, just as he attested in the morning when he presented a countenance which was always drawn but was then even more pale and deathly, and affirmed that the outcome of this day's battle, to whichever side the victory was granted, would totally destroy the kingdom of England. For he also declared that he would ruin all the partisans of the other side, if he emerged as the victor, predicting that his adversary would do exactly the same to the king's supporters if the victory fell to him. *Croyland Chronicle*

A LATE WARNING

The Tudor writers made much of the battle, embellishing accounts of it with many stories, like that concerning Richard's close ally, John duke of Norfolk:

which was warned by divers to refrain from the field in so much that the night before he should set forward toward the king, one wrote on his gate

Jack of Norfolke be not to bolde
For Dickon thy master is bought and solde

Edward Hall, *The Union of the two Noble Families of Lancaster and York (1550). More than either Vergil or More, this work of the lawyer Edward Hall was a eulogy of the Tudors.*

'A HORSE! A HORSE! MY KINGDOM FOR A HORSE'

In the course of battle Richard's horse was killed under him. He had earlier spurned all suggestions of flight from the field, even when the treachery of Sir William Stanley boded ill.

then to King Richard there came a Knight,
    & said, 'I hold it time for to flee;
for yonder Stanleys dints they be so wight,
    against them no man may dree.
here is thy horse at thy hand ready
    another day thou may thy worship win,
& for to reign with royalty,
    to wear the crown and be our king.'
he said, 'give me my battle axe in my hand,
    set the crown of England on my head so high!
for by him that shaped both sea and Land,
    King of England this day I will die!'

Hammond and Sutton, *Richard III. This quotation is from the metrical version of the very valuable* Ballad of Bosworth Field, *probably written before 1495.*

A COURAGEOUS PRINCE

There was general agreement about Richard's bravery at Bosworth from even the most hostile of writers:

There now began a very fierce battle between the two sides; [Henry] earl of Richmond with his knights advanced directly upon King Richard while the earl of Oxford, next in rank after him in the whole company and a very valiant knight, with a large force of French as well as English troops, took up his position opposite the wing where the duke of Norfolk was stationed. In the place where the earl of Northumberland stood, however, with a fairly large and well-equipped force, there was no contest against the enemy and no blows given or received in battle. In the end a glorious victory was granted by heaven to the earl of Richmond, now sole king, together with the priceless crown which King Richard had previously worn. As for King Richard he received many mortal wounds and, like a spirited and most courageous prince, fell in battle on the field and not in flight. *Croyland Chronicle*

A DESPERATE KING

At some point Richard decided to launch himself personally on Henry. A successful assault would have decided the day and Richard got near enough to kill Henry's standard-bearer. This Tudor account is unstinting in its praise of his end:

While the battle continued thus hot on both sides . . . king Richard understood, first by espials where earl Henry was afar off with small force of soldiers about him; then after drawing nearer he knew it perfectly by evident signs and tokens that it was Henry; wherefore, all inflamed with ire, he struck his horse with the spurs, and runneth out . . . against him. Henry perceived king Richard come upon him, and because all his hope was then in valiancy of arms, he received him with great courage. King Richard at the first brunt killed certain, overthrew Henry's standard, together with William Brandon the standard bearer, and matched also with John Cheney, a man of much fortitude, far exceeding the common sort, who encountered with him as he came, but the king with great force drove him to the ground, making way with weapons on every side. But yet Henry abode the brunt longer than ever his own soldiers would have weened, who were now almost out of hope of victory, when as lo, William Stanley with three thousand men came to the rescue: then truly in a very moment the residue all fled, and king Richard alone was killed fighting manfully in the thickest press of his enemies. *Polydore Vergil*

THE MYTH OF THE HAWTHORN BUSH

Amongst apocryphal stories concerning the crown, which claim greater notoriety than that of its hanging upon a thorn bush on Bosworth field? That Richard wore a crown that day is claimed by a number of sources and need not be questioned. No contemporary or sixteenth-century sources refer to it being found in the vicinity of a hawthorn bush. Vergil and the Croyland Chronicle *record it being found among the spoils. This leaves unexplained the origin of the Tudor badge of the crown and hawthorn which appears on Henry VII's tomb and in the window of his chapel in Westminster. The crowning of Henry by the traitor Stanley must also remain distinctly dubious.*

When the Lord Stanley saw the good will and gratuity of the people he took the crown of King Richard which was found amongst the spoil in the field and set it on the earl's head as though he had been elected king by the voice of the people.

Edward Hall

IGNOMINY

Where Henry V made reparation for his father's usurpation by transferring Richard II's remains from King's Langley to the Abbey and where Richard III himself had those of Henry VI moved from Chertsey Abbey to St George's Chapel, Windsor, no such act of reparation was ever performed for the last Plantagenet. Richard was buried in an unmarked grave and Henry VII managed £10 1s. for a coffin that became a horse-trough which, in turn, was broken and used for steps to the cellar of the White Horse Inn.

And Richard late king as gloriously as he in the morning departed from that town, so as irreverently was he that afternoon, brought into that town, for his body despoiled to the skin, and nought being left about him, so much as would cover his privy member, he was trussed behind a pursuivant called Norroy as a hog or an other vile beast, and all to besprung with mire and filfth, was brought to a church, in Leicester for all men to wonder upon, and was there lastly irreverently buried. And thus ended this man with dishonour as he that sought it, for had he continued still protector and have suffered the children to have prospered according to his allegiance and fidelity, he should have been honourably lauded over all, where as now his fame is darked and dishonoured as far as he was known, but God that is all mercifull forgive him his misdeeds.

Great Chronicle

CONTEMPORARY GRIEF

In the north, where he had been most loved and respected, York City Council entered in its minutes a lasting testimony to its true feelings on 23 August, 'the throne being vacant':

King Richard late mercifully reigning upon us was through great treason . . . piteously slain and murdered to the great heaviness of this city.

Hammond and Sutton, *Richard III*, quoting the York House Books

THE TUDORS

Henry VII

1485–1509

The period covered by the Tudors was one of change—'the new learning, the new prices, the new theology, the new world...'—but change did not come suddenly in 1485. Born in 1457, some three months after his father's death, Henry VII was a thoroughly medieval king. Through his mother, Lady Margaret Beaufort, whose influence on him was powerful, he could claim descent on the sinister side from John of Gaunt, third surviving son of Edward III. In religion, he was conventional, in government, masterful. Like usurpers before him, he was frequently troubled by rebellion and he was determined in the pursuit of security, wealth, and good order in the realm, though this rendered him harsh, rapacious, and miserly. Nevertheless, his statecraft brought a much needed stability to England.

DESCRIBED BY ONE WHO KNEW HIM

This is probably the best and most accurate description of the king:

His body was slender but well built and strong; his height above the average. His appearance was remarkably attractive and his face was cheerful, especially when speaking; his eyes were small and blue, his teeth few, poor and blackish; his hair was thin and white; his complexion sallow. His spirit was distinguished, wise and prudent; his mind was brave and resolute and never, even at the moments of greatest danger, deserted him. He had a most pertinacious memory. Withal he was not devoid of scholarship. In government he was shrewd and prudent, so that no-one dared to get the better of him through deceit or guile. He was gracious and kind and was as attentive to his visitors as he was easy of access. His hospitality was splendidly generous; he was fond of having foreigners at his court and he freely conferred favours on them. But those of his subjects who were indebted to him and who did not pay him due honour or who were generous only with promises, he treated with harsh severity.

Polydore Vergil, Anglica Historia, ed. D. Hay, Camden Society (1950)

HENRY'S ACCESSION PROPHESIED BY HENRY VI

Henry's biographer, Francis Bacon, published his account of the reign in 1622. Highly readable though it is, some of the stories cannot be taken at face value. A meeting between the two Henrys may well have taken place; otherwise the story is Bacon's own.

His worth may bear a tale or two, that may put upon him somewhat that may seem divine. When the lady Margaret his mother had divers great suitors for marriage, she dreamed one night, that one in the likeness of a bishop in pontifical habit did tender her Edmund earl of Richmond, the King's father, for her husband, neither had she ever any child but the King, though she had three husbands. One day when King Henry the sixth, whose innocency gave him holiness, was washing his hands at a great feast, and cast his eye upon King Henry [VII], then a young youth, he said: 'This is the lad that shall possess quietly that, that we now strive for.'

Francis Bacon, The Reign of King Henry the Seventh, ed. J. R. Lumby (Cambridge, 1881)

ESCAPE FROM BRITTANY, 1485

Henry demonstrated early on that he was not without panache. Nor, in the business of escape, was he without experience, for his departure from England in 1471 had been somewhat hasty. Edward IV had been displeased to find the young Tudor welcomed by Duke Francis of Brittany but failed to persuade the boy's host to extradite him. Richard III was even more anxious to secure Henry but at the eleventh hour Christopher Urswick, agent of Henry's mother, advised him to escape to France. A ruse brought Henry success.

Himself two days after, departing from Vannes, and accompanied with five servants only, feigned to go unto a friend, who had a manor not far off and, because a huge multitude of English people was left in the town, nobody suspected his voyage; but when he had journeyed almost five miles he withdrew hastily out of the highway into the next wood, and donning a serving man's apparel, he as a servant followed one of his own servants (who was his guide in that journey) as though he had been his master, and rode on with so great celerity, keeping yet no certain way, and he made no stay anywhere, except it were to bate his horses, before he had gotten himself to his company within the bounds of Anjou. *Polydore Vergil*

HENRY VII AND THE DUN COW

In common with the custom of victors Henry proceeded to London, his banners borne aloft to testify his success to all. Among these was one which bore a dun cow—a personal badge associated with Henry's earldom of Richmond.

In which time, that is to mean upon the xxvii day of August the king was received into London, the citizens being then again clothed in violet, and so brought unto Pauls where at the rood of the north dore he offered up three standards, whereof one was of the Arms of Saint George, the second a red firey dragon painted upon white and green sarcenet, and the third was a banner of tarteron bett with a dun cowe, and that done he was conveyed into the Bishop's palace and there lodged. *Great Chronicle*

PROPHECY OF HENRY'S CORONATION, 1485

Then at length, having won the goodwill of all men and at the instigation of both nobles and people, he was made king at Westminster on 31 October and called Henry, seventh of that name . . .

Thus Henry acquired the kingdom, an event of which foreknowledge had been possible both many centuries earlier and soon after his birth. For 797 years before, there came one night to Cadwallader, last king of the Britons . . . some sort of an apparition with a heavenly appearance; this foretold how long afterwards it would come to pass that his descendants would recover the land. This prophecy, they say, came true in Henry, who traced his ancestry back to Cadwallader [through his grandfather Owen Tudor]. The same prediction was made to Henry in his childhood by Henry VI. *Anglica Historia*

Union of the Red and White Roses

Though the phrase Wars of the Roses was not used in print until the mid-eighteenth century—causing some historians to scorn it—the idea was current in Shakespeare's day. The white rose was indeed one of the Yorkist badges, but Henry VI never actually used that of the red rose. Its association with the Tudors was, however, early and was used in the context of union.

In the year 1485 on the 22nd day of August the tusks of the Boar were blunted and the red rose, the avenger of the white, shines upon us.

Croyland Chronicle

It was too good for the bard to overlook its dramatic potential:

And then—as we have ta'en the sacrament—
We will unite the white rose and the red.
Smile, heaven, upon this fair conjunction,
That long have frowned upon their enmity.

Shakespeare, *Richard III*, v. viii

HENRY TUDOR AND ELIZABETH OF YORK

Henry VII married Elizabeth of York, Edward IV's eldest daughter, after the removal of the stigma of bastardy and the securing of a dispensation for the union, as the couple were cousins. The marriage had earlier been arranged by the two mothers, Lady Margaret Beaufort and Queen Elizabeth Woodville. Henry had promised to wed her in Rennes Cathedral as early as Christmas Day 1483, and finally married her on 18 January 1486.

This Margaret for want of health used the advice of a physician named Lewis, a Welshman born, who, because he was a grave man and of no small experience, she was wont oftentimes to confer freely withal, and with him familiarly to lament her adversity. And she, being a wise woman, after the slaughter of king Edward's children was known, began to hope well of her son's fortune, supposing that that deed would without doubt prove for the profit of the commonwealth, if it might chance the blood of king Henry the Sixth and of king Edward to be intermingled by affinity, and so two most pernicious factions should be at once, by conjoining of both houses, utterly taken away. *Polydore Vergil*

Margaret, therefore, during one of her talks with Lewis, prayed him to put the plan secretly to the queen,

for the Queen also used his head, because he was a very learned physician. Lewis nothing lingering spake with the queen . . . and declared the matter not as delivered to him in charge but as devised of his own head. The queen was so well pleased with this device, that she commanded Lewis to repair to the countess Margaret, who remained in her husband's house in London, and to promise . . . that she would do her endeavour to procure all her husband king Edward's friends to take part with Henry her son, so that he might be sworn to take in marriage Elizabeth her daughter after he shall have gotten the realm. Ibid.

ROYAL GRIEF, 1502

On 19 September 1486 Elizabeth presented her husband with a son and heir. A second son, Henry, followed on 28 June 1491 and a third, Edmund, in February 1499. In between came two daughters, Margaret and Mary, so that the future of the dynasty seemed secure. Edmund died in 1500 but far worse for his parents was the death of the heir, Prince Arthur, in April 1502, not quite five months after his marriage to the fifteen-year-old Catherine of Aragon. It was a severe test of the relationship between the king and queen.

When his Grace understood that sorrowful heavy tydings, he sent for the

Queene, saying that he and his Queene would take the painful sorrows together. After that she was come and saw the Kyng her Lord, and that naturall and paineful sorrowe, as I have heard saye, she with full great and constant comfortable words besought his Grace that he would first after God remember the weale of his own noble person, the comfort of his realme and of her. She then saied that my Lady his mother had never no more children but him only, and that God by his Grace had ever preserved him, and brought him where he was. Over that, howe that God had left him yet a fayre Prince, two fayre Princesses and that God is where he was, and we are both young ynoughe.

John Leland, De rebus Brittanicis collectanea, ed. T. Hearne, 6 vols. (1715), V, quoting a contemporary herald's report

Having thus comforted her husband Elizabeth retired and gave vent to her own great grief in private.

Then his Grace of true gentle and faithful love, in good hast came and relieved her, and showed her howe wise counsell she had given him before, and he for his parte would thanke God for his sonn, and would she should doe in like wise.

Ibid.

LADY MARGARET BEAUFORT

The bond between Henry and his mother remained unusually firm and permanent. For Margaret—woman of piety and benefactress of Cambridge colleges, notably St John's—fierce loyalty to her son channelled a natural capacity for political intrigue. For Henry, long exile overseas from 1471 to 1485 only strengthened his affections and he continued to rely on her help and advice after his accession. She recalled lovingly in a postscript to a letter from Calais:

this day of Saint Agnes's, that I did bring into this world my good and gracious prince, king and only beloved son.

Original Letters illustrative of English History, ed. H. Ellis, 11 vols. (1824–46), I

Henry's affections are revealed in a postscript of his own letter:

Madame, I have encumbered you now with this my long writing, but methinks that I can do no less, considering that it is so seldom that I do write, wherefore I beseech you to pardon me, for verily, madame, my sight is nothing so perfect as it has been, and I know well it will appear daily wherefore I trust that you will not be displeased, though I write not so often with mine own hand, for on my faith I have been three days ere I could make an end of this letter.

Ibid.

KING HENRY AND AN ASTROLOGER

Christopher Urswick, the source of this anecdote, became the king's almoner, after Henry's accession.

Erasmus tells a good story which he had from Urswick regarding the king about this time. Henry had been for some time in a declining state of health, and this had encouraged a saucy astrologer to foretell his death, and that it should happen before the year expired. The wise king had more mind to expose him than to punish him. So he sent to the man, and talked friendly with him, seeming not to know anything of his insolent prophecy. The king gravely asked him whether any future events could be foretold by the stars; 'Yes, Sir' (says the man) 'without all doubt.' 'Well, have you any skill in the art of foretelling?' The man affirmed that he had very good skill. 'Come then,' says the king, 'tell me where you are to be in the Christmas holidays that are now coming.' The man faltered at first, and then plainly confessed he could not tell where. 'Oh!' says the king, 'I am a better astrologer than you. I can tell where you will be,—in the tower of London,' and accordingly commanded him to be committed a prisoner thither. And when he had lain there till his spirit of divination was a little cooled, the king ordered him to be dismissed for a silly fellow.

Thomas A. Urwick, Records of the Family of Urwyk, Urswick or Urwick, ed. W. Urwick (privately printed, St Albans, 1893)

The Yorkist Shadow: Two Impostors

For all that he had defeated and killed Richard III, the new king had not eliminated the Yorkists. The rebellion was not long in coming. On 24 May 1487 in Christchurch cathedral, Dublin, a young boy presented as the son of George duke of Clarence, was crowned as Edward VI. The precise identity of the boy is uncertain, as is the very issue of impersonation, for Henry VII was unlikely ever to disclose the truth.

There was brought forward a certain sir William Simonds, a priest, twenty-eight years of age, as he asserted, who . . . publicly admitted and confessed that he himself abducted and carried across to places in Ireland the son of a certain organ-maker of the university of Oxford; and this boy was there reputed to be the earl of Warwick.

Michael Bennett, *Lambert Simnel and the Battle of Stoke* (Gloucester, 1987), quoting proceedings of a convocation of Canterbury. *In this volume written to commemorate the quincentenary of the rebellion, the author conveniently presents the contemporary evidence, some translated for the first time.*

Henry announced that the real Warwick was his prisoner in the Tower and called upon the pope:

As some of the prelates of Ireland, namely, the archbishop of Dublin, the archbishop of Armagh, and the bishops of Meath and Kildare, lent assistance to the rebels, and to a certain spurious lad, whom victory has now delivered into our hands, they pretending that the lad was the son of the late duke of Clarence, and crowning him king of England, we implore your Holiness to cite him as having incurred the censure of the church, and proceed against them at law. Ibid., quoting Henry's letter

In parliament, following the battle of Stoke, the impostor was described in more detail as:

one Lambert Simnel, a child of 10 years of age, son to Thomas Simnel, late of Oxford, joiner. Ibid., quoting the Rolls of Parliament

A French humanist in royal service, Bernard André, suggested other possibilities which were even more garbled:

The issue of cruel death of the sons of King Edward flaring up again, behold seditious men hatched another novel evil. In order to veil their plot in deceit, they maliciously put up a certain boy, lowly born, the son of either a baker or a shoemaker, as the son of Edward IV.

Ibid., quoting Bernard André, *Vita Henrici Septimi*

An Irish poem mocked the credulous archbishop of Dublin:

It is great pity that ye be deceived
By a false priest, that this matter began;
And that ye his child as a prince received
A boy, a lad, an organ-maker's son. Ibid.

Polydore Vergil was not a little confused:

Most recently, among other such enterprises, a certain lowborn priest called Richard, whose surname was Simons, a man as subtle as he was shameless, devised a crime of this fashion, by which he might disturb the peace of the kingdom. At Oxford, where he gave himself to study, he brought up a certain boy called Lambert Simnel. He first taught the boy courtly manners, so that if ever he should represent the lad as being born of royal stock, as he had resolved to do, people would the more easily credit it and have certain belief in his great creation. Some time later, since Henry VII, immediately on gaining power, had thrown Edward, the only son of the duke of Clarence, into the Tower of London, and since it was rumoured that Edward had been slain in that place, the priest Richard decided that the time had arrived when he might advantageously execute his planned villainy. He changed the boy's name and called him Edward . . . and forthwith departed with him to Ireland.

Ibid., quoting *Anglica Historia. Bennett's own conclusion is that Simnel was a pseudonym and that he was named for the cakes, which accords with the stories of his being a baker's son.*

A CULINARY END

Backing for the rebellion came from Edward IV's sister Margaret, dowager duchess of Burgundy, and an army of German mercenaries commanded by Martin Schwartz together with the Irish set sail for Lancashire. Leadership was provided, not by the young pretender but by John de la Pole, earl of Lincoln, nephew and probably intended heir of Richard III, as senior male representative of the Yorkist line, since 1484. Quite what would have happened regarding the succession had the rebels won is unclear but it might be supposed that Lincoln would have made his own bid for the crown in preference to the impostor.

The following day the king, having formed his whole force into three columns, advanced to the village of Stoke, halted below the earl's camp and, on the level ground there, offered battle . . . The Germans, fierce mountainmen, experienced in war, who were in the front line, yielded little to the English in valour; while Martin Schwartz their leader was not inferior to many in his spirit and strength. The Irish, on the other hand, though they fought most spiritedly, were nonetheless slain before the others, being according to their custom devoid of body armour . . . Of their leaders John, earl of Lincoln, Francis Lord Lovell, Thomas Boughton, the most bold Martin Schwartz and the Irish captain Thomas Geraldine were slain in that place . . . Lambert the false boy-king was indeed captured, with his mentor Richard: but each was granted his life, since the innocent lad was too young to have given offence, and since his mentor was a priest. Lambert is still alive to this very day, having been made trainer of the king's hawks; for some time before that he turned the spit and did other menial jobs in the royal kitchen. Ibid., quoting *Anglica Historia*

PERKIN WARBECK

The security of the realm was threatened a second time in 1491. As with the earlier conspiracy, much remains obscure. The pretender was hailed in Ireland as the younger son of Edward IV, Richard duke of York. Once again Margaret of Burgundy recognized him as her nephew. The affair dragged on for some years. In 1495, unswayed by any memory of the service rendered at Bosworth, Henry ordered the execution of Sir William Stanley for his involvement with Warbeck.

After abortive attempts to invade, Warbeck finally landed in Cornwall in September 1497 and was captured. Henry himself described the affair in a letter to Sir Gilbert Talbot, reserving especial venom for Margaret of Burgundy.

And not forgetting the great malice that the Lady Margaret of Burgundy beareth continually against us, as she showed lately in sending hither of a feigned boy [Simnel], surmising him to have been the son of the Duke of Clarence, and caused him to be accompanied with the Earl of Lincoln . . . whose end—blessed be God!—was as ye know well. And foreseeing now the perserverence of the same her malice by the untrue contriving eftsoons of another feigned lad called Perkin Warbeck, born at Tournay, in Picardy, which at his first [going] into Ireland called himself the bastard son of King Richard; after that the son of the said Duke of Clarence; and now the second son of our father, King Edward the Fourth, whom God assoile.

James Gairdner, *History of the Life and Reign of Richard the Third* and *The Story of Perkin Warbeck* (Cambridge, 1898), quoting Henry's letter

EXECUTIONS, 1499

Warbeck confessed to the king, who showed great leniency. His wife was sent to the queen and was well treated and Warbeck was kept unrestrained at court. The following year he tried to escape and was subsequently confined in the Tower. Over the ensuing months he became involved with Edward earl of Warwick and Henry finally decided to rid himself of both men. Bacon commented on Warwick's demise:

[Thus did] this winding ivy of a Plantagenet kill the true tree itself.

Francis Bacon

HENRY AND THE EARL OF KILDARE, 1496

After being cleared of alleged treason, Gerald FitzGerald, 8th earl of Kildare, was sent back to Ireland as the king's deputy. Someone is said to have remarked of Kildare to Henry: 'All England cannot rule yonder gentleman', to which Henry responded, 'No? Then he is meet to rule all Ireland.' Kildare was famous for his charm, sardonic humour and fiery temper:

The famous story told by Campion and Stanihurst that when he was being reprimanded for setting fire to the cathedral at Cashel, he apologized humbly to the king and said that he had only done this because he thought the archbishop was inside, is probably untrue (there is no record of his antagonism to Archbishop David Creagh), but again it is in character.

D. B. Quinn in *A New History of Ireland*, II, ed. Art Cosgrove (Oxford, 1987)

CONFOUNDED BY A MONKEY

Henry VII may well have had a pet monkey but whether or not it offended him as Francis Bacon suggested is unknown. The story may serve, nevertheless, to illustrate the sense of fear that some of the king's servants felt about this zealous ruler.

He was a Prince, sad, serious, and full of thoughts, and secret observations, and full of notes and memorials of his own hand, especially touching persons. As, whom to employ, whom to reward, whom to inquire of, whom to beware of, what were the dependencies, what were the factions, and the like; keeping, as it were, a journal of his thoughts. There is to this day a merry tale; that his monkey, set on as it was thought by one of his chamber, tore his principal notebook all to pieces, when by chance it lay forth: whereat the court, which liked not those pensive accounts, was almost tickled with sport.

Francis Bacon

AN OBITUARY, 1509

There was a general agreement about the many benefits derived from Henry's politic rule, even if he had been just a little greedy.

He well knew how to maintain his royal majesty and all which appertains to kingship at every time and in every place. He was most fortunate in war, although he was constitutionally more inclined to peace than to war. He cherished justice above all things; as a result he vigorously punished violence, manslaughter and every other kind of wickedness whatsoever. Consequently he was greatly regretted on that account by all his subjects, who had been able to conduct their lives peaceably, far removed from the assaults and evil doings of scoundrels. He was the most ardent supporter of our faith and daily participated with great piety in religious services . . . But all these virtues were obscured latterly by avarice, from which he suffered. This avarice is surely a bad enough vice in a private individual, whom it forever torments; in a monarch indeed it may be considered the worst vice since it is harmful to everyone and distorts those qualities of trustfulness, justice and integrity by which the State must be governed.

Anglica Historia

A FUNERAL ORATION

John Fisher, the saintly bishop of Rochester, confessor to Henry's mother, preached at the king's death in complimentary vein.

His politic wisdom in governance was singular, his wit always quick and ready, his reason pithy and substantial, his memory fresh and holding, his experience notable, his counsels fortunate and taken by wise deliberation, his speech gracious in diverse languages, his person goodly and amiable, his natural complexion of the purest mixture, his issue fair and in good number . . . his people were to him in as humble subjection as ever they were to king; his land many a day in peace and tranquility; his prosperity in battle against his enemies was marvellous; his dealings in time of perils and dangers was cold and sober with great hardiness.

The English Works of John Fisher, ed. J. E. B. Mayor, Early English Texts Society (reprinted, 1935)

Henry VIII

1509–1547

The word 'dissolute' clings to Henry, primarily for his six wives and the dissolution of the monasteries. Beyond that he was a Renaissance prince: acquisitive, cruel, fascinated by the arts, sport, fleets of ships. He saw no contradiction between his papal award of Defensor Fidei (*Defender of the Catholic Faith against Luther*) *and the English Reformation by which he, instead of the pope, became Head of the Church. Of the young resplendent Henry, Sir Thomas More said,* 'If a lion knew his own strength, hard were it for any man to rule him.' *His youthful beauty and mature corpulence were alike extreme.*

A REPRESSED YOUTH

According to the Spanish envoy called Fuensalida, who came to England in early 1508, the Prince of Wales was kept under such strict supervision that he might have been a young girl. He could not go out except through a private door which led into a park, and then only in the company of specially appointed persons. No one dared to approach him and speak to him. He spent his time in a room the only access to which was through the king's chamber and was so cowed that he never spoke in public except to answer a question from his father. [However] the same ambassador was soon reporting how the young Henry spent day after day at Richmond in his favourite, boisterous sport, tilting. True, even then his father was an onlooker—at least on occasion . . . the prince may not have been so thoroughly shut in as Fuensalida first supposed . . . Whatever the truth of the matter, a modern may well shake his head over this story of evident repression of an ebullient youth and conclude that it explains a good deal of the flamboyance and waywardness of the grown man.

J. J. Scarisbrick, *Henry VIII* (1968)

HENRY IN YOUTH FIGHTS BACK

The young Henry celebrated his resolve to have his own way—in verse:

For my pastance,
Hunt, song, and dance,
My heart is set!
Who shall me let?

H. Cam, *England Before Elizabeth* (1950), quoting Henry's poem

Impressions of a King

Historians do not lack for descriptions of Henry VIII left behind by his contemporaries. His size and the sense of majesty were clearly compelling.

A FOREIGNER'S VIEW

After dinner, we were taken to the King, who embraced us, without ceremony, and conversed for a very long while very familiarly, on various topics, in good Latin and in French, which he speaks very well indeed, and he then dismissed us, and we were brought back here to London . . .

His Majesty is the handsomest potentate I ever set eyes on; above the usual height, with an extremely fine calf to his leg, his complexion very fair and bright, with auburn hair combed straight and short, in the French fashion, and a round face so very beautiful, that it would become a pretty woman, his throat being rather long and thick. He was born on the 28th of June, 1491, so he will enter his twenty-fifth year the month after next. He speaks French, English, and Latin, and a little Italian, plays well on the lute and harpsichord, sings from book at sight, draws the bow with greater strength than any man in England, and jousts marvellously. Believe me, he is in every respect a most accomplished Prince; and I, who have now seen all the sovereigns in Christendom, and last of all these two of France and England in such great state, might well rest content.

English Historical Documents, V, 1485–1558, ed. C. H. Williams (1967), quoting letters written by a Venetian in 1515

The same year he wrote:

His Majesty came into our arbour, and, addressing me in French, said: 'Talk with me awhile! The King of France, is he as tall as I am?' I told him there was but little difference. He continued, 'Is he as stout?' I said he was not; and he then inquired, 'What sort of legs has he?' I replied 'Spare.' Whereupon he opened the front of his doublet, and placing his hand on his thigh, said "Look here! and I have also a good calf to my leg.' He then told me that he was very fond of this King of France, and that for the sake of seeing him, he went over there in person, and that on more than three occasions he was very near him with his army, but that he never would allow himself to be seen, and always retreated, which his Majesty attributed to deference for King Louis, who did not choose an engagement to take place; and he here commenced discussing in detail all the events of that war, and then took his departure . . .

After dinner, his Majesty and many others armed themselves *cap-à-pie*, and he chose us to see him joust, running upwards of thirty courses, in one of which he capsized his opponent (who is the finest jouster in the whole kingdom), horse and all. He then took off his helmet, and came under the windows where we were, and talked and laughed with us to our very great honour, and to the surprise of all beholders.

English Historical Documents, V, Quoting letters written by a Venetian

SIR THOMAS MORE AND HENRY VIII

For all the favour which Sir Thomas More enjoyed at the court and in the company of the king, he evidently had the measure of his lord:

And such entire favour did the king bear him that he made him chancellor of the duchy of Lancaster upon the death of Sir Richard Wingfield, who had that office before.

And for the pleasure he took in his company, would his grace suddenly sometimes come home to his house at Chelsea, to be merry with him; whither on a time, unlooked for, he came to dinner to him; and after dinner, in a fair garden of his, walked with him by the space of an hour, holding his arm about his neck. As soon as his grace was gone, I, rejoicing thereat, told Sir Thomas Moore how happy he was, whom the king had so familiarly entertained, as I never had seen him to do to any other except Cardinal Wolsey, whom I saw his grace once walk with, arm in arm. 'I thank our Lord, son,' quoth he, 'I find his grace my very good lord indeed, and I believe he doth as singularly favour me as any subject within this realm. Howbeit, son Roper, I may tell thee I have no cause to be proud thereof, for if my head could win him a castle in France (for then was there war between us) it should not fail to go.'

Ibid., quoting *The Lyfe of Sir Thomas Moore, Knight*

THE FIELD OF CLOTH-OF-GOLD, 7 JUNE 1520

A spectacular conference was held between Guisnes and Ardres on the boundary of the Calais marches and France, attended by Henry VIII and François I. It did more for prestige than peace.

When this day came, a great company having assembled from both sides which saluted their princes with great acclamation, the kings and a few only of their entourage entered a centrally-placed tent. There they first greeted each other like truly well-intentioned people, and then with evident satisfaction talked together in friendly fashion until evening. The foundation of friendship having thus been laid, Francis came the next day to salute the English queen, and Henry in his turn the French queen.

When they had dined they each returned home, meeting one another on the way. *Anglica Historia*

After this promising start things deteriorated:

For many days therefore everything resounded with happy voices, but it was possible to observe that not all the English viewed the French with happy minds . . . On account of this feeling King Francis easily saw that he and his subjects were poorly received. Taking advantage of a suitable opportunity for complaining, he is related to have said to Henry, 'I thank you, King Henry, most profoundly because I understand in certainty and (as they say) read in your face that you are truly in friendly agreement with me . . . But as for the other English, I am just as certain in my mind as through my observation that they are so far from liking us, that they even look upon us unwillingly . . . Henry neatly excused himself with a joke—without which he would have been unable to excuse himself. Ibid.

What was Henry's joke? A reference to the New Learning, perhaps? 'I see what your Majesty is trying to say: "Timeo Anglicos et dona ferentes—I fear the English even when they bring gifts."'

THE DOWNFALL OF THOMAS WOLSEY, 1529

As cardinal and organizer Wolsey was constantly celebrating High Mass on the Field of Cloth-of-gold. Nine years later he had fallen through failure to deliver from the pope the dissolution of Henry's marriage to Catherine of Aragon. On his death-bed Wolsey said:

'If I had served God as diligently as I have done the King, He would not have given me over in my grey hairs.'

George Cavendish, The Life and Death of Cardinal Wolsey, ed. R. S. Sylvester, Early English Texts Society (1959)

The Tragedy of Anne Boleyn, 1533–6

Of Henry's six wives, Anne Boleyn was assuredly his truest love. Their affair was passionate, as Henry's surviving letters to her reveal. Anne was intelligent and ambitious—like Queen Elizabeth Woodville in 1464, too good to be another of the king's mistresses.

My mistress and friend: I and my heart put ourselves in your hands, begging you to have them suitors for your good favour, and that your affection for them should not grow less through absence. For it would be a great pity to increase their sorrow since absence does it sufficiently, and more than ever I could have thought possible reminding us of a point in astronomy, which is, that the longer the days are the farther off is the sun, and yet the more fierce. So it is with our love, for by absence we are parted, yet nevertheless it keeps its fervour, at least on my side, and I hope on yours also: assuring you that on my side the ennui of absence is already too much for me: and when I think of the increase of what I must needs suffer it would be well nigh unbearable for me were it not for the firm hope I have of your steadfast affection for me. So, to remind you of that sometimes, and as I cannot be with you in person, I am sending you the nearest possible thing to that, namely, my picture set in a bracelet, with the whole device which you already know. Wishing myself in their place when it shall please you. This by the hand of

Your loyal servant and friend

H.Rex

English Historical Documents, V, quoting Wriothesley, *Chronicle*

And on another occasion:

No more to you at this present mine own darling for lack of time but that I would you were in my arms or I in yours for I think it long since I kissed you. Written after the killing of an hart at a xj. of the clock minding with God's grace tomorrow mightily timely to kill another: by the hand of him which I trust shortly shall be yours.

Henry R.

Ibid.

Nor did the letters lack intimacy:

Mine own sweetheart, these shall be to advertise you of the great elengenes [loneliness] that I find here since your departing, for I ensure you methinketh the time longer since your departing now last than I was wont to do a whole fortnight: I think your kindness and my fervents of love causeth it, for otherwise I would not have thought it possible that for so little a while it should have grieved me, but now that I am coming toward you methinketh my pains been half released . . . Wishing myself (specially an evening) in my sweetheart's arms, whose pretty dukkys [breasts] I trust shortly to kiss. Written with the hand of him that was, is, and shall be yours by his will.

H.R.

Ibid.

CELEBRATING THE DEATH OF QUEEN CATHERINE

Hall the chronicler narrated reprovingly that when Catherine of Aragon died 'Queen Anne wore yellow for mourning.' He suppressed the fact that Henry did the same:

When the news of her death at Kimbolton reached London, Henry—dressed from Head to toe in exultant yellow—celebrated the event with Mass, a banquet, dancing and jousting. Scarisbrick, *Henry VIII*

Nor was Anne Boleyn his only mistress during Catherine's life. He seduced Anne's sister Mary and also had a son by Bessie Blount, one of Catherine's ladies. The boy was created Duke of Richmond at six years old and given a household of his own with which to hold state at St James's Palace, which was built for him.

THE EXECUTION OF QUEEN ANNE BOLEYN, 19 MAY 1536

At eight in the morning Anne was brought to Tower Green, by the White Tower, where all the great of the land, from the Lord Chancellor to the aldermen, were seated on a platform. She had given Henry no male heir.

All these being on a scaffold made there for the execution, the said Queen Anne said as followeth: Masters, I here humbly submit me to the law, as the law hath judged me, and as for mine offences . . . God knoweth them, I remit them to God, beseeching him to have mercy on my soul; and I beseech Jesu save my Sovereign and master the King, the most goodliest, and gentlest Prince that is, and long to reign over you, which words she spake with a smiling countenance: which done, she kneeled down on both her knees, and said, To Jesu Christ I commend my soul and with that word suddenly the hangman of Calais smote off her head at one stroke with a sword: her body with the head was buried in the choir of the Chapel in the Tower. *Annals of John Stow*

HENRY'S UNCROWNED QUEEN

Henry's third and brief marriage achieved what he most desired. How long Jane Seymour might have enjoyed his favour after the birth of a son can only be guessed. What mattered was the final resolution of 'the king's private matter'.

Also the 20th. day of May [1536] the King was married secretly at Chelsea, in Middlesex, to one Jane Seymour, daughter to Sir John Seymour, knight, in the county of Wiltshire, late departed from this life, which Jane was first a waiting gentlewoman to Queen Katherine, and after to Anne Boleyn, late Queen, also; and she was brought to White Hall, by Westminster, the

30th. day of May, and there set in the Queen's seat under the canopy of estate royal.

Also the 4th. day of June, being Whitsuntide, the said Jane Seymour was proclaimed Queen at Greenwich, and went in procession, after the King, with a great train of ladies following after her, and also offered at mass as Queen, and began her household that day, dining in her chamber of presence under the cloth of estate . . .

This year, the 11th. day of October, anno 1537, and the 29th. year of the reign of King Henry the Eight, being Thursday, there was a solemn general procession in London, with all the orders of friars, priests, and clerks going all in copes, the mayor and aldermen, with all the crafts of the city, following in their liveries, which was done to pray for the Queen that was then in labour of child. And the morrow after, being Friday and the even of Saint Edward, sometime King of England, at two of the clock in the morning the Queen was delivered of a man child at Hampton Court beside Kingston. *English Historical Documents*, V, quoting Wriothesley, *Chronicle*

The son was quickly christened:

This year, the 25th. day of October, being Monday, the Prince was christened in the King's chapel at Hampton Court, the Archbishop of Canterbury and the Duke of Norfolk godfathers at the font, and my Lady Mary's grace, the King's daughter by Queen Katherine, godmother, and the Duke of Suffolk godfather at the confirmation, the Prince's name being Edward, proclaimed after his christening by the King of Heralds, 'Edward, son and heir to the King of England, Duke of Cornwall, and Earl of Chester.' The goodly solemnity of the lords and ladies done at the christening was a goodly sight to behold, every one after their office and degree; the Lady Elizabeth, the King's daughter, bearing the chrisom on her breast. Ibid.

Who remembered his mother?

This year, the 14 of October, being Wednesday, Queen Jane departed this life, lying in childbed, about 2 of the clock in the morning, when she had reigned as the King's wife, being never crowned, one year and a quarter.

Ibid.

HENRY AND THE CATHOLIC MARTYRS, 1535

John Fisher, bishop of Rochester, and Thomas More had opposed Henry's divorce. Pope Paul III, in order to strengthen Fisher's position with Henry, created him a cardinal, but he was executed on Tower Hill on 22 June.

Henry's agent in Italy, Sir Gregory di Casale, was horrified when he heard what the Pope had done. He was sure that it would infuriate Henry against Fisher. Ortiz was pleased that Fisher had been given the red hat, as he thought it was a great gesture against Henry; but he wrote to the Empress that he feared that before Fisher heard about it, God would give him the true red hat, the crown of martyrdom. The Pope was shaken when Casale told him how deeply Henry would resent what he had done, and that his well-meant gesture had endangered Fisher's life. He asked François to use his influence with Henry in Fisher's favour. François said that he would do all he could to save Fisher, but was not optimistic. . . . Casale's worst fears were realised. Henry was enraged when he heard that Fisher had been made a cardinal; he treated it as a challenge which he was bound to take up, and as an insult which he would avenge. He said that as Fisher had been given a cardinal's hat, he would cut off Fisher's head and send the head to Rome to have the hat put on it. Jasper Ridley, *Henry VIII* (1984)

On Tuesday 6 July 1535 More was informed that he would die before 9 a.m. He replied with alacrity:

'Master Pope, for your good tidings I most heartily thank you. I have been always much bounden to the king's highness for the benefits and honours that he hath still from time to time most bountifully heaped upon me . . . And . . . most of all . . . am I bound to his highness that it pleaseth him so shortly to rid me out of the miseries of this wretched world. And therefore will I not fail earnestly to pray for his grace, both here and also in another world.' Richard Marius, *Thomas More* (New York, 1984)

By royal order, More was instructed to be brief on the scaffold. His comment on the tottering structure is well known:

'I pray you, Master Lieutenant, see me safe up, and for my coming down, let me shift for myself.' Ibid.

SIGNS OF ROYAL UNPOPULARITY, 1537

When a Sussex man reported Henry's fall he added, 'It were better he had broken his neck.' He was called 'a mole who should be put down', 'a tyrant more cruel than Nero', 'a beast and worse than a beast'. It was said that 'Cardinal Wolsey had been an honest man if he had had an honest master', that Henry was 'a fool and my lord privy seal another'. 'Our king', said one malcontent, 'wants only an apple and a fair wench to dally with', while another told indignantly how Henry had spotted his woman while out riding near Eltham, 'grabbed her and taken her to his bed'. Things like this would never have been said . . . even five years ago. Scarisbrick, *Henry VIII*

THE DISSOLUTION OF THE MONASTERIES

How the king gave away abbey lands or possessions by free gift or lost them by gambling.

Herein take one story of many: Master John Champernoun, son and heir-apparent of Sir Philip Champernoun, of Modbury in Devon, followed the court, and by his pleasant conceits won good grace with the King. It happened, two or three gentlemen, the King's servants, and Mr Champernoun's acquaintance, waited at a door where the King was to pass forth, with purpose to beg of his Highness a large parcel of abbey-lands, specified in their petition. Champernoun was very inquisitive to know their suit, but they would not impart the nature thereof. This while out comes the King; they kneel down, so doth Mr Champernoun, being assured by an implicit faith, that courtiers would beg nothing hurtful to themselves; they prefer their petition, the King grants it; they render him humble thanks, and so doth Mr Champernoun. Afterwards he requires his share, they deny it; he appeals to the King, the King avows his equal meaning in the largesse. Whereupon, his companions were feign to allot this gentleman the priory of St Germain's, in Cornwall (valued at two hundred forty-three pounds and eight shillings of yearly rent; since, by him or his heirs, sold to Mr Eliot) for his partage. Here a dumb beggar met with a blind giver; the one as little knowing what he asked, as the other what he granted. Thomas Fuller, *The Church History of Britain* (1868)

TREACHEROUS KNIGHTS OF THE GARTER

Henry found time, during this hot and grim summer [1540], to pay attention to the affairs of the Order of the Garter, which always delighted him. The names of the knights of the most noble order were impressively inscribed in the books; but a growing number of them had been executed for high treason. The officials were uncertain whether their names should be erased from the list, as they so richly deserved; for the appearance of the books would be disfigured by a large number of erasures. The matter was referred to the King. He ordered that the names should remain on the list, but that opposite each name there should be added the words: *Vah, proditor* (Oh! a traitor!) Ridley, *Henry VIII*

HENRY'S SEX LIFE

Henry was neither ribald nor bawdy nor particularly lusty. In fact the King was exceedingly touchy about his sex-life, answering the Imperial Ambassador's argument that perhaps God had ordained the succession to remain in the female line by shouting three times over: 'Am I not a man like others?' Lacey Baldwin Smith, *Henry VIII. The Mask of Royalty* (1971)

The king was not amused by sixteenth-century bawdy:

There is a story that Henry enjoyed rhyming, and one day, while travelling on the river from Westminster to Greenwich to visit 'a fair lady whom he loved and lodged in the tower of the park', he challenged Sir Andrew Flamock to compose with him. The King wrote:

Within this tower
There lieth a flower
That hath my heart.

Exactly what Sir Andrew replied has been kept discreetly hidden, but a version of his answer appeared in one of the worst plays of almost any century:

Within this hour
She pist full sower
And let a fart.

Legend has it that the monarch was not amused and bid Flamock 'avant varlet' and begone. Ibid.

AN UNFAVOURABLE VIEW IN 1540

The French ambassador, Marillac, who was well placed to observe Henry and had known him over these later years, was not quite so enamoured as earlier writers—a true reflection perhaps of the decline of the king towards the end of his reign.

First, to commence with the head, this Prince seems tainted, among other vices with three which in a King may be called plagues. The first is that he is so covetous that all the riches in the world would not satisfy him. Hence the ruin of the abbeys, spoil of all churches that had anything to take, suppression of the knights of St John of Rhodes, from whom has been taken not only their ancient revenue, but the moveables which they had acquired which they have not been able to leave by will . . .

Thence proceeds the second plague, distrust and fear. This King, knowing how many changes he has made, and what tragedies and scandals he has created, would fain keep in favour with everybody, but does not trust a single man, expecting to see them all offended, and he will not cease to dip his hand in blood as long as he doubts his people. Hence every day edicts are published so sanguinary that with a thousand guards one would scarce be safe. Hence too it is that now with us, as affairs incline, he makes alliances which last as long as it takes for him to keep them.

The third plague, lightness and inconstancy, proceeds partly from the other two and partly from the nature of the nation, and has perverted the rights of religion, marriage, faith and promise, as softened wax can be altered to any form.

The subjects take example from the Prince, and the ministers seek only to undo each other to gain credit, and under colour of their master's good each attends to his own. For all the fine words of which they are full they will act only as necessity and interest compel them.

English Historical Documents, V, quoting Marillac's comments

A NINETEENTH-CENTURY VIEW OF HENRY VIII

Prince Pückler-Muskau was touring the British Isles in search of a rich wife and described to his ex-wife a visit to Hampton Court in 1826.

Most of the rooms in the palace are furnished just as they were in the time of William III a hundred and twenty years ago. The tattered chairs and tapestries are carefully preserved. Many interesting and excellent pictures adorn the walls . . . Let me mention only two splendid portraits: of Wolsey, the proud builder of this palace, and of Henry VIII, his treacherous master. Both are magnificent and entirely characteristic. You remember that fat lawyer, whom we had so much trouble in shaking off; his look of a wild beast, sensual, bloodthirsty—so far as one can be today—shrewd, crafty, full of intelligence and slyness, of boundless arrogance, and yet with an overwhelming tendency to baseness, and lastly entirely and frankly devoid of conscience—give the likeness of Henry a green frock coat with mother-of-pearl buttons, and you have his [the lawyer's] portrait to the life.

Pückler's Progress, ed. F. Brennan (1987)

LAST MONTHS

Obese and restricted by his hugely swollen legs, Henry died aged fifty-five. The fruits of his six marriages (two wives beheaded), were two princesses and one prince. It was Edward who succeeded him.

In March 1544—just as he was about to set out on his last campaign—the ulcer flared up once more and the fever returned. But in July of that year he crossed to Calais and rode a great courser to the siege of Boulogne. Though he was carried about indoors in a chair and hauled upstairs by machinery, he would still heave his vast, pain-racked body into the saddle to indulge his love of riding and to show himself to his people, driven by an inexorable will to cling to his ebbing life.

Scarisbrick, *Henry VIII*

Edward VI

1547–1553

Edward was short and slight with delicate colouring. Like his uncle Prince Arthur, he died of consumption at fifteen. But the stories about his childhood—and his reign was all childhood—are by no means those of a spirit in decline. Where his father had been ambivalent about Protestantism, Edward was a child-devotee.

THE CORONATION

The young king's precocious piety was already noticeable.

At Westminster the great church was preparing for the coronation; the choir was now made rich with hangings of Arras against the walls; the smoothed paving stones were laid with rushes. The trumpets blew as the three crowns in succession were set upon the young king's head. A ring of gold was placed upon his Grace's marrying finger. The choir sang the *Te Deum*; the organ played.

One can see him as he came up the aisle, a boy of ten. He was a quiet and studious child with a built-in distaste for all Church ceremonies. Alone of all his family, his clothes meant nothing to him. He probably hardly saw the light reflected from his jewellery. He did not like flattery; he was nourished on the Word of God.

They had made him ready for a gorgeous ceremonial. At the main door his horse was waiting, a patient beast, 'with a caparison of crimson satin embroidered with pearls and damasked gold'. King Edward had been dressed with splendour. He had a gown of cloth of silver with a girdle of white velvet wrought with Venice silk. The gown was powdered with rubies and diamonds. Pearls were scattered on the great white velvet cloak embroidered with Venetian silver. The king was small and slender and on his feet he wore white velvet buskins. The symbols of royalty were all about him—the insignia of the Chief Butler of England, the mantle, the rod of gold, the crown, two swords of state and a cap with the 'cyreillet'.

D. Mathew, *Lady Jane Grey: The Setting of a Reign* (1972)

Impressions of Edward VI

This precocity of Edward proved to be truly remarkable, so much so for his contemporaries that descriptions of him were extremely favourable, even eulogistic.

EDWARD'S ITALIAN DOCTOR AND ASTROLOGER

Of the excellent virtues and singular graces of king Edward, wrought in him by the gift of God, although nothing can be said enough in his commendation, yet, because the renowned fame of such a worthy prince shall not utterly pass our story without some grateful remembrance, I thought, in a few words, to touch some little portion of his praise, taken out of the great heaps of matter, which might be inferred. For, to stand upon all that might be said of him, it would be too long; and yet to say nothing, it were too much unkind. If kings and princes, who have wisely and virtuously governed, have found in all ages writers to solemnize and celebrate their acts and memory, such as never knew them, nor were subject unto them, how much then are we Englishmen bound not to forget our duty to king Edward: a prince, although but tender in years, yet for his sage and mature ripeness in wit and all princely ornaments, as I see but few to whom he may not be equal, so, again, I see not many, to whom he may not justly be preferred.

English Historical Documents, V, quoting John Foxe. *Foxe, the future Protestant martyrologist, preserved this well-known description by Girolamo Cardano when reporting on the king's health.*

EDWARD'S TUTOR

The books which you have written to the King's majesty, have been as acceptable to him as they deserved to be. A large portion of them I delivered to him myself, and am able therefore to inform you how kindly and courteously he received them, and how greatly he esteems them; and I can offer you my congratulations upon the subject. But since the King's majesty, debilitated by long illness, is scarcely yet restored to health, I cannot venture to make you any promise of obtaining a letter from him to yourself. But should a longer life be allowed him (and I hope that he may very long enjoy it) I prophesy indeed, that, with the Lord's blessing, he will prove such a king, as neither to yield to Josiah in the maintenance of true religion, nor to Solomon in the management of the state, nor to David in the encouragement of godliness. And whatever may be effected by nature or grace, or rather by God the source of both, whose providence is not even contained within the limits of the universe, it is probable that he will not only contribute very greatly to the preservation of the church, but also that he will distinguish learned men by every kind of encouragement. He has long since given evidence of these things, and has accomplished at this early period of his life more numerous and important objects, than others have been able to do when their age was more settled and matured.

Ibid., quoting Edward's tutor Sir John Cheke in a letter of 1553

ROGER ASCHAM ON EDWARD VI

The ability of our Prince equals his fortune, and his virtue surpasses both: or rather, as is fitting for a Christian to say, such is the manifold grace of God, that in eagerness for the best literature, in pursuit of the most strict religion, in willingness, in judgment, and in perseverance—the quality you most value in study—he is wonderfully in advance of his years. And in hardly any other particular do I consider him more fortunate than that he has had John Cheke as the instructor of his youth in sound learning and true religion. Latin he understands, and speaks, and writes with accuracy, propriety and ease. In Greek he has learned Aristotle's Dialectic, and now is learning his Ethics. He has made such progress in that language that he translates quite easily the Latin of Cicero's philosophy into Greek.

Ibid., quoting a description penned in 1550

A DIALOGUE ON HIS SISTER MARY'S CATHOLICISM

The Council had decided that Mary's disobedience over the new religion must be allowed.

The King's attendance was then requested. As Edward entered, the Lord Treasurer fell on his knees and told him that he and they and the realm were about to 'come to naught'. They must give way, pacify the Emperor, and let the Princess do as she desired; the bishops said that it might be done.

'Are these things so, my Lords?' said Edward, turning to them. 'Is it lawful by Scripture to sanction idolatry?'

'There were good kings, your Majesty,' they replied, 'who allowed the hill altars and yet were called good.'

'We must follow the example of good men,' the boy answered, 'when they have done well. We do not follow them in evil. David was good but David seduced Bathsheba and murdered Uriah. We are not to imitate David in such deeds as these. Is there no better Scripture?' The bishops could think of none.

'I am sorry for the realm, then,' the King said, 'and sorry for the danger that will come of it; I shall hope and pray for something better, but the evil thing I will not allow.'

J. A. Froude, *History of England*, IV (1893)

THE DOWNFALL OF THOMAS LORD SEYMOUR

When the young king heard from his uncle Lord Seymour, his mother Jane Seymour's brother, that his older uncle the Protector Somerset was getting too old for the job, Edward commented, 'It were better that he should die.' The Protector decided it were better that his treacherous brother should die, and Seymour sealed his own fate by trying to force his way back into Edward's confidence.

The withdrawal of the confidence which had been the basis of his designs drove the Admiral to the last fatal step. If Edward was not to be approached except in company, then there was nothing for it but to come to him after he had gone to bed. On the night of January 16th, 1549, Seymour, taking two of his servants and—final, incredible folly—armed with a pistol, let himself into the Private Garden and so reached the King's bedchamber without passing through the ante-chambers and passages. It had not occurred to him that Edward might have taken his own precautions. When everyone was asleep, the King had got out of bed and bolted the inner door on his own side, having put his little dog beyond the outer door. As soon as the Admiral and his men started fumbling with the lock, the dog . . . sprang up, barking furiously. Maddened, desperate, Seymour shot him.

As the report reverberated through the ante-chambers and galleries, yeomen, halberdiers and Gentlemen of the Privy Chamber came running. There stood the Admiral, the smoking pistol in his hand. To the torrent of questions he could only mumble: 'I wish to know whether His Majesty was safely guarded.' Within a few minutes he was under arrest, and next day in the Tower . . . The Protector seems to have been under the impression that his nephew was still attached to Seymour, and that the news of his treachery would come as a shock. If he had troubled to look at the Admiral's behaviour from Edward's point of view, he would have realised that no child is likely to forget or forgive the killing of a pet dog.

H. W. Chapman, *The Last Tudor King* (1958)

LEARNING ROYAL TEAM GAMES

Northumberland taught Edward to take an interest in hunting and clothes; Warwick devised games for the boy who preferred books to play.

Plans were made for him to lead a team of his Gentlemen of the Bedchamber against the same number of noblemen at archery, tilting and running at the ring in a series of matches. In his diary, Edward noted the results: 'The first day of the challenge at base, or running, the King won . . .' is followed by: 'I lost the challenge of shooting at rounds and won at rovers.' A month later the record of the semi-finals opens with: 'First came the King, sixteen footmen and ten horsemen in black silk coats pulled out with white taffety; then all the Lords, having three men likewise apparelled; and all gentlemen, their footmen in white fustian pulled out with black taffety. The other side came all in yellow taffety; at length the yellow band took it thrice in a hundred and twenty courses.' Then came disappointment—and some private criticism of the umpire: 'and my band touched often, which was counted as nothing, and took never, which seemed strange, and so was of my side lost'. Ibid.

A PROTESTANT DEATH-BED

Edward composed a prayer when he was too weak to write, but his attendants took it down:

Lord God, deliver me out of this miserable and wretched life, and take me amongst thy chosen; howbeit, not my will but Thy will be done. Lord I commit my spirit to Thee. O! Lord, thou knowest how happy it were for me to be with Thee: yet, for Thy Chosen's sake, send me life and health, that I may truly serve Thee. O! my Lord God, defend this realm from papistry, and maintain Thy true religion, that I and my people may praise Thy holy name, for Thy son Jesus Christ's sake. Amen.

J. G. Nichols, *Literary Remains of Edward VI* Roxburghe Club (1857)

EDWARD'S 'DEVISE' OF THE SUCCESSION

The duke of Northumberland, who had succeeded the beheaded Somerset as Protector, persuaded the dying king to will his crown to Lady Jane Grey, the duke's daughter-in-law. But in doing so Edward was thinking of Jane's Protestantism, not of her value on the political chessboard.

The dying king had a single wish, adhered to feverishly, that the Crown of England should continue to protect the Word of God.

. . . the matters of this world were fading from him. His tuberculosis was gaining fast as he now lay dying in his father's palace. He knew well that he had been a virtuous prince, who had spent his whole life in defending the True Reformed religion. He was a solemn child. He savoured sermons which, Sunday by Sunday, had laid before him, well-spiced with flattery, the duties of a Christian ruler. In some ways he was perhaps retarded. His whole life had been spent in innocence; now he would pass on his guardianship of Reformation values to a princess whose life was given to the study of the pure doctrine . . .

He knew that the pains of his complicated illness were by this time over. Now he stood before the Throne of Grace; he knew that he was due to enter at the Gates of Paradise. He died at nine o'clock of the evening of Thursday, 6 July 1553, a little earlier than had been expected.

Mathew, *Lady Jane Grey*

Jane was hardly older than Edward and his 'devise' spelt her demise.

The 'Nine-Day' Queen, 10–19 July 1553

Lady Jane Grey (Dudley) was proclaimed on 10 July by the Council, in accordance with Edward VI's will. When the country rose in favour of the direct royal line, the same Council proclaimed Mary Tudor nine days later. A proclamation against Mary, sent to Lord Northampton, showed that 'Jane the Quene', as she signed herself, possessed the worst possible advisers.

You will endeavour yourself in all things to the uttermost of your power, not only to defend our just title to the crown but also assist us to disturb, repel and resist, the feigned and untrue claim of the Lady Mary, bastard daughter of our great-uncle Henry the Eighth of famous memory.

Ibid.

THE EDUCATION OF QUEEN JANE

There is little time to form any impression of Lady Jane, but what view we have of her preserves a picture of a woman of scholastic ability.

Yet I cannot pass over two English women, nor would I wish, my dear Sturmius, to pass over anything if you are thinking about friends to be borne in mind in England, than which nothing is more desirable to me. One is Jane Grey, daughter of the noble marquis of Dorset. Since she had Mary, queen of France as grandmother she was related very closely to our King Edward. She is fifteen years of age. At court I was very friendly with her, and she wrote learned letters to me: Last summer when I was visiting my friends in Yorkshire and was summoned from them by letters from John Cheke that I should come to court, I broke my journey on the way at Leicester where Jane Grey was residing with her father. I was straightway shown into her chamber: I found the noble young lady reading (By Jupiter!) in Greek, Plato's Phaedo, and with such understanding as to win my highest admiration. She so speaks and writes Greek that one would hardly credit it. She has as tutor John Aylmer, one well versed in both tongues, and most dear to me for his humanity, wisdom, habits, pure religion, and many other bonds of the truest friendship. As I left she promised to write to me in Greek provided I would send her my letters written from the Emperor's court. I am awaiting daily a Greek letter from her: when it comes I will send it on to you immediately.

English Historical Documents, V, quoting Roger Ascham. *Ascham met Jane and described her in his letter of 1550, not knowing of course what her fate should be.*

Mary

1553–1558

In a more tolerant age 'Bloody' Mary could have been a noble mother superior, with her energy and kindliness. But for her in middle age (thirty-seven) and for England in mid-century everything went wrong. As Philip of Spain's wife, she was barren. As cousin of Philip's father the all-powerful Emperor, she was encouraged to make a holocaust of heretics, thus driving Protestant England into retaliatory fanaticism. She had a deep, carrying voice and in youth was said to be of 'pleasing' aspect.

BIRTH OF MARY

Mary's mother, Catherine of Aragon, had borne King Henry a son in 1514 who lived only a few days; this was the Queen's fourth unsuccessful confinement.

When, therefore, on 18 February 1516, the Queen was delivered of a daughter who lived and flourished, it was a matter for much more exuberant rejoicings than would normally have attended such an event. Henry treated the baby rather like a new toy, showing her off proudly to courtiers and diplomats; but he made it plain that her main significance was as a token of hope, '. . . by God's grace the sons will follow.'

D. M. Loades, *The Reign of Mary Tudor* (1979)

TWO HAPPY GLIMPSES

There was dressing-up and dancing for the young Princess.

Early in the spring of 1527, Princess Mary appeared as the principal character in a masque held at Greenwich Palace, for the entertainment of the French Ambassador. She and her ladies were disguised in Icelandic dresses, and were accompanied on the stage by six lords also dressed as Icelanders, who 'danced lustily with them'. At another banquet and masque given soon afterwards before the same Ambassador, the Princess was seen to issue from a cave with seven ladies, all 'apparelled after the Roman fashion'—in 'rich cloth of gold and crimson tinsel', *bendy*—that is, the dresses were striped in a slanting direction—a costume which would certainly have amazed the ancient Romans. 'Their hair was wrapped in coils of gold, and they wore bonnets of crimson velvet on their heads with pearls and precious stones, and the Princess and her seven ladies danced a ballet with eight lords.'

EXTRACT FROM PRINCESS MARY'S EXPENSE ACCOUNTS 1543

Item paid for two pair of gilt pots.
Item given to a poor woman of Hertford called mother Amnes.
Item paid for lute Strings.
Item paid for Feathers to Stuff cushions.
Item bringing a Chair for the king's [Henry VIII] new year gift.
Item to a Boy for little Fishes.

Privy Purse Expenses of the Princess Mary, ed. Frederick Madden (1831)

A DEMONSTRATION AGAINST QUEEN MARY AND THE OLD RELIGION, DECEMBER 1553

Mary's first Parliament reversed the religious laws in favour of Protestants in October 1553, restoring the Mass and making the Book of Common Prayer illegal.

In December, some Protestants succeeded in getting into Mary's presence-chamber at Whitehall and depositing there the corpse of a dog, with its head shaved like a priest's, its ears clipped and a rope around its neck. Mary sent a message to Parliament in which she warned her subjects that such acts would compel her to adopt harsh repressive measures. But the heresy statutes, under which heretics could be burned, were not re-enacted in this Parliament. Jasper Ridley, *The Life and Times of Mary Tudor* (1974)

The story of the dog-priest recalls an earlier story of Mary aged two. Seeing a Venetian organist at court dressed in a friar's habit, she called out, 'Priest! Priest!', a piece of infant precocity that was afterwards brought up against her.

MARY AND THE WYATT REBELLION, JANUARY, FEBRUARY 1554

In the first year of her reign there was a Protestant-inspired rising, led by Sir Thomas Wyatt, in protest at her marriage to Catholic Philip. The Queen's famous speech against it at the Guildhall (she was a good speaker) was an appeal to loyalty and love. The rising's suppression ushered in the epoch of blood and burnings.

'I am your Queen, to whom at my coronation, when I was wedded to the realm and laws of the same (the spousal ring whereof I have on my finger, which never hitherto was, nor hereafter shall be, left off), you promised your allegiance and obedience to me . . . And I say to you, on the word of a Prince, I cannot tell how naturally the mother loveth the child, for I was never the mother of any; but certainly, if a Prince and Governor may as naturally and earnestly love her subjects as the mother doth love the child, then assure yourselves that I, being your lady and mistress, do as earnestly and tenderly love and favour you. And I, thus loving you, cannot but think that ye as heartily and faithfully love me; and then I doubt not but we shall give these rebels a short and speedy overthrow.'

John Foxe, *The Actes and Monuments of these latter and perillous dayes (1563)*

AN INCIDENT IN THE REBELLION, 4 FEBRUARY 1554

This day Sir Nicholas Poynes, as it is said, being an assistant at the Tower, was with the queen to know whether they shoot off at the Kentishmen [with cannon], and so beat down the houses upon their heads. 'Nay,' said the queen, 'that were pity, for many poor men and householders are like to be undone there and killed. For' saith she, 'I trust, God willing,' saith she, 'that they shall be fought with tomorrow.'

The Chronicle of Queen Jane and of Two Years of Queen Mary, ed. J. G. Nichols, Camden Society (1850). *This is a contemporary but anonymous account.*

A STATE OF ALARM AT CATHOLIC ACTIVITIES

On Ash Wednesday that Wyatt was at Charing Cross did doctor Weston [a priest] sing mass before the Queen in harness under his vestments.

Ibid.

THE EXECUTION OF LADY JANE, 12 FEBRUARY 1554

Lady Jane Grey and her young husband Lord Guildford Dudley had lived in the Tower since Edward VI's death. As soon as the Wyatt rebellion collapsed, the two young prisoners were sent to the block; he first.

His carcase thrown into a cart, and his head in a cloth, he was brought to the chapel within the Tower, where the Lady Jane, whose lodging was in Partridge's house, did see his dead carcase taken out of the cart, as well as she did see him before alive on going to his death—a sight to her no less than death. By this time was there a scaffold made upon the green over against the White Tower, for the said Lady Jane to die upon . . . The said lady, being nothing abashed . . . with a book in her hand whereon she prayed all the way till she came to the said scaffold . . . First, when she mounted the said scaffold she said to the people standing thereabout: 'Good people, I am come hither to die, and by a law I am condemned to the same. The fact, indeed, against the queen's highness was unlawful, and the consenting thereunto by me: but touching the procurement and desire thereof by me or on my behalf, I do wash my hands thereof in innocency, before God, and the face of you, good Christian people, this day' and therewith she wrung her hands, in which she had her book. And then, kneeling down, she turned to Feckenham [the new dean of St Paul's] saying, 'Shall I say this psalm?' And he said, 'Yea.' Then she said the psalm of *Miserere mei Deus*, in English, in most devout manner, to the end. Then she stood up and gave . . . Mistress Tilney her gloves and handkercher, and her book to master Bruges, the lieutenant's brother; forthwith she untied her gown. The hangman went to her to help her therewith; then she desired him to let her alone, and also with her other attire and neckercher, giving to her a fair handkercher to knit about her eyes.

Then the hangman kneeled down, and asked her forgiveness, whom she gave most willingly. Then he willed her to stand upon the straw: which doing, she saw the block. Then she said, 'I pray you dispatch me quickly.' Then she kneeled down, saying, 'Will you take it off before I lay me down?' and the hangman answered her, 'No, madame.' She tied the kercher about her eyes; then feeling for the block said, 'What shall I do? Where is it?' One of the standers-by guiding her thereto, she laid her head down upon the block, and stretched forth her body and said: 'Lord, into thy hands I commend my spirit!' And so she ended. Ibid.

MARRIAGE—A DISTRESSFUL DUTY

Before marrying Philip of Spain in 1554 Mary told Simon Renard, one of the Emperor's representatives, that she considered herself at thirty-seven too old for Philip, aged twenty-six.

Mary then explained what was really on her mind: a man of twenty-six was likely to feel amorous, and this she would not like. She promised Renard that if she did marry Philip she would fall very deeply in love with him, because the Church commanded a wife to love her husband, but she would not do so out of any carnal desire. Mary knew that however much the idea of sex disgusted her, it was her duty to submit to it in order to have a child who would exclude [Protestant] Elizabeth from the throne; and this conflict between her duty and her instincts caused her great distress.

Ridley, *Life and Times of Mary*

FALSE PREGNANCIES, 1554 AND 1558

Twice at least Mary thought that the desired child was on the way. When Cardinal Pole the Papal Legate saluted her with the words used by the Angel Gabriel to the Virgin Mary—'Hail Mary full of grace!'—Queen Mary said she felt the child leap in her womb. And in the last year of her life she made a pathetic new will:

In the name of God Amen. I Mary by the Grace of God Queen of England, Spain, France, both Sicilies, Jerusalem and Ireland, Defender of the Faith, Archduchess of Austria, Duchess of Burgundy, Milan and Brabant,

Countess of Habsburg, Flanders and Tyrol, and lawful wife of the most noble and virtuous Prince Philip by the same Grace of God King of the said Realms and Dominions of England etc. Thinking myself to be with child in lawful marriage between my said dearly loved husband and Lord, Although I be at this present (thanks be unto Almighty God) otherwise in good health, yet foreseeing the great danger which by God's ordinance remain to all women in their travail of children, have thought good . . . to declare my last will & testament . . . *Privy Purse Expenses*

MARY'S IGNORANCE OF THE WORLD

One day Mary overheard her Lord Chamberlain say to her lady-in-waiting, Frances Neville, as he tickled her under the chin, 'My pretty whore, how dost thou?' A few minutes later Mary called for Frances to fix her farthingale and as the girl knelt to do so Mary said, 'God-a-mercy, my pretty whore.' When Mary explained to the deeply embarrassed Frances that she had only used the expression because she had just heard the Lord Chamberlain use it, Frances replied:

'My Lord Chamberlain is an idle gentleman, and we respect not what he saith or doth; but Your Majesty . . . doth amaze me either in jest or earnest to be called so by you. A whore is a wicked, misliving woman.'

Henry Clifford, *The Life of Jane Dormer* (1887)

Mary admitted that she had never heard the word whore before.

A FOREIGNER'S VIEW OF MARY IN 1557

She is of low rather than of middling stature, but, although short, she has not personal defect in her limbs, nor is any part of her body deformed. She is of spare and delicate frame, quite unlike her father, who was tall and stout; nor does she resemble her mother, who, if not tall, was nevertheless bulky. Her face is well formed, as shown by her features and lineaments, and as seen by her portraits. When younger she was considered, not merely tolerably handsome, but of beauty exceeding mediocrity. At present, with the exception of some wrinkles, caused more by anxieties than by age, which makes her appear some years older, her aspect, for the rest, is very grave. Her eyes are so piercing that they inspire not only respect, but fear in those on whom she fixes them, although she is very shortsighted, being unable to read or do anything else unless she has her sight quite close to what she wishes to peruse or to see distinctly. Her voice is rough and loud, almost like a man's, so that when she speaks she is always heard a long way off. In short, she is a seemly woman, and never to be loathed for ugliness, even at her present age, without considering her degree of queen. But whatever may be the amount deducted from her physical endowments, as much more may with truth, and without flattery, be added to those of her mind, as, besides the facility and quickness of her understanding, which comprehends whatever is intelligible to others, even to those who are not of her own sex (a marvellous gift for a woman), she is skilled in five languages, not merely understanding, but speaking four of them fluently, viz., English, Latin, French, Spanish, and Italian, in which last, however, she does not venture to converse, although it is well known to her; but the replies she gives in Latin, and her very intelligent remarks made in that tongue surprise everybody. Besides woman's work, such as embroidery of every sort with the needle, she also practises music, playing especially on the clavicorde and on the lute so excellently that, when intent on it (though now she plays rarely), she surprised the best performers, both by the rapidity of her hand and by her style of playing. Such are her virtues and external accomplishments. Internally, with the exception of certain trifles, in which, to say the truth, she is like other women, being sudden and passionate, and close and miserly, rather more so than would become a bountiful and generous queen, she in other respects has no notable imperfections; whilst in certain things she is singular and without an equal, for not only is she brave and valiant, unlike other timid and spiritless women, but so courageous and resolute that neither in adversity nor peril did she ever even display or commit any act of cowardice or pusillanimity, maintaining always, on the contrary, a wonderful grandeur and dignity, knowing what became the dignity of a sovereign as well as any of the most consummate statesmen in her service; so that from her way of proceeding and from the method observed by her (and in which she still perseveres), it cannot be denied that she shows herself to have been born of truly royal lineage.

English Historical Documents, V, quoting a description written by the Venetian ambassador, Giovanni Michieli

He observed a tendency

to a very deep melancholy, much greater than that to which she is constitutionally liable, from menstruous retention and suffocation of the matrix to which, for many years, she has been often subject, so that the remedy of tears and weeping, to which from childhood she has been accustomed, and still often used by her, is not sufficient; she requires to be blooded either from the foot or elsewhere, which keeps her always pale and emaciated. Ibid.

THE UNPOPULARITY OF MARY'S BURNINGS

The Spanish marriage and enforcement of Catholic rules like the celibacy of the clergy had already made Mary unpopular.

Two days ago, to the displeasure as usual of the population here, two Londoners were burned alive, one of them having been public lecturer in scripture, a person sixty years of age, who was held in great esteem. In a few days the like will be done to four or five more; and thus from time to time to many others who are in prison for this cause [heresy] and will not recant, although such sudden severity is odious to many people.

Loades, *Mary Tudor*, quoting the ambassador Giovanni Michieli

AN OPPRESSED MIND, 1558

Calais had been lost to France and King Philip was profoundly neglectful of his ailing wife. Mary's devoted ladies, noticing her melancholy, asked if King Philip was the cause.

'Indeed (said she) that may be one cause, but that is not the greatest wound that pierceth mine oppressed mind'; but what that was she would not express to them. Albeit afterward she opened the matter more plainly to Mistress Rise and Mistress Clarentius (if it be true that they told me, which heard it of Mistress Rise herself), who then being most familiar with her, told her that they feared she took thought for King Philip's departing from her. *Holinshed*

A TROUBLED HEART

'Not only that (said she) but when I am dead and opened you will find Calais lying in my heart.' Ibid.

DEATH OF MARY FROM FEVER

On 12 November, Suriam, from Brussels, reported correctly that she was on the point of death, and early on the morning of the 17th she died, with the full consolations of her faith and in the presence of faithful members of her household. During the last few days of her life, Mary had frequently lapsed into unconsciousness, and told those about her that she had been visited by delightful dreams of children 'like angels' . . . The moment of her actual death was almost unnoticed by those present, who subsequently reported that God had rewarded her faith and virtue with the most peaceful of ends. Reporting the news a fortnight later to his half-sister in Spain, Philip wrote '. . . the Queen my wife is dead. May God have received her into His Glory. I felt a reasonable regret for her death . . .'

Loades, *Mary Tudor*

Elizabeth I

1558–1603

Gloriana, Faery Queen, name-giver to a golden age of poets, statesmen and adventurers—the last of the Tudors conferred stardom on a royal line that had tarnished itself by her father's murder of her mother. The union of the Virgin Queen with her people was virtually an explicit substitute for the marriage she never made. She had auburn hair, white skin and hazel eyes, a striking combination.

A DISTURBED CHILDHOOD

Elizabeth was informed by her governess Katherine (Kat) Ashley of the execution of her mother Anne Boleyn and the resulting reduction in her status and style. The precocious child was four years old:

'Why, Governor, how hap it yesterday Lady Princess, and today but Lady Elizabeth?'

Chapman, *The Last Tudor King*

HER VIRGINITY THREATENED AT FIFTEEN

Lord Seymour (later beheaded) was secretly married to Catherine Parr, Henry VIII's widow and visited her in the same house in Chelsea where Elizabeth also lived.

Quite often, reflected the governess Katherine Ashley, pained by the recollections drawn out of her, Seymour would barge into her room of a morning before she was ready, and sometimes before she did rise, and if she were up he would bid her good morning and ask how she did, and strike her upon the back or buttocks familiarly . . . and sometimes go through to the maidens and play with them, and so forth. And if she were in her bed he would put open the curtains and bid her good morrow and make as though he would come at her, and she would go further in the bed so that he could not come at her.

Neville Williams, *Elizabeth Queen of England* (1967), quoting Burghley State Papers

THE CORONATION, 15 JANUARY 1559

With great solemnity the splendid host walked along the blue carpeting into the Abbey to see their queen crowned after the rites of her forefathers. One bizarre touch. This splendid carpet, provided at a cost of £145 to cover the route from the upper end of the Hall to the choir door of the Abbey, attracted the souvenir hunters. 'As Her Majesty passed the cloth was cut by those who could get at it', and the Duchess of Norfolk, who walked behind the Queen, was very nearly tripped up . . .

Much significance attaches to the order of service followed: it was the last coronation conducted in the Latin service of medieval times, and it was performed before Parliament met to settle the religion of the land and approve a new prayer book . . .

The service concluded with the saying of Mass by Bishop Oglethorpe. Both epistle and gospel were read in English as well as in Latin. Preceded by three naked swords and a sword in its scabbard, Elizabeth left her throne to make her offertory on her knees before the high altar, and to kiss the paten . . . Instead of returning to her throne, as the rubrics stated, she withdrew to a traverse (or pew secluded by a curtain) in St Edward's Chapel, where she remained till the consecration of the elements and the elevation of the host were completed . . . She absented herself from the choir during the consecration as a protest against Oglethorpe's elevation of the host according to the Roman rite. At the recent Christmas day Mass at the Abbey he had been ordered to omit the elevation, and when he refused the Queen left the service. The Bishop had agreed to crown the Queen, but he was adamant about the elevation. As a result Elizabeth could not but withdraw from the service while ritualistic practices so repugnant to her were performed. Ibid.

ROBERT DUDLEY VERSUS WILLIAM CECIL

Elizabeth was twenty-five when she came to the throne. As a romantic young woman she chose the handsome Lord Dudley, later Earl of Leicester, as her favourite. Her brilliant minister Cecil deplored this influence. The struggle for the queen, body and soul, was given a new twist by the violent death of Amy Robsart, Lady Dudley, on 8 September 1560, which freed Dudley to marry Elizabeth.

Three days before De Quadra [the Spanish Ambassador] wrote his account of his interview with Cecil, Dudley's wife, Amy Robsart, was found dead, with a broken neck, at the foot of a stone staircase in her house near Oxford.

Amy's death may have been an accident, it may have been suicide. It may even have been murder, though the evidence in the case tells rather against that verdict. De Quadra thought it was murder and so did the scandalmongers. That view of the matter found expression fifty years later on the London stage:

'The surest way to chain a woman's tongue is break her neck, a politician did it.'

De Quadra observed: 'Assuredly it is a matter full of shame and infamy . . . Likely enough a revolution may come of it. The Queen may be sent to the Tower and they may make a king of the Earl of Huntingdon, who is a great heretic, calling in a party of France to help them.'

De Quadra indubitably exaggerated the gravity of the situation, but it was grave enough to make Elizabeth pause . . . So far as Elizabeth was concerned, it was the issue between Elizabeth the woman and Elizabeth the Queen.

The situation was as bad as it could be in August and September. In October it improved. Cecil told De Quadra then that Elizabeth had decided not to marry Dudley, that she had told him so herself. This may have been so . . . It is fairly clear that by mid-October Cecil was back in his accustomed position by the Queen's side. It is also clear that her fervour had cooled. She had intended to raise Dudley to the peerage as Earl of Leicester, but the story ran in November that when the papers were presented to her for her signature she took a knife and slashed them to bits.

Conyers Read, *Mr Secretary Cecil and Queen Elizabeth* (1955)

Dudley was created Earl of Leicester four years later.

DAMAGING RUMOURS, 1560

In mid-June there was old mother Dowe of Brentwood embroidering odd tales she had picked up as she wandered about south-east Essex. Dudley, said she, had given the Queen a rich petticoat. 'Thinkest thou it was a petticoat?' chimed in a crony; 'No, no, he gave her a child, I warrant thee', and mother Dowe repeated this yarn in the next village she came to. Lord Robert and Elizabeth had played together and he was the father of her child. 'Why, she hath no child yet?' 'No,' said old Annie, 'if she hath not they have put one to making.' When arrested, the local magistrates wanted her tried in secret session to prevent the scandalous stories from reaching the public. Ten years later there were still folk rash enough to spread slanders that Elizabeth had had a child by Dudley, and some lost their ears for it.

Williams, *Elizabeth*

LORD BURGHLEY ON THE ROYAL SERVICE

Serve God by serving of the Queen, for all other service is indeed bondage to the devil.

Queen Elizabeth and Her Times, II, ed. Thomas Wright (1838), quoting Lord Burghley to his son

THE QUEEN CAUGHT A FEVER OR INFLUENZA, 1562

May it please your Honor, immediately upon the Queen's arrival here, she fell acquainted with a new disease, that is common in the town, called here the new acquaintance, which passed also through her whole court, neither sparing lord, lady, nor damoisel, not so much either French or English. It is a pain in their heads that have it, and a soreness in their stomachs, with a great cough that remaineth with some longer with other shorter time, as it findeth apt bodies for the nature of the disease. The Queen kept her bed six days. There was no appearance of danger, nor many die of the disease, except some old folks.

Ibid., quoting Thomas Randolf to William Cecil, the Queen's Lord Treasurer

ELIZABETH AND HER TWO CECILS, WILLIAM AND ROBERT

Sometimes the imperious queen was harsh toward her great minister William Cecil, Lord Burghley:

'I have been strong enough to lift you out of this dirt, and I am still able to cast you down again!' Paul Johnson, *Elizabeth I* (1974)

But more often she was encouraging, especially when he had gout:

'My lord, we make use of you, not for your bad legs, but for your good head.' Ibid.

and solicitous when he was on his death-bed, sending him medicine:

'I do entreat heaven daily for your longer life, else will my people and myself stand in need of cordials too. My comfort hath been in my people's happiness and their happiness in thy discretion.' Ibid.

William's son Robert, later Lord Salisbury, in turn became the Queen's powerful minister; he was small and a hunchback, his influence and size both being celebrated in a popular ballad of 1601:

Little Cecil trips up and down,
He rules both court and crown. Ibid.

THE QUEEN AND HER HERO, SIR WALTER RALEGH

Born about 1554, Ralegh was called by the biographer Aubrey tall, handsome, bold, and 'damnably proud'. Sir Robert Naunton (1563–1635), in his account of Elizabeth and her favourites called Fragmenta Regalia, *was specific about Ralegh's relations with the Queen:*

True it is, he had gotten the Queen's ear at a trice, and she began to be taken with his elocution, and loved to hear his reasons for her demands. And the truth is, she took him for a kind of oracle, which nettled them all.

Queen Elizabeth and Her Times, quoting *Fragmenta Regalia*

SIR WALTER RALEGH'S AMBITION TO SERVE

He was bred in Oriel College in Oxford; and thence coming to court, found some hopes of the Queen's favours reflecting upon him. This made him write in a glass window, obvious to the Queen's eye,

'Fain would I climb, yet fear to fall.'

Her Majesty, either espying or being shown it, did underwrite,

'If thy heart fail thee, climb not at all.'

Thomas Fuller, The Worthies of England (1662), ed. P. A. Nuttall (1890)

ELIZABETH'S ON–OFF MARRIAGE WITH THE DUC D'ALENÇON, 1581

The heir to the French throne, in order to further his suit with Elizabeth, sent an agent to her, one Simier, 'an artful man of an agreeable conversation' whom the favourite Leicester accused of gaining her affections by 'incantations and love potions'.

The quarrel went so far between Leicester and the French Agent that the former was suspected of having employed one Tudor, a bravo, to take away the life of his enemy; and the Queen thought it necessary, by proclamation, to take Simier under her own protection. It happened, that while the Queen was rowed in her barge on the Thames, attended by Simier, and some of her courtiers, a shot was fired which wounded one of her bargemen; but Elizabeth finding, upon enquiry, that the piece had been discharged by accident, gave the person his liberty, without further punishment. So far was she from entertaining any suspicion against her people as she was many times heard to say, 'That she would lend credit to nothing against them, which parents would not believe of their own children'.

John Nichols, *The Progresses of Queen Elizabeth* (1823)

AN ELABORATE FACE-SAVER

By 1581 Elizabeth no longer wanted to marry François, duc d'Alençon, despite Simier's endeavours, but she was too wise to humiliate him.

'François the Constant' still professed himself to be longing night and day to sleep in the great bed and show what a fine companion he could be. On

November 22nd, walking in the gallery at Whitehall, with the French Ambassador and other company, Elizabeth kissed him, drew a ring from her finger, and announced that she would marry him. Whether 'the force of modest love in the midst of amorous discourses' had carried her farther than she intended; whether, as was said, she spent the night among her weeping and wailing gentlewomen in doubts and cares, it is impossible to know. Probably not, for the promise to marry was made upon conditions which the French King was expected to refuse, and even should Henry III call Elizabeth's bluff, she could raise the terms still further. The announcement in the gallery saved Alençon's face.

J. E. Neale, *Queen Elizabeth* (1934)

THE QUEEN'S PROGRESSES

Lord Leicester writes to Lord Burghley of her famous Progress towards Kenilworth, 1575:

Even by and by her Majesty is going to the forest to kill some bucks with her bow, as she hath done in the park this morning. God be thanked she is very merry, and well disposed now. But at her first coming, being a marvellous hot day, at her coming hither not one drop of good drink for her, so well was she provided for, notwithstanding her oft telling of her coming hither. But we were fain to London with bottles, to Kenilworth, to divers other places, where ale was, her own here was such strong, as there was no man able to drink it, you had been as good to have drunk Malmsey, and yet was it laid in above three days before her Majesty came. It did put her very far out of temper, and almost all the company beside too; for none of us all was able to drink beer or ale here. Since, by chance, we have found drink for her to her liking; and she is well again, but I feared greatly two or three days some sickness to have fallen, by reason of this drink.

Queen Elizabeth and Her Times

TOUCHING FOR THE QUEEN'S EVIL: KENILWORTH, 1575

One day was set aside for queenly ceremony. Five young men were knighted, including Cecil's son Thomas, and afterwards Elizabeth received nine men and women afflicted with the 'king's evil', scrofula. These she attempted to heal, drawing on the curative power believed to inhere in her as queen. The ritual was one she carried out often. First she knelt in prayer, then, having purified herself, she 'pressed the sores and ulcers' of the sufferers, 'boldly and without disgust', confident that many of them would find the ministrations beneficial.

Carolly Erickson, *The First Elizabeth* (New York, 1983)

PRIVATE ENTERTAINMENT OF ROYALTY—ITS COST

The queen visited a wealthy citizen, Michael Hickes, at Ruckholt in 1597. Beforehand, Sir H. Maynard warned him not to overstrain his estate:

Some speech the Lord Chamberlain had with me touching your house, saying that he understood it was scant of lodgings and offices: whereupon I took occasion to tell his Lordship that it was true, and I conceived that it did trouble you that you had no convenient place to entertain some of her Majesty's necessary servants. His answer was, that you were unwise to be at any such charge, but only to leave the house to the Queen: and wished that there might be presented to her Majesty from your wife, some fine waistcoat or fine ruffle, or like thing which he said would be as acceptably taken as if it were of great price.

Erickson, *The First Elizabeth.*

ACADEMIC ENTERTAINMENT OF ROYALTY—ITS RISKS

The queen cut the performance (in translation) of a Sophoclean tragedy one evening at Cambridge, having had enough.

Her absence from the play disappointed a group of young men who had worked up a masque to be added as an epilogue to the main production, so they trailed her to Hinchinbrook, her next halt, and she allowed them to perform it for her in the hall. Their masque was in fact a burlesque on the Roman Mass, and one of the characters appeared as a dog bearing the host in his mouth. The Queen was so offended at this undergraduate prank in the worst possible taste that she left in the middle of the performance.

Williams, *Elizabeth*

The Queen's Parliamentary and Literary Gifts

Elizabeth may have been kept short in childhood, having 'neither gown, nor kirtle, nor petticoat, nor no manner of linen', but her education was by no means neglected, being in the hands of famous scholars like John Cheke and Roger Ascham. Hence the memorable style of her speeches. Rejecting parliament's urgent pleas for her marriage, Elizabeth showed that she believed it was her own business.

'As for my own part, I care not for death; for all men are mortal. And though I be a woman, yet I have as good a courage, answerable to my place, as ever my father had. I am your anointed Queen. I will never be by violence constrained to do anything. I thank God I am endued with such qualities that if I were turned out of the realm in my petticoat, I were able to live in any place in Christendom.'

THE TILBURY SPEECH, 1588

Order of the day to her troops drawn up to repel the duke of Parma's army in Armada year:

'My loving people, We have been persuaded by some that are careful of our safety, to take heed how we commit ourselves to armed multitudes, for fear of treachery; but I assure you, I do not desire to live in distrust of my faithful and loving people. Let tyrants fear. I have always so behaved myself that under God, I have placed my chiefest strength and good will in the loyal hearts and good will of my subjects; and therefore I am come amongst you, as you see, at this time, not for my recreation and disport, but being resolved, in the midst and heat of the battle, to live or die amongst you all; to lay for God, my kingdom, and for my people, my honour and my blood, even in the dust. I know I have but the body of a weak and feeble woman; but I have the heart and stomach of a King, and a King of England too, and think it foul scorn that Parma or Spain or any Prince of Europe, should dare to invade the borders of my realm; to which rather than any dishonour should grow by me, I myself will take up arms, I myself will be General, Judge and Rewarder of everyone of your virtues in the field.'

At the quatercentenary celebrations scholars suggested that although Elizabeth did address the army, the traditional speech given above must be regarded with suspicion.

THE QUEEN'S POETRY

Her belief in religion was poignantly expressed in a stanza said to have been quoted in answer to a question from her Catholic half-sister Mary I, and, according to tradition, composed by Elizabeth herself:

Hoc est corpus meum
As Christ willed it and spake it
And thankfully blessed and brake it
And as the sacred word doth make it
So I believe in it and take it
My life to give therefore
In earth to live no more.

Williams, *Elizabeth*

Her disbelief in love was equally poignantly expressed in a farewell stanza, 'On Monsieur's Departure':

I grieve, yet dare not show my discontent;
  I love, and yet am forced to seem to hate;
I dote, but dare not what I meant;
  I seem stark mute, yet inwardly do prate.
I am, and am not—freeze, and yet I burn,
Since from myself my other self I turn.

And again:

When I was fair and young and favour graced me,
Of many was I sought their mistress for to be,
But I did scorn them all and answered them therefore,
Go, go, go, seek some other where,
Importune me no more.

Williams, *Elizabeth*

THE QUEEN FORGIVES

This earl of Oxford [Edward de Vere], making of his low obeisance to queen Elizabeth, happened to let a Fart at which he was so abashed that he went to travell seven years. At his returne the queen welcomed him home and sayd, 'My lord, I had forgot the Fart'.

John Aubrey, Brief Lives, ed. Anthony Powell (1949)

ELIZABETH A FOCUS OF REJOICING—ARMADA YEAR, 1588

Bishop Goodman of Gloucester was a child of five when he saw the queen in London in November:

Suddenly there came a report to us (. . . much about 5 o'clock at night, very dark) that the Queen was gone to Council, and if you will see the Queen you must come quickly. Then we all ran; when the court gates were set open . . . the Queen came out in great state. Then we cried: 'God save your Majesty! God save your Majesty!' Then the Queen turned to us and said: 'God bless you all my good people!' Then we cried again: 'God save your Majesty!' The the Queen said again unto us: 'You may well have a greater prince, but you shall never have a more loving prince.' And so, looking one upon another a while, the Queen departed. This wrought such an impression upon us, for shows and pageants are ever best seen by torch-light, that all the way we did nothing but talk of what an admirable Queen she was, and how we would venture our lives to do her service.

Bishop Godfrey Goodman, The Court of James the First, ed. J. S. Brewer (1839). *Godfrey Goodman was a well-informed and moderate partisan of James I, and was at one time a prebendary of Westminster. Born in Wales in 1583, he was in continual trouble over debts and his high Anglican faith. He became bishop of Gloucester and died in 1656.*

'ITEM LOST FROM OUR BACK'

Daybooks were kept for the records of the Queen's Wardrobe of Robes. These books listed items of clothing given away by the Queen to her ladies or others, or jewels lost while on a Progress—'lost from Her Majesty's back.' The jewels were usually gold and diamond buttons or 'agletts', the gold and enamelled tags at the end of laces to make them easier to thread through eyelet holes. She once lost a curious brooch:

Lost from a Jewel (the 25 of February her Majesty being then at Wansted) called monster, having iii Fishes hanging at the end, one of these Fishes lost at Whitehall. [1584]

More usual was an item like the following:

Lost from off her majesty's Back the Viiith of November [1567] at Hampton Court, one Aglett of gold enamelled blue, set upon a Gown of purple velvet, the ground satin.

Her ladies would receive the grandest cast-offs as presents:

Item given by her Majesty's Commandment the ixth of December . . . 1580. One French gown of purple wrought velvet lined with purple satin with a broad garde [border] of purple velvet linen with purple taffeta. To the Lady Elizabeth Drury.

The queen's humbler servants also received gifts of clothes:

Item given by the queen her majesty's commandment and delivered the last of December [1562] to mistress Smytheson, her majesty's laundress, a French kirtle of russet satin edged with velvet and lined with russet taffeta.

One of the losses was due to the Oxford players:

Item there was occupied and worn at Oxford in a play before her majesty, certain of the Apparel that was late Queen Mary's in the charge of . . . Ralph Hope, at what time there was lost one forequarter of a Gown without sleeves of purple velvet with satin ground etc.

'Lost from Her Majesties Back', ed. Janet Arnold, The Costume Society (1980; spelling modernized)

THE QUEEN IN OLD AGE

Then, for the queen, she was ever hard of access, and grew to be very covetous in her old days: so that whatsoever she undertook, she did it to the halves only, to save charge . . . that the court was very much neglected, and in effect the people were very generally weary of an old woman's government. And this no doubt might be some cause of the Queen's melancholy, and that she should break out with such words as these: 'They have yoked my neck—I can do nothing—I have not one man in whom I can repose trust: I am a miserable forlorn woman.' *Bishop Godfrey Goodman*

ELIZABETH AND SHAKESPEARE

In 1587 the queen was infuriated to find that the captains in her army had failed to pay over what was owing to the troops:

From this experience she acquired a detestation of the captains which lasted for the rest of her life: it was one reason why she so much enjoyed Shakespeare's ridicule of Sir John Falstaff in *Henry IV* Parts One and Two. His combination of cowardice, greed, fraud and sloth seemed, to her, to present an accurate picture of what most, or at any rate many captains were like. Johnson, *Elizabeth I*

SHAKESPEARE AND THE DEATH OF ELIZABETH I

The old queen seems to have experienced Hamlet's fears—'To die, to sleep; / To sleep: perchance to dream: ay, there's the rub'—for on her death-bed she was said to have told Lady Scrope, a maid of honour,

'I saw one night my body exceedingly lean and fearful in a light of fire. Are you wont to see sights in the night?'

Frederick Chamberlin, *The Sayings of Elizabeth* (1923)

And she said to Lord Howard,

'If you were in the habit of seeing such things in your bed as I do when in mine, you would not persuade me to go there.' Ibid.

Nevertheless she also told the French ambassador:

'I am tired of living, with nothing to give content or anything to give pleasure.' Ibid.

And so it is that a modern historian, John Guy, sees Shakespeare writing 'To be, or not to be' as much for the dying Elizabeth as for Hamlet:

To die, to sleep—
No more, and by a sleep to say we end
The heartache and the thousand natural shocks
That flesh is heir to—'tis a consummation
Devoutly to be wished. Shakespeare, *Hamlet*, III. i

Such sentiments were equally applicable to the dusk of Tudor England.

When the bell tolled for the age of Gloriana, on 24 March 1603, Elizabeth had already lost her will to survive, Burghley was five years dead . . . Stability had begun to breed instability through structural decay.

John Guy, 'The Tudors', in *The Oxford Illustrated History of Britain*, ed. K. O. Morgan (Oxford, 1984)

Alas, Gloriana had changed into poor Yorick!

SAYINGS OF QUEEN ELIZABETH I

On hearing of the execution of Lord Seymour, 1549: 'This day died a man with much wit and very little judgement.'

On arriving at the Traitors' Gate of the Tower by command of Mary, 1554: 'Here lands as true a subject, being prisoner, as ever landed at these stairs.'

On hearing of her accession, 1558: 'Domino factum est istud et est mirabile in oculis nostris.' ['*This is the Lord's doing and it is marvellous in our eyes.*']

On marriage: 'I should call the wedding-ring the yoke-ring.'

To the French ambassador Fénelon: 'I think that, at the worst, God has not yet ordained that England shall perish . . .'

To Leicester, on (his excuses for) failing to subdue Cork: 'Blarney!'

On her descent from Henry VIII: 'Although I may not be a lioness, I am a lion's cub, and inherit many of his qualities . . .'

On receiving her first pair of silk stockings, 1559: 'I like silk stockings well, because they are pleasant, fine and delicate, and henceforth I shall wear no more cloth stockings.'

On Mountjoy's massacres in Ireland: 'I find that I sent wolves not shepherds to govern Ireland . . .'

Her last letter to the French King, 1603: 'All the fabric of my reign, little by little, is beginning to fall.'

Chamberlin, *The Sayings of Elizabeth*

Mary Queen of Scots and Queen Elizabeth I

Elizabeth on the birth of James VI and I, son of Mary Queen of Scots 1566:

'Alack, the Queen of Scots is lighter of a bonny son, and I am but of barren stock.'

Memoirs of Sir James Melville, ed. Francis Steuart (1929)

Always fearful of a Catholic plot to put Mary on the English throne, Elizabeth imprisoned her for life in England and finally had her put to death. They never met.

IMPRESSION OF MARY, IMPRISONED AT TUTBURY, 1569

For beside that she is a goodly personage, and yet in truth not comparable to our sovereign, she hath withal an alluring grace, a pretty Scottish accent, and a searching wit, clouded with mildness. Fame might move some to relieve her, and glory joined to gain might stir others to adventure much for her sake . . . My own affection by seeing the Queen's Majesty is doubled, and thereby I guess what sight might work in others. Her hair of itself is black, and yet Knollys told me that she wears her hair in sundry colours.

Queen Elizabeth and Her Times

SOME INVENTORIES OF MARY'S POSSESSIONS

These give an insight into her character at different periods: In the Will she made before the birth of her son James in 1566, she left to the Crown of Scotland, among other things, the 'Great Harry' jewel and a grand diamond cross; to her husband Henry Darnley, a

diamond ring enamelled in red. It is this with which I was married.

Joseph Robertson, *Inventories of Mary Queen of Scots* (Edinburgh, 1863)

An inventory of 1561 listed canvas for a bed for Nichola her female Fool; and an inventory made after her death included

An old black velvet gowne, broken . . . A payre of perfumed gloves . . . A little crown of thornes in golde enameled with a white sapphir at the end.

Ibid.

THE EXECUTION PERMITTED?

Parliament petitioned Elizabeth to have Mary executed, to which Elizabeth at length answered:

If I should say unto you that I mean not to grant your petition, by my faith I should say unto you more than perhaps I mean. And if I should say unto you I mean to grant your petition, I should then tell you more than is fit for you to know. And thus I must deliver you an answer answerless.'

THE EXECUTION PERFORMED AT FOTHERINGHAY CASTLE, 8 FEBRUARY 1587

The executioner that went about to pluck off her stockings, found her little dog crept under her coat, which being put from thence, went and laid himself down betwixt her head and body, and being besmeared with her blood, was caused to be washed, as were other things whereon any blood was. The executioners were dismissed with fees, not having any thing that was hers. Her body, with the head, was conveyed into the great chamber by the Sheriff, where it was by the chirurgeons embalmed until its interment.

Nichols, *Progresses of Elizabeth*

ELIZABETH RECEIVES THE NEWS IN LONDON

Despite the fact that she herself had signed the death warrant,

her countenance changed, her words faltered, and with excessive sorrow she was in a manner astonished, in so much as she gave herself over to grief, putting herself into mourning weeds and shedding abundance of tears. Antonia Fraser, *Mary Queen of Scots* (1969), quoting Camden's *Annals*

Only the remonstrances of her minister Cecil brought these 'theatricals' to an end.

ELIZABETH TO JAMES VI ON THE EXECUTION OF MARY QUEEN OF SCOTS, 1587

My Dear Brother: I would you know (though not felt) the extreme dolour that overwhelms my mind for that *miserable accident*, which, far contrary to my meaning, hath befallen. I have now sent this kinsman of mine [Robert Carey], whom, ere now, it hath pleased you to favour, to instruct you truly of that, which is irksome for my pen to tell you.

I beseech you—that as God and many *moe* know how innocent I am in this case—so you will believe me, that if I had bid aught, I would have abided by it. I am not so base-minded, that the fear of any living creature, or prince, should make me afraid to do that [which] were just, or, when done, to deny the same. I am not of so base a lineage, nor carry so vile a mind . . . if I had meant it, I would never lay it on others' shoulders; no more will I *not* damnify myself that thought it not.

Chamberlin, *The Sayings of Elizabeth*

The most interesting question today is whether Elizabeth was justified in deceiving herself.

POPE SIXTUS V ON ELIZABETH'S EXECUTION OF MARY

'What a valiant woman. She braves the two greatest kings by land and sea . . . It is a pity that Elizabeth and I cannot marry: our children would have ruled the whole world.' Williams, *Elizabeth*

THREE CENTURIES AFTER

Dean Stanley discovers Mary's tomb, 1867:

It was discovered that Mary shared her catacomb with numbers of her descendants, including her grandson Henry, Prince of Wales, who died before his prime, her granddaughter Elizabeth of Bohemia, the Winter Queen, and her great-grandson Prince Rupert of the Rhine, among the most romantic of all the offshoots of the Stuart dynasty. Most poignant of all were the endless tiny coffins of the royal children who had died in infancy: here were found the first ten children of James II, and one James Darnley, described as his natural son, as well as the eighteen pathetic babies born dead to Queen Anne, and her sole child to survive infancy, the young Duke of Gloucester.

Finally the coffin of the Queen of Scots herself was found, against the north wall of the vault, lying below that of Arabella Stuart, that ill-fated scion of the royal house who had been the child-companion of Mary's captivity. The coffin itself was of remarkable size, and it was easy to see why it had been too heavy to carry in procession at Peterborough Cathedral at the first burial. But so securely had the royal body been wrapped in lead at the orders of the English government on the afternoon of the execution, that the casing had not given way in the slightest, even after nearly 300 years. The searcher felt profoundly moved even by the inanimate spectacle. No attempt was made to open it now. 'The presence of the fatal coffin which had received the headless corpse at Fotheringhay,' wrote Dean Stanley, 'was sufficiently affecting without endeavouring to penetrate further into its mournful contents.' The vault was thus reverently tidied, the urns rearranged, and a list made of the contents. But the Queen's own coffin was left untouched, and the little children who surrounded her were not removed.

Fraser, *Mary Queen of Scots*

THE STUARTS

Though Mary Stuart Queen of Scots never reached the throne of England, her descendants did. She herself was a great-granddaughter of Henry VII through his daughter Margaret Tudor, whom he astutely married to the Scots king. Mary's son James was a Stuart also through his father Henry Stuart, the murdered Lord Darnley. Elizabeth I, in one of her speeches against marriage, had hoped that Almighty God would send the royal line an heir 'that may be a fit governor, and per adventure more beneficial to the realm than such offspring as may come of me'. It is doubtful if this was the Almighty's thought in sending James I.

James I

(JAMES VI OF SCOTLAND)
1603–1625

The first of the Stuart Kings of England, born 1566, was said to be 'the wisest fool in Christendom'. Part of his 'wisdom' concerned the subordination of parliament to the king and of the Church to the bishops. He produced a wisecrack: 'No bishops, no king.' In total disagreement the Puritans set sail in the Mayflower *for a bishop-free America. The fool in James steered him towards unpopular male favourites. But in his native Scotland he had got the better of hostile forces, even supernatural ones. He was said to have boorish manners, a loud voice, and the body of 'an old young man'—though 'not delicate'.*

KING JAMES MENACED BY WITCHCRAFT IN SCOTLAND, *C.*1590

Francis earl of Bothwell, nephew of Mary Queen of Scots' third husband James Bothwell, was known as 'the Wizard Earl' and reputed to have ordered the destruction of the king by witchcraft. His uncle had murdered James's father, Lord Darnley, in 1567.

About this time many witches were taken in Lothian, who deposed concerning some design of the earl of Bothwell's against his Majesty's person . . . Especially a renowned midwife called Amy Simson affirmed, That she, in company with nine other witches, being convened in the night beside Prestonpans, the devil their master being present.

A body of wax, shapen and made by the said Amy Simson, wrapped within a linen cloth, was first delivered to the devil; who, after he had pronounced his verdict, delivered the said picture to Amy Simson, and she to her next neighbour, and so everyone round about, saying, 'This is King James VI, ordered to be consumed at the influence of a nobleman, Francis earl Bothwell.' Afterward again at their meeting by night in the kirk of North-Berwick, where the devil, clad in a black gown, with a black hat upon his head, preached unto a great number of them out of the pulpit, having like light candles round about him.

The effect of his language was to know what hurt they had done ... what success the melting of the picture had, and such other vain things. And because an old silly poor ploughman, called Gray Meilt, chanced to say, That nothing ailed the King yet, God be thanked; the Devil gave him a great blow. Thus divers among them entered in reasoning, marvelling that all this devilry could do no harm to the King, as it had done to divers others. The devil answered, 'Il est un homme de Dieu, Certainly he is a man of God,' and does no wrong wittingly, but he is inclined to all godliness, justice and virtue; therefore God hath preserved him in the midst of many dangers. Now after that the devil had ended his admonitions, he came down out of the pulpit, and caused all the company come kiss his arse: which they said was cold like ice; his body hard like iron.

Memoirs of Sir James Melville

Gunpowder, Treason, and Plot, 1605

A BLOW-BY-BLOW CONTEMPORARY ACCOUNT

On the 5th November we began our Parliament, when the King should have come in person, but he refrained through a practice but that morning discovered. The plot was to have blown up the King at such a time as he should have been set in his royal throne, accompanied with his children, nobility, and commoners, and assisted with all the bishops, judges, and doctors; at one instant and blast to have ruined the whole state and kingdom of England; and for the effecting of this there was placed under the parliament-house, where the King should sit, some thirty barrels of powder with good store of wood, faggots, pieces, and bars of iron. How this came forth is sundry ways delivered ... But howsoever certain it is that upon a search lately made on Monday night in the vault under the parliament chamber before spoken of, one Johnson was found with one of those close lanterns, preparing the train against the next morrow, who being brought into the galleries of the court, and there demanded if he were not sorry for his so foul and heinous a treason, answered, that he was sorry for nothing, but that the act was not performed . . . Some say that he was servant to one Thomas Percy; others that he is a Jesuit, and had a shirt of hair next to his skin. But he was carried to the Tower on Tuesday following, whither the lords were to examine him . . .

When Johnson was brought to the King's presence, the King asked him how he could conspire so hideous a treason against his children, and so many innocent souls, which never offended him? He answered that it was true; but a dangerous disease required a desperate remedy. He told some Scots that it was his intent to have blown them back again into Scotland . . .

Since Johnson's being in the Tower he beginneth to speak English; and yet he was never upon the rack [he soon was], but only by his arms upright . . . Johnson's name is now turned into Guy Vaux, alias Faux.

Bishop Godfrey Goodman

THE KING'S REACTION TO THE PLOT

James persuaded himself that it was he alone who had discovered the plot. (In fact it was revealed by an informer to the government.) The king was therefore in a state of elation and self-congratulation, but at the same time terror-stricken. The Venetian ambassador noted:

The King . . . does not appear nor does he take his meals in public as usual. He lives in the innermost rooms with only Scotsmen about him.

D. Harris Willson, *King James VI and I* (1956), quoting the ambassador

James's infant daughter, Princess Mary, died in 1607, and was buried in Henry VII's Chapel, Westminster Abbey.

The following year Princess Mary died aged two and a half, described by her father as 'a most beautiful infant'. As she lay dying she is reputed to have repeated, 'I go, I go—Away I go,' and again, 'I go, I go.' She is buried next to her sister and commemorated by a stiff little figure in a black dress lying on one elbow, a lion at her feet, and looking far older and worldly wise than her two and a half years. For some time afterwards James was wont to say, with heavy and tortuously theological wit, that he 'would not pray *to* the Virgin Mary . . . but *for* the Virgin Mary'.

Olivia Bland, *The Royal Way of Death* (1986)

JAMES ON THE KING'S DIVINE RIGHT

The state of Monarchy is the supremest thing upon earth: for Kings are not only God's lieutenants, and sit upon God's throne, but even by God himself they are called Gods.

Roy Strong, *Van Dyck: Charles I on Horseback* (1972), quoting the king's speech to parliament on 21 March 1609

JAMES AND THE HOUSE OF COMMONS

This house of parliament did not fit in with James's ideas on royal rule:

I am surprised that my ancestors should ever have allowed such an institution to come into existence.

The Royal Favourites

After the decline of the favourite Somerset, the queen and Prince Charles arranged a strategem by which George Villiers, their own candidate, should be knighted.

Upon St George's Day, the Queen and the Prince being in the bedchamber with the King, it was so contrived that Buckingham [George Villiers] should be in some nearness to be called in upon any occasion; and when the Queen saw her own time, he was called in. Then did the Queen speak to the Prince to draw out the sword and to give it to her; and immediately with sword drawn she kneeled to the King and humbly beseeched his Majesty to do her that special favour as to knight this noble gentleman whose name was George, for the honour of St George, whose feast he now kept. The King at first seemed to be afeard that the Queen should come to him with a naked sword, but then he did it very joyfully; and it might very well be that it was his own contriving, for he did much please himself with such inventions. *Bishop Godfrey Goodman*

James was with his lord treasurer when his 'gentlemen waiters' announced that dinner was served. The king took no notice.

The gentlemen came the second time and told his majesty that the time was far spent, and that dinner was upon the table: still the King had business with my lord and came not. The gentlemen came again and told his majesty that his meat was grown cold, and they would carry it back again unless he came as soon as they were gone back. My lord told the King that he did wish they would eat up all the meat and leave him the reversion, for so they had done with his estates; they had culled out all the best things and left him to live upon projects and fee-farms. The King then went to dinner and caused his carver to cut him out a court-dish, that is, something of every day, which he sent him as part of his reversion; so much was the King taken with that conceit. Ibid.

James and his favourite Buckingham exchanged many loving letters. One from Buckingham began 'Dear Dad and Gossip' and ended 'Your most humble slave and dog, Steenie'. Postscript:

Even as I was sending this, Kate [Buckingham's wife, Katherine Manners] and I received another present from you, for which we give you our humble thanks. Your presents are so great, we cannot eat them so fast as they come.

Bishop Godfrey Goodman

Buckingham often signed 'Steenie'. James began and ended a letter: 'Sweet Heart . . . God bless thee and me! James R.'

James wrote to his son Charles and Buckingham when they were abroad together:

My Sweet Boys: . . . I have no more to say, but that I wear Steenie's picture in a blue ribbon under my waistcoat, next my heart. And so God bless you both, & send you a joyful & happy return . . . your dear dad & true friend.

Ibid.

James wrote again:

My Sweet Babie: . . . I pray God that, after a happy conclusion there, ye may both make a comfortable & happy return in the arms of your dear dad. James R.

Ibid.

TWO ACCOUNTS OF THE ROYAL FAVOURITES

And these . . . his favourites or minions . . . like burning-glasses, were daily interposed between him and the subject, multiplying the heat of oppressions in the generall opinion, though in his own he thought they screened them from reflecting on the Crowne: Through the fallacy of which maxime his son came to be ruined . . . Now, as no other reason appeared in favour of their choyce but handsomenesse, so the love the King shewed was as amorously conveyed as if he had mistaken their sex, and thought them ladies; which I have seene Sommerset and Buckingham labour to resemble, in the effiminatenesse of their dressings; though in W[horeson] lookes and wanton gestures, they exceeded any part of woman kind my conversation did ever cope withall.

Robert Ashton, *James I by His Contemporaries* (1969), quoting Francis Osborne, *Traditional Memoyres on the Raigne of James I* (first printed 1811)

At Royston and Newmarket
He'll hunt till he be lean.
But he hath merry boys
That with masks and toys
Can make him fat again.

Buckingham became one of these 'merry boys', and a major reason why he kept his hold on the King's affections was that he made him laugh. James may seem a pathetic creature in twentieth-century eyes, but as far as his subjects were concerned he was the King, and they treated him with respect not untinged with awe. How refreshing, therefore, it must have been to James when he found in his young favourite someone who instinctively knew how to breach the wall of majesty when the occasion was opportune and to treat the sovereign like a human being. From time to time he adopted a bantering, mock-chiding tone that is best described as cheeky. When he was in Spain, for instance, in 1623, he reprimanded the King for being too mean in supplying his son with jewels, and gave his 'poor and saucy opinion what will be fittest more to send'. Then, in a postscript, he listed the animals which he was despatching to James as a gift and promised to 'lay wait for all the rare colour birds that can be heard of'. But the gift was not unconditional, for 'if you do not send your baby [Charles] jewels enough I'll stop all other presents. Therefore look to it!' There was always a risk, of course, in using such peremptory language to the King, but Buckingham knew his man. James was delighted and sent him hearty thanks for 'thy kind drolling letter'.

R. Lockyer, 'An English Valido? Buckingham and James I', in *For Veronica Wedgwood: These Studies in Seventeenth Century History*, eds. R. Ollard and P. Tudor-Craig (1986)

TOUCHING FOR THE KING'S EVIL—A PLOY

He was a King in Understanding, and was content to have his Subjects ignorant in many things: as in curing the *King's-Evil*, which he knew a Device, to aggrandise the Virtue of Kings, when Miracles were in fashion; but he let the World believe it, though he smil'd at it in his own Reason, finding the Strength of the Imagination a more powerful Agent in the Cure, than the Plaisters his Chirugions prescrib'd for the Sore.

Ashton, *James I*, quoting Arthur Wilson, *Life and Reign of James I* (1719)

JAMES'S MELANCHOLY

An attempt to cure the king's melancholy made by the favourite Buckingham and his mother badly miscarried:

But our king receiving so many Delays and Dissatisfactions from *Spain* and *Rome*, they begot him so much Trouble and Vexation, that . . . press'd upon his Natural Temper some Fits of Melancholy, which those about him with facetious Mirth, would strive to mitigate; And having exhausted their Inventions or not making use of such as were more pregnant, the Marquis and his Mother (instead of Mirth) fell upon Profaneness, thinking with that to please him . . . For they caused *Mrs Aspernham*, a young Gentlewoman of the Kindred to dress a Pig like a Child, and the old Countess, like a Midwife, brought it into the King in a rich Mantle. *Turpin* that married one of the Kindred . . . was dress'd like a Bishop in his Sattin Gown, Lawn Sleeves, and other Pontifical Ornaments, who (with the *Common Prayer* Book) began the Words of Baptism, one attending with a silver Basin of Water for the Service; the King hearing the Ceremonies of Baptism read, and the squeaking Noise of that Brute he most abhorred, turned himself to see what Pageant it was; and finding *Turpin's* Face, which he well knew, dress'd like a Bishop; and the Marquis, whose Face he most of all loved, stand as a Godfather; he cried out, *Away for shame, what Blasphemy is this?* and turning away with a Frown, he gave them Pause to think, that such ungodly Mirth would rather increase than cure his Melancholy.

Ibid.

'A Grace by Ben Johnson, Extempore, before King James'

. . . And God blesse every living thing
That lives, and breath's, and loves the King.
God bless the Councell of Estate,
And Buckingham, the fortunate.
God blesse them all, and keepe them safe,
And God blesse me, and God blesse Raph.

The king was mighty enquisitive to know who this Raph was. Ben told him 'twas the drawer at the Swanne tavernne, by Charing-cross, who drew him good Canaric. For this drollerie his majestie gave him an hundred poundes.

John Aubrey, Brief Lives

KING JAMES'S 'WIT'

At a consultation at Whitehall, after queen Elizabeth's death how matters were to be ordered and what ought to be donne, Sir Walter Ralegh declared his opinion, 'twas the wisest way for them a group or cabal of which Ralegh was a member to keep the government in their own hands . . . It seems there were some of this caball who kept not this so secret but that it came to king James's ears; who, where the English noblesse mett and received him, being told upon their presentment to his majesty their names, when Sir Walter Ralegh's name was told 'Ralegh' said the king 'On my soule, mon, I have heard *rawly* of thee.'

Ibid.

SAYINGS OF JAMES I

On smoking: 'A custom loathsome to the eye, hateful to the nose, harmful to the brain, dangerous to the lungs, and in the black, stinking fume thereof, nearest resembling the horrible Stygian smoke of the pit that is bottomless.'

On the King's rule: 'I will govern according to the common weal, but not according to the common will.'

On the Dean of St Paul's poetry: 'Dr Donne's verses are like the peace of God; they pass all understanding.'

On the Bodleian Library, Oxford: 'Were I not a king, I would be a University man. And if it were so that I must be a prisoner, if I might have my wish, I would have no other prison than this library, and be chained together with these good authors.'

To his son Prince Henry in the Basilikon Doron, *1599*: 'You are a little God to sit on his throne and rule over other men.'

DEATH-BED OF JAMES I, 1625

Early in the year James suffered from fever with convulsions, while staying at his palace of Theobalds near Hatfield in Hertfordshire. 'I shall never see London again', he said sadly. Buckingham and his mother sent in a new medicine to cure him; it made him worse. When James found Buckingham's mother kneeling by his bed and crying out that she had been accused of poisoning him,

'Poisoning me?' said he; and with that, turning himself, swooned.

He died on 27 March surrounded by complacent archbishops, bishops, and chaplains;

without pangs or convulsions at all, Solomon slept.

Willson, *King James*, quoting contemporary commentators

In fact 'Solomon' had a severe stroke two days before the end and died in the distressing throes of dysentery.

Charles I

1625–1649

Fortunately Charles did not inherit his father's personal tastes; unfortunately he absorbed James's absolutism. It was tragic that a man of so much family affection and artistic feeling should have believed it necessary to vindicate illusory sovereign rights by steps that led to civil war. Parliament sullied its victory: for the Puritans' reign of terror—striking off the heads of exquisite medieval statues—foreshadowed the fate of Charles himself. After a feeble start he grew into an athletic young man. The lifelong impediment in his speech may have accounted for his shyness, and this in turn for the concentration on his own perceptions.

QUEEN HENRIETTA MARIA

When Henrietta Maria came to England as Charles I's bride, the French clergy bade her become a missionary of the Catholic faith:

The Queen's ideas of missionary activity were somewhat elementary, and consisted of breaking into an Anglican service in the royal household with a pack of beagles and interrupting the preacher with hunting noises.

H. R. Trevor-Roper, *Archbishop Laud* (1940), from Salvetti's *Newsletters*

A PURITAN WOMAN'S VIEW

The court of Charles and his wife was compared to that of James I. The tone was distinctly condemnatory, but is the product of prejudice.

The face of the Court was much chang'd in the change of the King, for King Charles was temperate and chaste and serious; so that the fooles and bawds, mimicks and Catamites of the former Court grew out of fashion, and the nobillity and courtiers, who did not quite abandon their debosheries, had yet that reverence to the King to retire into corners to practise them. Men of learning and ingenuity in all arts were in esteeme, and receiv'd encouragement from the king, who was a most excellent judge and a greate lover of paintings, carvings, gravings, and many other ingenuities less offensive than the bawdry and prophane abusive witt which was the only exercise of the other Court. But as in the primitive times it is observed that the best emperors were some of them stirr'd up by Sathan to be bitterest persecutors of the Church, so this King was a worse encroacher upon the civill and spirituall liberties of his people by farre than his father.

He married a papist, a French lady of a haughty spiritt, and a greate witt and beauty, to whom he became a most uxorious husband. By this means the court was replenish't with papists ... the Puritans more than ever discountenanc'd and persecuted ... The example of the French king was propounded to him, and he thought himselfe no Monarch so long as his will was confin'd to the bounds of any law. But knowing that the people of England were not pliable to an arbitrary rule, he plotted to subdue them to his yoke by forreigne force ...

But above all these [Archbishop Laud and the Earl of Strafford] the King had another instigator of his owne violent purpose, more powerfull than all the rest, and that was the Queene ... who, growne out of her childhood, began to turne her mind from those vaine extravagancies she liv'd in at first to that which did lesse become her, and was more fatall to the kingdom, which never is in any place happie where the hands that are made only for distaffes affect the management of Sceptres. If anyone object the fresh example of Queen Elizabeth, let them remember the felicity of her reigne was the effect of her submission to her masculine and wise Councellors; but wherever male princes are so effeminate to suffer women of foreigne birth and different religions to intermeddle with the affairs of State, it is alwayes found to produce sad desolations; and it hath been observed that a French Queene never brought any happinesse to England.

Mrs Lucy Hutchinson, Memoirs of the Life of Colonel Hutchinson, ed. James Sutherland (Oxford, 1973). *Lucy Hutchinson, born 1620, was said by her husband to be 'above the pitch of ordinary women'. She wrote his life after he died in 1664, having defended Nottingham Castle for the parliament. He was one of the regicides.*

CHARLES MAKES AN ENEMY

King Charles I had complaint against [Henry Martin] for his wenching. It happened that Henry was in Hyde-parke one time when his majestie was there, goeing to see a race. The king espied him, and sayd aloud, 'Let that ugly rascall be gonne out of the parke, that whore-master, or else I will not see the sport.' So Henry went away patiently, *sed manebat alta mente repostum* [but the sarcasm remained deep within him]. That sarcasme raysed the whole countie of Berks against him [the king] ... shortly after he was chosen knight of the shire for that county ... and proved a deadly enemy to the king.

John Aubrey, Brief Lives

A ROYAL BREACH OF PRIVILEGE

In the prelude to the civil war, as parliament claimed more power for itself, Charles attempted in person to arrest five Members of the House of Commons, 4 January 1642.

The next day in the afternoon, the king, attended only by his own guard, and some few gentlemen, who put themselves into their company in the way, came to the House of Commons; and commanding all his attendants to wait at the door, and to give offence to no man; himself, with his nephew, the Prince Elector, went into the House, to the great amazement of all: and the Speaker leaving the chair, the king went into it; and told the House, 'he was sorry for that occasion of coming to them; that yesterday he had sent his sergeant at arms to apprehend some, that, by his command, were accused of high treason; whereunto he expected obedience, but instead thereof he had received a message. He declared to them, that no king of England had been ever, or should be, more careful to maintain their privileges, than he would be; but that in cases of treason no man had privilege; and therefore he came to see if any of those persons, whom he had accused, were there; for he was resolved to have them, wheresoever he should find them: and looking then about, and asking the Speaker whether they were in the House, and he making no answer, he said, he perceived the *birds were all flown*, but expected they should be sent to him, as soon as they returned thither and assured them in the word of a king, that he never intended any force, but would proceed against them in a fair and legal way' . . . They took very little notice of the [king's] accusing the members; but the king's coming to the House, which had been never known before . . . was looked upon as the highest breach of privilege that could possibly be imagined.

Selections from Clarendon's History of the Rebellion, ed. G. Huehns (Oxford, 1978)

CHARLES ON HIS DEVOTION TO TRUTH AND RIGHT, 10 JANUARY 1642

I will rather choose to wear a crown of thorns with my Saviour, than to exchange that of gold, which is due to me, for one of lead.

John Ganden, Eikon Basilike: The Portraiture of his Sacred Majesty in his Solitudes and Sufferings, ed. P. A. Knachel (Ithaca, NY, 1966)

Thomas Wentworth, First Earl of Strafford, was impeached by the Long Parliament for corrupting the king and setting him against them. When Strafford, a prisoner in the Tower, heard that his friend the king had agreed, albeit under duress, to the bill of attainder, he quoted bitterly:

Put not your trust in princes nor in the sons of men, for in them there is no salvation.

After Strafford's execution on Tower Hill, Charles was to agree, sadly, that his friend had spoken the truth.

CHARLES AT HIS TRIAL

The king refused either to plead or even to acknowledge the court.

One thing was remark'd in him by many of the Court [including Colonel Hutchinson who was one of the Commissioners and signed the death warrant], that when the bloud spilt in many of the battles where he was in his owne person, and had caus'd it to be shed by his owne command, was lay'd to his charge, he heard it with disdainfull smiles, and lookes and gestures which rather expresst sorrow that all the opposite party to him had not bene cutt off than that any were; and he stuck not to declare in words that no man's blood spilt in this quarrell troubled him but only one, meaning the Earle of Strafford. *Lucy Hutchinson*

The Execution of Charles I

On 29 January 1649, the day before his execution, Charles said farewell to two of his children, Elizabeth aged thirteen (later queen of Bohemia) and Henry duke of Gloucester aged eight. His eldest sons and heirs, Charles and James, had fled abroad.

Both children immediately fell on their knees, Elizabeth crying bitterly. The King raised them to their feet, and drawing them aside—for they were not alone—spoke first to his daughter. He had much of importance to say to her that he could say to no one else. He was anxious, not without cause, about the relations between his two eldest sons, between whom there was much adolescent jealousy. She was to tell 'her brother James, whenever she should see him, that it was his father's last desire, that he should no more look upon Charles as his eldest brother only, but be obedient unto him as his sovereign.' The princess was crying so much that he could not be sure that she was taking it in. 'Sweet heart, you will forget this,' he said. She shook her head. 'I shall never forget it whilst I live,' and she promised to write it down.

It is thus from the account that she set down that night that we know what passed between them . . . (Here followed his instructions to her not to 'torment' herself for him for his death would be glorious, being for the 'laws and liberties of this land, and for maintaining the true Protestant Religion'. He recommended to her certain books which would ground her against Popery. He said he had forgiven his enemies and hoped God and his family would forgive them also. He sent a message of faithful love to his wife.)

After reading over what she had written, the Princess noticed an omission and added a postscript:

Further, he commanded us all to forgive these people, but never to trust them, for they had been most false to him and to those that gave them power, and he feared also to their own souls; and desired me not to grieve for him, for he should die a martyr; and that he doubted not but the Lord would settle his throne upon his son; and that we should be all happier than we could have expected to have been if he had lived.

He said less to the Duke of Gloucester, and in the simplest possible language for it was important the child should understand. The unity of the family and legal descent of the Crown might depend on this:

'Mark, child, what I say,' said the King, taking his son on his knee, 'they will cut off my head, and perhaps make thee a king: but mark what I say, you must not be a king so long as your brothers Charles and James do live; for they will cut off your brothers' heads when they can catch them, and cut off thy head too, at last: and therefore I charge you, do not be made a king by them.'

The child who, all the time his father spoke, had 'looked very steadfastly upon him', now said with great firmness: 'I will be torn in pieces first.'

This answer greatly pleased the King. He had little more to say and every reason, both for his own and the children's sake, not to prolong the interview. He gave them the casket and most of his remaining jewels to take away, keeping back only a few personal things and the George [Order of St George], cut in a single onyx, that he intended to wear on the scaffold. Then he kissed and blessed them both and sent them away.

Soldiers on guard, and spectators outside the gates of St James's who saw the children leave, predicted that the Princess would die of grief, and within a day or two the newspapers were reporting that she had actually done so.

C. V. Wedgwood, *The Trial of Charles I* (1964)

The morning of the execution, 30 January 1649:

Between five and six o'clock he awoke, drew back the bed curtain and called to Herbert [Thomas Herbert his attendant] who had fallen into an uneasy and restless sleep. 'I will get up,' said the King, 'I have a great work to do this day . . . Herbert,' he said, 'this is my second marriage day; I would be as trim to-day as may be, for before to-night I hope to be espoused to my blessed Jesus.'

The bitter January frost was still unbroken and the King, anxious that he might not feel the cold, put on two shirts so that he would not shiver when he came to prepare for the block and so give an impression of fear. 'I fear not death. Death is not terrible to me. I bless my God I am prepared.'

Wedgwood, *The Trial of Charles I*

The king was taken to Whitehall, where he waited for hours instead of minutes as expected, there having been some hold-up apparently over the executioners, who finally appeared on the black-draped scaffold wearing disguises.

There had been difficulties too among the officers in charge of the proceedings some hours before the King reached Whitehall. The three to whom the death-warrant had been directed, Hacker, Hunks and Phayre, had to sign the order for the execution itself . . . Voices are raised in argument: Hunks—the mistakenly named Hercules Hunks—has lost his nerve. No, he will not sign. Cromwell shouts at him; he is 'a froard, peevish fellow'. Colonel Axtell appears in the doorway, and speaks: 'Colonel Hunks, I am ashamed of you; the ship is now coming into harbour and will you strike sail before we come to anchor?' Hunks did not sign. Ibid.

The three names—Hunks, Hacker and Axtell—seem to be too good to be true, unless from Restoration drama. One of the sentences that Charles spoke on the scaffold referred to Strafford, whose execution he himself had ordered:

'An unjust sentence that I suffered to take effect, is punished now by an unjust sentence on me.'

Another sentence referred to the axe. He was speaking about the affairs of the Church and the king:

He broke off short, for one of the officers on the scaffold happened by accident to touch the axe. 'Hurt not the axe,' said the King, 'that may hurt me.'

After his words to the small group around were finished:

The King stood for a moment raising his hands and eyes to Heaven and praying in silence. Then slipped off his cloak and lay down with his neck on the block. The executioner bent down to make sure that his hair was not in the way, and Charles, thinking that he was preparing to strike, said, 'Stay for the sign.'

'I will, an' it please Your Majesty,' said the executioner. A fearful silence had now fallen on the little knot of people on the scaffold, on the surrounding troops, and on the crowd. Within a few seconds the King stretched out his hands and the executioner on the instant and at one blow severed his head from his body.

A boy of seventeen, standing a long way off in the throng, saw the axe fall. He would remember as long as he lived the sound that broke from the crowd, 'such a groan as I never heard before, and desire I may never hear again.'

Ibid., with quotes from the Diaries and Letters of Philip Henry

CHARLES'S FIRST RESTING PLACE

The body of the king had been brought to Windsor by night and deposited on a long table in the Deanery, where it remained until it was interred in St George's Chapel. I have often looked at the table which Dean Albert Bailey had discovered being used in the servant's quarters. I believe it had been used to brush the clothes of former Deans! What a strange 'catafalque' for a deceased sovereign!

Princess Marie-Louise, *My Memories of Six Reigns* (1956)

THE OPENING OF CHARLES'S TOMB

In 1813, during the construction of George III's tomb at Windsor, workmen accidentally broke through the wall of the Henry VIII vault. The Prince Regent was informed and agreed to allow the vault—opposite the eleventh Knight's stall on the Sovereign's side—to be opened so that 'a doubtful point in [Clarendon's] History might be cleared up' by a comparison of its contents with contemporary accounts of Charles's burial. In the presence of the Prince Regent and Sir Henry Halford, the King's Physician, the coffins were examined; the two more ancient were virtually left alone, but Halford's description of the third, and its contents, is of considerable interest.

On removing the pall, a plain leaden coffin, with no appearance of ever having been inclosed in wood, and bearing an inscription, KING CHARLES, 1648, in large legible characters, on a scroll of lead encircling it, immediately presented itself to the view. A square opening was then made in the upper part of the lid, of such dimensions as to admit a clear insight into its contents. These were, an internal wooden coffin, very much decayed, and the Body, carefully wrapped up in cere-cloth, into the folds of which a quantity of unctuous or greasy matter, mixed with resin, as it seemed, had been melted, so as to exclude, as effectually as possible, the external air. The coffin was completely full; and, from the tenacity of the cere-cloth, great difficulty was experienced in detaching it successfully from the parts which it enveloped. Wherever the unctuous matter had insinuated itself, the separation of the cere-cloth was easy; and when it came off, a correct impression of the features to which it had been applied was observed in the unctuous substance. At length, the whole face was disengaged from its covering. The complexion of the skin of it was dark and discoloured. The forehead and temples had lost little or nothing of their muscular substance; the cartilage of the nose was gone; but the left eye, in the first moment of exposure, was open and full, though it vanished almost immediately: and the pointed beard, so characteristic of the period of the reign of King Charles, was perfect. The shape of the face was a long oval; many of its teeth remained; and the left ear, in consequence of the interposition of the unctuous matter between it and the cere-cloth, was found entire.

It was difficult, at this moment, to withhold a declaration, that, notwithstanding its disfigurement, the countenance did bear a strong resemblance to the coins, the busts, and especially to the pictures of King Charles I by Vandyke, by which it had been made familiar to us. It is true, that the minds of the Spectators of this interesting sight were well prepared to receive this impression; but it is also certain, that such a facility of belief had been occasioned by the simplicity and truth of Mr Herbert's Narrative, every part of which had been confirmed by the investigation, so far as it had advanced: and it will not be denied that the shape of the face, the forehead, an eye, and the beard, are the most important features by which resemblance is determined.

When the head had been entirely disengaged from the attachments which confined it, it was found to be loose, and, without any difficulty was taken up and held to view. It was quite wet, [Halford believed this liquid to be blood] and gave a greenish tinge to paper and to linen, which touched it. The back part of the scalp was entirely perfect, and had a remarkably fresh appearance; the pores of the skin being more distinct, as they usually are when soaked in moisture; and the tendons and ligaments of the neck were of considerable substance and firmness. The hair was thick at the back part of the head, and, in appearance, nearly black. A portion of it, which has since been cleaned and dried, is of a beautiful dark brown colour. That of the beard was a redder brown. On the back part of the head, it was [no?] more than an inch in length, and had probably been cut so short for the convenience of the executioner, or perhaps by the piety of friends soon after death, in order to furnish memorials of the unhappy King.

On holding up the head, to examine the place of separation from the body, the muscles of the neck had evidently retracted themselves considerably; and the fourth cervical vertebra was found to be cut through its substance, transversely, leaving the divided portions perfectly smooth and even, an appearance which could have been produced only by a heavy blow, inflicted with a very sharp instrument, and which furnished the last proof wanting to identify King Charles the First.

In the matter of the hair Halford's account and statements by contemporary observers conflict: at the time of his trial and execution Charles's hair is described as being grey and, so far as we know, neither Charles nor the embalmers removed any back hair. It might indeed have simply disappeared as 'memorials'. As to the face itself, however, Halford is more definite and allowing, as he did, for actual discoloration, disfigurement and a readiness to project a preconceived image, the century and a half since the Restoration had otherwise effected no radical change in the idealised 'portraiture' of Van Dyck and *Eikon Basilike*. Research in both artistic technique and historiography has since suggested further revision, but the popular tradition survives: those Regicides who discovered, with their faces towards Whitehall, that no crown was cut off with Charles's head, would today have reason, in the longevity of Charles the Martyr, to doubt their other belief about stone dead having no fellow.

A. A. Mitchell, 'Charles the First in Death', *History Today* (1966)

Discussion of the opening of Charles I's coffin was renewed in the Sunday Telegraph *by Vivian Nolan on 6 September 1987, and the threads were finally tied up by an archivist on the 20th.*

We in the school of St Peter in York have a special interest in the recent correspondence regarding the burial of King Charles, as it was one of our old boys, Sir Thomas Herbert, who made the arrangements for the funeral. The body was embalmed, it is said, and the head stitched in place by Thomas Fairfax's surgeon. The body was then placed in a vault with the bodies of Henry VIII and Jane Seymour.

As earlier correspondents have written, the vault was opened in 1813 and various relics removed. Mr C. H. Keeling is correct in assuming that these were placed in a casket which is still in existence. It is, in fact, where it should be—on the coffin of King Charles. It was returned by Sir Henry St John Halford to the Prince of Wales who decided, with the permission of Queen Victoria, to replace the remains, and this was done privately on December 13, 1888.

The wooden casket, encased in oak and lead, with an engraved inscription giving details of the contents, was lowered through a small opening made in the floor of the chapel so that it rested on the velvet pall which still covered the coffin.

In addition to the vertebra removed in 1813 and turned into a salt cellar, a workman removed a finger bone from the body of Henry VIII and used it to make a knife handle! *Sic transit gloria.*

J. V. Mitchell,
Honorary Archivist, St Peter's School, York

Sunday Telegraph, 20 September 1987

Charles II

1660–1685

With his saturnine features, set off by a black periwig, and powerful athletic frame, Charles was never confined to the lightweight image of a 'merry monarch'. But his happy escape from capture and joyful Restoration, not to mention his wit and mistresses, more than accounted for the sobriquet. His keen intellect delighted in science and he was an unstoppable talker.

A MERRY YOUTH

Three stories indicate the Prince's lightness of humour. First when he was eleven, he refused to take some medicine which he was given. His mother, at Newcastle's request, wrote to reprimand him. His reply was to advise Newcastle that he himself would improve his health by not relying on too much physic. Secondly, when he and Newcastle played at butts together and his governor had the better of him, he remarked 'What, my lord, have you invited me to play the rook [sharper] with me?' Lastly, when Charles was in Oxford during the civil war the Earl of Berkshire was once incited to 'hit him on his head with his staff' because he observed the Prince to be laughing during service time in church and exchanging pleasantries with the ladies seated near him. Maurice Ashley, *Charles II* (1971)

The Escape

After the royalists' defeat by Cromwell's soldiers at the battle of Worcester on 3 September 1651, Charles hid in an oak tree with Colonel Carlis (or Carelesse), on his melodramatic bid for safety. Charles himself told the story of his escape twice over to the diarist Samuel Pepys: once on board the Royal Charles *on 23 May 1660 when sailing triumphantly home to England; again at Newmarket in October 1680. Many others told the great saga, including a hunted priest, Father Huddleston, who helped Charles. First, Pepys in 1660:*

Upon the Quarter-deck he fell in discourse of his escape from Worcester. Where it made me ready to weep to hear the stories that he told of his difficulties that he had passed through. As his travelling four days and three nights on foot, every step up to the knees in dirt, with nothing but a green coat and a pair of country breeches on and a pair of country shoes, that made him so sore all over his feet that he could scarce stir.

Charles II's Escape from Worcester: A Collection of Narratives Assembled by Samuel Pepys, ed. W. Matthews (1967)

Fr. Huddleston added a point about his shoes:

His shoes were old, all slasht for the ease of h[is] feet and full of gravell, with little rowlls of pa[per] between his toes; which he said he was advised to, to keep them from galling. Ibid.

Pepys continues:

Yet he was forced to run away from a miller and other company that took them for rogues.

His sitting at table at one place, where the master of the house, that had not seen him in eight years, did know him but kept it private; when at the same table there was one that had been of his own Regiment at Worcester, could not know him but made him drink the Kings health and said that the King was at least four fingers higher then he [Charles was six-foot two at least] . . . In another place, at his Inn, the master of the house, as the King was standing with his hands upon the back of a chair by the fire-side, he kneeled down and kissed his hand privately, saying that he would not ask him who he was, but bid God bless him whither that he was going. Then the difficulty of getting a boat to get into France, where he was fain to plot with the master thereof to keep his design from the four men and a boy (which was all his ship's company), and so got to Feckam [Fécamp] in France.

At Roane [Rouen] he looked so poorly that the people went into the rooms before he went away [to Paris], to see whether he had not stole something or other. *The Diary of Samuel Pepys*, I

Charles did not tell Pepys the Royal Oak story during this first account. The incidents from all sources have been brought together here in a modern re-telling by Richard Ollard. The date was 6 September.

A hue and cry there certainly was: both the house and the woods surrounding it were likely to be searched during the day that was just coming on.

For the moment, however, the coast was clear. The two Penderels, together with Colonel Carlis, went back into the wood to fetch Charles. Breakfast consisted of bread and cheese. As a special luxury William Penderel's wife made the King a posset of thin milk and small beer . . . More to the point, she warmed some water to bathe his feet while Carlis pulled off his shoes and stockings. Both were sopping wet and the shoes were full of gravel. As there was not another pair in the house that would fit him—not that these did—Mrs Penderel put some hot cinders in them to dry him . . . It was now high time to be taking cover. Colonel Carlis had already selected a tall oak, standing by itself with a good all-round view, whose lower branches had been lopped and whose top bushed out so thickly as to render anyone in it invisible from below. Carlis and the King climbed into it by means of William Penderel's wood-ladder. Provisions—the inevitable bread and cheese and small beer—were passed up to them, together with a couple of pillows, as Charles had now been three nights without sleep, and the ladder was withdrawn. Sure enough the military were soon upon the scene. 'While we were in this tree,' Charles told Pepys, 'we see soldiers going up and down, in the thicket of the wood, searching for persons escaped, we seeing them now and then peeping out of the wood. This did not prevent him from falling into a profound slumber, his head on Carlis's arm which after a time became numb. This put the Colonel into a quandary. If he were to speak to the King loud enough to wake him he risked discovery. So he very sensibly woke him by pinching him.

Richard Ollard, *The Escape of Charles II* (1966)

Meanwhile there were incidents down below, including a visit by a Cromwellian officer to the house demanding the king and saying there was £1,000 reward on his head. Charles and Carlis came down out of the oak as soon as it got dark, and when the king heard of the reward he showed his dismay.

To people as poor as the Penderels a thousand pounds was unimaginable wealth . . . Colonel Carlis took the bull by the horns and told the King in front of them all that, 'if it were one hundred thousand pounds, it were to no more purpose, and that he would engage his soul for their truth.' This was the only occasion on which he allowed for a moment his face to betray him.

Ibid.

The king had his hair cut with William Penderel's shears, having earlier been made to rub his hands along the inside of the chimney to blacken his face. Later, in the west country, he was given boiled walnut juice to stain his white skin.

Meanwhile the King passed a most uncomfortable night in a hiding-place between two walls which was not long enough for him to lie down in. Still, it was secure enough.

Ibid.

THE RESTORATION 1660

The country had shown itself in favour of a restored parliament and restored monarchy. Pepys, a 26-year-old clerk in the Exchequer, was also in the household of his cousin Edward Mountagu, naval commander, and therefore crossed from Holland on the king's ship.

23 [May]. That done [renaming the ships: for instance. *Nazeby* into *Charles*] . . . we weighed Ancre, and with a fresh gale and most happy weather we set sail for England—all the afternoon the King walking here and there, up and down (quite contrary to what I thought him to have been), very active and stirring.

25. [The disembarkation] I went . . . and one of the King's footmen, with a dog that the King loved (which shit in the boat, which made us laugh and me think that a King and all that belong to him are but just as others are) went in a boat by ourselfs; and so got on shore . . . upon the land at Dover. Infinite the Croud of people and the gallantry of the Horsmen, Citizens, and Noblemen of all sorts.

The Diary of Samuel Pepys, I. *Samuel Pepys (1633–1703), secretary of the Admiralty and MP, was a great servant of the navy. His frank and brilliant diaries revealed himself and his times from 1660 to 1669.*

THE MARRIAGE OF CHARLES AND THE CATHOLIC PRINCESS, CATHERINE OF BRAGANZA

In order to get round any difficulties with the Papacy there was no formal marriage by proxy in Portugal but Catherine of Braganza sailed to England . . . in May 1662. On 28 May Charles wrote to Clarendon from Portsmouth, where he had travelled to meet his bride, telling him he was glad he was not called upon to consummate the marriage the previous night for he was sleepy 'and matters would have gone very sleepily'.

I can now give you an account of what I have seen abed [he added], which in short is, her face is not so exact as to be called a beauty though her eyes are excellent good, and not anything in her face that can in the least shock one, on the contrary she hath as much agreeableness in her looks altogether as ever I saw, and if I have any skill in physiognomy, which I think I have, she must be as good a woman as ever was born; her conversation as much as I can perceive is very good for she has wit enough and the most agreeable voice . . . In a word I think myself very happy.

How Charles discovered the quality of her conversation is obscure since the Queen spoke neither French nor English: perhaps they conversed in broken Spanish. Catherine had been brought up in a nunnery; her ladies were unprepossessing and their heavy native skirts or farthingales made them a laughing stock at the sophisticated English Court . . . In private he is supposed to have remarked that he thought they had brought him a bat instead of a woman.

Maurice Ashley, *Charles II*

THE SOUL OF BARBARA CASTLEMAINE, 1663

Charles's mistress, Barbara Villiers, Countess of Castlemaine and Duchess of Cleveland, converted to Catholicism from ostentatious Anglicanism.

When quizzed on this *volte-face*, King Charles remarked with spirit that he never concerned himself with the souls of ladies, but with their bodies, in so far as they were gracious enough to allow him.

Antonia Fraser, *King Charles II* (1979)

THE MISTRESSES AND THE ACTRESSES

Lady Castlemaine [Barbara Villiers] was a termagant. Charles was little concerned over her infidelities but mocked them. When she took up with the handsome John Churchill, the future Duke of Marlborough, and Charles found them together in Barbara's apartments, he said to Churchill, 'Go, you are a rascal, but I forgive you because you do it to get a living.' Lady Castlemaine could not stand the King's sarcasm. She told him once that it very ill became him to reproach the one woman in England who least deserved it; that he never ceased to pick quarrels with her since his low tastes had first declared themselves; that to gratify his base desires he needed only stupid geese like Stuart and Wells and that little slut of an actress [Nell Gwyn] he had recently taken up with.

Floods of angry tears accompanied these storms; after which, taking on the part of Medea, 'she would close the scene by threatening to massacre her children and burn the palace over his head'. As for Frances Stuart, she constantly provoked the King by increasing his ardour 'without diminishing her virtue by making the final sacrifice' . . .

It is said that it was when the King was at Tunbridge Wells in the summer of 1668 that he first met two actresses both of whom were to become his mistresses, 'Moll' Davis and Nell Gwyn. Moll Davis was a singer and dancer who appeared at the Duke's theatre . . . She sang a ballad 'My lodging is on the cold ground' which greatly impressed the King. The ballad 'raised the fair songstress from her bed on the cold ground to the royal bed'. . . . Pepys thought her 'the most impertinent slut' but he admired Nell Gwyn, a comedienne who appeared at the Theatre Royal, Drury Lane. Nell was generally popular. She is reputed to have said 'I was but one man's whore, though I was brought up in a bawdy house to fill strong waters for the guests.' . . . Another of his mistresses, Winifred Wells, was a Maid-of-Honour to the Queen. Gramont said she had the 'carriage of a goddess and the physiognomy of a dreamy sheep'. Charles treated all his mistresses generously. He arranged for the upkeep and ennoblement of their children. He does not appear to have practised birth control. In all he was to have thirteen illegitimate children, eight sons and five daughters.

Ashley, *Charles II*

The Great Fire of London, 1666

The diarists Pepys and Evelyn both described the king's energy. Charles's courage in personally helping to douse the fire in the City, until his clothes and face were soaked and blackened, won him immense popularity, as did the purse of 100 guineas he distributed among groups of fire-fighters. Pepys wrote on 2 September 1666:

Having stayed [in a boat by the Tower], and in an hour's time seen the fire rage every way, and nobody to my sight endeavouring to quench it, but to remove their goods, and leave all to the fire; and having seen it get as far as the Steeleyard, and the wind mighty high and driving it into the city, and everything, after so long a drougth, proving combustible, even the very stones of churches, and among other things, the poor steeple by which pretty Mrs [Horsley] lives . . . taken fire in the very top and there burned till it fall down—I to White-hall with a gentleman with me who desired to go off from the Tower to see the fire in my boat—to White-hall, and there up to the King's closet in the chapel, where people came about me and I did give them an account [which] dismayed them all; and word was carried in to the King, so I was called for and did tell the King and [James] Duke of York what I saw, and that unless his Majesty did command houses to be pulled down, nothing could stop the fire. They seemed much troubled, and the King commanded me to go to my Lord Mayor from him and command him to spare no houses but to pull down before the fire every way . . . At last met my Lord Mayor in Canning Streete, like a man spent, with a hankercher about his neck. To the King's message, he cried like a fainting woman, 'Lord, what can I do? I am spent! People will not obey me. I have been pull[ing] down houses. But the fire overtakes us faster than we can do it.'

The Diary of Samuel Pepys, VII

John Evelyn wrote:

The burning still rages; I went now on horse back, and it was now gotten as far as the Inner Temple; all Fleetestreete, old baily, Ludgate hill, Warwick Lane, Newgate, Paules Chaine, Wattlingstreete now flaming and most of it reduc'd to ashes, the stones of Paules flew like granados, the Lead mealting downe the streetes in a streame, and the very pavements of them glowing with fiery rednesse, so as nor horse nor man was able to tread on them, and the demolitions had stopped all the passages, so as no help could be applied; the Easter[n] Wind still more impetuously driving the flames forewards: Nothing but the almighty power of God was able to stop them, for vaine was the help of man: on the fift it crossed towards White-hall, but ô the Confusion was then at that Court: It pleased his Majestie to command me among the rest to looke after the quenching of fetter-lane end, to preserve (if possible) that part of Holborn, whilst the rest of the Gent: tooke their several posts, some at one part, some at another, for now they began to bestirr themselves, and not 'til now, who 'til now had stood as men interdict, with their hands a crosse, and began to consider that nothing was like to put a stop, but the blowing up of so many houses, as might make a [wider] gap, than any had yet ben made by the ordinary method of pulling them downe with Engines; This some stout Seamen proposd early enought to have saved the whole Citty, but some tenacious and avaritious Men, Aldermen etc. would not permitt, because their houses must have ben [of] the first . . . It is not indeed imaginable how extraordinary the vigilance and activity of the King and Duke was, even labouring in person, and being present to command, order and encourage Workemen; by which he shewed his affection to his people and gained theirs.

The Diary of John Evelyn, ed. J. Bowle (Oxford, 1983). *John Evelyn (1620–1706), was a royalist, an Anglican, secretary of the Royal Society founded by Charles II, and a friend of Pepys.*

CHARLES AND LORD PEMBROKE—A QUAKER 'OF A SORT'

Pepys told a story which illustrates the king's scepticism. The editors of Pepys's diary add a footnote showing the king's wit. 4 April 1668:

By and by the King comes out, and he did easily agree to what we moved . . . And then to talk of other things; about the Quakers not swearing, and how they do swear in the business of a late election of a Knight of the Shire of Hartfordshire in behalf of one they have a mind to have—and how my Lord of Pembroke will now and then, he says he hath heard him at the tennis-Court, swear to himself when he loses. And told us what pretty notions my Lord Pembroke hath of the first chapter of Genesis—how Adam's sin was not the suckeing (which he did before) but the swallowing of the apple; by which the contrary elements begun to work in him and to stir up evil passions—and a great deal of such fooleries, which the King made mighty mockery at.

Footnote: In 1665 Pembroke had told the King that the end of the world would come that year, and bade him prepare for it. Whereupon the King had offered him seven years' purchase for his manor of Wilton, but Pembroke replied, 'No and please your Majesty it shall die with me'.

The Diary of Samuel Pepys, IX

TOUCHING FOR THE KING'S EVIL, 28 MARCH 1684

There was so greate and eager a concourse of people with their children, to be touch'd of the Evil, that six or seven were crush'd to death by pressing at the Chirurgeon's door for tickets. *John Evelyn*

CHARLES ON HIS PAGE AND LATER MINISTER, SIDNEY GODOLPHIN

'He is never in the way, and never out of the way.'

CHARLES'S MEDICAL ATTENTION ON HIS DEATH-BED

Evelyn heard that the king was ill on 4 February 1685. Charles died two days later. The throne passed to his brother, James duke of York.

I went to Lond, hearing his Majestie had ben the moneday before surpriz'd in his bed chamber with an Apoplectical fit, and so, as if by Gods providence, Dr King (that excellent chirurgeon as well as Physitian) had not ben accidentally present to led him bloud with his lancet in his pocket) his Majestie had certaincly died that moment, which might have ben of direfull consequence, there being no body else with the King save this doctor and one more, as I am assured: It was a mark of the extraordinary dexterity, resolution, and presentnesse of Judgment in the Doctor to let him bloud in the very paroxysme, without staying the coming of other physitians, which regularly should have ben don, and the not doing so, must have a formal pardon as they tell me: This rescued his Majestie for that instant, but it prov'd onely a reprieve for a little time; he still complain'd and was relapsing and often fainting and sometimes in Epileptical symptoms 'til Wednesday, for which he was cupp'd, let bloud againe in both jugularies, had both vomit and purges etc: which so relieved him, that on the Thursday hops of recovery were signified in the publique Gazett; but that day about noone the Physitians conjectur'd him somewhat feavorish; This they seem'd glad of, as being more easily alaied, and methodicaly to be dealt with, than his former fits, so as they prescrib'd the famous Jesuits powder; but it made his Majestie worse; and some very able Doctors present, did not think it a feavor, but the effect of his frequent bleeding, and other sharp operations used by them about his head: so as probably the powder might stop the Circulation, and renew his former fitts, which now made him very weake: Thus he pass'd Thursday night with greate difficulty, when complaining of a paine in his side, the[y] drew 12 ounces more of blood from him, this was by 6 in the morning on friday, and it gave him reliefe, but it did not continue; for being now in much paine and strugling for breath, he lay doz'd, and after some conflicts, the Physitians desparing of him, he gave up the Ghost at halfe an houre-after Eleaven in the morning, being the 6 of Feb: in the 36t yeare of his reigne, and 54 of his age.

John Evelyn

James II

1685–1688

Coming to the throne at fifty-one, James duke of York had none of his brother Charles II's flair, not even in the pursuit of women. The duke of Buckingham once pointed the difference between the brothers: 'The King could see things if he would; the duke would see things if he could.' Exile was familiar to him: first to escape Cromwell; then because of his unpopularity as an active Catholic who tried to rule without Parliament; lastly when the 'Glorious Revolution' enabled him to concentrate henceforth on religion rather than rule. James was not unlike his brother Charles in appearance, though fair instead of dark and moderately, instead of immensely, tall.

PRINCE JAMES AS LORD HIGH ADMIRAL, 1665

James was in command at the battle of Lowestoft, aboard the Royal Charles. *Things went badly in the morning, when several of his commanders were killed, including his great friend Charles Berkeley and one of his Gentlemen of the Bedchamber. 'James himself was splashed by their noble blood as he stood upon his quarter-deck.' But things were to change dramatically:*

By two o'clock in the afternoon the fire from the Dutch ships began to slacken . . . Half an hour later James was able to turn the tables on his opposite number. A lucky shot from the *Royal Charles* struck the magazine of Obdam's flagship which blew up killing the Dutch commander-in-chief and 400 of his men. By four o'clock the battle had degenerated into a confused rout. James ordered his fireships to be clapped on to the Dutch warships as they became entangled with one another. As Cornelius van Tromp, who was with the Dutch fleet, reported, it 'got into such confusion that they all ran away from the enemy before the wind', losing a dozen of their ships captured or burnt. James gave instructions for a chase, which was not abandoned until about nine in the evening when it was growing dark. Thus the battle lasted for the best part of eighteen hours. It was a definite victory for the English as the Dutch losses were heavier than was normal in seventeenth-century warfare. At eleven o'clock at night James, exhausted by the bloody battle, retired to bed, fully dressed with a quilt thrown over him. He slept the sleep of a conqueror . . .

[Parliament] voted a further £1,250,000 for the continuation of the naval war and gave James £120,000 'in token of the great sense they had of his conduct and bravery' at Lowestoft. Maurice Ashley, *James the Second* (1971)

THE ESCAPE OF JAMES DUKE OF YORK

As the period of his youth coincided with the extraordinary social upheavals which began with the civil wars and ended only with the Restoration, he had passed through adventures enough to do credit to a hardened veteran. Taken prisoner as a mere boy during the Civil Wars, he was confined in St James's Palace, from which, assisted by Colonel Bamfield, he effected his escape in 1648 by an exceedingly clever ruse. One night, after supper, he went to play at 'hide and seek' with his brother and sister, at which game for the past fortnight he had practised assiduously, 'and had used to hide himself in places so difficult to find' that it usually took the other children half-an-hour to search for him. On this particular night, the Duke having first taken the precaution to shut up in his sister's room a little dog that was wont to follow him, crept out of the palace through a back door, met there the trusty Colonel Bamfield, proceeded with him in a coach as far as Salisbury House, stepped out and bolted down a side street to the river. There they took boat and got out a little lower down, proceeding to the house 'of one, Loe, a Surgeon, where they found Mrs Murray, who had women's cloths in readiness to disguise the Duke.' John Beresford, *Gossip of the Seventeenth and Eighteenth Centuries* (1923)

THE PLAGUE OF 1665

James duke of York and Anne Hyde, his duchess, daughter of Lord Clarendon, fled first to Salisbury and then to Oxford.

27 July. So despatched all my business . . . and so we stayed and saw the King and Queene set out toward Salsbury—and after them, the Duke and Duchesse—whose hands I did kiss. And it was the first time I did ever or did see anybody else kiss her hand; and it was a most fine white and fat hand; and it was pretty to see the young pretty ladies dressed like men; in velvet coats, caps and ribbands, and with laced bands just like men—only, the Duchesse herself it did not become.

The Diary of Samuel Pepys, VI. *The fashion for plush hunting caps and red vests had been set by the queen in 1662.*

DEATH OF THE DUCHESS OF YORK

Anne died of breast cancer in 1671. She had become a Roman Catholic, though the bishop of Oxford who visited her on her death-bed did not know it.

The Bishop spoke but little and fearfully. He happened to say he hoped she continued still in the truth: upon which she asked, 'What is truth?'; and then, her agony increasing, she repeated the word '*Truth, truth,*' very often, and died in a few minutes.

Bishop Burnet, *The History of My Own Times*, I (Oxford, 1823). *Gilbert Burnet (1643–1715) was a Scot who disliked Charles II and supported William of Orange, entering England from Holland in his train in 1688. He became bishop of Salisbury as a reward and later governor of New York.*

JAMES AS KING

Mary of Modena, his second wife, was queen. Would he turn over a new leaf?

The king did, some days after his coming to the crown, promise the queen and his priests, that he would see Mrs Sidley no more, by whom he had some children. And he spoke openly against lewdness, and expressed a detestation of drunkenness. He sat many hours a day about business with the council, the treasury, and the admiralty. It was upon this said, that now we should have a reign of action and business, and not of sloth and luxury, as the last was. Mrs Sidley had lodgings in Whitehall: orders were sent to her to leave them. This was done to mortify her; for (as she was naturally bold and insolent) she pretended that she should now govern as absolutely as the duchess of Portsmouth [one of Charles II's mistresses] had done: yet the king still continued a secret commerce with her. And thus he began to reign with fair appearances. Ibid.

The Birth of the 'Old Pretender'

Young James was the king's only legitimate son and a Catholic like both his parents.

The Queen was with child. Before the end of October 1687 the great news began to be whispered . . . The great body of the nation listened with mingled derision and fear. There was indeed nothing very extraordinary in what had happened . . . As, however, five years had elapsed since her last pregnancy, the people, under the influence of that delusion which leads men to believe what they wish, had ceased to entertain any apprehension that she would give an heir to the throne. On the other hand, nothing seemed more natural and probable than that the Jesuits should have contrived a pious fraud. . . . A suspicion, not indeed well founded, but by no means as absurd as is commonly supposed, took possession of the public mind. The folly of some Roman Catholics confirmed the vulgar prejudice. They spoke of the auspicious event as strange, and miraculous . . .

One fanatic announced that the Queen would give birth to twins, of whom the elder would be king of England, and the younger Pope of Rome. Mary could not conceal the delight with which she heard this prophecy, and her ladies found that they could not gratify her more than by talking of it. The Roman Catholics would have acted more wisely if they had spoken of the pregnancy as a natural event, and if they had borne with moderation their unexpected good fortune . . .

The birth followed on James's imprisonment of seven Anglican bishops.

Scarcely had the gates of the Tower been closed on the prisoners when an event took place which increased the public excitement. It had been announced that the Queen did not expect to be confined till July. But, on the day after the Bishops had appeared before the Council, it was observed that the King seemed to be anxious about her state. In the evening, however, she sate playing cards at Whitehall till near midnight. Then she was carried in a sedan to St James's Palace, where apartments had been very hastily fitted up for her reception. Soon messengers were running about in all directions to summon physicians and priests, Lords of the Council, and Ladies of the Bedchamber.

Lord Macaulay, *History of England* (1898)

THE WARMING-PAN RUMOUR

There was a rumour that a child had been smuggled into Mary's bed inside a warming-pan who was then passed off as her son.

There, on the morning of Sunday, the tenth of June, a day long kept sacred by the too faithful adherents of a bad cause, was born the most unfortunate of princes, destined to seventy-seven years of exile and wandering, of vain projects, of honours more galling than insults, and of hopes such as make the heart sick. The calamities of the poor child had begun before his birth. The nation over which, according to the ordinary course of succession, he would have reigned, was fully persuaded that his mother was not really pregnant. By whatever evidence the fact of his birth had been proved, a considerable number of people would probably have persisted in maintaining that the Jesuits had practised some skilful sleight of hand; and the evidence, partly from accident, partly from gross mismanagement, was really open to some objections. Many persons of both sexes were in the royal bedchamber when the child first saw the light; but none of them enjoyed any large measure of public confidence. Of the Privy Councillors present, half were Roman Catholics; and those who called themselves Protestants were generally regarded as traitors to their country and their God. Many of the women in attendance were French, Italian and Portuguese. Of the English ladies some were Papists, and some were the wives of Papists . . . The Princess Anne was, of all the inhabitants of the island, the most deeply interested in the event. Her sex and her experience qualified her to act as the guardian of her sister's [Mary's] birthright and her own. She had conceived strong suspicions . . .

In this temper Anne had determined to be present and vigilant when the critical day should arrive. But she had not thought it necessary to be at her post a month before that day, and had, in compliance, it was said, with her father's advice, gone to drink the Bath waters. Sancroft, whose great place made it his duty to attend, and on whose probity the nation placed entire reliance, had a few hours before been sent to the Tower by James. The Hydes were the proper protectors of the rights of the two Princesses. The Dutch Ambassador might be regarded as the representative of William, who, as first prince of the blood and consort of the King's eldest daughter, had a deep interest in what was passing. James never thought of summoning any member, male or female, of the family of Hyde; nor was the Dutch Ambassador invited to be present.

Posterity has fully acquitted the King of the fraud which his people imputed to him. But it is impossible to acquit him of folly and perverseness such as explain and excuse the error of his contemporaries. Ibid.

DEPARTURE AND ARRIVAL

James II fled his country, leaving London and the crown to William of Orange. John Evelyn noted the event:

18 The Pr: comes to St. James, fills W-hall (the King taking barge to Gravesend at 12 a Clock) with Dut[c]h Guard . . . All the world go to see the Prince at St Jamess where is a greate Court, there I saw him and severall of my Acquaintance that come over with him: He is very stately, serious, and reserved: The Eng: souldiers etc. sent out of Towne to distant quarters: not well pleased: Divers reports and opinions, what all this will end in; Ambition and faction feared . . .

24 The King passes into France, whither the queen and child wer gon a few days before. *John Evelyn*

JAMES IN IRELAND

James tried to hold Ireland against William with the help of French troops. He was defeated at the Battle of the Boyne on 1 July 1690. Reaching the safety of Dublin, he encountered Lady Tyrconnel and informed her:

'Madam, your countrymen have run away.'

To which Lady Tyrconnel retorted:

'Sire, your majesty seems to have won the race.'

William III and Mary II

1689–1702 and 1689–1694

William was a grandson of Charles I through his daughter Mary, who married the Stadtholder of the Dutch Republic. He was therefore both nephew and son-in-law to James II. It was with a Dutch fleet and army that William landed at Torbay in Devon (quickly to be joined by many English) and with a French–Irish army that James II failed to defeat William at the Boyne. Parliament accepted William as a way of restoring their authority. William accepted the English crown as a way of safeguarding the Netherlands against France. Mary was an outgoing beauty, 4½ inches taller than the withdrawn William, who may have had homosexual inclinations; but the marriage did not fail. The only sport William enjoyed was hunting, perhaps because he could go it alone.

WILLIAM AND MARY AS CHILDREN

As nephew of James II Prince William of Orange was Mary's first cousin. Pepys saw William at The Hague in 1660 and Mary nine years later in London.

About 10 at night the Prince comes home, and we found an easy admission. His attendance very inconsiderable as for a prince. But yet handsome, and his tutor a fine man and himself a very pretty boy.

. . . I did see the young Duchess [Mary], a little child in hanging sleeves, dance most finely, so as almost to ravish me, her airs were so good . . .

The Diary of Samuel Pepys, I and IX

THE CHARACTER OF WILLIAM

The prince had been much neglected in his education: for all his life long he hated constraint. He spoke little. He put on some appearance of application: but he hated business of all sorts. Yet he hated talking, and all house games, more. This put him on a perpetual course of hunting, to which he seemed to give himself up, beyond any man I ever knew: but I looked on that always as a flying from company and business . . . He had no vice, but of one sort [a reference to his favourites Portland and Albemarle] in which he was very cautious and secret. He had a way that was affable and obliging to the Dutch. But he could not bring himself to comply enough with the temper of the English, his coldness and slowness being very contrary to the genius of the nation.

Burnet, *History*

Lord Dartmouth added a footnote on William's vice:

Bishop Burnet told me, if I lived to read his History, I should be surprised to find he had taken notice of King William's vices; but some things, he said, were too notorious for a faithful historian to pass over in silence.

Princess Mary's romantic passion for her best friend, Lady Frances Apsley (later Lady Bathurst) was affected by the love literature of her day; it resulted in some extraordinary letters.

St James's ten o'clock

Of all things in the world I long to see my dear dear dear Aurelia [Frances] to tell her the cause of my neglect which I have not time to do in this letter only to assure her I shall always be your most affectionate friend and dutiful wife.

Mary Clorine

If you do not come to me some time today dear husband [sic] that I may have my belly full of discourse with you I shall take it very ill . . . If you come you will mightily oblige your faithful wife,

Mary Clorin [sic]

I have written it in such a hand that I believe you cannot read it pray burn it & send word whether you can or no by the bearer.

for Mrs Apsley

Letters of Two Queens, ed. B. Bathurst (1924)

A CRISIS IN THE MARRIAGE OF WILLIAM AND MARY

The royal couple had been married in England on 4 November 1677. Prince William's long-standing association with Betty Villiers, which Princess Mary had come to accept, faced a dangerous moment in October 1685 when it became public. The first step was when Mary's father James II, armed with information given him by spies in Mary's household, brought to his daughter's notice the (already known) humiliations of her position. William then caught the spies by intercepting a packet of their letters being sent to England through the ambassador Skelton. After reading these letters William sent for Mary.

What was said between them was overheard by no one, but Daniel de Bourdon, who was close friends with one of the companions of a lady-in-waiting, makes a plausible guess at it. Once alone with Mary, the Prince asked, 'whether she was aware that people were attempting to destroy their unity and to drive her to a scandalous separation by making her believe falsehoods.' She replied that, 'although if the truth were to be told, she had

had reason not to be too happy for some time now, she had nevertheless shut up her sorrow in her heart, without having told anyone of it, and that if someone had been touched and moved to intervene, it had not been her doing, she knew very well how to suffer in silence.'

'Well,' said the Prince, 'you protect thus those of your household who dare to sow dissension between us, and who by false reports, make you doubt of my fidelity to you.' 'Have I not reason to do so?' she asked. 'No,' said the Prince, 'and I swear to you by all that is most sacred that what has caused you pain was simply a distraction, that there has been no crime [adultery]; but there are servants of yours, for whom I have done everything I could, who betray me, and if you believe the oath I make before God, never to violate the trust I swore to you, you must abandon them to my just indignation.'

The Princess—still according to Bourdon— was completely disarmed by his assurances, burst into tears and threw herself into his arms, protesting that she had no knowledge and no part in the correspondence of her servants, and telling him to act as he thought fit. The Prince then interviewed the culprits privately . . . and ordered them abruptly to pack their cases and be ready to leave in two hours. They were forbidden to write or see their mistresses again, and were shipped off to England.

H. and B. van der Zee, *William and Mary* (1973), quoting the memoirs of Bourdon

HOW THE REIGN CAME TO BE THAT OF 'WILLIAM AND MARY'

One of Mary's ministers offered to talk parliament round into setting her alone on the throne.

She made him a very sharp answer; she said she was the Prince's wife, and would never be other than what she should be in conjunction with him and under him; and that she would take it extremely unkindly, if any, under a pretence of their care for her would set up a divided interest between her and the Prince.

Burnet, *History*

As for the prince, he totally rejected the idea of being prince consort, saying that unless he were king, on an equality with the queen, he would go back to Holland.

Mary's happy face on ascending her deposed father's throne met with some criticism.

I saw the new Queene and King, so proclaim'd the very next day of her coming to White-hall, Wednesday 13. Feb. with wonderfull acclamation and general reception, Bonfires, bells, Gunns etc: It was believed that they both, especialy the Princesse, would have shewed some (seeming) reluctancy at least, of assuming her Fathers Crowne and made some Apologie, testifying her regret, that he should by his misgovernment necessitat the Nation to so extraordinary a proceeding, which would have shewed very handsomly to the world, (and according to the Character give[n] of her piety etc) and consonant to her husbands first Declaration, that there was no intention of Deposing the King, but of Succoring the Nation; But, nothing of all this appeared; she came into W-hall as to Wedding, riant and jolly, so as seeming to be quite Transported: rose early on the next morning of her arrival, and in her undresse (as reported) before her women were up; went about from roome to roome, to see the Convenience of White-hall: Lay in the same bed and appartment where the late Queene lay: and within a night or two, sate downe to play at Basset, as the Q. her predecessor us'd to do: smiled upon and talked to every body; so as no manner of change seem'd in Court, since his Majesties last going away, save that the infinite crowds of people thronged to see her, and that she went to our prayers: This carriage was censured by many: she seemes to be of a good nature, and that takes nothing to heart whilst the Pr: her husband had a thoughtfull Countenance, is wonderfull serious and silent, seemes to treate all persons alike gravely . . . *John Evelyn*

WILLIAM III'S ALLEGED DISCOURTESY TO HIS SISTER-IN-LAW PRINCESS ANNE

I believe I could fill as many sheets, as I have already written, with relating the brutalities that were done to the Prince and Princess George and Anne in that reign. The King was indeed so ill-natured and so little polished by education, that neither in great things nor in small had he the manners of a gentleman. I shall give you an instance of his worse than vulgar behaviour at his own table, when the Princess dined with him.

It was in the beginning of his reign, and when she was with child of the Duke of Gloucester. There happened to be a plate of pease [sic], the first that had been seen that year. The King, without offering the Princess the least share of them, eat them every one up himself. Whether he offered any to the Queen, I cannot say but he might do that safely enough, for he knew, she durst not touch them. The Princess confessed, when she came home, she had so much mind to the pease, that she was afraid to look at them, and yet could hardly keep her eyes off them.

The Conduct of the Dowager Duchess of Marlborough, from her first coming to Court to the year 1710, ed. N. Hooke (1742)

A reviewer signing himself 'Britannicus' *lambasted the* Conduct *for its bitterness, explaining why William and Mary disliked its author Sarah Churchill, duchess of Marlborough (born Jennings):*

It seems their Majesties thought she *assumed too much* in directing the Princess as she did, and did not care to fall under the Tuition of *her* and *her Lord* [John Churchill, Duke of Marlborough]; and for this reason King *William* is not allowed to have so much as the *Manners* of a *Gentleman*, and Queen *Mary* is said *to have wanted Bowels.*

THE MOLEHILL THAT BECAME A MOUNTAIN

The death of William on 8 March (old style) 1702 followed a fall. Mary had died from smallpox eight years before.

On 20 February Bidloo [William's devoted Dutch doctor] was suddenly sent for in the evening to Kensington, and found the King with his arm in a sling. The King told him: 'I was riding this afternoon in the Park near Hampton Court, and I was urging the horse into a gallop when she fell on her knees. I tried to pull her up by the reins, but she fell first forward and then sideways, and I fell on my right shoulder on the ground. It was odd, because it was level ground.'

The horse Sorrel, which William had been riding for the first time, had in fact stumbled on a mole-hill, and the King had broken his collar-bone in his fall. His surgeon Dr Ronjat set it at Hampton Court, but William insisted on returning to Kensington that evening as he had planned, and during the long coach-ride it had been jolted out of place again. Bidloo had to reset it, and afterwards William went to bed and slept perfectly soundly, apparently unaffected by the accident.

Next day he felt well enough to work as usual, and thought no more of his fall until on 27 February they took the bandages off and discovered that the fracture had not mended and was slightly swollen: his right hand and arm—the same side as the fracture—looked odd and puffy too. William refused to take much notice of this and of his growing weakness, dictated letters as usual.

van der Zee, *William and Mary*

By 3 March he was a very sick man but he refused the powders, herbal decoctions, and juleps prescribed by his doctors.

William refused them all, and on 6 March he was so weak that he could no longer keep his food down but vomited often. New medicines were tried, like powdered crabs' eyes and pearled julep, and sal volatile to revive him . . . and when the doctors begged the King to eat a little to keep his strength up, he replied with a touch of impatience: 'Believe me, gentlemen, I know particularly well that forced feeding does me no good.'

Ibid.

His beloved Albermarle arrived at the bedside from Holland on 7 March, to be told by the King, 'Je tire vers ma fin.' *Did Albermarle send for his rival favourite*

Portland deliberately too late to hear the king speak again? William could only take Portland's hand and feebly press it to his heart.

Then his head fell back on his shoulder, and 'shutting his eyes, he expired with two or three soft Gasps' . . . The mourning in England was nothing like as deep and prolonged as had been that for Queen Mary, and the English Jacobites invented their toast 'To the little gentleman in black velvet'—the mole on whose hill Sorrel had stumbled and whose fall had broken the King's collar-bone and health.

Ibid.

Was it because their father James II had had syphilis as duke of York that neither Mary nor Anne his daughters was able to produce heirs to the throne?

Upon the whole matter the Duke was often ill: the children were born with ulcers, or they broke soon after: and all his sons died young and unhealthy. This has, as far as anything whatsoever that could be brought in the way of proof, prevailed to create a suspicion that so healthy a child as the pretended Prince of Wales could neither be his, nor be born of any wife with whom he lived long. The violent pain that his eldest daughter had in her eyes, and the gout which has so early seized our present Queen [Anne], are thought the dregs of a tainted original. Upon which, Willis, the great physician, being called to consult for one of his sons, gave his opinion in these words, *Mala stamina vitae*; which gave such offence that he was never called for afterwards.

Burnet, *History*

Anne

1702–1714

The last of the Stuarts, Anne was more consciously English than any of her Stuart predecessors, certainly far more so than her Hanoverian successors. Shy, conscientious, stout, short-sighted, gouty, she was to be described as 'one of the smallest people ever set in a great place'. In her name great land victories were to be won, the like of which had not been known since Crécy and Agincourt. Her staunch, simple Protestantism helped her to bear the sorrow of seventeen pregnancies and sixteen dead babies. Even the beloved William duke of Gloucester died in 1701, perhaps of hydrocephalus, aged eleven. Anne was homely; her age heroic. As a girl she had rosy cheeks, and always a cheerful smile.

LADY ANNE AND LORD MULGRAVE

Neither Pepys nor Evelyn, the great seventeenth-century diarists, noted the birth of a second daughter to James duke of York. But at seventeen Anne attracted the attentions of a courtier, and though the affair was exaggerated it affected her future.

The villain of this melodrama was John Sheffield, Lord Mulgrave, a favourite of Charles II, a bachelor and, at thirty-five, Lady Anne's senior by eighteen years. The nature of Mulgrave's 'soe briske attempts upon the Lady Anne' remains obscure. Mulgrave claimed that his crime was 'only ogling', but all independent accounts agreed that he had written letters 'intimating too near an address to her'. When this correspondence was discovered and Mulgrave was banned from court in November 1682, London gossips concluded that Mulgrave had seduced Lady Anne, 'so far as to spoil her marrying to any body else, and therefore the town has given him the nickname of King John'. Although Mulgrave was temporarily exiled to Tangiers, his later career suggests that his intimacy with Lady Anne was vastly overrated by gossip . . .

The Mulgrave affair contained serious portents for the future. First, it hastened negotiations for Lady Anne's marriage [to Prince George of Denmark] . . . Secondly, it underlined—and perhaps contributed to—a growing divergence between Lady Anne and her sister. The Princess of Orange's letter to their mutual friend, Frances Apsley, is revealing: while deploring the fate of 'my pore sister' and conventionally, if unconvincingly, defending Lady Anne ('not but that I believe my sister very innocent'), the

Princess of Orange strongly hinted that her sister lacked discretion and good judgement: 'I am so nice upon the point of reputation that it makes me mad she should be exposed to such reports, & now what will not this insolent man say being provokt.' No correspondence between the royal sisters is extant for this period, but it is likely that some hint of her sister's priggish attitude was conveyed to Lady Anne.

Finally, and most importantly, the Mulgrave affair marked the flowering of the friendship between Lady Anne and Sarah Jennings, a friendship which would influence the destiny of Europe.

Edward Gregg, *Queen Anne* (1980)

PRINCESS ANNE'S CONSORT: PRINCE GEORGE OF DENMARK

Removed from his native country—and it was taken for granted that he and Anne should live in England—and from the life of action to which he was accustomed, he never managed to carve out a proper role for himself, but remained always in the background, occupying himself by making model ships and important only because he was Anne's husband. His remarkable appetite and fondness for the bottle were the most striking things about him, and they quickly ruined his looks and sapped whatever energy he once had, turning him into a gross, rather ridiculous figure; the King's much-quoted opinion of him, 'I've tried him drunk and I've tried him sober but there's nothing in him,' was only one of a long line of Court jokes at his expense. It was Mulgrave, Anne's erstwhile admirer, who later unkindly suggested that his fits of asthma were due to the fact that he was forced to breathe hard lest he should be taken for dead and removed for burial.

Gila Curtis, *The Life and Times of Queen Anne* (1972)

Nevertheless Anne as queen admired her consort even more than before.

For Anne proposed to make him not only King Consort in England but, if the Dutch could be persuaded to agree, Stadtholder in the Netherlands as well. Nobody else, in either England or the Netherlands, shared her enthusiasm, but it was only after much tactical dissuasion that she gave up the idea. Even then it was only to appoint him Generalissimo of all her forces and Lord High Admiral of the Fleet, which was scarcely more acceptable. By this time bovine George had long since cast aside all ambitions—it was reported that he loved only 'his news, his bottle, and the queen'.

Ibid.

The passionate friendships of Anne with women have been endlessly discussed. Sarah Jennings, when Dowager Duchess of Marlborough, described how she came to enter Princess Anne's service on her marriage to Prince George of Denmark, in 1683.

Her first lady of the bedchamber, Lady Clarendon, said Sarah, 'looked like a mad-woman, and talked like a scholar'.

The Princess had a different taste. A friend was what she most coveted: and for the sake of friendship (a relation which she did not disdain to have with me) she was fond even of that *equality* which she thought belonged to it. She grew uneasy to be treated by me with the form and ceremony due to her rank; nor could she bear from me the sound of words which implied in them distance and superiority. It was this turn of mind, which made her one day propose to me, that whenever I should happen to be absent from her, we might in all our letters write ourselves by feigned names, such as would import nothing of distinction of rank between us. Morley and Freeman were the names which her fancy hit upon; and she left me to chuse by which of them I would be called. My frank, open temper naturally led me to pitch upon Freeman, and so the Princess took the other; and from this time Mrs Morley and Mrs Freeman began to converse as equals, made so by affection and friendship.

The Conduct of the Dowager Duchess of Marlborough

Sarah told the story of Anne's escape from her father James II when she heard he was coming to London, during the 'Glorious Revolution' of 1688.

She sent for me, told me her distress, and declared, *That rather than see her father she would jump out at window*. This was her very expression.

A little before, a note had been left with me, to inform me where I might find the bishop of London (who in that critical time absconded) if her Royal Highness should have occasion for a friend. The Princess, on this alarm, immediately sent me to the bishop. I acquainted him with her resolution to leave the court, and to put herself under his care. It was hereupon agreed, that, when he had advised with his friends in the city, he should come about midnight in a hackney coach to the neighbourhood of the Cockpit [Anne's home in Whitehall] in order to convey the Princess to some place where she might be private and safe.

The Princess went to bed at the usual time to prevent suspicion. I came to her soon after; and by the back-stairs which went down from her closet, Her Royal Highness, my Lady Fitzharding, and I, with one servant, walked to the coach, where we found the bishop and the earl of Dorset. They conducted us that night to the bishop's house in the city and the next day to my lord Dorset's at Copt-hall. From thence we went to the earl of Northampton's and from thence to Nottingham, where the country gathered about the Princess; nor did she think herself safe, till she saw that she was surrounded by the Prince of Orange's friends. Ibid.

ATTEMPT TO HUMILIATE PRINCESS ANNE IN CHURCH

This is another story against William and Mary by Sarah. She and Anne hated William, referring to him as 'Caliban'.

Another foolish thing, that was done by the same advice, as I suppose, was sending to the minister of St James's church, where the Princess used to go . . . to forbid them to lay the text upon her cushion, or take any more notice of her than of other people. But the minister refusing to obey without some order from the crown in writing, which they did not care to give, that noble design dropt. Ibid.

QUEEN ANNE'S CORONATION, 23 APRIL 1702

It was a cruel blow to her that, weighted down by all this magnificence, she also happened to be suffering from a severe attack of gout, and therefore gained the dubious distinction of being the only English monarch who had to be carried to her coronation. And so it was that Anne, carried by the Yeomen of the Guard in an open sedan chair with a low back, over which her six-yard train could pass to her ladies behind, arrived in Westminster Abbey shortly after eleven o'clock. The ceremony took over five hours, and for a gouty invalid must have been an exhausting ordeal. The Archbishop of York, whom Anne preferred as being more High Church than Canterbury, preached the sermon, and then she was finally crowned late in the afternoon by the Archbishop of Canterbury, who also expressed the rather tactless hope that she would 'leave a numerous posterity to rule these kingdoms' . . .

Two weeks later, on 4 May, England declared war against France. 'It means I'm growing old when ladies declare war on me,' joked Louis [XIV] when he was told. But his levity was misplaced. Curtis, *Queen Anne*

The era of Marlborough and Blenheim was being ushered in.

ANNE HEARS THE NEWS OF BLENHEIM

Anne received the news at Windsor while, legend has it, she was playing at dominoes seated in one of the great bay windows overlooking the park. It had taken eight days for Marlborough's messenger, Colonel Parke, to ride back across Europe and then wait for the wind to blow him to England with his momentous message from the Duke [of Marlborough]. It was scribbled to his wife on the back of a tavern bill: 'I have not time to say more, but to beg you will give my duty to the Queen, and let her know her army has had a glorious victory . . .' Anne, with tears of joy running down her cheeks, gave Parke a miniature of herself and a thousand guineas in reward. In London copies of Marlborough's note were struck off the presses in their hundreds and the people went mad with joy . . . It was a land victory such as England had not known since Agincourt . . . Anne would preside over England's rising fortunes for a decade to come; but never again would the same unalloyed taste of victory sweep over the nation and bind the people together as it did for a rare moment after the Battle of Blenheim.

Curtis, *Queen Anne*.

ANNE AS PEACEMAKER IN PARLIAMENT

But the moment of national rejoicing did not last for long and when Parliament reassembled in the autumn of 1705, for all Anne's pleas for unity and concord, the two parties still found plenty to quarrel about. It was partly in the hope that her presence might induce members to act more moderately that, in this session, Anne revived Charles II's practice of attending debates in the House of Lords in person, sitting either on the throne or, in the colder months, on a bench beside the fire.

Ibid.

Anne was a moderate Tory, but in her later years the Tories found her as obstinate as the Whigs did.

In February 1711 Anne celebrated her forty-sixth birthday at St James's Palace. For the occasion she wore a gown of green flowered satin embroidered with gold, and attended a special performance of 'an Italian dialogue in Her Majesty's praise set to excellent music by the famous Mr Handel . . .' Reflecting the political changes of the previous year, the entertainment was an almost exclusively Tory affair and most of the Whigs boycotted the Palace. The Tory ladies, however, put on a good show. Some were 'scarce able to move under the load of jewels', and it was said that such a splendid Court had not been seen since 1660.

Outward display, however, could not hide from those around her that the Queen, ageing prematurely and rapidly, was growing not only infirm but incapable. Her chief doctor and close friend, Sir David Hamilton, felt compelled to warn her of the dangers of her 'disquiets and uneasiness', while her ministers, needing decisions met only with prevarication. Harley found that there was 'no other remedy but to let Her Majesty take her own time which never failed to be the very longest that the nature of the thing would suffer her to defer it.'

Ibid.

Queen Anne eventually transferred her favours from Sarah Marlborough to her bedchamber woman, Abigail Hill, Sarah's cousin, who had just married Mr Masham. Sarah told how she found out.

I went presently to the Queen and asked her, *why she had not been so kind as to tell me of my cousin's marriage*, expostulating with her upon the point, and putting her in mind of what she often used to say to me out of Montaigne, *that it was no breach of promise of secrecy to tell such a friend anything, because it was no more than telling it to one's self.* All the answer I could obtain from Her Majesty was this, *I have a hundred times bid Masham tell it you, and she would not.*

The conduct both of the Queen and of Mrs Masham convinced me that there was some mystery in the affair, and thereupon I set myself to enquire as particularly as I could into it. And in less than a week's time, I discovered that my cousin was become an absolute favourite . . . that Mrs Masham came often to the Queen, when the Prince was asleep, and was generally two hours every day in private with her. Ibid.

'QUEEN ANNE'S DEAD!'

After much suffering from gout, Anne died on 1 August 1714 of erysipelas. Her physician, John Arbuthnot, told his friend Dean Swift: 'I believe sleep was never more welcome to a weary traveller than death was to her.' She had survived the well-meant but horrific remedies of her doctors (bleeding, blistering, emetics, the shaving of her head) for several days, but five days before her release, when asked how she was, she replied, 'Never worse. I am going.' There were premature rumours that she had 'gone' two days before the event. It seems no more than just that her death should become proverbial: George Colman the Younger introduced into one of his plays the line:

Lord help you! Tell 'em Queen Anne's dead.

THE HANOVERIANS

The best one can say of them qua *Royal House is that they improved as they went along, sloughing the boorishness that descended with them on the British throne. Thackeray showed how easy it was to lay on the ridicule with a trowel. Yet Britain under the Hanoverians increased in power and parliament preferred a German Protestant whose heart was in Hanover to a British Catholic 'Pretender' whom they refused to call 'James III'. The Act of Settlement (1701) turned the hereditary spotlight on to the Electress Sophia of Hanover, granddaughter of James I. Unfortunately this gifted woman died barely two months before Queen Anne. Her far from gifted son succeeded.*

THE ELECTRESS SOPHIA AND HER SON'S MISTRESS

George fell in love with Mlle Schulenberg [his mother's maid-of-honour] ... One evening when she was in waiting behind the Electress's chair at a ball, the Princess Sophia, who had made herself Mistress of the language of her future subjects, said in English to Mrs Howard (afterwards Countess of Suffolk) then at her Court, 'Look at that Malkin [scarecrow], & think of her being my Son's passion.' Mrs Howard, who told me the story, protested she was terrified, forgetting that Mlle Schulenberg did not understand English.

Reminiscences written by Mr Horace Walpole in 1788 for the amusement of Miss Mary and Miss Agnes Berry, ed. Paget Toynbee (Oxford, 1924)

George I

1714–1727

This reign of a foreigner had its advantages. It enabled a 'prime minister' to emerge and preside over the king's government; and it presented Sir Robert Walpole, a great peacetime statesman, with his opportunity.

GEORGE I AND HIS WIFE

George I, while electoral prince, had married his Cousin the Princess [Sophia] Dorothea, only child of the Duke of Zell; a match of convenience to reunite the dominions of the family. Though She was very handsome, the Prince who was extremely amorous, had several Mistresses, which provocation & his absence in the army of the Confederates probably disposed the Princess to indulge some degree of coquetry. At that moment arrived at Hanover the famous and beautiful Count Königsmark.

Horace Walpole, *Reminiscences*

Her youth (she was married at sixteen), beauty, and state of neglect encouraged the count to make advances which she accepted, though without 'transgressing her duty'. Her ladies betrayed her to the prince who, in a flaming passion, ordered the count to leave the country next day.

The Princess, surrounded by Women too closely connected with her Husband, and consequently enemies of the Lady they injured, was persuaded by them to suffer the count to kiss her hand before his abrupt departure; he was actually introduced by them into her bedchamber the next morning before She rose. From that moment He disappeared, nor was it known what became of him, till on the death of George I, on his son the new King's first journey to Hanover, some alterations in the palace being ordered by him, the body of Königsmark was discovered under the floor of the Electoral Princess's dressingroom, the Count having probably been strangled there the instant he left her, & his body secreted there.

Ibid.

The unfortunate Sophia Dorothea lived in seclusion at Ahlden for the rest of her life, leaving the way clear for the two beloved Germans he brought with him to England.

THE 'MAYPOLE' AND THE 'ELEPHANT-AND-CASTLE'

One of the German favourites was immensely tall and thin, the other hugely fat; hence their nicknames in England. The fat one, Mme Kilmansegge, was created countess of Darlington by George I and her compatriot, the Schulenberg, became duchess of Kendal.

Lady Darlington, whom I saw at my mother's in my infancy, and whom I remember by being terrified at her enormous figure, was as corpulent & ample as the Duchess was long & emaciated. Two fierce black eyes, large & rolling beneath two lofty arched eyebrows, two acres of cheeks spread with crimson, an ocean of neck that overflowed & was not distinguished from the lower part of her body, and no part restrained by stays—no wonder that a child dreaded such an ogress, and that the mob of London were highly diverted at the importation of so uncommon a seraglio!

Ibid.

NO HARM IN PERKS

One of the favourites was jeered at by a mob. 'Goot people,' she called, 'why you abuse us? We come for all your goots!' A voice from the crowd shouted back: 'Yes, damn ye, and for all our chattels too!' A German servant of the King's was shocked by the looting and asked permission to return to Hanover, where people were honest and careful. George: 'Bah! It is only English money. Steal like the rest!'

The Duke of Windsor, 'My Hanoverian Ancestors', ed. J. Bryan III, unpublished

THE 1715 REBELLION IN FAVOUR OF THE 'OLD PRETENDER'

James, son of James II and his second wife Mary of Modena, was a Catholic and so excluded from the succession. The 'Fifteen' was put down with harshness and the king was criticized for ostentatiously going to a ball on the day that Lord Derwentwater and Lord Kenmuir, James's supporters, were executed. He was said to have thrown Lady Nithsdale to the floor when she came to plead for her husband. However, when Lord Nithsdale escaped from the Tower dressed as a woman George merely remarked:

'It is the best that a man in his situation could have done.'

Lord Nottingham, Lord Privy Seal, lost his job when he pleaded for mercy towards the Jacobite peers. Horatio Walpole, brother of Robert Walpole, 'an honest diplomat at The Hague', put the king's point of view arguing his need to placate his Whig ministers. Walpole wanted all seven captured noblemen executed.

The conduct of Lrd Nottingham & his brother in relation to the condemn'd Lrds, is unaccountable; to plead for mercy in favour of the most stubborn Rebels, and at the same time to cast the Odium of Cruelty upon the mildest & best of Princes; is what I do not understand; I don't doubt but their removal is look'd upon here by men of Sense as necessary to preserve union in his Majtys counsels and steadiness in his government; I am wth the greatest respect & affection Yr Lrdps &c—H. Walpole.

An Honest Diplomat at the Hague: The Private Letters of Horatio Walpole, 1715–1716, ed. John J. Murray (Bloomington, Ind./The Hague, 1955; spelling modernized)

TRIBUTE TO GEORGE BY HORACE WALPOLE

On one of his journeys to Hanover his coach broke. At a distance in view was a chateau of a considerable German nobleman. The King sent to borrow assistance. The possessor came, conveyed the King to his house, & begged the honour of his Majesty's accepting a dinner, while his carriage was repairing—and while dinner was preparing, begged leave to amuse his Majesty with a collection of pictures, which he had formed in several tours to Italy—but what did the King see in one of the rooms but an unknown portrait of a person in the robes & with the regalia of the Sovereigns of Great Britain. George asked whom it represented? The nobleman replied with much diffident but decent respect that in various journeys to Rome he had been acquainted with the Chevalier de St Georges [the Old Pretender], who had done him the honour of sending him that picture. 'Upon my word,' said the King instantly, 'it is very like to the Family.' It was impossible to remove the embarrassment of the proprietor with more good breeding.

Horace Walpole, *Reminiscences*

THE DUKE OF WINDSOR AND THE JACOBITES

In certain circles a favourite toast was 'The King' drunk with a wine glass held above a glass of water, to signify 'The King over the water'.

Toward the end of a dinner at a great English house I noticed that I alone had been given a finger bowl. My hostess explained that for more than two hundred years the family had taken this precaution against an expression of Jacobite sentiment in the presence of Hanoverian royalty.

Windsor, 'My Hanoverian Ancestors'

ANTAGONISM OF SOVEREIGN AND HEIR—A HANOVERIAN VICE

Towards the end of September 1716 the smouldering quarrel between the king and the prince of Wales flared up.

The King thought that the Prince had threatened the life of the Duke of Newcastle at the christening of the Prince's son. The incident had all the quality of high farce. The King, to show his authority, had insisted that the traditional right of the Lord Chamberlain to be a godparent of the Prince's child should be honoured. This infuriated his son, who detested all his father's servants. Fury once felt needed expression and at the christening he grabbed Newcastle's arm and said: 'Rascal, I find you out.' The Prince's accent was never very clear and Newcastle's intelligence rarely exact. Bewildered and confused, Newcastle understood the Prince to have said: 'I fight you.' He rushed back to the King in an irrepressible state of excitement and said that his life had been threatened. The King called a cabinet meeting; ministers of State were sent to interrogate the Prince, who called Newcastle a liar, and this promptly led the King to place his son under what was virtually close arrest. The ministers began to talk of *Habeas Corpus*; the Prince, with perhaps memories of his mother's long imprisonment, began to express contrition. The King would not listen to his son, but the fears of his ministers worked on him sufficiently to call off the Yeomen of the Guard. However, he relieved his feelings by expelling his son from St James's Palace but insisted on his grandchildren remaining; in future he himself would be responsible for their education. The Princess naturally followed her husband into exile; they were devoted to their little daughters whom they had left behind and after a time paid a clandestine visit to them. Immediately they had a sharp rebuke from the King.

Monsieur Coke

You will say from me that I find it very bad that he came to St James without my permission, I being there, and that he and the Princess must in future ask permission of me when they want to come and see the children, which will be granted once a week.

J. H. Plumb, *Sir Robert Walpole*, I (1956). *The letter is translated from George's execrable French.*

THE (TEMPORARY) RECONCILIATION

There was national rejoicing over a reconciliation three and a half years later. This is how it was celebrated in Axminster. Mrs Elizabeth Molesworth, lady-in-waiting to the princess, writes to Mrs Henrietta Howard, the prince's mistress:

I suppose you have had no small share in the joy this happy reconciliation has occasioned. I heartily congratulate you upon it. Mr Molesworth [her husband] testified his zeal at the expense of his sobriety; for he was not satisfied to make his men drunk, but got drunk himself, and it was no fault of his that I was not so too; in short, he celebrated the news in a manner that alarmed the country people, for after he had made them ring the bells all day, in the evening he made his troop draw up before his lodging and he at the head of them, and began the king and prince's healths together, and then the princess, and after, the rest of the royal family; at every health he made his troop fire round a volley of shot: he invited several gentlemen to pledge these healths, and when they had done they threw the glasses over their heads. When this was done he carried them all with him to drink a bowl of punch. As to his men, after they had dispatched a barrel of ale they thought themselves not glad enough, and he, to make them so, went amongst them and gave them money to finish in wine. He is at present a little disordered with that night's work.

Letters of Henrietta Howard, Countess of Suffolk (1824)

After the temporary reconciliation the prince heard his father begin to grumble as he took his leave: 'Votre conduite! Votre conduite!'

Grant Robertson, in England under the Hanoverians *and Ragnhild Hatton, in* George I *both mention a plot to kidnap the prince—John Lord Hervey's memoirs speak of it: a letter was found by the prince among the king's papers after his death, in which Sunderland suggested kidnapping the prince and transporting him to the plantations. George refused; but that a courtier could suggest such a thing shows the extent of the king's known dislike of his son.*

SAYINGS OF GEORGE I

He greatly enjoyed the company of Dean Lockyer. Once when the Dean had just returned from a visit to Rome the king asked him jocularly whether he had 'succeeded in converting the Pope'. The Dean replied: 'No, your Majesty. His Holiness has most excellent Church preferment and a most desirable bishopric, and I had nothing better to offer him.'

George is said to have made a single bow to intellect when a German nobleman congratulated him on becoming king of England: 'Congratulate me rather on having Newton for a subject in one country and Leibniz in the other.'

It was doubtful whether even this salute was sincere, since George did nothing to satisfy Leibniz's wish to visit England. His famous remark made in a guttural accent was more characteristic of his attitude to the intellect and arts: 'I hate all Boets and Bainters.'

His craze for agricultural improvement is said to have prompted him to ask a notorious question of his minister: 'How much would it cost him to close St James's Park to the public and use it to grow turnips?' *The answer was:* 'Only three crowns, Sire.'

The same story is also told of Queen Anne who learnt that it would cost her two crowns to enclose the park in her garden.

THE YOUNG HORACE WALPOLE'S LONGING TO SEE THE KING

The night but one before the King began his last journey, my Mother carried me at Ten at night to the apartment of the Countess of Walsingham on the ground-floor towards the garden at St James's, which opened into that of . . . the Duchess of Kendal [Mme Schulenberg]:

Notice being given that the King was come down to supper, Lady Walsingham took me alone into the Duchess's ante-room, where we found alone the King and her. I knelt down, & kissed his hand, he said a few words to me, & my conductress led me back to my Mother.

The person of the King is as perfect in my memory as if I saw him but yesterday. It was That of an elderly Man rather pale & exactly like to his pictures & coins; not tall, of an aspect rather good than august, with a dark tye wig, a plain coat, waistcoat and breeches of snuff-coloured cloth, with stockings of the same colour, & a blue ribband over all.

So entirely was He my object, that I do not believe I once looked at the Duchess; but as I could not avoid seeing her on entering the room, I remember that just beyond his Majesty stood a very tall, lean, ill-favoured old Lady . . .

My childish loyalty & the condescension in gratifying it, were, I suppose, causes that contributed very soon afterwards to make me shed a flood of tears for that Sovereign's death, when with the other Scholars of Eton College I walked in procession to the proclamation of the Successor.

Horace Walpole, *Reminiscences*

A DOUBLE DEATH

Princess Sophia Dorothea died on 13 November 1726. Towards the end of her life, George had offered to liberate and reinstate her. According to court legend she refused:

'If what I am accused of is true, I am unworthy of his bed; and if it is false, he is unworthy of mine.'

George left her unburied for over six months and then at last decided to travel to Hanover where the body lay. Walpole tells a good story of his departure and its cause.

I have said that the disgraced Princess died but a short time before the King. It is known that in Queen Anne's time there was much noise about French Prophets. A Female of that vocation (for we know from Scripture that the gift of prophecy is not limited to one gender) warned George I to take care of his Wife, as he would not survive her a year. That Oracle was probably dictated to the French Deborah [prophetess] by the Duke and Duchess of Zell, who might be apprehensive lest the Duchess of Kendal should be tempted to remove entirely the obstacle to her conscientious union with their Son-in-law. Most Germans are superstitious, even such as have few other impressions of Religion. George gave such credit to the denunciation, that on the eve of his last departure he took leave of his Son and the Princess of Wales with tears, telling them he should never see them more. It was certainly his own approaching Fate that melted him, not the thought of quitting for ever two persons he hated. He did sometimes so much justice to his Son as to say 'Il est fougueux [fiery], mais il a de l'honneur.'—as for Queen Caroline, to his confidants he termed her 'Cette Diablesse Madame la Princesse'. Ibid.

Flouting all decorum, the king had gone to a play at the Haymarket on the day that he heard of his ex-wife's death and later set out for her burial accompanied by the 'Maypole'. But feeling unwell and anxious about the prophecy, he parted from his mistress before reaching Osnabrück and went on alone. Before arriving at the city of his birth, he had a massive stroke in his coach and died in the room in which he had been born sixty-seven years before.

EPILOGUE: A GHOSTLY VISITANT TO THE 'MAYPOLE'

In a tender mood he [George I] promised the Duchess of Kendal that if She survived him, & it were possible for the Departed to return to this World, he would make her a visit. The Duchess on his death so much expected the accomplishment of that engagement, that a large raven, or some black fowl flying into one of the windows of her Villa at Isleworth, She was persuaded it was the soul of her departed Monarch so accoutred, & received & treated it with all the respect & tenderness of Duty, till the Royal Bird or She took their last flight.

Horace Walpole, *Reminiscences*

George II

1727–1760

The second of the Georges marked a political change that had begun after Queen Anne. The monarch reigned but parliament ruled—or did so more and more. This produced a paradox. George II, himself unremarkable, presided over Britain's most sensational expansion under Pitt the Elder, Robert Clive, and James Wolfe. The king's defects in no way diminished the country's glories, and the lines written by Walter Savage Landor were unfair:

George the First was always reckoned
Vile, but viler George the Second.

At least George loved his clever wife as well as his mistresses. And he was responsible for one of the rare Hanoverian witticisms. To someone who complained that General Wolfe was mad, George retorted,

'Then I wish he would bite some other of my generals!'

George II was neatly built, fair with bulging blue eyes and a rich complexion, receding forehead and prominent nose. Though greedy he was not dishonest.

GEORGE II'S BOASTED ENGLISHNESS

On coming to the throne George said before addressing parliament:

'I have not a drop of blood in my veins which is not English.'

However, the end of the sentence was later quoted as 'dat is not English'; and as prince of Wales he had shown little admiration for the country he was to head.

He thought there were no manners out of Germany. Sarah Marlborough once coming to visit the Princess Caroline, whilst Her Royal Highness was whipping one of the roaring Royal children, 'Ah!' says George, who was standing by, 'you have no good manners in England, because you are not properly brought up when you are young.' He insisted that no English cook could roast, no English coachman could drive: he actually questioned the superiority of our nobility, our horses, and our roast beef!

W. M. Thackeray, *The Four Georges*, (1860). *It must be remembered that Thackeray's comments are not those of a contemporary historian but of a brilliant nineteenth-century satirist.*

THE CORONATION OF QUEEN CAROLINE

The dress of the Queen on this occasion was as fine as the accumulated riches of the City and suburbs could make it, for besides her own jewels (which were a great number and very valuable), she had on her head and on her shoulders all the pearls she could borrow of the ladies of quality at one end of the town, and on her petticoat all the diamonds she could hire of the Jews and jewellers at the other, so that the appearance and the truth of her finery was a mixture of magnificence and meanness not unlike the *éclat* of royalty in many other particulars when it comes to be nicely examined and its sources traced to what money hires or flattery lends.

John Lord Hervey, *Memoirs of the Reign of George II*, ed. J. W. Croker (1884)

J. W. Croker put a footnote to this waspish report, quoting from Walpole's Reminiscences:

There was some little excuse for this. 'At the death of Queen Anne, such a clearance had been made of her Majesty's jewels, or the new King George I had so instantly distributed them among his German favourites, that Lady Suffolk told me Queen Caroline never obtained of the late Queen's jewels but one pearl necklace.'

Horace Walpole added that the queen's coronation petticoat was so stiff with jewels that a draw-string had to be attached to pull it up and down like a curtain as she knelt.

CAROLINE'S INFLUENCE ON GEORGE II, 1733

She always at first gave in to all his notions, though never so extravagant, and made him imagine any change she wrought in them to be an afterthought of his own.

Hervey

QUEEN CAROLINE AND THE PRINCESS ROYAL'S WEDDING, 1734

The Prince of Orange was a less shocking and less ridiculous figure in this pompous procession and at supper than one could naturally have expected such an Aesop, in such trappings and such eminence, to have appeared. He had a long peruke like hair that flowed all over his back, and hid the roundness of it; and as his countenance was not bad, there was nothing very strikingly disagreeable about his stature.

But when he was undressed, and came in his nightgown and nightcap into the room to go to bed, the appearance he made was as indescribable as the astonished countenances of everybody who beheld him. From the shape of his brocaded gown and the make of his back, he looked behind as if he had no head, and before as if he had no neck and no legs. The Queen [Caroline] in speaking of the whole ceremony next morning alone with Lord Hervey, when she came to mention this part of it, said, 'Ah! mon Dieu! when I saw the monster enter, to lie with my daughter, I thought I should faint . . .' What seems most extraordinary was, that from the time of their being married till they went out of England, Lord Hervey . . . said she always behaved to him as if he had been an Adonis. Ibid.

The king had in fact given his daughter permission to reject her hunchbacked suitor; she answered that she would have him even if he were a baboon. 'Well, then, there is baboon enough for you', said George II.

COVETOUSNESS

Sir Robert Walpole told Lord Hervey that he had had the King's orders to buy one hundred lottery tickets for his Majesty to send to her [Mme Walmoden, his mistress] . . . He also told Lord Hervey that the King, to save making the £1000 disbursement out of his privy purse, had ordered him to charge the purchase-money of these tickets in the secret service—adding, that his Majesty, like all other covetous people, fancied always that he gave less when he gave out of a stock which, though equally his own, was money he had never fingered. Ibid.

As prince of Wales, George II once tried to marry Lady Diana Spencer.

Money he soon wanted; old Sarah Duchess of Marlborough, ever proud and ever malignant, was persuaded to offer her favourite granddaughter, Lady Diana Spencer, afterwards Duchess of Bedford, to the Prince of Wales, with a fortune of an hundred thousand pounds. He accepted the proposal, & the day was fixed for their being secretly married at the Duchess's lodge in the great park at Windsor. Sir Robert Walpole got intelligence of the project, prevented it, and the secret was buried in silence. Horace Walpole, *Reminiscences*

He gave little in charity, and the only present [Prime Minister] Walpole ever had from him was said to have been a diamond with a flaw in it.

J. M. Rigg, 'George II', in *Dictionary of National Biography* (1908)

Evidence of Unpopularity

The king was once in a storm at sea. Somebody asking, two or three days after the tempest, 'How the wind was now for the king?' was answered, 'Like the nation—against him.' Ibid.

During one of George's foreign tours, a broken-down old horse was turned loose in the streets of London; the animal bore a placard 'Let nobody stop me—I am the King's Hanover equipage going to fetch His Majesty and his whore to England.' . . . When the Government tried to reduce the consumption of gin, mobs stormed round the royal coach screaming 'No Gin! No King!' John Clarke, *George II*

THE FIRST PRODUCTION OF *THE BEGGARS' OPERA*

It was a spectacular success with all but the government and King. George II forbade the Duchess of Queensberry from coming to Court as a punishment for her having raised subscriptions for the printing of Gay's sequel, *Polly*. She replied to the King: 'The Duchess of Queensberry is surprised and well pleased that the King and Queen have given her so agreeable a command as forbidding her the Court, where she never came for diversion, but to bestow a very great civility on the King and Queen. She hopes that by so unprecedented an order as this the King will see as few as he wishes at Court, particularly such as dare to think or speak the truth.' A. Calder-Marshall, *The Grand Century of the Lady* (1976)

GEORGE'S OBSESSION WITH HANOVER INCREASED HIS UNPOPULARITY

During one of his absences a notice appeared on the gates of St James's Palace: 'Lost or strayed out of this house, a man who has left a wife and six children on the parish.' The reward for finding him was four shillings and sixpence, 'nobody judging him to deserve a crown [five shillings]'. Another public notice posted at the Royal Exchange read: 'It is reported that his Hanoverian Majesty designs to visit his British dominion for three months in the spring.' Windsor, 'My Hanoverian Ancestors'

The Mistresses

They did nothing for his popularity, though he hoped they publicized his virility and freedom from his wife's influence. But the contemporary jingle still flourished:

You may strut, dapper George, but 'twill all be in vain;
We know 'tis Queen Caroline, not you that reign.

The duke of Windsor relates that he first met Queen Caroline in a nursery rhyme:

Queen, Queen Caroline
Dipped her nose in turpentine,
Turpentine to make it shine,
Poor Queen Caroline.

The poor queen was also said to dip her rheumatic legs in freezing mud in order to accompany George on his walks. Two of her ladies for whom George lusted were Miss Mary Bellenden and Mrs Henrietta Howard, who became the countess of Suffolk after Prince George became king.

Miss Bellenden by no means felt a reciprocal passion. The Prince's gallantry was by no means delicate; & his avarice disgusted her. One evening sitting by her he took out his purse & counted his money. He repeated the numeration a second time: the giddy Bellenden lost her patience & cried out, 'Sir, I cannot bear it! If you count your money any more, I will go out of the room.' The chink of the gold did not tempt her more than the person of his Royal Highness.

Miss Bellenden married John Campbell, afterwards 4th Duke of Argyll. Mrs Howard was the intimate friend of Miss Bellenden, had been the confidante of the Prince's passion, and on Mrs Campbell's eclipse, succeeded to her friend's post of favourite—but not to her resistance.

From the steady decorum of Mrs Howard I should conclude that she would have preferred the advantages of her situation to the ostentatious éclat of it: but many obstacles stood in the way of total concealment; nor do I suppose that love had any share in the sacrifice she made of her virtue. She had felt poverty, and was far from disliking power. Mr Howard was probably as little agreeable to her as he proved worthless. The King, though very amorous, was certainly more attracted by a silly idea he had entertained of gallantry being becoming, than by a love of variety; & he added the more egregious folly of fancying that his inconstancy proved he was not governed; but so awkwardly did he manage that artifice, that it but demonstrated more clearly the influence of the Queen. With such a disposition secrecy would by no means have answered his Majesty's view: yet the publicity of the intrigue was especially owing to Mr Howard, who far from ceding his wife quietly, went one night into the quadrangle of St James's, and vociferously demanded her to be restored to him before the Guards and other audience. Being thrust out, he sent a letter to her by the Archbishop of Canterbury, retaining her, and the Archbishop by *his* instructions consigned the summons to the Queen, who had the malicious pleasure of delivering the letter to her rival . . .

She was discreet without being reserved: & having no bad qualities, & being constant to her connections she preserved uncommon respect to the end of her life; & from the propriety & decency of her behaviour was always treated as if her virtue had never been questioned; her friends even affecting to suppose that her connection with the King had been confined to pure friendship—unfortunately his Majesty's passions were too indelicate to have been confined to platonic love for a woman who was deaf [she went deaf early in life]—sentiments he had expressed in a letter to the Queen, who however jealous of Lady Suffolk, had latterly dreaded the King's contracting a new attachment to a younger rival, & had prevented Lady Suffolk from leaving the court as early as she had wished to. 'I don't know', said his Majesty, 'why you will not let me part with an old deaf woman of whom I am weary.'

. . . *she* was elegant, her lover, the reverse and most unentertaining, and void of confidence in her. His motions too were measured by etiquette and the clock. He visited her every evening at nine; but with such dull punctuality, that he frequently walked about his chamber for ten minutes with his watch in his hand, if the stated minute was not arrived.

But from the Queen she tasted more positive vexations. Till she became Countess of Suffolk (1731), she constantly dressed the Queen's head, who delighted in subjecting her to such servile offices, though always apologising to *her good Howard.* Often her Majesty had more complete triumph. It happened more than once, that the King coming into the room while the Queen was dressing, has snatched off her handkerchief [covering her bosom], & turning rudely to Mrs Howard, has cried 'Because you have an ugly neck yourself, you hide the Queen's!'—oh! that you had seen that royal neck! Since the days of Homer, who admired the cow-like eyes of Juno, never I believe were seen dugs that would have assorted so well with the delineation of that quadruped.

It is certain that the King always preferred the Queen's person to that of any other woman; nor ever described his idea of beauty, but he drew the picture of his wife. They always went to bed on his return from Hanover as soon as he came out of the drawing-room, as at all times they did after dinner.

Horace Walpole, *Reminiscences*

THE COMING OF AMELIA SOPHIA, COUNTESS VON WALMODEN

Queen Caroline's instinct to keep a hold of Lady Suffolk for fear of finding something worse was correct. Lord Hervey relates that when George II fell in love with this young and beautiful German he told his wife: 'You must love the Walmoden, for she loves me.' But instead of loving the Walmoden, Caroline made a determined approach to Lady Suffolk, begging her not to leave the court just because the king had tired of her in favour of Walmoden. Afterwards she reported the interview to Hervey.

'I told her', said the Queen, 'that she and I were not of an age to think of these sort of things in such a romantic way; and said, "My good lady Suffolk, you are the best servant in the world, and, as I should be most extremely sorry to lose you, pray take a week to consider of this business and give me your word not to read any romances in that time, and then I dare say you will lay aside all thought of doing what, believe me, you will repent . . ." '

Hervey

Hervey added that the queen told him Lady Suffolk had hitherto received £2,000 a year rising to £3,200 from George II,

besides several little dabs of money both before and since he came to the Crown.

Finally, Caroline changed tactics and wrote to George suggesting he bring the Walmoden back to England with him rather than staying so long away in Hanover to see her. George wrote back delighted, extolling the virtues of his mistress to his wife.

Beside his womanizing, the king had a coarse line in rudeness to men. In 1735 he criticized Hervey's friendship with Bishop Hoadley:

'A pretty fellow for a friend!' said the King, turning to Lord Hervey. 'Pray what is it that charms you in him? His pretty limping gait' (and then he acted the Bishop's lameness), 'or his nasty stinking breath?—phaugh!—or his silly laugh, when he grins in your face for nothing, and shows his nasty rotten teeth? . . .'

Ibid.

It is fair to add that this was not wholly gratuitous but comes in the context of an attack on the bishop as an unpleasant hypocrite, preaching from a Bible he did not believe and taking money from a government he criticized.

THE ALTAR AND THE TABLES

Gambling was the rage in the eighteenth century and George II saw no incompatibility between the two festivals that closely followed each other. After all, gold, which George adored, featured at both.

Wednesday, Jan. 5, 1731. This being the Twelfth Day . . . Their Majesties, the Prince of Wales, and the three eldest Princesses, preceded by Heralds etc., went to the Chapel Royal, and heard Divine Service. The King and Prince made the Offerings at the altar, of Gold, Frankincense and Myrrh, according to Custom. At night, their Majesties etc., played at Hazard, for the benefit of the Groom Porter [an officer of the Lord Steward's department of the Royal Household in charge of gambling], and 'twas said the King won 600 Guineas, and the Queen 360, Princess Amelia 20, Princess Caroline, the Earl of Portmore and the Duke of Grafton, several thousands.

The Gentleman's Magazine, first number

THE KING AND QUEEN AGREED IN HATING THE PRINCE OF WALES

Prince Frederick (Fretz) was the father of George III and died before he could succeed to the throne.

'My God,' says the Queen, 'popularity always makes me sick; but *Fretz's* popularity makes me vomit.' *Hervey*

Caroline also said:

'Our first-born is the greatest ass, the greatest liar, the greatest canaille and the greatest beast in the whole world and we heartily wish he was out of it.' Clarke, *George II*

George II suggested:

Fretz might be a 'Wechselbag' or changeling. *Hervey*

DEATH OF QUEEN CAROLINE, 1737

In Horace Walpole's opinion George II 'loved Queen Caroline's little finger more than he loved Lady Suffolk's whole body'. George refused to allow Fretz to say goodbye to his mother and Caroline is said to have been thankful: 'At least I shall have one comfort in having my eyes eternally closed. I shall never see that monster again!' Caroline also refused the consolation of the Church of England: she was a learned lady who loved books including theology, but had to read them in her closet as the sight of a book made George see red.

The Queen's chief study was divinity; and she had rather weakened her faith than enlightened it. She was at least not orthodox; & her confidante Lady Sundon, an absurd and pompous simpleton, swayed her countenance towards the less-believing clergy. The Queen however was so sincere at her death, that when Archbishop Potter was to administer the sacrament to her, she declined taking it, very few persons being in the room. When the prelate retired, the courtiers in the anteroom crowded round him crying, 'My Lord, has the Queen received?' His Grace artfully eluded the question, only saying most devoutly, 'Her Majesty was in a heavenly disposition'—and the truth escaped the public.

Horace Walpole, *Reminiscences*

One of her last acts was to hand back the ruby ring George had given her at their coronation with the words, 'Naked I came to you and naked I go from you.' Hervey describes the last scene of all, preceded by its moment of farce:

'It is not necessary to examine whether the Queen's reasoning was good or bad in wishing the King, in case she died, should marry again:—it is certain she did wish it . . . and gave it now as her advice to him when she was dying—upon which his sobs began to rise and his tears to fall with double vehemence. Whilst in the midst of this passion, wiping his eyes, and sobbing between every word, with much ado he got out the answer, 'Non, j'aurai des Maîtresses.' To which the Queen made no other reply than, 'Ah! mon Dieu! cela n'empêche pas.' . . .

About ten o'clock on Sunday night—the King being in bed and asleep on the floor at the feet of the Queen's bed, and the Princess Emily in a couch-bed in a corner of the room—the Queen began to rattle in the throat; and Mrs Purcel giving the alarm that she was expiring, all in the room started up . . . All she said before she died was 'I have now got an asthma. Open the window.' Then she said, '*Pray*.' Upon which the Princess Emily began to read some prayers of which she repeated ten words before the Queen expired. The Princess Caroline held a looking-glass to her lips, and finding there was not the least damp upon it, cried, 'Tis over!' and said not one word more . . . The King kissed the face and hands of the lifeless body several times, but in a few minutes left the Queen's apartment and went to that of his daughters, accompanied only by them.

Hervey

THE BATTLE OF DETTINGEN, 1743

The king's courage on this occasion, and his 'tenderness' towards his dying wife, regained for him some of his long-lost popularity.

He was the last British monarch to lead his troops into battle.

The Gentleman's Magazine reported that Lord Carteret had sent a despatch to the Duke of Newcastle which began: '. . . His Majesty was all the Time in the Heat of the Fire; but is in perfect Health . . .' The King's battle-cry had been patriotic enough: 'Now, boys, now for the honour of England; fire and behave bravely and the French will soon run.' The King was on foot for earlier in the action his horse had bolted with him to the rear. He walked back, saying, 'I can be sure of my own legs. They will not run away with me.' The men cheered their 'Little Captain' and drove the French off.

Windsor, 'My Hanoverian Ancestors'

PROTOCOL AND THE PRETENDERS

Four years after the 1745 Rebellion when Bonnie Prince Charlie, the Young Pretender, was defeated, Lord Chesterfield, putting on the voice of Polonius, advised his son how to treat this awkward family if he should run into them while abroad.

5 September 1749. You will, in many parts of Italy, meet with numbers of the Pretender's people . . . especially at Rome, and probably the Pretender himself. It is none of your business to declare war on these people; as little as it is your interest, or, I hope, your inclination, to connect yourself with them; and, therefore, I recommend to you a perfect neutrality. Avoid them as much as you can with decency and good manners; but, when you cannot, avoid any political conversations or debates with them: tell them that you do not concern yourself with political matters—that you are neither a maker nor a deposer of Kings—that, when you left England, you left a King in it, and have not since heard either of his death, or of any revolution that has happened, and that you take Kings and kingdoms as you find them; but enter no farther into matters with them, which be of no use, and might bring on heat and quarrels. When you speak of the old Pretender, you will call him only the Chevalier de St George, but mention him as seldom as possible. Should he chance to speak to you at any assembly (as I am told he sometimes does to the English), be sure that you seem not to know him, and answer him civilly, but always either in French or in Italian; and give him, in the former, the appellation of *Monsieur*, and in the latter of *Signore*. Should you meet with the Cardinal of York [second son of the Old Pretender], you will be under no difficulty, for he has, as Cardinal, an undoubted right to *Eminenza* . . . never be drawn into any altercation with them about the imaginary rights of their King as they call him.

Letters of Lord Chesterfield, ed. J. B. Bradshaw (1926)

A MASKED BALL IN HANOVER—WITH THACKERAY'S SATIRICAL COMMENTS

'At night, supper was served in the gallery with three great tables, and the King was very merry. After supper dancing was resumed, and I [Lord Hervey] did not get home till five o'clock by full daylight to Hanover. Some days afterwards we had, in the opera-house at Hanover, a great assembly. The King appeared in a Turkish dress; his turban was ornamented with a magnificent *agrafe* of diamonds; the Lady Yarmouth [Walmoden] was dressed as a sultana . . .'

So, while poor Caroline is resting in her coffin, dapper little George, with his red face and white eyebrows and goggle-eyes, at sixty years of age, is dancing a pretty dance with Madame Walmoden, and capering about dressed up like a Turk! For twenty years more [actually only sixteen], that little old Bajazet went on in this Turkish fashion, until the fit came which choked the old man, when he ordered the side of his coffin to be taken out, as well as that of poor Caroline's who had preceded him, so that his sinful old bones and ashes might mingle with those of the faithful creature. O strutting Turkey-cock of Herrenhausen! O naughty little Mahomet! in what Turkish paradise are you now, and where be your painted houris?

Thackeray, *The Four Georges*, quoting *Hervey*

A LAMENT FOR THE DECEASED GEORGE II

Parson Porteous, poet, prose-writer and afterwards bishop, composed a threnody which included many ridiculous lines, such as the suggestion that George

Saw in his offspring all himself renewed;
The same fair path of glory still pursued

and also saw young Prince George (III)

Blend all his grandsire's virtues [George II's virtues] with his own,
And form their mingled radiance for the throne—
No farther blessing could on earth be given—
The next degree of happiness was—heaven!

Thackeray commented:

If he had been good, if he had been just, if he had been pure in life, and wise in council, could the poet have said much more? It was a parson who came and wept over this grave, with Walmoden sitting on it, and claimed heaven for the poor old man slumbering below. Here was one who had neither dignity, learning, morals, nor wit—who tainted a great society by a bad example; who, in youth, manhood, old age, was gross, low, and sensual; and Mr Porteous, afterwards my Lord Bishop Porteous, says the earth was not good enough for him, and that his only place was heaven! Bravo, Mr Porteous! The divine who wept these tears over George the Second's memory wore George the Third's lawn.

Thackeray, *The Four Georges*

George III

1760–1820

Young George ascended the throne at twenty-two, well-intentioned but ill-prepared for a reign that was to face George Washington, the French Revolution, and Napoleon. Sadly, by the time of Waterloo his conversations were carried on with creatures not of this world, including angels. He was right in thinking that the king should not be the prisoner of one political party—the Whigs—but misguided in trying to be above party.

He was of medium height and not unattractive but his mannerisms—fast talk and abundant gestures—worked against regal dignity, despite a fine voice. Today the main interest and argument revolve around his three bouts of 'madness'. Were they mental, due to the effect of wrangles with his eldest son and ministers? Or was a disease responsible, like porphyria, or lead poisoning caused by the vessels which held his favourite sauerkraut and lemonade? Whatever the truth, the symptoms were those of insanity.

DEATH OF GEORGE'S FATHER, 'POOR FRED', 1751

The teenage prince was genuinely unhappy when his father finally succumbed to a blow from a cricket ball. Putting his hand on his heart he said:

'I feel something here, just as I did when I saw the two workmen fall from the scaffold at Kew.'

Stanley Ayling, *George the Third* (1972)

INFLUENCE OF PRINCE GEORGE'S MOTHER

In the year of George III's death, 1820, John Nichols's 'Recollections' launched the famous story of his mother having continually admonished her son as a child, 'George, be a king!' This is said to have encouraged him in his later obstinacy. Modern scholars reject this interpretation, arguing that if the princess ever used the words she was simply telling young George to sit up, take his elbows off the table, be a man, be a king.

HOSTILITY OF GEORGE II TOWARDS HIS GRANDSON

Calling attention to the traditional hostility of the Hanoverians towards their heirs, Lord Chesterfield observed that George II's dislike of his son had been repeated in the case of his grandson:

There is nothing new under the sun, nor under the grandson either.

Letters of Lord Chesterfield, I

LOVE AFFAIRS LEGENDARY AND REAL, 1761

In his first year as monarch, still a bachelor, George is said to have tried to 'be a king' by seducing a 'Fair Quaker' named Hannah Lightfoot, much as his royal ancestor had seduced 'Fair Rosamund'. The Royal Archives at Windsor do not support this story or the further tale of a bastard daughter. But George's love for Lady Sarah Lennox in summer 1761 in response to her advances is mentioned by Walpole:

She appeared every morning at Holland House, in a field close to the great road (where the King passed on horseback) in a fancy habit making hay.

Horace Walpole, *Reminiscences*

Thackeray wrote that she was

making hay at him. Thackeray, *The Four Georges*

The above version of the story does not tally with Lady Sarah's family tradition. According to this, George proposed to her through an intermediary, her cousin Susan Fox-Strangways, and next time he met Sarah, wrote her son Henry Napier, drew her 'into a recess in one of the large windows,' *and asked her* 'what she thought of his proposal'.

'Tell me, for my happiness depends on it!' 'Nothing, Sir,' was my mother's reply, upon which he left her abruptly, exclaiming pettishly 'Nothing comes of nothing.'

Ayling, *George the Third*

MARRIAGE TO PRINCESS CHARLOTTE OF MECKLENBURG-STRELITZ, 1761

George got married before anyone else could make hay at him or of him. Charlotte's nose was not her best point having a marked upward tilt.

Though later popular, the people when they first saw her shouted 'Pug! Pug! Pug!' Charlotte, puzzled, said, 'Vat is dat they do say—*poog*? Vat means *poog*?' The Duchess of Ancaster, to whom she had addressed the question replied 'It means 'God bless Your Royal Highness.'

Windsor, 'My Hanoverian Ancestors'

It was considered a bad omen when the great diamond dropped out of the king's crown between St James's and Westminster on his wedding day and was later seen as heralding the loss of America; in much the same way the falling of the Maltese Cross from George V's bier was thought to be ominous for the reign of Edward VIII.

THE CORONATION—THROUGH A SCHOOLBOY'S EYES, SEPTEMBER 1760

George and Charlotte were crowned a fortnight after their wedding. Hickey was eleven years old. His father paid 50 guineas for one of the boxes high up in the Abbey known as 'nunneries'; it held twelve members of family and friends. They set off in the family coach from St Albans Street at midnight but did not reach the Abbey till 7 a.m., owing to the chaos of carriages, some aiming for the Abbey, others for Westminster Hall:

This created much confusion and running against each other, whereby glasses and panels were demolished without number, the noise of which accompanied by the screeches of the terrified ladies, was at times truly terrific. *Memoirs of William Hickey*, ed. Peter Quennell (1960)

They found a hot breakfast awaiting them in their 'nunnery', after which young Hickey amused himself running up and down the long gallery till noon, when notice was given that their Majesties' procession had started:

Exactly at one they entered the Abbey, and we had a capital view of the whole ceremony. Their Majesties . . . being crowned, the Archbishop of Canterbury mounted the pulpit to deliver the sermon; and, as many thousands were out of the possibility of hearing a single syllable, they took the opportunity to eat their meal, when the general clattering of knives, forks and plates, and glasses that ensued, produced a most ridiculous effect, and a universal burst of laughter followed. Ibid.

SIMPLE TASTES

They met, and they were married, and for years they led the happiest simplest lives surely ever led by married couple . . . They had the simplest pleasures . . . little country dances, to which a dozen couples were invited, and where the honest King would stand up and dance for three hours at a time to one tune; after which delicious excitement they would go to bed without any supper (the Court people grumbling sadly at that absence of supper), and get up quite early the next morning, and perhaps the next night have another dance . . .

He was a patron of the arts, after his fashion; kind and gracious to the artists whom he favoured, and respectful to their calling. He wanted once to establish an Order of Minerva for literary and scientific characters; the knights were to take rank after the Knights of the Bath, and to sport a straw-coloured ribbon and a star of sixteen points. But there was such a row among the *literati* as to the persons who should be appointed, that the plan was given up, and Minerva and her star never came down amongst us.

Thackeray, *The Four Georges*

Describing George III's cultural interests Thackeray implied wrongly that the king was a cultural philistine:

The theatre was always his delight. His bishops and clergy used to attend it, thinking it no shame to appear where that good man was seen. He is said not to have cared for Shakespeare or tragedy much; farces and pantomimes were his joy; and especially when a clown swallowed a carrot or a string of sausages, he would laugh so outrageously that the lovely Princess by his side would have to say, 'My gracious monarch, do compose yourself.' But he continued to laugh, and at the very smallest farces, as long as his poor wits were left him.

Ibid.

The King and the Arts

George discoursed with the writer Fanny Burney on Shakespeare and refused to succumb to the idolatry of the Bard so characteristic of the period.

'Was there ever,' cried he, 'such stuff as a great part of Shakespeare? Only one must not say so! But what think you?—What?—Is there not sad stuff? What?—What?'

'Yes, indeed, I think so, sir, though mixed with such excellencies, that—'

'Oh!' cried he, laughing good-humouredly, 'I know it is not to be said! But it's true. Only it's Shakespeare, and nobody dare abuse him.'

Then he enumerated many of the characters and parts of plays that he objected to; and when he had run them over, finished with again laughing, and exclaiming,

'But one should be stoned for saying so.'

Diary and Letters of Madame d'Arblay [Fanny Burney], II, ed. by her Niece [Charlotte Barrett] (7 vols, 1842–6)

George's brother Prince William of Gloucester said to the great Edward Gibbon, on being presented with the second volume of his Decline and Fall of the Roman Empire, *'Another damned, thick, square book! Always scribble, scribble, scribble, eh, Mr Gibbon?' In contrast, George collected a fine library of thick books in the Queen's House (Buckingham Palace), which was to form the nucleus of the British Library. The following is an account of his obtaining a rare item for his collection. Sir John Fenn described in the Preface to his third volume of the famous fifteenth-century* Paston Letters *how George III showed an interest.*

During their continuance in that repositary [the library of the Society of

Antiquaries] it was intimated to the Editor [Fenn] that the King had an inclination to inspect and examine them; they were immediately sent to the Queen's Palace, with an humble request from the Editor, that, if they should be thought worthy of a place in the Royal Collection, His Majesty would be pleased to accept them; to this request a most gracious answer was returned, and they are now in the Royal Library.

The Paston Letters and Papers of the Fifteenth Century, ed. Norman Davis, 2 vols. (Oxford, 1971–6), I

The answer was indeed gracious: Fenn was summoned to the levee at St James's on 23 May 1787, where he

had the honour of presenting to his Majesty (bound in three volumes) the original Letters, of which he had before presented a printed copy, and was forthwith knighted.

Ibid., quoting from Fenn's Preface and article in *Morning Chronicle*, 24 May 1787

Nor must the tribute of Dr Johnson be forgotten, after being presented to George III in 1767. The full story, as narrated by Boswell, begins by explaining that Dr Johnson frequently visited the Queen's House to read in the library. Hearing this, George III expressed a wish to meet the great man on his next visit. Johnson was deep in a book by the fireside when Mr Barnard, the librarian, suddenly announced his Majesty:

'Sir, here is the King.' Johnson started up and stood still. His Majesty approached him, and at once was courteously easy . . .

His Majesty . . . mentioning his having heard that the Doctor had been lately at Oxford, asked him if he was not fond of going thither. To which Johnson answered, that he was indeed fond of going to Oxford sometimes, but was likewise glad to come back again. The King then asked him what they were doing at Oxford. Johnson answered, he could not much commend their diligence, but that in some respects they were mended, for they had put their press under better regulations . . .

His Majesty enquired if he was then writing any thing. He answered, he was not, for he had pretty well told the world what he knew, and must now read to acquire more knowledge. The King . . . then said 'I do not think you borrow much from any body.' Johnson said, he thought he had already done his part as a writer. 'I should have thought so too (said the King), if you had not written so well.'—Johnson observed to me, upon this, that 'No man could have paid a handsomer compliment; and it was fit for a King to pay. It was decisive.' When asked by another friend, at Sir Joshua Reynolds's, whether he made any reply to this high compliment, he answered, 'No, Sir. When the King has said it, it was to be so. It was not for me to bandy civilities with my Sovereign.'

Boswell's Life of Johnson, ed. R. W. Chapman (3rd edn; Oxford, 1979)

The king and the doctor proceeded to cover a wide range of subjects, all raised by his Majesty—the work of libraries, literary journals, reviews, and individual scholars—including a recent publication:

His Majesty then asked him what he thought of Lord Lyttelton's *History*, which was then just published. Johnson said, he thought his style pretty good, but that he had blamed Henry II rather too much. 'Why (said the King), they seldom to these things by halves.' 'No, Sir (answered Johnson), not to Kings.' . . .

During the whole of this interview, Johnson talked to his Majesty with profound respect, but still in his firm, manly manner, with a sonorous voice, and never in that subdued tone which is commonly used in the levee and in the drawing-room. After the King withdrew, Johnson shewed himself highly pleased with his Majesty's conversation and gracious behaviour. He said to Mr Barnard, 'Sir, they may talk of the King as they will; but he is the finest gentleman I have ever seen.' Ibid.

'Farmer George'

The young king started off well with his declaration:

'Born and educated in this country, I glory in the name of Britain.'

His public appeal was badly dented by his involvement with unpopular ministries but restored by his blameless private life and by his often comical love of the countryside and country pursuits.

On one occasion he played the part of King Alfred, and turned a piece of meat with a string at a cottager's house. When the old woman came home, she found a paper with an enclosure of money, and a note written by the Royal pencil: 'Five guineas to buy a jack' [device for automatically turning a roasting spit]. It was not splendid, but it was kind and worthy of Farmer George. One day, when the King and Queen were walking together, they met a little boy—they were always fond of children, the good folk—and patted the little white head. 'Whose little boy are you?' asked the Windsor uniform. 'I am the King's beefeater's little boy,' replied the child. On which the King said, 'Then kneel down, and kiss the Queen's hand.' But the innocent offspring of the beefeater declined this treat. 'No,' said he, 'I won't kneel, for if I do, I shall spoil my new breeches.' The thrifty King ought to have hugged him and knighted him on the spot.

Thackeray, *The Four Georges*

Cartoons showed 'Farmer George' hobnobbing with a pigman, and asking a farm labourer's wife how the apple gets inside the dumpling. The incident was celebrated in verse:

The Apple-Dumplings and a King

(An old widow has finished preparing some dumplings for the pot when the King arrives—)

In tempting row the naked dumplings lay,
When lo! the monarch in his usual way,
Like lightening spoke, 'What's this? what's this? what? what?'
'No!' cried the staring monarch with a grin,
'How? how? the devil got the apple in?'

Peter Pindar (John Wolcot), *Poems* (1791)

THE CHARMS OF WINDSOR

The king liked watching boys playing cricket or flying kites in the Park. Charles Knight, son of a Windsor bookseller, wrote:

Many a time had he bidden us good morning when we were hunting for mushrooms in the early dew and he was returning from his dairy to his eight o'clock breakfast. Every one knew that most respectable and amiable of country squires, and His Majesty knew every one.

John Brooke, *King George the Third* (1972)

George expressed his dislike of London:

'I certainly see as little of London as I possibly can, and am never a volunteer there.' Ibid.

George as the first commuter:

Even in the 1790s when he was approaching his sixtieth year, after a long and tiring day at Court, he would take a hasty dinner at the Queen's House [Buckingham Palace], mount his horse, and ride the twenty miles to Windsor. London was his place of business but Kew and Windsor were his homes. King George was the first commuter. Ibid.

GEORGE'S INTEREST IN TECHNOLOGY

As well as 'Farmer George' he was the 'Button-maker' because he used a lathe to make a set of ivory buttons. He was fascinated by the largest telescope of the great astronomer Sir William Herschel at Slough and paid £4,000 for its construction. The scientist's sister, Caroline, noted on 17 August 1787:

One anecdote of the old tube . . . I must give you. Before the optical parts were finished, many visitors had the curiosity to walk through it, among the rest King George III, and the Archbishop of Canterbury, following the King, and finding it difficult to proceed, the King turned to give the hand, saying, 'Come, my Lord Bishop, I will show you the way to Heaven!'

Memoir and Correspondence of Caroline Herschel (1750–1848), ed. Mrs John Herschel (1876)

GEORGE III ON PROOF-READING

The king had some valuable advice for the diarist and author Fanny Burney.

He laughed; and enquired who corrected my proofs?

'Only myself,' I answered.

'Why, some authors have told me that they are the last to do that work for themselves. They know so well by heart what ought to be, that they run on without seeing what is. They have told me, besides, that a mere plodding head is best and surest . . . and that the livelier the imagination, the less it should be trusted to.

Madame d'Arblay, VI

REX V. WILKES, 1763

The imprisonment of Wilkes for an article in the North Briton, *allegedly libelling the king, produced the slogan against George, 'Wilkes and Liberty!' Years after Wilkes's release on Habeas Corpus and £1,000 damages, followed by his election as Lord Mayor, he and the king met affably at court. When Princess Amelia took Lady Mary Coke, her lady-in-waiting, to see the great feast she was giving for the haymakers on her Gunnersbury estate, she was upset by untimely mention of the affair.*

There was a table for fifty; 'tis amazing how soon the great pieces of roast & boiled beef were despatched; there were besides six or seven of the largest baked puddings I ever saw. several dishes of beans & bacon, pease, cabbage, &c.; the ale & beer came in pales, & went off as quick as the eatables. Most of them behaved with propriety, but one man having drunk too freely, all decency & gratitude subsided; he raised his voice & drank, 'Wilks [*sic*] & liberty for ever.' The Princess turn'd about & said, 'I cou'd cry'; but the Man, being reproved by one of the servants, grew very loyal & . . . joined with the others in singing 'God Save our Noble King &c.'

Letters and Journals of Lady Mary Coke, III (Bath, 1970)

A HUMOROUS CONCLUSION

After Wilkes turned Courtier, and attended the Levees, the King asked him how his friend Sergeant Glynn was. The Sergeant had been very intimate for many years with Wilkes, had been engaged with him in many of his seditioning Transactions, and employed for him as his Counsel in all his important Westminster Hall Trials and Transactions. 'My Friend, Sir,' says Wilkes to the King, 'he is no Friend of mine, he loves Sedition and Licentiousness, which I never delighted in. In fact, Sir, I have nothing to do with such a Man. He was a Wilkite, which I never was.' The King said The confidence and humour of the Man made him forget at the Moment his Impudence.

After Wilkes's Conversion and this conversation with George the 3rd, Wilkes dined with his Son, George the 4th, then Prince of Wales; after dinner Toasts were proposed, not the healths of Persons, but as was often done in those days, Sentiments. Wilkes gave as his Toast 'God Save the King'. 'God bless me' said the Prince, 'how long have you been so loyal as to give that Toast?' 'Ever since I became acquainted with your Royal Highness' replied Wilkes.

Lord Eldon's Anecdote Book, ed. A. L. J. Lincoln and R. L. McEwen (1960)

THE BOSTON 'TEA-PARTY', 1773

When the rebellious American colonists tipped British tea into Boston harbour as a protest against the tea duty, the king believed that stubbornness alone would prevent the 'tea-party' turning into an orgy. After the 'colonists' had won their War of Independence, George showed his more attractive side. He said to John Adams, Washington's successor as President:

I wish you, Sir, to believe, that it may be understood in America, that I have done nothing in the late contest but what I thought myself indispensably bound to do by the duty which I owed my people. I will be very frank with you. I was the last to consent to the separation; but the separation having been made, and having become inevitable, I have always said, as I say now, that I would be the first to meet the friendship of the United States as an independent power.

His bonhomie was rewarded:

A wealthy American once said: 'They say King George is a very honest fellow; I should like to smoke a pipe with him.'

Brooke, *George the Third*

A MAD ATTACK ON THE KING, 1786

His carriage had just stopped at the garden-door at St James, and he had just alighted from it, when a decently-dressed woman, who had been waiting for him some time, approached him with a petition. It was rolled up, and had the usual superscription—'For the King's most excellent Majesty.' She presented it with her right hand; and, at the same moment that the King bent forward to take it, she drew from it, with her left hand, a knife, with which she aimed straight at his heart!

The fortunate awkwardness of taking the instrument with the left hand made her design perceived before it could be executed; the King started back, scarce believing the testimony of his own eyes; and the woman made a second thrust, which just touched his waistcoat before he had time to prevent her; and at that moment one of the attendants, seeing her horrible intent, wrenched the knife from her hand.

'Has she cut my waistcoat?' cried he, in telling it—'Look! for I have had no time to examine.'

Thank Heaven, however, the poor wretch had not gone quite so far. 'Though nothing', added the King, in giving his relation, 'could have been sooner done, for there was nothing for her to go through but a thin linen, and fat.' While the guards and his own people now surrounded the King, the assassin was seized by the populace, who were bearing her away, no doubt to fall the instant sacrifice of her murtherous purpose, when the King, the only calm and moderate person then present, called aloud to the mob, 'The poor creature is mad!—Do not hurt her! She has not hurt me!'

Madame d'Arblay, III

CONGRATULATIONS ON HM'S ESCAPE

A loyal Address was presented at the Sheldonian Theatre, Oxford:

When the address was ended, the King took a paper from Lord Harcourt, and read his answer. The King reads admirably; with ease, feeling, and force, and without any hesitation. His voice is particularly full and fine. I was very much surprised by its effect . . . Next followed music: a good organ, very well played, antheme-d and voluntary-ed us for some time. The Vice-Chancellor and Professors next begged for the honour of kissing the King's hand.

The king noticed that the Vice-Chancellor would not be capable of walking backwards down the steps after the ceremony:

He therefore dispensed with being approached to his seat, and walked down himself into the area, where the Vice-Chancellor kissed his hand, and was imitated by every Professor and Doctor in the room.

Notwithstanding this considerate good-nature in his Majesty, the sight, at times, was very ridiculous. Some of the worthy collegiates, unused to such ceremonies, and unaccustomed to such a presence, the moment they had kissed the King's hand, turned their backs to him, and walked away as in any common room; others, attempting to do better, did still worse, by tottering and stumbling, and falling foul of those behind them; some, ashamed to kneel, took the King's hand straight up to their mouths; others, equally off their guard, plumped down on both knees, and could hardly get up again; and many, in their confusion, fairly arose by pulling his Majesty's hand to raise them.

Madame d'Arblay, III

The queen's ladies, including Fanny Burney, had a worse time than the dons. Having stood for five hours with nothing to eat but a little chocolate consumed in turn behind a screen of colleagues, they were at last given fruit and bread in Christ Church; but even here the queen suddenly appeared and they all had to crush their crusts back into their pockets. Though devoted to their Majesties personally, Fanny never approved of court protocol. She resented being rung for by a bell and once told her sister-in-law that when in waiting one must choke rather than cough or sneeze, and if a pin happened to run into one the only thing to do was to bite the inside of one's cheek and swallow the piece; never, never show pain. This fits with the story of a request to Queen Charlotte that a pregnant and exhausted lady, who was holding Princess Charlotte at her christening, might sit for a moment, and the Queen's reply,

'She may stand! She may stand!'

The King's First Attack of Illness, 1788

Fanny Burney was appointed joint keeper of the queen's robes in 1786. To her we owe the horrific account of the king's breakdown in October. At the height of the attack, the terrified queen told Lady Harcourt that the king's eyes looked like blackcurrants and that the foam poured from his mouth.

25 Oct.: I had a sort of conference with his Majesty, or rather I was the object to whom he spoke, with a manner so uncommon, that a high fever alone could account for it. A rapidity, a hoarseness of voice, a volubility, and an earnestness—a vehemence, rather—it startled me inexpressibly; yet with a graciousness exceeding even all I ever met with before—it was almost kindness! Heaven—Heaven preserve him! The Queen grows more and more uneasy. She alarms me sometimes for herself, at other times she has a sedateness that wonders me still more.

29 Oct.: The King is very sensible of the great change there is in himself, and of [the queen's] disturbance at it. It seems, but Heaven avert it! a threat of a total break up of the constitution. This, too, seems his own idea. I was present at his first seeing Lady Effingham on his return to Windsor this last time. 'My dear Effy,' he cried, 'you see me, all at once, an old man.' . . . Lady Effingham, in her well-meaning but literal way, composedly answered, 'We must all grow old, sir; I am sure I do.'

5 Nov.: O dreadful day! My very heart has so sickened in looking over my memorandums, that I was forced to go to other employments. I will not, however, omit its narration. 'Tis too interesting ever to escape my own memory, and my dear friends have never yet had the beginning of the thread which led to all the terrible scenes of which they have variously heard . . . At noon the King went out in his chaise, with the Princess Royal, for an airing. I looked from my window to see him; he was all smiling benignity, but gave so many orders to the postillions, and got in and out of the carriage twice, with such agitation, that again my fear of a great fever hanging over him grew more and more powerful. Alas! how little did I imagine I should see him no more for so long—so black a period! . . . Only Miss Planta (the youngest of the Queen's ladies) dined with me. We were both nearly silent: I was shocked at I scarcely knew what, and she seemed to know too much for speech . . . Meanwhile, a stillness the most uncommon reigned over the whole house. Nobody stirred; not a voice was heard; not a step, not a motion. I could do nothing but watch, without knowing for what: there seemed a strangeness in the house most extraordinary.

At seven o'clock Columb [a page] came to tell me that the music was all forbid, and the musicians ordered away!

This was the last step to be expected, so fond as his Majesty is of his Concert, and I thought it might have rather soothed him.

Madame d'Arblay, II

Later on Fanny heard the ghastly details from the courtier Fairley.

O my dear friends, what a history! The King, at dinner, had broken forth into positive delirium, which long had been menacing all who saw him most closely; and the Queen was so overpowered as to fall into violent hysterics. All the Princesses were in misery, and the Prince of Wales had burst into tears. No one knew what was to follow—no one could conjecture the event.

Ibid.

GREVILLE'S REPORT

The king had agreed that the queen should sleep in a separate room provided it was next door to him. The equerry Robert Greville reported an incident during the night of 5 November 1788:

In the Night H::M:: found an opportunity of getting out of his Bed, and taking a Candle. He open'd the Queen's Apartment, and coming gently to the bedside, He open'd the Curtains and looked in—Miss Goldsworthy who slept in the Queen's Room and Her Majesty were much alarmed—The King looking Earnestly at the Queen said, 'I will confess the truth, I thought you had deceived Me, and that you was not here.'

Diaries of Robert Fulke Greville, ed. F. M. Bladon (1930)

FANNY'S DIARY CONTINUED: 5 NOVEMBER

Some time after one o'clock she was sent for by the queen.

My poor Royal mistress! Never can I forget her countenance—pale, ghastly pale she looked; she was seated to be undressed . . . her whole frame was disordered, yet she was still and quiet . . . I gave her some camphor julep, which had been ordered her by Sir George Baker [the king's doctor]. 'How cold I am!' she cried, and put her hand on mine; marble it felt! and went to my heart's core!

Madame d'Arblay, II

The following day Fanny rose at 6 a.m. in an awful state of suspense. She heard men's voices in the queen's little dressing-room. They had sat up all night and Fanny was struck by the horror on their faces. She was then summoned to the queen.

O what a scene followed! What a scene was related! The King, in the middle of the night, had insisted upon seeing if his Queen was not removed from the house; and he had come into her room, with a candle in his hand, opened the bed-curtains, and satisfied himself she was there, and Miss Goldsworthy (one of her ladies) by her side. This observance of his directions had much soothed him; but he stayed a full half-hour, and the depth of terror during that time no words can paint. The fear of such another entrance was now so strongly upon the nerves of the poor Queen, that she could hardly support herself . . .

The King—the Royal sufferer—was still in the next room, attended by Sir George Baker and Dr Heberden, and his pages, with Colonel Goldsworthy occasionally, and as he called for him. He kept talking unceasingly; his voice was so lost in hoarseness and weakness, it was rendered almost inarticulate; but its tone was still all benevolent—all kindness—all touching graciousness.

Fanny had to carry the best account she could manage to the queen.

Nothing could be so afflicting as this task; even now, it brings fresh to my ear his poor exhausted voice. 'I am nervous,' he cried; 'I am not ill, but I am nervous: if you would know what is the matter with me, I am nervous. But I love you both (his doctors) very well; if you would tell me true: I love Dr Heberden best, for he has not told me a lie. Sir George has told me a lie—a white lie, he says, but I hate a white lie! If you will tell me a lie, let it be a black lie!'

The Princesses asked to be allowed to see their mother.

She burst into tears, and declared she could neither see them, nor pray, while in this dreadful situation, expecting every moment to be broken in upon, and quite uncertain in what manner, yet determined not to desert her apartment, except by express direction from the physicians. Who could not tell to what heights the delirium might rise; there was no constraint, no power; all feared the worst, yet none dared take any measures.

Ibid.

FANNY'S HEALTH V. THE KING'S HEALTH, 1789

The king and court had been removed to Kew to escape the crowds at Windsor. For the sake of her health, Fanny was advised to walk around Kew Gardens, while the king, beginning to recover, was safely away in Richmond. This she did on 2 February; Fanny was desperately short-sighted.

I had proceeded, in my quick way, nearly half the round, when I suddenly perceived, through some trees, two or three figures. Relying on the instructions of Dr John Willis [son of the king's notorious 'mad doctor' Francis Willis], I concluded them to be workmen and gardeners; yet tried to look sharp, and in so doing, as they were less shaded, I thought I saw the person of his Majesty! Alarmed past all possible expression, I waited not to know more, but turning back, ran off with all my might. But what was my terror to hear myself pursued!—to hear the voice of the King himself loudly and hoarsely calling after me, 'Miss Burney! Miss Burney!'

I protest I was ready to die. I knew not in what state he might be at the time; I only knew the orders to keep out of his way were universal . . .

The steps still pursued me, and still the poor hoarse and altered voice rang in my ears: more and more footsteps resounded frightfully behind me—the attendants all running, to catch their eager master, and the voices of the two Doctors Willises loudly exhorting him not to heat himself so unmercifully.

Heavens, how I ran! I do not think I should have felt the hot lava from Vesuvius—at least not the hot cinders—had I so run during its eruption. My feet were not sensible that they even touched the ground.

Soon after, I heard other voices, shriller, though less nervous, call out 'Stop! Stop! Stop!'

I could by no means consent: I knew not what was purposed, but I recollected fully my agreement with Dr John that very morning, that I should decamp if surprised and not be named. . . . Still, therefore, on I flew; and such was my speed, so almost incredible to relate or recollect, that I fairly believe no one of the whole party could have overtaken me, if these words, from one of the attendants, had not reached me, 'Doctor Willis begs you to stop!'

'I cannot! I cannot!' I answered, still flying on, when he called out 'You must, ma'am; it hurts the King to run.' . . .

When they were within a few yards of me, the King called out 'Why did you run away?'

Shocked at a question impossible to answer, yet a little assured by the mild tone of his voice, I instantly forced myself forward, to meet him . . .

The effort answered: I looked up, and met all his wonted benignity of countenance, though something still of wildness in his eyes. Think, however, of my surprise, to feel him put both his hands round my two shoulders, and then kiss my cheek!

I wonder I did not really sink, so exquisite was my afright when I saw him spread out his arms! Involuntarily, I concluded he meant to crush me: but the Willises, who had never seen him till this fatal illness, not knowing how very extraordinary an action this was from him, simply smiled and looked pleased, supposing, perhaps, it was his customary salutation!

I believe, however, it was but the joy of a heart unbridled, now, by the forms and proprieties of established custom and sober reason. To see any of his household thus by accident, seemed such a near approach to liberty and recovery, that who can wonder it should serve rather to elate than lessen what yet remains of his disorder! *Madame d'Arblay*, II

THE WILLIS TREATMENT

Dr the Revd Francis Willis and his two sons, Dr John and Dr Robert, submitted George III to the full rigours of eighteenth-century medicine, including the strait-jacket and tying him to his bed. The king detested Willis and occasionally got the better of him. A courtier recorded a dialogue between the king and the clerical physician:

'Sir, your dress and appearance bespeaks you of the Church, do you belong to it?'

'I did formerly, but lately I have attended chiefly to physick.'

'I am sorry for it. You have quitted a profession I have always loved, and you have embraced one I most heartily detest . . .'

'Sir, Our Saviour Himself went about healing the sick.'

'Yes, yes, but He had not £700 [a year] for it.'

On another occasion the king said to the same courtier, with a wink, that the doctor was 'a great rascal' and added:

'Tricking in love and physick you know is all fair.'

Nor was he more complimentary about his regular physician: once he picked out a knave from a pack of cards and wrote on the back:

'Sir Richard Warren Bart First Physician to the King.'

Diaries of Robert Fulke Greville

But the doctors had the last word:

Dr Willis remained firm, and reproved him in nervous and determined language, telling him he must control himself otherwise he would put him in a strait waistcoat. On this hint Dr Willis went out of the room and returned directly with one in his hand . . . The King eyed it attentively and alarmed at the doctor's firmness of voice and procedure began to submit . . .

Ibid.

THE LEGEND OF THE OAK TREE

One of the most persistent legends about the king's illness in 1788 is the story that he got out of his coach in Windsor Great Park and shook hands with an oak tree under the impression it was Frederick the Great.

This legend derives from a pamphlet, *History of the Royal Malady by a page of the Presence*, published in 1789. From a note opposite the title page it has been assumed that the author was one Philip Withers. There was no one of this name in the Royal Household in 1789. Brooke, *George the Third*

THE ROYAL STRIKE

When the king's illness recurred—in 1801—he at first refused to be supervised by the Willises again but, with the connivance of the queen, they virtually kidnapped him and kept him in the White House at Kew. Here he remained in isolation from his family, yet still carrying out the royal functions of corresponding with ministers, giving the royal assent to Acts of parliament and creating at least one peerage. Finally he contemplated industrial action. This threat of a royal strike succeeded and the king duly visited his family. He had told Lord Chancellor Eldon, who visited him on 19 May, that:

he had taken a solemn determination, that unless he was that day allowed to go over to the house where the Queen and his family were, no earthly consideration should induce him to sign his name to any paper or to do one act of government whatever. Brooke, *George the Third*

A LIBEL IN *THE TIMES*, 1789

Owing to the well-known hostility between the king and his two eldest sons, The Times *suggested that the sons' rejoicing at their father's recovery was humbug:*

The Royal Dukes, and the leaders of opposition in general, affect to join with the friends of our amiable Sovereign, in rejoicing on account of his Majesty's recovery. But the insincerity of their joy is visible. Their late unfeeling conduct will for ever tell against them; and contradict the artful professions they may think it prudent to make.

It argues infinite wisdom in certain persons, to have prevented the Duke of York from rushing into the King's apartment on Wednesday. The rashness, the Germanick severity, and insensibility of this young man, might have proved ruinous to the hopes and joys of a whole nation.

The Times: Past Present Future (1985)

Following this press attack on the royal dukes, they sued John Walter, founder–owner of The Times, *who was convicted, fined £50 and sent to Newgate gaol for two years. The story behind this drastic punishment was devious. After George III's recovery, Walter accepted an under-the-counter government salary of £300 a year to publish paragraphs favourable to the king's party. The two 'libellous' paragraphs had been secretly sent to Walter by the Secretary of the Treasury. While in gaol he continued to receive his £300 salary as well as £250 from the prince of Wales as conscience money; a year later he received a gift of another £250 from prime minister William Pitt's secret service fund. So there was humbug, but not on one side only.*

A SECOND MAD ATTACK ON THE KING, 1800

James Hadfield shot at the King at Drury Lane theatre when Sheridan was manager and Michael Kelly singing in The Marriage of Figaro. 'Never shall I forget His Majesty's coolness. The whole audience was in an uproar. The King on hearing the report of the pistol retired a pace or two, stopped, and stood firmly for an instant, then came forward to the front of the box, put his opera glass to his eye, and looked round the house without the smallest appearance of alarm or discomposure.' Michael Kelly, who wrote this account, finished the performance by singing a new verse of God save the King which Sheridan had written for the occasion:

From every latent foe,
From the assassin's blow,
God save the King! . . .

Brooke, *George the Third*

SAYINGS OF GEORGE III

On some drawings by William Blake: 'What—what—what! Take them away, take them away!'

On Burke's Reflections on the French Revolution: 'A good book, every gentleman ought to read it.'

Asking Burney how she wrote her book: 'But what?—what?—how was it? . . . How came you—how happened it—what?—what?'

To his architect Wyatt: 'Six hours sleep are enough for a man, seven for a woman and eight for a fool.'

To his gardener at Kew who was packing up a basket of plants for Dr Willis: 'Get another basket, Eaton, at the same time, and pack up the doctor in it, and send him off at the same time.'

SAYINGS ON AND BY GEORGE III'S CHILDREN

George III had fifteen children by Queen Charlotte.

Wellington on the nine sons: 'They are the damnedest millstones about the necks of any government that can be imagined.'

Ernest duke of Cumberland, the fifth of George's nine sons: 'Nothing in my eyes is so terrible as a family party.'

Princess Sophia, the fifth of George's six daughters, to the prince of Wales on the Windsor 'nunnery', as the girls called themselves because they married late or not at all: 'Poor old wretches as we are, *four old cats, four old wretches a dead weight* upon you, *old lumber to the country*, like *old clothes*. I wonder you do not vote for putting us *in a sack* and *drowning us* in the *Thames*.'

Charles Greville on the daughters: 'They were secluded from the world, mixing with few people, their passions boiling over, and ready to fall into the hands of the first man whom circumstances enabled to get at them.'

A scandalous rumour was buried by Wellington (George IV's executor) that Princess Sophia had been 'got at' by her brother the duke of Cumberland. In fact she had an illegitimate son by General Garth, a courtier. She explained her pregnancy as dropsy, cured by roast beef. Greville always said the Hanoverians needed plenty to eat and drink to keep them fit.

THE KING'S FINAL BREAKDOWN, 1811

George had been blind since 1805 and the death of his youngest and favourite daughter Princess Amelia was a blow from which he did not recover. Lord Byron wrote to Robert Dallas, 25 August 1811:

The newspapers seem much disappointed at his Majesty's not dying, or doing something better. I presume it is almost over.

Byron's Letters and Journals, II, ed. L. A. Marchand (1973)

Yet the king was to haunt Windsor Castle for another eight or nine years, a cross between a bearded Old Testament prophet and King Lear, while his unfavourite son operated the Regency. Fanny Burney reported on 11 May 1813:

The beloved King is in the best state possible for his present melancholy situation; that is, wholly free from real bodily suffering, or imaginary mental misery, for he is persuaded that he is always conversing with angels.

Madame d'Arblay, VI

THE KING'S DEATH

I see the good old King is gone to his place—one can't help being sorry—though blindness—and age and insanity are supposed to be drawbacks—on human felicity—but I am not at all sure that the latter at least—might not render him happier than any of his subjects.

Byron's Letters, VIII: Lord Byron to John Murray

Byron's 'The Vision of Judgement' also dealt ironically with the king's death:

He died—but left his subjects still behind,
One half as mad—and 'tother no less blind.

THE FUNERAL IN ST GEORGE'S CHAPEL

The service was at seven in the evening. Mrs Arburthnot and her husband, a Cabinet official, had visited the coffin earlier in the day. She considered it

most splendid, of purple velvet almost covered with gold richly embossed & with immensely massive gold handles . . . but I cannot help feeling that the profusion of gold & *glitter* takes away from those feelings of awe with which we usually contemplate the closing scene of human existence.

The Journal of Mrs Arbuthnot, I, ed. C. Bamford and the 7th duke of Wellington (1950)

Of the service itself she wrote:

The chaunting, & indeed the whole service, was most impressive & affecting, & when the coffin was lowered into the vault Ld Winchilsea broke his staff of office & threw it in & the Garter King at Arms attempted to proclaim the style; but the poor old man, who is 92, was so much affected that his voice was quite inaudible.

And thus has sunk into an honoured grave the best man & the best King that ever adorned humanity; and it is consoling to the best feelings of the human heart that *such* a sovereign was followed to his last home by countless thousands of affectionate subjects drawn to the spot by no idle curiosity to view the courtly pageant, but to pay a last tribute of respect & to shed the tear of affection & gratitude over the grave of him who, for sixty long years, had been The Father of his people! Ibid.

Mrs Arbuthnot was right about George III's popularity at his death, due to his scandal-free personal life, piety and simplicity, and the pathos of his long illness. She admired him as much as she detested his successor George IV.

George IV

1820–1830

The Georges seemed to have ended with the fourth of that name. Walter Savage Landor finally put paid to them:

When from earth the Fourth descended,
God be praised, the Georges ended.

In a sense he was right, for the chasm between the fourth and fifth Georges was such as to make them appear of different breeds. Filial and marital loyalty, sense of duty, personal integrity, and self-discipline—all were wanting in George IV. Even in his twenties he was stout and florid, but his many drawing-room graces—singing, dancing, witty conversation—suggested a paragon of deportment. He was to die of over-indulgence and yet he had once been the 'First Gentleman in Europe', Prince Charming, and Prinny, THE Prince.

PERDITA AND FLORIZEL

The actress Mary Robinson became his mistress soon after he was seventeen and she twenty-one. He promised her a bond for £20,000 when he came of age but dropped her long before. She died in poverty. Meanwhile they called themselves, romantically, Prince Florizel and Perdita. Their first meeting was at Kew, George being accompanied by his brother Frederick duke of York.

It was agreed that she would dine with Lord Malden at the inn on the island between Kew and Brentford. They were to watch for the signal of a handkerchief being waved from the opposite shore, then they were to take a boat for the landing place by the iron gates of the Palace. They arrived there safely and the Prince, with Frederick at his side, ran down the avenue to meet them. But there was only time for the exchange of a few words when, as the moon rose, the noise of people approaching from the Palace disturbed them and she and Lord Malden had to run back to the boat.

Christopher Hibbert, *George the Fourth* (1972)

THE COMPULSIVE SPENDER

He spent over £20 a week on cold cream and almond paste, perfumed almond powder and scented bags, lavender water, rose water, elder flower water, jasmine pomatum and orange pomatum, eau de cologne, eau romaine, Arquebusade, essence of bergamot, vanilla, eau de miel d'Angleterre, milk of roses, huile antique and oil of jasmine. He bought them all in huge quantities—perfumed powder was delivered in amounts of up to £33 at a time; toothbrushes came by the three dozen. But then he bought almost everything in huge quantities: in need of a few walking sticks, he bought thirty-two in one day. Ibid.

HANOVERIAN HOSTILITY

Once prime minister North tried to reconcile the prince to his father. Besides being recklessly extravagant and sunk in debt, the prince supported the Whig opposition, chiefly to annoy the Tory king; for when George became Prince Regent he ditched his political friends.

When George IV was Prince of Wales, Lord North, then Minister, had made himself a Party, at the Prince's desire, to reconcile the King and the Prince relative to some Matter, which caused some uneasy Feelings between them. Lord North succeeded, and called on the Prince to inform him of that, and addressed to this effect. Now, my dear Prince, do in future conduct yourself differently—Do so for God's Sake, do so for your own Sake, do so for your excellent Father's Sake, do so for the sake of that good natured Man Lord North, and don't oblige him again to tell your good Father so many Lies, as that good natured Man has been obliged to tell him this Morning. *Lord Eldon's Anecdote Book*

BETROTHAL BY TRICKERY, 1784

The prince had fallen madly in love with Maria Fitzherbert, a beautiful and charming widow but a Catholic. After repeated failures, he hit on a plan to win her. His physician, Dr Keate, dashed up to her house in a carriage, saying the Prince had stabbed himself: come at once!

She found him in his private apartments on the ground floor which overlooked the garden and St James's Park. He was pale and covered with blood which issued from a wound in his side, the wound which his friends assured her had been self-inflicted by the Prince in consequence of her cruelty. According to the Prince, he had fallen upon his sword. According to another account he had stabbed himself with a dagger. According to a third, he had tried to shoot himself, but hit the head of his bed instead; the pistol had been taken away from him; he then possessed himself of a table knife and drove it into his side . . . The theory has also been put forward that the Prince had simply been 'blooded' by Keate to relieve the violence of his passion, and he had dabbled the blood about his clothes to make himself more interesting to his beloved.

W. H. Wilkins, *Mrs Fitzherbert and George IV* (1905)

Mrs Fitzherbert, 'deprived almost of consciousness,' agreed to save his life by permitting him to put a ring on her finger. The prince borrowed one from the duchess of Devonshire, who had accompanied her, and she went back to Devonshire House where a deposition was signed by her.

MARRIAGE IN SECRET, 1785

The marriage took place on 15 December, just over a year after the betrothal, between the Protestant heir to the throne and the Catholic widow. When Mrs Fitzherbert's Catholic friend, Mrs Jerningham, eventually heard about it, she wrote:

Mrs Fitzherbert has, I believe, been married to the Prince. But it is a very hazardous undertaking, as there are two acts of Parliament against the validity of such an Alliance: concerning her being a subject and her being a Catholick. God knows how it will turn out.

The Jerningham Letters, I, ed. Egerton Castle (1896)

In order that the illegal marriage should remain secret, a Church of England clergyman had to be found and bribed. The Revd Robert Burt was given £500 down and preferment. The ceremony took place at 6 p.m. in Mrs Fitzherbert's drawing-room at Park Street, Park Lane, according to the rites of the Church of England. (Some fifteen years later the marriage was pronounced canonical by the Pope, so that by 1800 Maria began living with George and beginning what she afterwards called 'the eight happiest years' of her life.)

There were two witnesses, Maria's uncle Sir Henry Errington and her brother Jack Smythe. The prince slipped into the house under cover of the December darkness. The doors were locked.

Years later the signatures of the nervous witnesses were cut off the marriage certificate at their request, and the document itself was placed first in Coutts's Bank and afterwards in the Royal Archives.

When rumours of the marriage began to circulate, George's Whig friend, Charles James Fox, had to lie to Parliament on his behalf, stating:

on direct authority [the report is] a base and scandalous calumny.

Wilkins, *Mrs Fitzherbert*

The Brunswick Bride, 1795

George needed a legal marriage to settle his debts. Lord Malmesbury was sent to negotiate the marriage with Caroline of Brunswick, George's German cousin. He observed some ominous traits in the Princess's otherwise amiable character.

21 January 1795: Princess Caroline has a tooth drawn—she sends it down to me by her page—very nasty and indelicate.

6 March: I had two conversations with the Princess. One on the toilette, on cleanliness, and on delicacy of speaking. On these points I endeavoured, as far as was possible for a *man*, to inculcate the necessity of great and nice attention to every part of dress, as well as to what was hid, as to what was seen. (I knew she wore coarse petticoats, coarse shifts, and thread stockings, and these never well washed, or changed often enough.) I observed that a long toilette was necessary, and gave her no credit for boasting that hers was a 'short' one.

Diaries and Correspondence of James Harris First Earl of Malmesbury, ed. by his grandson (1844)

Lady Hester Stanhope, Pitt's niece, was another to notice the 'short' toilette:

She did not know how to put on her own clothes, . . . putting on her stockings with the seam before, or one of them wrong side outwards.

Henry Colburn, *Memoirs of Lady Hester Stanhope* (1845)

Malmesbury goes on to describe the arrival of the bride and her reception:

I immediately notified the arrival to the King [George III] and Prince of Wales; the last came immediately. I, according to the established etiquette, introduced . . . the Princess Caroline to him. She very properly, in consequence of my saying to her it was the right mode of proceeding, attempted to kneel to him. He raised her (gracefully enough), and embraced her, said barely one word, turned round, retired to a distant part of the apartment, and calling me to him, said, 'Harris, I am not well; pray get me a glass of brandy.' I said, 'Sir, had you not better have a glass of water?'—upon which, he, much out of humour, said, '*No*; I will go directly to the Queen,' and away he went. The Princess, left during this short moment alone, was in a state of astonishment; and, on my joining her, said, 'Mon Dieu! est-ce que le Prince est toujours comme cela? Je le trouve très gros, et nullement aussi beau que son portrait.' *Malmesbury Diaries*

CAROLINE'S ACCOUNT OF THE WEDDING-NIGHT

Years afterwards she prepared her own story to be published if the bill to divorce her got through to the House of Commons. It did not. But society got wind of the story.

It is a sort of Journal which she kept when she first married, & gives the whole account of her marriage & the King's treatment, & quantities of notes which passed between them during the first year. She says in it that the King was so drunk the night he married that, when he came into her room, he was obliged to leave it again; and he remained away all night and did not return again till the morning; that he then obliged her to remain in bed with him & that that is the *only time* they were together as husband & wife.

Journal of Mrs Arbuthnot, I

Another story says that he collapsed into the bedroom grate and remained there till dawn. At any rate their only child, Princess Charlotte, was conceived.

'PRINCELY' BEHAVIOUR

At Mrs Vaneck's assembly last week, the Prince of Wales, very much to the honor of his polite and elegant Behaviour, measured the breadth of Mrs Vaneck behind with his handkerchief, and shew'd the measurement to most of the Company.

G. W. E. Russell, *Collections and Recollections* (1898), quoting from a contemporary 'unpublished diary'

THE PRINCE OF WALES AND HIS BROTHER

One of his [the Prince of Wales's] stories was about the huge size of the penis of one of his Royal brothers—a fact which he had discovered one night while riding with him in a carriage. His brother had felt the need to relieve himself: when he did so out of the carriage window, the water flowed as from a fountain and the driver urged the horses forward to escape what he thought was a rainstorm!

A Persian at the Court of King George 1809–1810, trans. and ed. Margaret Morris Clarke (1988)

PRINNY'S QUARREL WITH BEAU BRUMMELL

Having been society's fashionable trend-setter and intimate of 'Prinny', Brummell fell out of favour. But after winning £20,000 at whist at White's Club, he was invited again to Carlton House, the prince's luxurious London home.

At the commencement of the dinner, matters went smoothly, but Brummel, in his joy at finding himself with his old friend, became excited and drank too much wine. His Royal Highness—who wanted to pay off Brummell for an insult he had received at Lady Cholmondeley's ball, when the beau, turning towards the Prince, said to Lady Worcester, 'Who is your fat friend?'—had invited him to dinner merely out of a desire for revenge. The Prince therefore pretended to be affronted with Brummell's hilarity, and said to his brother, the Duke of York, who was present, 'I think we had better order Mr Brummell's carriage before he gets drunk.' Whereupon he rang the bell, and Brummell left the royal presence. This circumstance originated the [apocryphal] story about the beau having told the Prince to ring the bell.

Reminiscences of Captain Gronow, ed. John Raymond (1964)

THE DELICATE INVESTIGATION, 1806

This is what they called the inquiry into Princess Caroline's alleged adultery with various gentlemen while separated from the prince and living at Blackheath. One of the witnesses was William Cole, a footman:

Mr Bidgood's wife has lately told him [said Cole], that Fanny Lloyd told her, that Mary Wilson had told Lloyd, that one day, when she went into the Princess's room, she [the Princess] and Sir Sidney Smith [the Admiral] were *in the fact*: that she [Mary Wilson] immediately left the room, and fainted at the door.

Thea Holme, *Prinny's Daughter* (1976)

Caroline was exculpated—with a warning.

'DREADFUL ACCIDENTS' AT THE OPENING OF CARLTON HOUSE TO THE PUBLIC, 1811

After the grand fête for 3,000 VIPs on 19 June, the Prince Regent planned to open it to successive batches of the 'well-dressed' public on 24, 25, and 26 June.

The method adopted was to let in about 200 at a time . . . This was repeated every half-hour [from eleven] till three, by which time the number in front, extending from Carlton House to Haymarket on the one side, and to St James's Street on the other, could not be less than thirty-thousand. Most of these were females, whose screams and shrieks became so distressing that the gate was opened; which instead of giving relief, increased the evil, for many were thrown down, and trampled upon by those behind. One lady had her leg broken, and others were carried away apparently dead. Even such as were fortunate enough to escape personal injury, suffered in their dress; and few of them could leave Carlton House, until they had obtained fresh garments.

Revd G. N. Wright, *William IV*, 2 vols. (1902), II

The Duke of Clarence was given the invidious task of declaring the gates shut.

A FAT ADONIS, 1812

Leigh Hunt, poet and critic, was sent to gaol for two years for calling the recently appointed Prince Regent a 'fat Adonis of fifty'. He was visited in prison by Byron; and another poet and friend, Charles Lamb, mocked the Prince in anonymous verse:

Not a fatter fish than he
Flounders round the polar sea.
See his blubbers—at his gills
What a world of drink he swills . . .

Every fish of generous kind
Scuds aside or shrinks behind;
But about his presence keep
All the monsters of the deep . . .
By his bulk and by his size,
By his oily qualities,
This (or else my eyesight fails),
This should be the Prince of Whales.

A KINDLY WHALE AT HIS MARINE PAVILION, BRIGHTON

To one young lady who had given some slight offence he showed that he was displeased by bowing to her curtly as she arrived and giving her nothing more than a 'little parting shake of the hand' when she left. The next morning when she arrived she made him a curtsy 'perhaps rather more brave, more low and humble than usual' (meaning—'I beg your pardon dear foolish, beautiful Prinny for making you take the pet'). Immediately he held out his hand, and all was forgiven.

Hibbert, *George the Fourth*

A KILLER WHALE AT THE MARINE PAVILION

And now I have one more story of the bacchanalian sort, in which Clarence and York, and the very highest personage of the realm, the great Prince Regent, all play parts. The feast took place at the Pavilion at Brighton, and was described to me by a gentleman who was present at the scene. In Gillray's caricatures, and amongst Fox's jolly associates, there figures a great nobleman, the Duke of Norfolk . . . celebrated for his table exploits. He had quarrelled with the Prince, like the rest of the Whigs; but a sort of reconciliation had taken place; and now, being a very old man, the Prince invited him to dine and sleep at the Pavilion, and the old Duke drove over from his Castle of Arundel with his famous equipage of grey horses, still remembered in Sussex.

The Prince of Wales had concocted with his Royal brothers a notable scheme for making the old man drunk. Every person at table was enjoined to drink wine with the Duke—a challenge which the old toper did not refuse. He soon began to see that there was a conspiracy against him; he drank glass for glass; he overthrew many of the brave. At last the First Gentleman of Europe proposed bumpers of brandy. One of the Royal brothers filled a great glass for the Duke. He stood up and tossed off the drink. 'Now,' says he, 'I will have my carriage, and go home.' The Prince urged upon him his previous promise to sleep under the roof where he had been so generously entertained. 'No,' he said; he had had enough of such hospitality. A trap had been set for him; he would leave the place at once and never enter its doors more. The carriage was called, and came; but, in the half-hour's interval, the liquor had proved too potent for the old man; his host's generous purpose was answered, and the Duke's old grey head lay stupefied on the table. Nevertheless, when his post-chaise was announced, he staggered to it as well as he could, and stumbling in, bade the postillions drive to Arundel. They drove him for half-an-hour round and round the Pavilion lawn; the poor old man fancied he was going home. When he awoke that morning he was in bed at the Prince's hideous house at Brighton . . . I can fancy the flushed faces of the Royal Princes as they support themselves at the portico pillars, and look on at old Norfolk's disgrace; but I can't fancy how the man who perpetrated it continued to be called a gentleman.

Thackeray, *The Four Georges*

THOUGH THE PRINCE'S PUBLIC FACE WON ADMIRATION, HIS PRIVATE FACE WAS CRITICIZED

Prinny is exactly in the state one would wish; he lives only by protection of his visitors. If he is caught alone, nothing can equal the execrations of the people who recognise him . . . All agree that Prinny will die or go mad. He is worn out with fuss, fatigue, and *rage*. He came to Lady Salisbury on Sunday from his own dinner beastly drunk, whilst her guests were all perfectly sober. It is reckoned very disgraceful in Russia for the higher orders to be drunk. He already abuses the Emperor lustily, and his (the Emperor's) waltzing with Lady Jersey, the Prince's mistress, last night at Lady Cholmondeley's would not mend his temper, and in truth he only stayed five minutes, and went off sulky as a bear, whilst everybody else stayed and supped and were as merry as could be.

The Creevey Papers, I, ed. Sir H. Maxwell (1904): Creevey to Mrs Creevey, 14 June 1814

DEDICATION OF JANE AUSTEN'S *EMMA* TO HRH THE PRINCE REGENT

On 15 November 1815 Jane wrote to the Revd James Stanier, the Prince Regent's chaplain and a great fan of hers:

Sir: I must take the liberty of asking you a question. Among the many flattering attentions which I received from you at Carlton House on Monday last, was the Information of my being at liberty to dedicate any future work to HRH the P.R. without the necessity of any solicitation on my part . . . I intreat you to have the goodness to inform me how such a Permission is to be understood, & whether it is incumbent on me to shew my sense of the Honour, by inscribing the Work now in the Press, to HRH. *Jane Austen's Letters*, ed. R. W. Chapman (2nd edn, 1952)

Stanier wrote that it was not 'incumbent' on her to dedicate Emma *but would be very welcome.*

He wrote again on 27 March 1816 from Brighton Pavilion that His Royal Highness the Prince Regent thanked her for 'the handsome copy you sent him of your last excellent novel'—handsome no doubt because of the Dedication—and added that many of the nobility at Brighton 'have paid you the just tribute of their praise.'

FIRST VISIT TO BRIGHTON AS KING

March 1820: The King is gone down to Brighton, much better & enjoying the society of Lady Cunningham [mis-spelt because the famous Lady Conyngham was until now little known]: The reigning favorite. The Duke of Wellington told me that, a few days ago, a dispatch of one of the foreign ministers was seen at the Office in which was the sentence, 'Le Prince Regent d'Angleterre âgé de soixante-cinq ans, a quitté la Marquise de Hertford, âgée de soixante-cinq ans, pour devenir amoureux fou de la Marquise de Cunningham, âgée de cinquante ans.' Pretty ideas foreign powers must have of our gracious Sovereign!! *Journal of Mrs Arbuthnot*, II

The king and his Tory government detested Queen Caroline but they could not stop her returning to England from abroad to claim her rightful place on the throne, after the death of George III.

The Queen came to the Ground after the Review was over, and evidently to try how she would be received by the Officers & Troops, and I am happy to say that the experiment failed compleatly. I met her in Hounslow, which at the time was full of Officers, Troops & people; and they literally took no notice of her. I passed close by her at a Gallop; & what is very extraordinary, after I had passed and had taken no notice of her, there was not a Man in the Street who did not pull off his Hat to me!

Wellington and His Friends, ed. 7th Duke of Wellington (1965), quoting a letter of Wellington to Mrs Arbuthnot, 6 August 1820

Caroline was to become more popular as the coronation approached, and signs of disaffection among some regiments caused Wellington anxiety. Owing to a failure to provide barracks for all, some were billeted with radical tradesmen, and came under the influence of the West End pro-Caroline mobs, who were joined by trade unionists from Wapping.

CAROLINE COMES ON 'TRIAL', 1820

She entered the House of Lords in August to hear the debates—prolonged and damaging (to all parties)—on the Bill of Pains and Penalties, George IV's doomed attempt to get a divorce from his wife.

She popped all at once into the House, made a *duck* to the throne, another to the Peers and a concluding jump into the chair which was placed for her. Her dress was black figured gauze, with a good deal of trimming, lace, &c: her sleeves white, and perfectly episcopal; a handsome white veil, so thick as to make it very difficult to me, who was as near to her as anyone, to see her face; . . . with a few straggling ringlets on her neck, which I flatter myself from their appearance were not her Majesty's own property.

She squatted into her chair with such a grace that the gown is at this moment hanging over every part of it—both back and elbows.

The Creevey Papers, I, 17 August 1820

The accusations of adultery against Caroline, together with her habit of dropping off in the House during these broiling summer weeks, provoked a popular rhyme:

Her conduct at present no censure affords,
She sins not with courtiers but sleeps with the Lords.

According to some of the evidence on her 'sins', during a cruise she would sleep in a tent on deck with her majordomo and take a bath with him being in the cabin if not in the bath with her, thus inspiring another popular couplet:

The Grand Master of St Caroline has found promotion's path.
He is made both Knight Companion and Commander of the Bath.

Lady Bessborough had seen her at a ball in Italy wearing 'a girl's white, frock-looking dress, but with shoulder, back and neck [bosom] quite low (disgustingly so) down to the middle of her stomach', a wreath of pale pink roses crowned by a black wig and fierce-looking black painted eyebrows.

Sightseers from other parts of London coming to see the queen on trial caused yet more verse:

And who were your company—heigh ma'am; ho ma'am? . . .
We happened to drop in
With *gemmen* from Wapping . . .

Roger Fulford, *The Trial of Queen Caroline* (1967)

A blunder by the queen's counsel (he likened her to the 'woman taken in adultery' who was told by the Saviour to 'go and sin no more'—the defence's whole point being that she had never committed adultery) produced the witty quatrain:

Most gracious queen, we thee implore
To go away and sin no more;
Or if that effort be too great,
To go away at any rate.

But the wittiest comment was by Queen Caroline herself, who is said to have remarked that the only time she committed adultery was when she went to bed with 'Mrs Fitzherbert's husband'.

After a majority of only nine for its third reading, the bill was never presented to the Commons, and Creevey summed up for Caroline's friends:

The Bill is gone, thank God! to the devil. *The Creevey Papers*, I

THE CORONATION—BY AN INSIDER, 19 JULY 1821

6 May 1821: The Queen has written to Lord Liverpool [prime minister] to say she means to go to the Coronation, desires to have ladies of high rank appointed to hold her train, & wishes to know *what dress* His Majesty wd desire her to wear!! The impudence of this woman is beyond belief. It wd have been well to have sent her word to appear in a white sheet.

July 1821: 'The King returned to the Hall [Westminster] about 5 o'clock, when the Earl Marshal, the Duke of Wellington (High Constable) & Lord Anglesea (High Steward) rode up the Hall with the first course, & backed out again. They came in again with the second course, & the two latter with the Champion. It was very well done; the Duke of Wellington rode a white arabian who backed most perfectly. There were a great many services done, caps given & returned, falcons presented by the Duchess of Atholl. The peers drank to the King, & he in turn to the peers & his good people, & the whole concluded with 'God save the King', sung by the choristers, & chorused by the whole assembly. After the riding was over, the people had been allowed to crowd into the body of the Hall & only a small space was kept open at the foot of the steps, & it is not possible to describe any thing finer than the scene was, the galleries all standing up waving their hats & handkerchiefs & shouting, 'God bless the King!' Altogether it was a scene I would not have missed seeing for the world, & shall never again see so fine a one.

The King behaved very indecently; he was continually nodding & winking at Ly Conyngham & sighing & making eyes at her. At one time in the Abbey he took a diamond brooch from his breast & looking at her, kissed it, on which she took off her glove & kissed a ring she had on!!! Any body who could have seen his disgusting figure, with a wig the curls of which hung down his back, & quite bending beneath the weight of his robes & his 60 years would have been quite sick. *Journal of Mrs Arbuthnot*, II

'CORONATION SOLILOQUY'

Put into the mouth of the re-incarnated 'Sun King' by the satirical poet Leigh Hunt.

Yes, my hat, Sirs,
Think of that, Sirs,
    Vast and plumed and Spain-like,
See my big,
Grand robes; my wig
    Young, yet lion-mane like.
Glory! Glory!
I'm not hoary,
    Age it can't come o'er me;
Mad, Grave, gazing on the grand man,
    All alike adore me.

DEATH OF QUEEN CAROLINE

Caroline died on 7 August 1821, just nineteen days after her frustrated attempt to be crowned, some said of misery, others of an internal obstruction. She was buried in Brunswick, not without further violence, for the London mob insisted on her coffin being carried through the City on its way to the port of Harwich and some of them were shot down in the government's futile attempt to resist them.

QUEEN CAROLINE'S MEMORIAL

She desires that on her coffin may be inscribed 'Caroline [of Brunswick], the *injured* Queen of England' . . . It is observed that she says *injured*, not innocent, and that no clergyman attended her in her last moments.

The Croker Papers 1808–1857, ed. Bernard Pool (1967): letter, 11 August 1821

GEORGE IV RECEIVES THE NEWS OF NAPOLEON'S DEATH ON ST HELENA

Napoleon died on 5 May 1821 and George heard of it while he was waiting at Anglesey to know the result of Caroline's illness, before beginning his Royal Progress in Dublin. Both Lord Holland and Sir William Fraser tell the story, in slightly different words, of George mistaking the news of Napoleon's death for that of his wife. The news was given by a courtier:

'I have, Sir, to congratulate you: your greatest enemy is dead.'
'Is *she*, by God?'

Lord Holland, Further Memoirs of the Whig Party 1807–1821, ed. Lord Stavordale (1905)

GEORGE IV'S MILITARY FANTASIES

His imaginary victories were not entirely ignominious, since as prince of Wales he had been thwarted by his father in a strong wish to serve his country on the field of battle. In later years he persuaded himself that he had led the German charge at Salamanca disguised as General Brock, and had headed the famous attack of the 'Tenth or Prince Regent's Own Royal Hussars' at Waterloo. He would often shout down the dinner table to the duke of Wellington for corroboration: 'Was that not so?' To which the tactful duke would reply, 'I have often heard your Majesty say so.'

THE KING'S STATE VISIT TO SCOTLAND—AND SIR WALTER SCOTT, 1822

You know how, when George IV came to Edinburgh, a better man than he [Walter Scott] went on board the Royal yacht to welcome the King to his kingdom of Scotland, seized a goblet from which His Majesty had just drunk, vowed it should remain for ever as an heirloom in his family, clapped the precious glass in his pocket, and sat down on it and broke it when he got home . . . Thackeray, *The Four Georges*

CHANTREY AND THE WIG

The famous sculptor Francis Chantrey had worked as a youth in a grocer's shop and wondered how the king, as his sitter, would treat him:

'Now, Mr Chantrey,' he said, 'I insist upon you laying aside everything like restraint, both for your own sake and for mine; do here, if you please, just as you would if you were at home.' Chantrey began to prepare the clay, and the King, after watching for some time, suddenly took off his wig, and, holding it at arm's length, said, 'Now, Mr Chantrey, which way shall it be, with the wig or without it?' Harold Armitage, *Francis Chantrey* (1915)

Chantrey sensibly decided for the wig. The story is remarkably like one about Queen Elizabeth II: when sitting to an artist for her portrait, she asked him on the second day, 'With the teeth or without?' meaning, with her mouth shut or not.

VISIT OF THE COMPOSER ROSSINI TO BRIGHTON

After a brief conversation, 'which seems to have left a very agreeable impression upon Rossini', the King invited him to hear his band; and, taking him by the arm, he led him into the music room. Rossini, he said, would now hear some music which might not be to his liking. 'But I have only chosen the first piece,' added the King. 'After that the band will play whatever you wish.' The first piece must have been more or less to Rossini's taste, for it was the overture to his own opera *La Gazza Ladra.* Rossini had already discovered which were the King's favourite pieces. He now asked for them, and pointed out 'their characteristic beauties'. And, since it was to be an evening of graceful gestures and reciprocated compliments, Rossini finally told his host that he had never heard *God Save the King*, except on the piano, and that he would like to hear it performed by his excellent band. The King was evidently gratified . . .

Two apocryphal stories remain: it is said that Rossini once accompanied 'the vocal efforts' of George IV himself; the King, a bass, got into the wrong key, but Rossini continued to play as though nothing untoward had happened. 'It was my duty', he explained, 'to accompany your Majesty. I am ready to follow you wherever you may go.' The other story is less graceful: at a grand concert at St James's Palace, the King said, 'Now Rossini, we will have one piece more, and that shall be the *finale*.' 'I think, Sir,' replied Rossini, 'we have had music enough for one night.' And he made his bow. Joanna Richardson, *George IV: A Portrait* (1966)

ROYAL ENTERTAINMENT

26 July 1821—the King's Drawing Room: The King had a Drawing Room. Everybody was fresh presented, which was a mere whim of the King & a great bore, as he is not remarkably agreeable to kiss.

Journal of Mrs Arbuthnot, II

6 July 1825: I was at a ball at St James's on the 4th. It was very magnificent, all the Royal Family were there & every body dressed as fine as they could be. The supper was *very bad*, positively I could not get any thing to eat, & it was a *standing up* supper, which is not very Royal, I think, the King himself ate standing. There were no French wines, bad fruit, no hot meat or soups; in short, an individual wd not have ventured to give so bad a supper. I suppose it is Lady Conyngham's economy. She is the most avaricious woman in the world and, I have understood, considers dining out as *tant de gagné upon the weekly bills.* Ibid.

GEORGE IV AND CATHOLIC EMANCIPATION, 1829

The right of Catholics to sit in Parliament was introduced by Wellington and Peel in the teeth of opposition from their own Tory party. The king was a prime opponent. As he gradually gave way he was heard to mutter gloomily,

'Arthur [Wellington] is the King of England, O'Connell King of Ireland, and I am Canon of Windsor.'

His opposition was understandable, since his father had threatened to go mad again if Pitt touched the 'Catholic Question', and Peel himself had changed his mind. What was strange, and typical of George's fantasies, was his assertion, when the battle was almost won that the Emancipation Bill was 'his measure,' and that

he knew he should carry it at a canter, although the Duke was very nervous!! *Journal of Mrs Arbuthnot*, II

WELLINGTON'S ACCIDENT IN HYDE PARK

1 June 1829: The Duke fell from his horse the other day at the Review in consequence of having on his head the extravagant Grenadier cap which the King has thought fit to order & with which, in a high wind, it is impossible to balance yourself; and the King, at the ball, called up Lord Anglesey & said to him, 'Pray, Anglesey, is it your custom to fall from your horse at the head of your regiment?' Ld A, of course, repeated this to every body with delight. Ibid.

Both the king and Anglesey were hostile to Wellington over Catholic Emancipation, though from opposing political standpoints.

The King's Last Weeks, 1830

As he grew older a kind of torpor enveloped him, until, at the end of his reign, he was called at six or seven in the morning, breakfasted in bed, transacted what business his Ministers could induce him to transact, still in bed, read every newspaper all through, got up in time for dinner at six, and retired to bed again between ten and eleven. In the night would often ring his bell forty times, and, though a watch hung by his side, he would not make the effort of turning his head to look at it, but would ring for a Page to tell him the time: similarly, he would not even stretch out his hand for a glass of water. Osbert Sitwell, *Left Hand Right Hand*, I (1945)

THE KING'S BREAKFAST, 9 APRIL 1830

I heard of the King this morning. What do you think of his breakfast yesterday morning for an Invalid? A Pidgeon and Beef Steak Pye of which he eat two Pigeons and three Beefsteaks. Three parts of a Bottle of Mozelle, a Glass of Dry Champagne, two Glasses of Port & a Glass of

Brandy! He had taken Laudanum the night before, and again before this breakfast, again last night and again this Morning!

Wellington and His Friends, quoting a letter of Wellington to Mrs Arbuthnot

DECLINE AND DEATH, 26 JUNE 1830

23 April 1830: His mode of living is really beyond belief. One day last week, at the hour of the servants' dinner, he called the Page in & said, 'Now you are going to dinner. Go down stairs & cut me off just such a piece of beef as you wd like to have yourself, cut from the part you like the best yourself, & bring it me up.' The Page accordingly went and fetched him an enormous quantity of roast beef, all of which he ate, & then slept for 5 hours.

One night he drank two glasses of hot ale & toast, three glasses of claret, some strawberries!! and a glass of brandy. Last night they gave him some physic and, after it, he drank three glasses of port wine & a glass of brandy. No wonder he is likely to die. But they say he will have all these things & nobody can prevent him. I dare say the wine wd not hurt him, for with the Evil (which all the Royal Family have) it is necessary, I believe, to have a great deal of high food, but the mixture of *ale* and strawberries is enough to kill a horse.

29 June 1830: The poor King died on the 26th at ½ past 3 in the morning, apparently without pain. The Duke saw him two days before and did not think he wd have died so soon; but a blood vessel in his stomach burst. He put his hand up to his breast, exclaimed, 'Good God, what do I feel? This must be death!' & died in a few minutes. *Journal of Mrs Arbuthnot*, II

THE KING'S LYING-IN-STATE

16 July 1830: It was in one of the old State Rooms in the Castle. The coffin was very fine and a most enormous size. They were very near having a frightful accident for, when the body was in the leaden coffin, the lead was observed to have bulged very considerably & in fact was in great danger of bursting. They were obliged to puncture the lead to let out the air & then to fresh cover it with lead. Rather an *unpleasant operation*, I shd think, but the embalming must have been very ill done. Ibid.

A BOW AND A GRIN

Madame Tussaud has got King George's coronation robes: is there any man now alive who would kiss the hem of that trumpery? He sleeps since thirty years: do not any of you, who remember him, wonder that you once respected and huzza'd and admired him? . . . The Sailor King who came after George was a man: the Duke of York was a man, big, burly, loud, jolly, cursing, courageous. But this George, what was he? I look through all his life, and recognise but a bow and a grin. I try and take him to pieces, and find silk stockings, padding, stays, a coat with frogs and a fur collar, and star and blue ribbon, a pocket-handkerchief prodigously scented, one of Truefitt's best nutty-brown wigs reeking with oil, a set of teeth and a huge black stock, underwaistcoats, more underwaistcoats, and then nothing.

Thackeray, *The Four Georges*, giving Thackeray's famous dismissal of George IV

And Wellington, as executor, distributed the king's wardrobe among the Pages of the Backstairs. He found less than nothing when he went through the late king's papers:

6 July 1830: The Duke told us he had been examining the King's papers, that there was nothing but volumes of love letters, chiefly from Ly Conyngham, some *foul copies* of his own to Ly Conyngham descriptive of the most furious passion, trinkets of all sorts, quantities of women's gloves, dirty snuffy pocket handkerchiefs with old faded nosegays tied up in them; in short, such a collection of trash he had never seen before. He said he thought the best thing would be to burn them all. *Journal of Mrs Arbuthnot*, II

William IV

1830–1837

Prince William Henry duke of Clarence, third son of George III, acceded to the throne at the age of sixty-five. It was too late to change his character, which had been formed in the Navy and continued to swing between basic decency and eccentric buffoonery. He fought in a naval battle before ascending the throne, something that was not repeated until Prince Bertie (George VI) served at the battle of Jutland. Like his brother George IV with the bill for Catholic Emancipation, William IV gave his royal assent to a historic bill he hated, the Great Reform Bill, but with better grace. In youth he looked a typical, good-natured sailor-boy; in maturity he was affable, garrulous, undignified, with a curious pear-shaped head; in later life he was the spitting image of 'a respectable old admiral'.

NORMAL IN THE NAVY

Amidst the welter of colourful anecdote which surrounds the Prince's naval days, it is by now impossible to sift the literally true from the embellished, the embellished from the invented. When William was asked his name on his first day aboard did he really reply: 'My father's name is Guelph and you are welcome to call me William Guelph'? When he was behaving with undue self-confidence did one of his fellow midshipmen really say to him: 'Avast there, my hearty, the son of a whore here is as good as the son of a king'? The latter at least sounds improbably picturesque, and yet something close to it must have been said a dozen times. Prince William joked, shouted, cursed, grumbled like every other boy. He fought Lieutenant Moody of the Royal Marines, and after peace had been restored, shook him by the hand with the rather patronising remark: 'You are a brave fellow, though you are a Marine.' He drank and gambled whenever the chance was given him—which was not very often—did his work with reasonable consicentiousness, got into trouble from time to time. His life, in fact, was almost entirely normal—in that, of course, lay its abnormality.

Philip Ziegler, *King William IV* (1971)

PRINCE WILLIAM HENRY AT THE SIEGE OF GIBRALTAR AGED FOURTEEN, 1780

A conference took place between Admiral Digby and the Spanish Admiral Don Juan who visited Digby's ship. William's most recent biographer, Philip Ziegler, quotes the story of this meeting but doubts whether such words were ever on the lips of the proud Spaniard.

During the conference between the two Admirals, the Prince retired; and when it was intimated that Don Juan wished to return [to his ship], his Royal Highness appeared in the character of a midshipman, and respectfully informed the Admiral that the barge was ready. The Spaniard, astonished to see the son of a monarch acting as a warrant officer, could not help exclaiming, 'Well does Great Britain merit the empire of the sea, when the humblest stations in her navy are filled by princes of the blood!'

Wright, *William IV*, I

PRINCES AT A MASKED BALL

There is a story of a masked ball in the autumn of 1785 at which two of the guests came to blows. They were marched off by the guard and unmasked. 'Aye, William, is it you?' 'Aye, George, is it you?' And the embarrassed guards hurriedly released their princely prisoners. Ziegler, *William IV*

WILLIAM IN HANOVER, 1785

Denied the company of aristocratic or even middle-class women, he was forced to perform

'with a lady of the town against a wall or in the middle of the parade . . .'

He told his brother the Prince of Wales that he loathed

'this damnable country, smoaking, playing at twopenny whist and wearing great thick boots. Oh, for England and the pretty girls of Westminster; at least to such as would not clap or pox me every time I fucked.'

Ibid., quoting a letter of April 1785

ON THE HALIFAX STATION, NOVA SCOTIA 1787/8

Prince William Henry, as he was officially called, commanded the Andromeda *and became friends with a lively young officer, William Dyott, who was also on the station. The prince observed moderation as to drink but not as to women. Dyott first met him in October, and decided he was like George III only better looking—about 5 feet 7 or 8 inches tall, and fair.*

Wednesday morning: I met him walking in the street by himself. I was with Major Vesey of the 6th regiment. His Royal Highness made us walk with him; he took hold of my arm, and we visited all the young ladies in the town . . . [He] dislikes drinking very much [but on this first acquaintance he drank] near two bottles of Madeira.

A month later: He would go into any house where he saw a pretty girl, and was perfectly acquainted with every house of a certain description in the town.

6 November: . . . he told me I must dine with him before he sailed, and that we would have a small snug party and none of the great people.

A dance on the 7th: The last dance before supper . . . his Royal Highness, Major Vesey, and myself, and six very pretty women danced Country Bumpkin for near an hour. He . . . joked me all the evening on our party at dinner; but I must say I never in my life saw him the smallest degree lose his dignity or forget his princely situation.

The 'mutual passion' between the Prince and the Governor's wife was of course not unprincely.

His character is, where he takes a liking he is very free, but always guarded, and if ever any man takes the smallest liberty he cuts instantly.

Friday the 9th after dining on HRH's ship: When we first came on shore he was very much out indeed, shouted and talked to every person he met.

The last dinner before HRH sailed: He did not drink himself; he always drinks Madeira [not the claret provided]. He took very good care to see everybody fill, and he gave twenty-three bumpers without a halt. In the course of my experience I never saw such fair drinking. When he had finished his list of bumpers, I begged leave . . . to stand upon our chairs with three times three . . . I think it was the most laughable sight I ever beheld, to see the Governor, our General, and the Commodore, all so drunk they could scarce stand on the floor, hoisted upon their chairs with each bumper in his hand; and the three times three cheers was what they were afraid to attempt for fear of falling . . . HRH saw we were all pretty well done and he walked off. There were just twenty dined, and we drank twenty-three bottles of wine. *Dyott's Diary*, I, ed. R. W. Jeffery (1907)

AN IRONIC COLLEAGUE, 1787

He once jeered at a fellow captain for being the son of a schoolmaster from Hackney and asked why the captain had not followed the same career as his father. 'Why, sir,' retorted the admirable Captain Newcombe, 'I was such a stupid, good-for-nothing fellow, that my father could make nothing of me, so he sent me to sea.' Ziegler, *William IV*

THE SAILOR HOME FROM THE SEA

Then he said he had been making acquaintance with a new Princess, one he did not know nor remember—Princess Amelia. 'Mary, too,' he said, 'I had quite forgot; and they did not tell me who she was; so I went up to her, and, without in the least recollecting her, she's so monstrously grown, I said, 'Pray, ma'am, are you one of the attendants?'

Madame d'Arblay, V, May 1789

A TYPICAL ADVANCE BY THE DUKE OF CLARENCE, 1814

At the end of a party at the Brighton Pavilion, William was asked to take the place of the Prince Regent and hand the unaccompanied Princess Lieven to her carriage.

He was, as usual after dinner, a little lively and unsteady on his legs. He walked slowly, and, after having put me in the carriage, and, as they were going to raise the carriage-step, he pushed the footman roughly, got into the carriage without a hat, and ordered them to drive on. All this was done so quickly that I had no time to stop it, but I felt very ill at ease. Hardly was he in the carriage than he said:

'Are you cold, Madame?'
'No, Monseigneur.'
'Are you warm, Madame?'
'No, Monseigneur.'

(His conversation always began like that.)

'Permit me to take your hand.' (This was an extra.)
'It is needless, Monseigneur!'

But this did not prevent him from taking my hand. Fear seized me, for he was evidently drunk. With the other hand I hastened to lower the carriage window as a precautionary measure. As I did not want to use it, I soon racked my brains for something to distract his attention. I have said that he was very stupid, very ignorant of everything. He took no interest in anything, great affairs preoccupied him not at all. He had only one fixed idea in politics—Hanover.

Princess Lieven, *Unpublished Diary*

Princess Lieven raised the topic of Hanover; it worked. He dropped her hand and poured out 'a torrent of great words' on Hanover for the rest of the way.

LIFE WITH MRS DOROTHY JORDAN

William fell madly in love with the beautiful actress, set up house with her at Bushey and by her had ten children. When George III first heard about the liaison he asked:

'Hey, hey:—what's this—what's this. You keep an actress, they say.' 'Yes, sir.' 'Ah, well, well; how much do you give her, eh?' 'One thousand a year, sir.' 'A thousand, a thousand; too much; too much! Five hundred quite enough! Quite enough!' Brian Fothergill, *Dorothy Jordan* (1965)

The success of her career compared with his provoked a satirical rhyme:

As Jordan's high and mighty squire
  Her playhouse profits deigns to skim,
Some folks audaciously enquire
  If *he* keeps *her* or *she* keeps *him*.

Anon.

THE END OF MRS JORDAN

Under pressure from his mother, William parted from Mrs Jordan in 1811. Though he showed her nothing but kindness, giving her a generous allowance, stories circulated of the prince's meanness in supposedly requesting repayment.

Enough wit was left her, however, to signify her refusal by sending him a playbill with this notice: 'Positively no money refunded after the curtain has risen.' Windsor, 'My Hanoverian Ancestors'

THE ROYAL MATRIMONIAL MARATHON

After the death of Princess Charlotte in childbirth, William duke of Clarence and Edward duke of Kent decided to get married and breed for the succession. William married Princess Adelaide of Saxe-Meiningen in 1818, but their first two daughters died in infancy and their twin daughters were stillborn. Meanwhile the dukes of Cambridge and Cumberland had also married and joined Clarence and Kent in the race for the heir to the throne. The competition caused further satire:

Yoics! the Royal sport's begun!
I' faith, but it is glorious fun,
For hot and hard each Royal pair
Are at it hunting for the heir.

Peter Pindar, *Poems*

WILLIAM'S ACCESSION ANNOUNCED TO HIM

He was roused from bed to receive the news and composedly returned to bed, 'in order,' he explained, 'to enjoy the novelty of sleeping with a queen'. Windsor, 'My Hanoverian Ancestors'

FIRST IMPRESSIONS OF THE NEW KING

29 June 1830: The King came immediately to town. The Privy Council were sworn in & he went thro' all the necessary ceremonies on the 26th & 28th, & each night returned to Bushey, where he now is. He announced at once that he shd change the uniform of the Guards and put all the cavalry into scarlet except *the Blues*, said he shd make two Field Marshals (two old women, Sir Samuel Hulse & Sir Alured Clarke), talked of making Lord Combermere an Earl and a Privy Councillor & some other follies; but I dare say he will be easily put off them &, if he don't go mad, will do very well . . .

21 July 1830: The King is somewhat *wild* and talks & shews himself too much. He walked up St James's Street the other day quite alone, the mob following him, & one of the common women threw her arms round him & kissed him.

Journal of Mrs Arbuthnot, II

WILLIAM AND THE REFORM OF PARLIAMENT

Lord Grey, the Whig prime minister, requested the king to dissolve parliament, as a preliminary to a general election and victory for Reform. For the only time in his life the agitated king responded with verse:

I consider Dissolution
Tantamount to Revolution.

But when he heard that a diehard peer was out to defeat Reform by preparing a motion to obstruct his prerogative, he promptly ordered his carriage to go to parliament. The startled Lord Albemarle, Master of the Horse, received a messenger urging him to hurry while still at breakfast.

'Lord bless me! is there a revolution?

'Not at this moment, but there will be if you stay to finish your breakfast.'

Lord Albemarle, *Fifty Years of my Life* (1877)

Albemarle left his breakfast, the royal horses were harnessed, the crown was rushed to the Palace from the Tower of London and the king's robes were snatched back from a studio where they were being officially painted. Meanwhile uproar was growing in both Houses. The Tory Sir Robert Peel refused to be silent though cannon already boomed the king's approach; a diehard, Lord Londonderry, flourished a whip at a Whig duke while Lord Mansfield egged him on and five other peers tried to drag him off. The king asked Brougham, his lord chancellor, what all the din could be.

'If it please your Majesty, it is the Lords debating.'

Works of Henry Lord Brougham, III (1872)

WILLIAM IV

With his crown crooked and precariously balanced on his head, King William IV dissolved the last unreformed parliament.

EFFECTS OF 'REFORM' ON ROYALTY

13 June 1831: We returned to town for the ball on the 13th, but the balls at St James's are now like bear gardens. All sorts of people are invited and, instead of its being an honor to be invited, as it was in George the 4th's time, people now don't go if they happen to be tired or unwilling. In short, the Queen is treated like any other lady & very soon wd, I shd think, be even worse off. The Ministers put no sort of check upon the licentiousness of the public Press, and there are penny publications every day in which the King and Queen are called *Mr & Mrs* Guelph! And meetings at a place called the Rotunda, where the most blasphemous & republican doctrines are preached without any hindrance. If such poison is permitted to be daily & hourly administered, I don't see how other consequences can follow than total demoralisation & ruin. *Journal of Mrs Arbuthnot*, II

CHANGING TIMES

On his first visit to the Palace in the reign of William and Adelaide, Charles Fulke Greville, Clerk of the Privy Council, looks before and after: before to George IV and after to the as yet unknown Victoria.

What a *changement de decoration*; no longer George the 4th, capricious, luxurious, and misanthropic, liking nothing but the society of listeners and flatterers, with the Conyngham tribe and one or two Tory Ministers and Foreign Ambassadors; but a plain, vulgar, hospitable gentleman, opening his doors to all the world, with a numerous family and suite, with a frightful Queen and a posse of bastards, originally a Whig Minister, and no foreigners, and no Toad-eaters at all. Nothing more different, and looking at him one sees how soon this act will be finished, and the scene be changed for another probably not less dissimilar. Queen, bastards, Whigs, all will disappear, and God knows what replace them.

The Greville Memoirs 1814–1860, I, 5 June 1831, ed. Lytton Strachey and Roger Fulford (1938)

SCANDAL ATTACKS EVEN THE BLAMELESS QUEEN ADELAIDE

Jonathan Peel told me yesterday morning that L[ady] A[lice] Kennedy had sent word to his wife that the Queen is with child; if it be true, and a queer thing if it is, it will hardly come to anything at her age, and with her health; but what a difference it would make! Ibid., III, 20 January 1835

THE ALLEGED FATHER

Lord Howe was soon being named as the father:

Munster told me the day before yesterday that he heard of the Queen's being with child on the day of the Lord Mayor's dinner; that She is now between two and three months gone. Of course there will be plenty of scandal. Alvanley proposes that the Psalm '*Lord, how* wonderful are thy works' should be sung. It so happens, however, that Howe has not been with the Court for a considerable . . . time.

The Greville Memoirs, 25 January 1835

SAYINGS OF WILLIAM IV

To a shy boy he met at a party: 'Come, we are both boys, you know.'

On a dinner party in 1786: 'We sat down thirty people. Few got up sober, for we were at the table and bottle seven hours and a half.'

On books: 'I know no person so perfectly disagreeable and even dangerous as an author.'

William rebuked Cumberland in a famous toast: 'The land we live in, and let those who don't like it, leave it!'

On being rescued from a mob: 'Oh, never mind all this: when I have walked about a few times they will get used to it, and will take no notice.'

On the Tory defeat, 1835: 'I will have no more of these sudden changes. The country shan't be disturbed in this way, to make my reign tumble about, like a topsail sheet-block in a breeze.'

DEATH-BED OF WILLIAM IV

A few days before Waterloo Day, 18 June 1837, the King said, 'Doctor, I know I am going but I should like to see another anniversary of the battle of Waterloo. Try if you cannot tinker me up to last out that day.' The doctor did so and on the morning of the day the King said, 'I know I shall never live to see another sunset.' The doctor replied, 'I hope your Majesty will live to see many.' The King ended the dialogue with the words 'Oh, that is quite another thing.' He died two days later.

Ziegler, *William IV*

HIS OBITUARY

He was not a man of talent or of much refinement . . . But he had a warm heart, and it was an English heart.

The Times, 20 June 1837

VICTORIA AND HER DESCENDANTS

Victoria

1837–1901

Victoria gave her name to an age that was remarkable both for material changes and superficially immutable 'values'. The Victorian Age saw the transition from a perfected coaching system to 'miracles' of modern transport and communication, two more Reform Acts and the need, amid so much wealth, to combat poverty. The queen called the only British–European war of her reign (the Crimean War) 'unnecessary' but entered strongly into the spirit of colonial wars. Fatherless at eight months old and widowed at forty-two, she bore nine children despite ambiguous feelings about motherhood, and denounced women's rights—as well as their wrongs. Free from racial prejudice and from class snobbery all her life, she also began life without prudery but, partly under her husband Prince Albert's influence, gradually adopted 'Victorian' sexual taboos. Ascending the throne at eighteen, this diminutive woman—she did not grow to quite five feet tall—was to survive seven attempts on her life, reign longer than any other British monarch and become Britain's first and last Queen-Empress.

VICTORIA'S PARENTS

Her father was Edward duke of Kent, fourth son of George III.

He was the baldest of the whole family and Sheridan suggested that this was because grass did not grow upon deserts. The Duke remarked when he heard this joke, 'If Sheridan means that I haven't genius, I can tell him that such a gift would have been of small value to a Prince, whose business it is to keep quiet. I am luckier in having, like my country, a sound constitution.' *Life of Queen Victoria, The Times*, 1901

Lord Melbourne was to tell Victoria that her father was, like George IV,

very agreeable [and unlike William IV] he had none of that talking; and was much more posé [calm]. Queen Victoria's Journal, Royal Archives

Her mother was Princess Victoria of Saxe-Coburg, sister of Prince Leopold

afterwards king of the Belgians, and of Prince Albert's father. She was the widow of the prince of Leiningen, by whom she had two children, Charles and Feodore, Victoria's half-brother and half-sister. The young duchess of Kent could not speak English and after her marriage had to make a public speech of thanks from a phonetical script:

Ei hoeve tu regrétt, biing *aes yiett* so littl cônversent in thie Inglisch lênguetsch, uitsch obleitschës—miy, tu seh, in *averi fiú words*, theat ei em môhst grêtful for yur congratuleschens end gud uishes, end *heili*, flatterd, bei yur allucheon to mei brother. Ibid.

THE BIRTH OF PRINCESS VICTORIA

She was brought into the world at Kensington Palace on 24 May 1819 by a 'medical lady' of whom her father the duke of Kent wrote proudly:

The Medical Lady whom the Duchess's family and herself wish to attend her, in conjunction with Dr Wilson . . . not only is an accoucheuse, but also practises as a Physician in all Ladies' complaints, having gone through the regular course of Anatomy, Physics, etc., at Gottingen.

Elizabeth Longford, 'Queen Victoria's Doctors' in *A Century of Conflict*, ed. Martin Gilbert (1966)

Ironically, the future queen was to abominate 'medical ladies', especially those who studied anatomy alongside their male colleagues in the medical schools.

The duke also boasted of the baby's vigour:

The little one is rather a pocket Hercules, than a pocket Venus.

Cecil Woodham-Smith, *Queen Victoria: Her Life and Times* (1972)

VICTORIA STANDS FIRE LIKE A SOLDIER'S DAUGHTER

The duke of Kent and his family were staying at Woolbrook Cottage, Sidmouth, for the sake of its usually mild climate. But the duke caught a feverish cold during the exceptionally savage winter and died on 23 January 1820 of 'cupping and bleeding', the regular medical treatment of the day. A month before his death, when Victoria was aged seven months an accident nearly carried off his daughter also.

About four in the afternoon of 28th December 1819, the Duchess was sitting in the drawing-room with the infant Princess, when, to her terror, a shot shattered the window. She was '*most* exceedingly' alarmed. The culprit turned out to be an apprentice boy called Hook, who was in the road, firing at birds and using an unnecessarily heavy type of shot, swan shot. After this agitating incident the Duke and Duchess of Kent showed themselves at their best. Within half-an-hour Captain Conroy wrote on the Duke's behalf to a Mr George Cornish, probably the local magistrate, asking him to 'adopt some measures for the prevention of such an occurrence. But, their Royal Highnesses desire me most particularly to request that the Boy may not be punished, they only interfere to prevent the thing happening again.' Ibid.

EARLIEST MEMORIES

At the age of fifty-three the queen recalled one of her childhood phobias:

I had a great horror of *Bishops* on account of their wigs and *aprons*, but recollect this being partially got over in the case of the then Bishop of Salisbury . . . by his kneeling down and letting me play with his badge of Chancellor of the Order of the Garter. With another Bishop, however, the persuasion of showing him my 'pretty shoes' was of no use.

Letters of Queen Victoria, I, ed. A. C. Benson and Lord Esher, 9 vols. (1907–30)

As a young queen she told her prime minister Lord Melbourne about her bishop phobia—how she 'hated' them as 'a very *little girl' (Journal, Royal Archives)—and indeed the phobia lasted all her life. During her Diamond Jubilee she attended an ecclesiastical party at Lambeth Palace; on the way home she said to her lady-in-waiting, Lady Lytton:*

'A very ugly party. I do not like bishops.'
Lady Lytton (shocked): 'Oh, but your dear Majesty likes *some* bishops . . .'
'Yes, I like the man but *not* the bishop.'

Lady Lytton's Court Diary, ed. Mary Lutyens (1961)

IMPRESSIONS OF VICTORIA AS A CHILD

Lord Albemarle was lord-in-waiting to the duke of Sussex, Victoria's paternal uncle, who also lived in Kensington Palace:

One of my occupations of a morning, while waiting for the Duke, was to watch from the window the movements of a bright little girl, seven years of age. It was amusing to see how impartially she divided the contents of the watering pot between the flowers and her own little feet . . . She wore a large straw hat, and a suit of white cotton; a coloured fichu round the neck was the only ornament she wore. The young lady I am describing was Princess Victoria . . . Albemarle, *Fifty Years of my Life*

Other, unnamed observers remembered the little princess as a decidedly spirited child with a mind of her own. She was called Drina during her first years, a shortening of Alexandrina, after the Tsar Alexander of Russia.

Mama: When you are naughty you make me and yourself very unhappy.

Drina: No Mama, not *me*, not myself, but *you*!

One morning before lessons began her tutor asked Mama if she had been good.

Mama: Yes, she has been good this morning, but yesterday there was a little storm.
Drina: Two storms—one at dressing and one at washing.

On another day her music master had to correct her:

Master: There is no royal road to music, Princess. You must practise like everybody else.
Victoria (shutting the piano with a bang): There! you see there is no *must* about it.

Nor was Victoria prepared to abandon the royal road with her playmates. One day she noticed that little Lady Jane Elliot was about to play with the royal toys:

'You must not touch those, they are mine; and I may call you Jane, but you must not call me Victoria.' *Cornhill Magazine*

PRINCESS VICTORIA AND 'UNCLE KING'

In 1872 Queen Victoria wrote down an account of her first visit to George IV at Royal Lodge, Windsor. She was seven:

When we arrived at the Royal Lodge the King took me by the hand, saying: 'Give me your little paw.' He was large and gouty but with a wonderful dignity and charm of manner. *Letters of Queen Victoria*, I

The duke of Wellington described the same occasion:

Virginia Water, 2 August 1826: The King was very drunk, very blackguard, very foolish, very much out of temper at times, and a very great bore! In short, I never saw him so bad as yesterday. The little Princess is a delightful Child. She appeared to please the King.

Wellington and His Friends, quoting a letter of Wellington to Mrs Arbuthnot

It was not so pleasant to kiss his face, which was covered with grease-paint. Her great-grandson, Prince David, the future duke of Windsor, was to feel equally ambivalent about giving his 'little paw' to his great-grandmother Queen Victoria:

Seventy-five years later, when *I* was seven, I was taken in to see *her* in the Long Gallery at Windsor, and she gave me her fat, boneless little hand, and I remember *my* distaste! Windsor, 'My Hanoverian Ancestors'

'I WILL BE GOOD'

On 11 March 1830, two months before her eleventh birthday, Princess Victoria was made aware, dramatically, of her great inheritance. Having just opened 'Howlett's Tables' of the Kings and Queens of England to begin her history lesson with her German governess the Baroness Lehzen, she suddenly noticed that an extra page had been slipped into the book.

'I never saw that before,' exclaimed Victoria.
'No, Princess. It was not thought necessary that you should.'

The new page brought the royal family tree up to date, showing that only her 'Uncle King' and her Uncle William duke of Clarence stood between herself and accession. And she probably knew that George IV was on his death-bed.

'I am nearer to the throne than I thought.'

She then shed a few tears and told Lehzen that whereas some children might boast of the splendour, they would not realize the difficulties. Lifting up the forefinger of her right hand she said:

'I will be good.'

She went on to explain what she meant by being 'good':

'I understand now why you urge me so much to learn, even Latin. . . . but you told me Latin is the foundation of the English grammar and of all the elegant expressions; and I learnt it, as you wished it, but I understand all better now'

—and she put her right hand into Lehzen's repeating solemnly,

'I will be good.'

Baroness Lehzen to Queen Victoria, 2 December 1867, Royal Archives

A good girl who learnt her lessons was what she meant, but the thought at the back of her mind that some time she would become a good queen made the words memorable.

AN ATTACK OF TYPHOID

She was seriously ill while staying at Ramsgate, aged sixteen. But it was not so much the illness as the struggle it caused within her mother's household that made the event traumatic. She believed that Sir John Conroy, her mother's majordomo, was out to gain control of herself as well as of her mother. (*Wellington and Greville supposed they were lovers.*) *On her side she could count on Uncle Leopold and her beloved Lehzen; against her the ambitious knight, who took advantage of her weakness to demand, in advance, the position of private secretary when she came to the throne. Though too ill to keep her journal, she persistently refused to sign the document he put before her. After she had become queen she described her victory to Lord Melbourne, her prime minister:*

Victoria: I resisted in spite of my illness.
Melbourne: What a blessing! Journal, Royal Archives

THE FIRST RAILWAY TRAIN, 1837

8 February: We went to see the railroad near Hersham, and saw the steam carriage pass with surprising and startling quickness, striking sparks as it flew along the railroad, enveloped in clouds of smoke and making a loud noise. It is a curious thing indeed!' Ibid.

'*I AM QUEEN*', 20 JUNE 1837

Up to the very last minute Victoria's mother and half-brother, Prince Charles of Leiningen, tried to coerce the princess into accepting what they called Conroy's 'system'—in other words his domination. But the princess shut herself all day in her room and refused to speak to them while her uncle William IV lay dying. She was sleeping as always in her mother's bedroom when the news came that William had died.

I was awoke at 6 o'clock by Mamma who told me that the Archbishop of Canterbury and Lord Conyngham were here and wished to see me. I got out of bed and went into my sitting room (only in my dressing gown) and *alone*, and saw them. Lord Conyngham (the Lord Chamberlain) then acquainted me that my poor Uncle, the King, was no more, and had expired at 12 minutes past 2 this morning and consequently that *I* am *Queen*. *The Girlhood of Queen Victoria*, I, extracts from Victoria's Journal, ed. Viscount Esher, 2 vols. (1912)

It was said that the moment Conyngham reached the word 'queen' she did not wait for anything else but shot out her hand for him to kiss. Her bed was at once moved into a room of her own.

Curiosity About the Queen

In accordance with their system, Conroy and the duchess had maintained control of Victoria by keeping her out of the public eye. Three hard-bitten men of the world described the queen's first Council on 21 June 1837.

Croker the diarist wrote:

I cannot describe to you with what a mixture of self-possession and feminine delicacy she read the paper. Her voice, which is naturally beautiful, was clear and untroubled; and her eye was bright and calm, neither bold nor downcast, but firm and soft. There was a blush on her cheek . . . and certainly she *did* look as interesting and handsome as any young lady I ever saw. *The Croker Papers 1808–57*

Charles Greville, clerk to the Council, wrote:

She looked very well, and though so small in stature, and without any pretension to beauty, the gracefulness of her manner and the good expression of her countenance give her on the whole a very agreeable appearance, and with her youth inspire an excessive interest in all who approach her, and which I can't help feeling myself. *The Greville Memoirs*, III

The Duke of Wellington, a Tory like Croker, was heard to say:

'She not merely filled her chair, she filled the room.'

All three were soon to become less sentimental about the young queen.

SPECULATION ABOUT THE NEW QUEEN

Scarcely more than a month after Victoria's accession the Tories, as exemplified by the duke of Wellington, were regretting the death of the old king:

. . . partly on account of the dreadful uncertainty attending the Govt. of a Young Lady of 18.

I must say of Her however that everything that has appeared of Her from the Moment she appeared in the Council at Kensington is excellent. I know no more. She is surrounded by Whiglings Male and female; and no body knows any thing except Gossip . . .

It is said that the Young Queen is a Person of Character with a Will of Her own. She can have no Knowledge However even to the amount of that which the late King had.

Wellington's Private Correspondence, Roxburghe Club (1952). *This unfinished letter from Wellington is to an unnamed correspondent on 26 July 1837.*

THE QUEEN AND HER LIBERAL PRIME MINISTER, 1837

Lord Melbourne sees her every day for a couple of hours, and his situation is certainly the most dictatorial, the most despotic, that the world has ever seen. Wolsey and Walpole were in strait waistcoats compared to him.

The Croker Papers. Melbourne was then aged fifty-eight.

A CYNICAL VIEW OF THE YOUNG QUEEN, 1838

I was much amused at seeing our young Victoria playing the popular to her people . . . She passed this house [Brooks's Club in St James's] in state—

four royal carriages and an escort of Horse Guards. The mother had judiciously chosen a chariot for herself and daughter, so they were both visible to all. The young one was rather too short to nod quite above the door, but she was always at it as well as she could. The Croker Papers

THE QUEEN AT A PERFORMANCE OF *KING LEAR*, 1838

It was not pleasant to see her, when Macready's 'Lear' was fixing all other hearts and eyes, chattering to the Lord Chamberlain, and laughing, with her shoulder turned to the stage. I was indignant, like a good many other people: but, in the fourth act, I saw her attention fixed; and then she laughed no more. She was interested like the rest of the audience; and, in one way, more than others. Probably she was the only person present to whom the play was entirely new to her, in as much as she was not previously aware that King Lear had any daughters.

Harriet Martineau, *Autobiography* (1877)

In fact the duchess of Kent had presented her daughter with a copy of 'King Lear' on her nineteenth birthday—but perhaps she did not read it.

The Coronation

Harriet Martineau, the writer and philosopher, attended the ceremony in a mood of rational criticism. She was an unaccompanied spinster aged thirty-five.

I was quite aware that it was an occasion (I believe the only one), on which a lady could be alone in public without impropriety . . . Except for a mere sprinkling of oddities, every body was in full dress. In the whole assemblage, I counted six bonnets [rather than lace, pearls, combs, tiaras, on the head] . . . The throne, an armchair with a round back, was covered, as was its footstool, with cloth of gold, stood on an elevation of four steps, in the centre of the area . . . From a quarter to seven [peers and their ladies] arrived faster and faster . . . Old hags, with their dyed or false hair drawn to the top of the head, to allow the putting on of the coronet, had their necks and arms bare and glittering with diamonds: and those necks were so brown and wrinkled as to make one sick; or dusted over with white powder which was worse than what it disguised . . . The younger were as lovely as the aged were haggard . . . Prince Esterhazy, crossing a bar of sunshine, was the most prodigious rainbow of all. He was covered with diamonds and pearls . . . While he was thus glittering and gleaming, people were saying . . . that he had to redeem those jewels from pawn, as usual, for the occasion . . .

In order to see the enthroning, I stood on the rail behind our seats . . .

every moment expecting that the rail would break. Her small dark crown looked pretty, and her mantle of cloth of gold very regal. She herself looked so small as to appear puny. The homage was as pretty a sight as any; trains of peers touching her crown, and then kissing her hand. It was in the midst of that process that poor Lord Rolle's disaster sent a shock through the whole assemblage. It turned me very sick. The large, infirm old man was held up by two peers, and had nearly reached the royal footstool when he slipped through the hands of his supporters, and rolled over and over down the steps, lying at the bottom coiled up in his robes. He was instantly lifted up; and he tried again and again, amidst shouts of admiration of his valour. The Queen at length spoke to Lord Melbourne, who stood at her shoulder, and he bowed approval; on which she rose, leaned forward, and held out her hand to the old man, dispensing with his touching of the crown . . . A foreigner in London gravely reported to his own countrymen, what he entirely believed on the word of a wag, that the Lords Rolle held their title on the condition of performing the feat at every Coronation . . .

The enormous purple and crimson trains, borne by her ladies, dressed all alike, made the Queen look smaller than ever . . .

It was a wonderful day; and one which I am glad to have witnessed; but it had not the effect on me which I was surprised to observe in others. It strengthened instead of relaxing my sense of the unreal character of monarchy in England . . . There was such a mixing up of the Queen and the God, such homage to both, and adulation so alike in kind and degree that, when one came to think of it, it made one's blood run cold. Ibid.

THE QUEEN'S OWN COMMENTS, EXTRACTED FROM HER JOURNAL

Thursday, 28 June: I was awoke at four o'clock by the guns in the Park, and could not get much sleep afterwards on account of the noise of the people, bands, &c., &c. Got up at 7 feeling strong and well.

On the crowds: Their good-humour and excessive loyalty was beyond everything, and I really cannot say *how* proud I feel to be the Queen of *such* a *Nation*.

After the Enthronement, Homage and Communion: I then again descended from the Throne, and repaired with all the Peers bearing the Regalia, my Ladies and Train-bearers, to St Edward's Chapel, as it is called; but which, as Lord Melbourne said, was more *unlike* a Chapel than anything he had ever seen; for, what was *called* an *Altar* was covered with sandwiches, bottles of wine, &c. The Archbishop came in and *ought* to have delivered the Orb to me, but I had already got it. There we waited for some minutes; Lord Melbourne took a glass of wine, for he seemed completely tired; the Procession being formed, I replaced my Crown (which I had taken off for a few minutes), took the Orb in my left hand and the Sceptre in my right, and thus *loaded* proceeded through the Abbey, which resounded with cheers, to the first Robing-room . . . And here we waited for at least an hour, with *all* my ladies and train-bearers; the Princesses went away about half an hour before I did; the Archbishop had put the ring on the wrong finger [actually it had been made for the wrong finger, the fifth instead of the fourth], and the consequence was that I had the greatest difficulty to take it off again—which I at last did with great pain. At about ½ p. 4 I re-entered my carriage, the Crown on my head and Sceptre and Orb in my hand, and we proceeded the same way as we came—the crowds if possible having increased. The enthusiasm, affection and loyalty was really touching, and I shall ever remember this day as the *proudest* of my life. I came home at a little after 6—really *not* feeling tired.

The Girlhood of Queen Victoria, I, ed. Lord Esher (1912)

Certainly not too tired to run upstairs immediately and give her spaniel Dash his bath.

FURTHER SCENES INSIDE THE ABBEY

The Earl of Albemarle, Treasurer of the Household, knelt with a knightly grace to present the gold nugget of one-pound's weight, which the Queen was to put into the offertory plate; and afterwards it was with a lordly air that he scattered among the crowds in the aisles handfuls of commemorative medals in gold and silver. This item in the proceedings, which will doubtless be omitted from future coronations in deference to the more respectful modern notions as to the sanctity of a church, caused much turbulent scrimmaging. His Majesty's judges, in their robes of scarlet and ermine, stretched forth their hands with decorous langour to try to seize some of the flying mementoes, but of course disdained to stoop and pick up anything from the floor; the Aldermen of London, less proud, sprawled over the flags in their furred gowns and grabbed one another by the sleeves in their rude scramble for the pieces . . . The Turkish Ambassador caused much diversion by his absolute bewilderment at the magnificence of the spectacle presented to his gaze. He was so wonder-struck that he could not walk to his place; but stood as if he had lost his senses, and kept muttering, 'All this for a woman!'

Life of Queen Victoria, *The Times*

A PROBLEM OF WEIGHT, 1838

King Leopold and Melbourne both urged Victoria to walk if she did not want to get fat, to which she objected that when she walked she got stones in her shoes. 'Have them made tighter,' said Lord M. while Uncle [Leopold] dragged in poor Princess Charlotte again, saying that she had died through not walking enough. In the end Victoria and Melbourne had quite a set-to about it. She complained of feeling morbid in St George's Chapel, Windsor; all her relations were buried there and she would go there too. 'Do more walking,' advised Melbourne, 'My feet swell,' retorted Victoria. 'Do more then!' Victoria: 'No!' Melbourne: 'Yes!' Victoria: 'Donna Maria [Queen of Spain] is so fat and yet she took such exercise.' This silenced Lord M. Journal, Royal Archives

GREVILLE'S FIRST DINNER-PARTY AT BUCKINGHAM PALACE, 1840

Just before dinner the queen entered the round room next to the gallery with the duchess of Kent, preceded by the Chamberlain and six ladies. She shook hands with the women 'and made a sweeping bow to the men', and went directly into dinner. She talked merrily to her neighbours; then left; a quarter of an hour later the men joined her and 'huddled about the door in the sort of half-shy, half-awkward way people do'. . . . The queen advanced, talked to each.

Greville then gives his sarcastic account of their 'deeply interesting' dialogue:

Q. 'Have you been riding to-day Mr. Greville?'
G. 'No, Madam, I have not.'
Q. 'It was a fine day.'
G. 'Yes, Ma'am, a very fine day.'
Q. 'It was rather cold though.'
G. (like Polonius) 'It *was* rather cold, Madam.
Q. 'Your sister, Ly. Francis Egerton, rides I think, does not She?'
G. 'She does ride sometimes, Madame.'
(A pause, when I took the lead though adhering to the same topic.)
G. 'Has your Majesty been riding to-day?'
Q. (with animation). 'O, yes, a very long ride.'
G. 'Has your Majesty got a nice horse?'
Q. 'O, a very nice horse.' *The Greville Memoirs*, IV, 11 March 1840

THE BEDCHAMBER PLOT, MAY 1839

To the young queen's consternation, her beloved Lord Melbourne's Whig government was about to be superseded by Sir Robert Peel and his Tories. For the first time she behaved unconstitutionally by refusing Peel's request to replace her Whig ladies of the bedchamber with Tories; thus by her royal act of defiance she prolonged Lord M.'s premiership. She described to Lord M. her interview with his defeated rival:

I said I could *not* give up *any* of my Ladies, and never had imagined such a thing. He asked if I meant to retain *all*. '*All*,' I said. 'The Mistress of the Robes and the Ladies of the Bedchamber?' I replied, 'All' . . . I never saw a man so frightened.

She added to Melbourne that Peel tried to reach a compromise by saying that he needed to change only the most prominent ladies–

'to which I replied *they* were of more consequence than the others, and that I could *not* consent and that it had never been done before. He said I was a Queen Regnant, and that made the difference. 'Not here,' I said—and I maintained my right.

After consulting Wellington, Peel delivered his ultimatum: either she must accept some Tory ladies or he would not serve as prime minister. Again she wrote to Lord M.:

I was calm and very decided and I think you would have been pleased to see my composure and great firmness. The Queen of England will not submit to such trickery. Keep yourself in readiness for you may soon be wanted.

Charles Greville heard a surprisingly accurate report of the interview on the court grapevine, which he reproduced in his diary on 9 May. He has the queen say to Melbourne:

'Do not fear that I was not calm and composed. They wanted to deprive me of my Ladies, and I suppose that they would deprive me next of my dressers and my housemaids; they wish to treat me like a girl, but I will show them that I am Queen of England.'

Greville then analysed the queen's attitude in the crisis:

The simple truth in this case is that the Queen could not endure the thought of parting with Melbourne, who is everything to her. Her feelings which [Greville wrote originally, 'are *sexual* though She does not know it, and'] are probably not very well defined to herself, are of a strength sufficient to bear down all prudential considerations and to predominate in her mind with irresistible force. *Letters of Queen Victoria*, I; and *The Greville Memoirs*

The Royal Engagement, 1839

Prince Albert of Saxe-Coburg and Gotha, born three months after Victoria, his first cousin, and now just twenty, had visited Windsor already in 1836 together with his elder brother Prince Ernest. But this time it was different. Her Journal showed that she had fallen in love with him from the moment of his arrival.

10 October: At ½ p. 7 I went to the top of the staircase and received my 2 dear cousins Ernest and Albert—whom I found grown and changed, and embellished. It was with some emotion that I beheld Albert—who is *beautiful*.

11 October: They remained some little time in my room and really are charming young men; Albert really is quite charming, and so excessively handsome, such beautiful blue eyes, an exquisite nose, and such a pretty mouth with delicate moustachios and slight but very slight whiskers; a beautiful figure, broad in the shoulders and a fine waist.

15 October: Saw my dear Cousins come home quite safe from the Hunt, and charge up the hill at an immense pace . . . At about ½ p. 12 I sent for Albert; he came to the Closet where I was alone, and after a few minutes I said to him, that I thought he must be aware *why* I wished them to come here—and that it would make me *too happy* if he would consent to what I wished (to marry me). We embraced each other and he was *so* kind, *so* affectionate. I told him I was quite unworthy of him—he said he would be very happy 'das Leben mit dir zu zubringen', and was so kind, and seemed so happy, that I really felt it was the happiest brightest moment of my life.

The Girlhood of Queen Victoria, II

1–4 November: He was *so* affectionate, *so* kind, *so* dear, we kissed each other again and again and he called me 'Liebe Kleine. Ich habe dich so lieb, ich kann nicht sagen wie.' ['Darling little one. I love you so much, I can't express how much.']. . . . Oh! what *too* sweet delightful moments are these!! . . . We sit so nicely side by side on that little blue sofa; no two Lovers could ever be happier than we are! . . . He took my hands in his, and said my hands were so little he could hardly believe they *were* hands, as he had hitherto *only* been accustomed to handle hands like Ernest's.

Journal, Royal Archives

When the time came for declaring her intended marriage to the Council, the queen expressed her feelings:

When she saw the Dss. of Gloster in town, and told her she was to make her declaration the next day, the Dss. asked her if it was not a nervous thing to do. She said, 'Yes; but I did a much more nervous thing a little while ago.' 'What was that?' 'I proposed to Prince Albert.'

The Greville Memoirs, IV, 26 November 1839

A DISPUTE OVER PRINCE ALBERT'S PRECEDENCE

The queen wished her husband to receive the place of honour immediately after herself (as Queen Elizabeth II accorded to Prince Philip). Instead the Tories in

1840 planned to give Prince Albert the lowest possible precedence. Greville asked the Tory Lord Ellenborough:

'What are you going to do about the precedence?' To which he said, 'O, give him the same which P[rince] G[eorge] of D[enmark] had: place him next before the A. Bishop of Canterbury.' I said, 'That will by no means satisfy her'; at which he tossed up his head, and with an expression of extreme contempt said 'Satisfy her? What does that signify?'

The Greville Memoirs, IV, 31 January 1940

Queen Victoria showed her 'dissatisfaction' or rather fury by writing in her Journal on 2 February 1840:

Poor dear Albert, how cruelly are they ill-using that dearest Angel! Monsters! you Tories shall be punished. Revenge, revenge!

Journal, Royal Archives

She tried to prevent the Tory Wellington from being invited to their wedding but failed. However, he was later to become her trusted family adviser.

The Wedding Day, 10 February 1840

After entering the date in her Journal, the queen's first words on the great day were:

'the last time I slept alone. Ibid.

She sent an encouraging note to Albert, who had slept under her roof, contrary to protocol:

Dearest . . . how are you today, and have you slept well? . . . Send one word when you, my most dearly loved bridegroom, will be ready. Thy ever faithful, Victoria R. Ibid.

Then, after breakfast, again contrary to protocol:

I saw Albert for the *last* time *alone* as my *Bridegroom*. Ibid.

The marriage was at the Chapel Royal, St James's.

At ½ p. 12 I set off, dearest Albert having gone before. I wore a white satin gown with a very deep flounce of Honiton lace, imitation of old. I wore my Turkish diamond necklace and earrings and Albert's beautiful sapphire brooch. Ibid.

The Honiton lace was commissioned long before the queen's engagement, either late in 1838 or early in 1839 and probably not as a wedding dress at all, but to help 200 'poor women' from among the many unemployed lace-makers in Devon. The Queen found her twelve bridesmaids, dressed in white with white roses, waiting for her. They had been locked up for ninety minutes to prevent them from straying. When the service began the Queen was both joyful and impressed:

The Ceremony . . . *ought* to make an everlasting impression on every one who promises at the Altar to *keep* what he or she promises.

She particularly liked the simple way

in which we were called 'Victoria, wilt thou have &c . . .' & 'Albert, wilt thou &c . . .'

Her happiest moment was when Albert put on the ring. When Agnes Strickland, contemporary author of 'Victoria from Birth to Bridal', commented on the touch of 'melancholy' in the queen's smile after the marriage, Victoria scribbled in the margin of the book, 'Not melancholy—only joy!'

WEDDING GOSSIP

She was given away by her uncle of Sussex, of whom a wag of the times said, 'The Duke of Sussex is always ready to give away what does not belong to him.' The marriage service was conducted according to the rubric of the Church of England, the Archbishop having dutifully waited upon Her Majesty beforehand, to know if the promise 'to obey' was to be omitted, but she replied that she wished 'to be married as a woman, not as Queen'. When Prince Albert solemnly repeated the words, 'With all my worldly goods I thee endow,' it was observed by some that the bride gave him an arch smile. Sarah A. Tooley, *The Personal Life of Queen Victoria* (1896)

Before the huge wedding breakfast, Victoria had exactly half an hour sitting alone with Albert on a sofa in the Palace. She gave him his *ring and he said they must never have a secret which they did not share. Twenty years later on 10 February 1863 she wrote in her Journal,*

And we never did. Journal, Royal Archives

They left for their four-day honeymoon at Windsor at about four o'clock, the queen in a white silk dress trimmed with swansdown and bonnet with orange blossom:

I & Albert alone, which was SO delightful. Ibid.

Charles Greville sniffed at their unimpressive going-away chariot:

They went off in a very poor and shabby style. Instead of the new chariot in which most married people are accustomed to dash along, they were in one of the old travelling coaches, the postilions in undressed liveries, and with a small escort, three other coaches with post horses following.

The Greville Memoirs

But contrary to Greville's sour account, they also had a large escort of spontaneous well-wishers in gigs and on horseback who galloped alongside them all the way to Windsor.

On arrival, Victoria was for once more exhausted than Albert. Even his kisses failed to revive her and she lay all evening on the sofa with a sick headache,

but ill or not, I NEVER NEVER spent such an evening!!! My DEAREST DEAREST DEAR Albert sat on a footstool by my side, & his excessive love & affection gave me feelings of heavenly love & happiness, I never could have *hoped* to have felt before! He clasped me in his arms, and we kissed each other again and again! His beauty, his sweetness and gentleness—really how can I ever be thankful enough to have such a *Husband*! . . . to be called by names of tenderness, I have never yet heard used to me before—was bliss beyond belief! Oh! this was the happiest day of my life! May God help me to do my duty as I ought and be worthy of such blessings!

Journal, Royal Archives

MARRIED BLISS

12 February: Already the 2nd. day since our marriage . . . I *feel* a purer more unearthly feeling than I ever did . . . We sat in my large sitting room; he at one table, and I at another, and we both *tried* to write, I my journal, and Albert a letter, but it ended always in talking.

13 February: Got up at 20 m. to 9. My dearest Albert put on my stockings for me. I went in and saw him shave; a great delight for me.

23 February (*signing state papers*): Albert helped me with the blotting paper when I signed.

Ibid.

CARPING CRITICISM

Greville discovered that the bridal pair had got up rather early on 11 February after their first night, and drew the wrong conclusion. He also criticized them for a short, sociable honeymoon, with 'no interval for retirement, no native delicacy characteristic of an English woman'.

Married on Monday, she collected an immense party on Wednesday, and She sent off in a hurry for Clarence Paget to go down and assist at a ball or rather a dance which she chose to have at the Castle last night. This is a proceeding quite unparalleled . . . It was much remarked too that she and Prince Albert were up very early on Tuesday morning walking about, which is very contrary to her former habits. Strange that a bridal night should be so short; and I told Lady Palmerston that this was not the way to provide us with a Prince of Wales. *The Greville Memoirs*, IV, February 1840

Actually they were not up till 8.30 and did not go out walking till noon.

THE FIRST PREGNANCY

Contrary to Greville's gloomy prediction, Victoria was expecting her first child very soon after her marriage. Dr Locock, the royal accoucheur, had his first interview with his royal patient a few weeks before the birth, and he passed on an account to his friend Lady Mahon, who passed it on to her friend Mr Arbuthnot, who passed it on in a letter to his friend the duke of Wellington:

At the commencement of [the interview] Locock says he felt shy & embarrassed; but the Queen very soon put him at his ease.

Every Medical observation which he made, & which perhaps might bear two significations, was invariably considered by Her Majesty in the least delicate sense. She had not the slightest reserve & was always ready to express Herself, in respect to Her present situation, in the very plainest terms possible.

She asked Locock whether she should suffer much pain. He replied that some pain was to be expected, but that he had no doubt Her Majesty would bear it very well. 'O yes,' said the Queen, 'I can bear pain as well as other People.'

It was a subject going so near the wind of delicacy, that I [Mr Arbuthnot] could do no more than listen without asking questions. A good deal was told me by Lady Mahon to the same effect; but the results of the whole was that Locock left Her Majesty without any very good impressions of Her; & with the certainty that She will be very ugly & enormously fat. He says that Her figure now is most extraordinary. She goes without stays or anything that keeps Her shape within bounds; & that She is more like a barrel than anything else. Longford, *Queen Victoria's Doctors*

Locock went on to tell his confidante that there would be nobody at the delivery except himself, Prince Albert and a maid. Lady Mahon said that no doubt the queen would be very relieved at this privacy,

upon which he [Locock] remarked he verily believed from Her manner that as to delicacy, She would not care one single straw if the whole world was present. Ibid.

Locock and Prince Albert then retired to the Prince's apartment.

The Prince remarking upon what the Queen had said with regard to pain, told Locock that he did not think she could bear pain well at all, and that he expected she would make a great *Rompos*. Longford, *Queen Victoria's Doctors*

HUSBAND NOT MASTER, 1840

Apart from the fear of a 'Rompos' during his wife's labour, Prince Albert had anxieties about his own position. In May he wrote to his friend Prince William of Löwenstein:

In my home life I am very happy and contented; but the difficulty of filling my place with proper dignity is that I am only the husband, and not the master in the house. Roger Fulford, *The Prince Consort* (1949)

Some conflict on this was inevitable, but for the time being fortunately her preoccupation with maternity encouraged the queen to leave more and more decisions to her husband: for instance, dealing with a terrifying experience, as well as with the appalling lack of efficiency and security in the Palace.

THE FIRST ASSASSINATION ATTEMPT, 10 JUNE 1840

Prince Albert gave a graphic account of the event:

We had hardly proceeded a hundred yards from the Palace when I noticed, on the footpath on my side, a little mean-looking man holding something towards us; and before I could distinguish what it was, a shot was fired, which almost stunned us both, it was so loud and fired barely six paces from us. Victoria had just turned to the left to look at a horse, and could not therefore understand why her ears were ringing . . . The horses started and the carriage stopped. I seized Victoria's hands and asked if the fright had not shaken her, but she laughed at the thing. I then looked again at the man, who was still standing in the same place, his arms crossed, and a pistol in each hand. His attitude was so affected and theatrical it quite amused me. Suddenly he again pointed his pistol and fired a second time. This time Victoria also saw the shot, and stooped quickly, drawn down by me. Prince Albert's Memorandum, Royal Archives

Then someone on the footpath seized him and their attendants closed in, the crowd shouting, 'Kill him! Kill him!' Albert continued:

I called to the postilion to go on and we arrived safely at Aunt Kent's. From thence we took a short drive through the Park, partly to give Victoria a little air, and partly to show the public we had not . . . lost all confidence in them.

Ibid.

The man with the pistols turned out to be a weak-minded waiter, aged about eighteen, who was sent to Hanwell asylum, though Melbourne suspected him of being a French revolutionary and others of being a hired assassin paid by Victoria's wicked uncle the duke of Cumberland, now king of Hanover. The bullet holes were found in the Palace garden wall by a young bystander, John Everett Millais, the future painter.

BIRTH OF THE PRINCESS ROYAL

There was no 'Rompos' when the queen's first child was born on 21 November 1840, but a certain degree of disappointment. No one was present during labour except Prince Albert, Dr Locock and Mrs Lilly the midwife, though there was the necessary bevy of bishops and ministers next door to witness that the royal birth had nothing to do with a warming-pan. After the delivery Dr Locock said to the queen:

'Oh, Madam, it is a Princess.'
'Never mind, the next will be a Prince.'

The queen was right. The Princess Royal, first known as Pussy, then as Vicky, was followed within the year by Bertie, the prince of Wales and future Edward VII. Victoria was furious at two pregnancies so soon and so close together and was later to warn Vicky against the 'Schattenseite' (shadow side) of marriage.

THE BOY JONES

The case of an intruder at Buckingham Palace showed the need for Albert's reform of his deplorably run home. A nightly ration of wine, for instance, was still provided by the queen for the man who had stood guard over George III; but no one guarded the little Princess Royal.

On 2 December 1840, Mrs Lilly, the Queen's monthly nurse, heard a stealthy noise in the Queen's sitting-room shortly after one in the morning and summoned a page, who found a boy rolled up under the sofa, on which the Queen only three hours previously had been sitting . . . The intruder was at once recognised as 'the boy Jones' who on a former occasion had . . . contrived to enter the Palace. He alleged he could obtain an entrance into the Palace whenever he pleased, by getting over the wall on Constitution Hill and creeping through one of the windows. When asked why he entered the apartment of Her Majesty he replied 'I wanted to know how they lived at the Palace. I was desirous of knowing the habits of the people, and I thought a description would look very well in a book.' . . . He declared 'that he had sat upon the throne, that he saw the Queen and heard the Princess Royal squall'. He had slept under one of the servants' beds and helped himself to 'eatables' at night, 'leaving his finger prints on the stock for the soup'. Enquiry by Home Office officials revealed him to be Edmund Jones, aged 17 . . . On his first appearance he had been declared insane and discharged, this time he was committed to the House of Correction . . . and put on the treadmill. He affected an air of great consequence and repeatedly requested the police 'to behave towards him as they ought, to a gentleman who was anxious to make a noise in the world'.

Woodham-Smith, *Queen Victoria*

After his third visit to the palace in March 1841 he was given another spell of the treadmill and then sent to sea, where his conduct was eventually 'good'.

THE FIRST ROYAL TRAIN JOURNEY—TO EDINBURGH

The Town Council of Edinburgh, unable to imagine that a Sovereign could keep early hours, had not assembled to meet her Majesty when she entered the city at about nine in the morning. Startled by the blare of trumpets as they wended their way leisurely to the place of meeting, the Lord Provost and Councillors were soon seen hurrying, breathless, at the tail of the Royal procession, with their gowns ballooning behind them, and the jibes of their sarcastic fellow-townsmen ringing in their ears.

Life of Queen Victoria, *The Times*

THE BIRTH OF BERTIE, PRINCE OF WALES, 9 NOVEMBER 1841

After the usual statement the bulletin ran thus: 'Her Majesty and the Prince are perfectly well.' When this was shown to the Queen by Prince Albert, previous to its publication, she said, with a laugh, 'My dear, this will never do.' 'Why not?' asked the Prince. 'Because', replied the Queen, 'it conveys the idea that you were confined also.' Prince Albert was a little dumbfounded, but the bulletin was altered to, 'Her Majesty and the *infant* Prince are perfectly well.'

Tooley, *Personal Life of Victoria*

DISCIPLINING VICKY

The baby Princess Royal was rather like her mother in certain respects.

Her mother had to exercise severe discipline to keep her in order. For example, when Dr Brown, of Windsor, entered the service of Prince Albert, the little princesses, hearing their father address him as 'Brown', used the same form of speech. The Queen corrected them, and told them to say 'Dr Brown'. All obeyed except 'Vicky', who was threatened with 'bed' if she transgressed again. Next morning when the Doctor presented himself to the royal family, the young Princess, looking straight at him, said, 'Good morning, Brown!' Then, seeing the eyes of her mother fixed upon her, she rose and, with a curtsy, continued, 'and good night, Brown, for I am going to bed,' and she walked resolutely away to her punishment.

Ibid.

DOMESTIC BLISS AT OSBORNE, 1844

The children again with us, & such a pleasure & interest! Bertie & Alice [born 1843] are the greatest friends & always playing together.—Later we both read to each other. When I read, I sit on a sofa, in the middle of the room, with a small table before it, on which stand a lamp & candlestick, Albert sitting in a low armchair, on the opposite of the table with another small table in front of him on which he usually stands his book. Oh! if I could only exactly describe our dear happy life together!

Journal, Royal Archives

CURIOSITIES OF CAMBRIDGE, 1843

Until the 1890s when Cambridge modernized its drainage system, all the sewage went into the river, so that walking along the backs was not the pleasantest of experiences. Victoria accompanied Albert when he received an honorary degree from Cambridge University:

There is a tale of Queen Victoria being shown over Trinity by the Master, Dr Whewell, and saying, as she looked down over the bridge: 'What are those pieces of paper floating down the river?' To which with great presence of mind, he replied: 'Those, ma'am, are notices that bathing is forbidden.'

Gwen Raverat, *Period Piece: A Cambridge Childhood* (1952)

POMP AND CIRCUMSTANCE

An Irish novelist gave her impression of the State Opening of Parliament in 1844 by Queen Victoria:

Enter the crown and cushion and sword of state and mace—the Queen, leaning on Prince Albert's arm. She did not go up the steps to the throne well—caught her foot and stumbled against the edge of the footstool, which was too high. She did not seat herself in a decided, queenlike manner, and after sitting down pottered too much with her drapery, arranging her petticoats. That footstool was much too high! her knees were crumpled up, and her figure, short enough already, was foreshortened as she sat, and her drapery did not come to the edge of the stool: as my neighbour Miss Fitzhugh whispered, 'Bad effect.' However and nevertheless, the better half of her looked perfectly ladylike and queenlike; her head finely shaped, and well held on her shoulders with her likeness of a kingly crown, that diadem of diamonds. Beautifully fair the neck and arms; and the arms moved gracefully, and never too much. I could not at that distance judge of her countenance, but I heard people on the bench near me saying that she looked 'divinely gracious'.

Dead silence: more of majesty implied in that silence than in all the magnificence around. She spoke, low and well: 'My lords and gentlemen, be seated.' Then she received from the lord in waiting her speech, and read: her voice, perfectly distinct and clear, was heard by us ultimate auditors; it was not quite so fine a voice as I had been taught to expect; it had not the full rich tones nor the varied powers and inflections of a perfect voice. She read with good sense, as if she perfectly understood, but did not fully or warmly feel, what she was reading. It was more a girl's well-read lesson than a Queen pronouncing her speech. She did not lay emphasis sufficient to mark the gradations of importance in the subjects, and she did not make pauses enough.

F. A. Edgeworth, *A Memoir of Maria Edgeworth with Selections from her Letters*, 3 vols. (1867)

VICTORIA'S FIRST SEA BATHE, 1847

'Drove down to the beach [at Osborne] with my maid & went into the bathing machine, where I undressed & bathed in the sea (for the first time in my life), a very nice bathing woman attending me. I thought it delightful till I put my head under the water, when I thought I should be stifled.'

Longford, *Victoria R.I.*, quoting Queen Victoria's Journal

In future the queen sponged her face on the beach before her dip and then, head erect, 'plunged about' in the ocean.

THE SABBATH

Queen Victoria was not a sabbatarian but she kept a sense of propriety.

One Sunday afternoon Princess Thora [her granddaughter] came to the Queen and asked whether she and two of the maids-of-honour might play tennis. Grandmama's reply was, 'Yes, so long as you pick up the balls yourself. Being Sunday, I do not think it right to make others work for your amusement.'

Princess Marie-Louise, *My Memories of Six Reigns* (1956)

THE QUEEN AND 'PAM'—'PILGERSTEIN'—ALIAS LORD PALMERSTON

A love–hate relation with the queen began with her fear of Palmerston's 'unscrupulous dexterity' (which included his entering uninvited a lady-in-waiting's bedroom at Windsor) and ended in her lamenting his death:

He had often worried and distressed us, though as Pr. Minister he had behaved *very well.*

Journal, Royal Archives

Albert clashed with him continually over foreign policy in which both were intensely interested. The royal couple slated him in German, privately calling him 'Pilgerstein' (pilgrim palmer-stone) and referring to his alleged blunders as 'bocks'. His worst 'bock' according to Albert was not showing the Palace his foreign office despatches. Soon after the famous 1850 example of Pilgerstein's 'gunboat diplomacy' known as the Don Pacifico affair, Albert sent for him and accused him of withholding dispatches from the queen or changing them after she had approved them. Pilgerstein immediately resorted to 'unscrupulous dexterity' by breaking down and sobbing out,

'What have I done?' Ibid.

Albert was 'dreadfully fagged' by the interview but Pilgerstein was soon up to his old 'bocks' again. Even as Home Secretary in 1853 Palmerston thought of nothing but the Eastern Question, involving Turkey and Russia.

When asked by the Queen whether he had news about the strikes which were agitating the north of England he is reported to have answered absently: 'No, Madam, I have heard nothing; but it seems certain that the Turks have crossed the Danube.'

Kingsley Martin, *The Triumph of Lord Palmerston* (1924)

THE YEAR OF REVOLUTIONS, 1848

While European royalties were trembling and thrones tottering, the British couple were merely wrestling with a publisher.

The most severe trial undergone by Victoria and Albert during that turbulent year was the news—by anonymous letter—that prints of their etchings, the private record of their home life, had been surreptitiously made and were to be published and sold to the public. In one of the few cases ever brought by a member of the Royal Family as a private citizen against another, the Prince Consort took out an injunction against William Strange the publisher. When Strange appealed, in February 1849, the injunction was upheld, establishing a legal precedent that a work of art cannot be published without the artist's consent.

Marina Warner, *Queen Victoria's Sketchbook* (1979)

THE CRYSTAL PALACE, 1851

Albert was responsible for the organization of the Great Exhibition, a task which provided a very acceptable outlet for his energies.

Queen Victoria visited the Crystal Palace Exhibition almost daily from its opening until she went to Osborne towards the end of July, getting up early, arriving before 10 a.m., and systematically working her way through every section. The French courts she found beautiful beyond description, the American machinery 'inventive' but 'not entertaining' (later the 'cotton machines' from Bradford and Oldham won her unstinted applause); only the Prussian and Russian sectors were contemptibly thin. Among individual items she selected for special notice the Indian pearls, the Sheffield Bowie knives made exclusively for America, the Chubb locks, the electric telegraph and a machine for making fifty million medals a week . . . She feasted on anecdotes relayed by the Foreign Secretary Lord Granville:

Visitor to Exhibition pointing to a block of alum:
'What's that?'
Well-informed friend: 'That's a statue of Lot's wife.'

Another visitor describing the statue of St Michael casting out Satan: 'The Queen and the Pope.'

Victoria's summary of the Great Exhibition:

'It was the *happiest, proudest* day in my life, and I can think of nothing else. Albert's dearest name is immortalised with this *great* conception, *his* own, and my *own* dear country *showed* she was *worthy* of it.'

The Great Exhibition marked the halfway point in Queen Victoria's married life, the summit of Albert's career and the climax of early Victorian England. Longford, *Victoria R.I.*

THE QUEEN'S FIRST CHLOROFORM, 1853

Chloroform was used for the first time in the royal palace for the birth of Prince Leopold. It was not only science's gift to the queen but also the queen's gift to the nation, for her use of it helped to dispel the myth that pain in childbirth was woman's divinely appointed destiny.

Sir James Clark to Dr Simpson of Edinburgh: The Queen had chloroform exhibited to her during her last confinement . . . It was not at any time given so strongly as to render the Queen insensible, and an ounce of chloroform was scarcely consumed during the whole time. Her Majesty was greatly pleased with the effect, and she certainly never has had a better recovery.'

The queen described it more lyrically: Dr Snow gave that blessed Chloroform & the effect was soothing, quieting & delightful beyond measure. Ibid.

PRESENTING MEDALS AFTER THE CRIMEAN WAR, 1855

When the queen heard that the soldiers refused to give up their medals for engraving in case they did not get back the actual one she had presented, she was profoundly moved and touched—

so much so that Mrs Norton, Melbourne's old lady friend, afterwards questioned Lord Panmure the War Minister about Her Majesty's emotion:

Mrs Norton: Was the Queen touched?
Lord Panmure: Bless my soul, no! She had a brass railing before her, and no one could touch her. [This was not true.]
Mrs Norton: I mean, was she moved?
Lord Panmure: Moved! she had no occasion to move. Ibid.

The Queen and Her Daughters

THE PRINCESS ROYAL TO MARRY THE CROWN PRINCE OF PRUSSIA

The Prussians wished the wedding to take place in Berlin. Victoria was furious.

The assumption of its being *too much* for a Prince Royal of Prussia to *come* over to marry *the Princess Royal of Great Britain* IN England is too absurd, to say the least . . . Whatever may be the usual practice of Prussian Princes, it is not *every* day that one marries the eldest daughter of the Queen of England. Ibid.

ANOTHER DETERMINED DAUGHTER

'Baby' Beatrice, born in 1857 and Victoria's youngest daughter, was eyeing a rich iced pudding at luncheon when she was only two years old.

Victoria: Baby musn't have that, it's not good for Baby.
Baby (*helping herself*): But she likes it, my dear. Ibid.

MARRIAGE AND PREGNANCY

Queen Victoria at times wrote frankly and discouragingly to her daughter Vicky on these subjects. Even the results—babies—had not much to be said for them aesthetically.

Now to reply to your observation that you find a married woman has much more liberty than an unmarried one; in one sense of the word she has,—but what I meant was—in a physical point of view—. . . aches—and sufferings and miseries and plagues—which you must struggle against—and enjoyments etc. to give up . . . you will feel the yoke of a married woman! Without that—certainly it is unbounded happiness—if one has a husband one worships! It is a foretaste of heaven . . . I had 9 times for 8 months to bear with those above-named enemies and real misery . . . and I own it tried me sorely; one feels so pinned down—one's wings clipped—in fact . . . only half oneself—particularly the first and second time. This I call the 'shadow side'.

Dearest Child: Private Correspondence of Queen Victoria and the Crown Princess of Prussia, 1858–61, ed. Roger Fulford (1964), 24 March 1858

What you say of the pride of giving life to an immortal soul is very fine, dear, but I own I cannot enter into that; I think much more of our being like a cow or a dog at such moments; when our poor nature becomes so very animal and unecstatic . . . Ibid., 15 June 1858

Vicky's eldest son William, the future Kaiser, arrived on 27 January 1859 after a difficult birth.

Poor dear darling! I pitied you so! It is indeed too hard and dreadful what we have to go through and men ought to have an adoration for one, and indeed do everything to make up, for what after all they alone are the cause of! Ibid., 9 March 1859

I am shocked to hear baby leaves off his caps so soon; I hope however only in the nursery, for they look so frightful to be seen without caps. In the nursery it is wholesome but it is not pretty. Ibid., 16 April 1859

The queen's niece Princess Ada of Schleswig-Holstein was again for the third time 'in that most charming situation' (pregnant) after being married less than three years.

I positively think those ladies who are always enceinte quite disgusting; it is more like a rabbit or a guinea-pig than anything else and really it is not very nice. There is Lady Kildare who has two a year, one in January and one in December—and always is so, whenever one sees her! Ibid., 15 June 1859

The queen was considering the marriage of her second daughter, Princess Alice.

'. . . all marriage is such a lottery—the happiness is always an exchange—though it may be a very happy one—still the poor woman is bodily and morally the husband's slave. That always sticks in my throat. When I think of a merry, happy, free young girl—and look at the ailing, aching state a young wife generally is doomed to—which you can't deny is the penalty of marriage.' Ibid.

Marie, the queen's daughter-in-law married to Prince Alfred, had just had a baby:

She nurses the child—which will enchant you. As long as she remains at home—and does not publish the fact to the world—by taking the baby everywhere and can do it well—which they say she does now—I have nothing to say (beyond my unfortunately—from my earliest childhood—totally insurmountable disgust for the process). Ibid., 27 October 1874

A RIDING ACCIDENT

Though in theory the queen lamented the freedom which a young wife lost, she disapproved of those who remained free.

A dreadful thing has happened near here. Poor Lady Charles Ker . . . one of those great riding, hunting ladies . . . had a frightful fall taking a fence on Monday last—fractured her skull and is still alive but that is all! . . . May it be a warning to many of those fast, wild young women who are really unsexed. And to the husbands, fathers and brothers too who allow their wives, daughters and sisters to expose themselves in such an unfeminine way. In other respects the poor young thing was very quiet and not very strong—but imagine her going down alone to hunt while her husband was walking about in London! Ibid.

WEIGHT-WATCHING

Victoria was already worried at the age of nineteen about weighing 8 stone, 13 pounds. The problem was worse at the age of forty and the earliest fears had been forgotten. She wrote to her daughter Vicky:

I did not tell you that the other day when we were going down Craig na-Ban—which is very steep, and rough, Jane Churchill fell and could not get up again (having got her feet caught in her dress) and Johnny Brown (who is our factotum and really the perfection of a servant for he thinks of everything) picked her up like une scène de tragédie and when she thanked him, he said 'Your Ladyship is not so heavy as Her Majesty!' which made us laugh very much. I said 'Am I grown heavier do you think?' 'Well, I think you are,' was the plain spoken reply. So I mean to be weighed as I always thought I was light. Ibid., 26 September 1859

THE GREAT EXPEDITIONS

The queen loved touring in the Highlands incognito—and being recognized. The first great expedition was on 4 September 1860 to an inn in Fettercairn. They were served a 'very fair dinner' by 'a ringleted woman' as the gillies, Brown and Grant, were too 'bashful'; ('bashful' meant intoxicated in Brown's case). The queen found the roast lamb 'good', the cranberry tart 'good' but the potatoes 'very good'.

The people were very amusing about us . . . They told Jane [Shackle, the wardrobe maid] 'Your lady gives no trouble.' . . . At about ten minutes to ten o'clock we started in the same carriages [embellished with the royal monogram] . . . evidently 'the murder was out' for all the people were in the street, and the landlady waved her pocket-handkerchief, and the ringleted maid (who had curlers in the morning) waved a flag from the window. *Leaves from the Journal of Our Life in the Highlands*, ed. Sir Arthur Helps (1868)

The last great expedition, 8 October 1861, was made to a much poorer area.

Unfortunately there was hardly anything to eat, and there was only tea, and two miserable starved Highland chickens, without any potatoes! No pudding, and no *fun* . . . Ibid.

Death of the Prince Consort, 14 December 1861

Prince Albert was already suffering from the beginnings of his mortal illness when he made his last effort to intervene in political affairs. Wearing his wadded dressing-gown with crimson velvet collar, he crept to his sitting-room at 7 a.m., lighted his German green-shaded lamp (there was no fire) and set about redrafting the brusque message which the Foreign Secretary was about to send to the American Confederates, after they had seized two Southern envoys from the British vessel Trent. *His conciliatory draft made negotiations possible and probably saved the two countries from war.*

His doctors finally diagnosed his illness as typhoid (though according to some retrospective diagnoses it was cancer). As soon as the anguished queen felt able to resume her journal she described the death-bed. All Albert's last words, as recorded by Victoria, were spoken in German, showing that in death he reverted to the language of his happy youth, though normally, the queen said, they spoke no more German than English to each other.

Two or three long but perfectly gentle breaths were drawn, the hand clasping mine, & (oh! it turns me sick to write it) *all all* was over . . . I stood up, kissing his dear heavenly forehead & called out in a bitter agonising cry, 'O! my dear Darling!' & then dropped on my knees in mute, distracted despair, unable to utter a word or shed a tear! Journal, Royal Archives

Dean Wellesley, who was present, wrote afterwards:

We heard her loud sobs as she went off to her solitary room.

Sixty years later the historian Lytton Strachey was to give a twist of melodrama to his account of the same scene:

She shrieked—one long wild shriek that rang through the horror-stricken Castle—and understood that she had lost him for ever.

Lytton Strachey, *Queen Victoria* (1921)

THE WIDOW AT WINDSOR

At first the queen's mourning and seclusion were total. To some of her ministers, for whom George III was a living memory, it seemed that she might go mad. When Lord Clarendon mentioned the possibility of a change of government, she tapped her forehead crying,

'My reason, my reason.'

Lord Clarendon's Letters to the Duchess of Manchester, ed. A. L. Kennedy (1956)

Politics without Albert to advise her were unendurable.

One day she sent for the Duchess of Sutherland, and, leading her to the Prince's room, fell prostrate before his clothes in a flood of weeping, while she adjured the Duchess to tell her whether the beauty of Albert's character had ever been surpassed. Strachey, *Victoria*

She even thought in a distraught moment of ending her own life, as she recalled in a letter to her daughter the German Empress in October 1888, when Vicky had just lost her husband:

I too wanted once to put an end to my life *here*, but a *Voice* told me for *His* sake—no, 'Still Endure'. Longford, *Victoria R.I.*

'Still Endure' was to become her motto.

A press attempt was made in 1864 to bounce the queen out of her seclusion. On 1 April (appropriate date) The Times *wrote a leader on a bogus report that the royal seclusion was at last coming to an end:*

Her Majesty's loyal subjects will be very well pleased to hear that their Sovereign is about to break her protracted seclusion . . . before long the whole Court will recover from its suspended animation . . . We are not a people to take much on trust, or to conceive that to be real which does not meet the eye. *The Times*

Six days later the queen published her reply in The Times:

The Queen heartily appreciates the desire of her subjects to see her, and whatever she *can* do to gratify them in this loyal and affectionate wish she *will* do . . .

[She emphasizes that she cannot take on extra ceremonial] More the Queen *cannot* do; and more the kindness and good feeling of her people will surely not expect from her. *The Times*, 7 April 1864

Shortly before the article in The Times, *posters had appeared outside Buckingham Palace:*

These commanding premises to be let or sold, in consequence of the late occupant's declining business.

Later it was ironically suggested that the palace should be made into a home for fallen women.

Benjamin Disraeli, a favourite prime minister and author of Sybil, Coningsby *and many other novels, helped Queen Victoria to emerge from her seclusion by awakening her interest in politics, particularly in her becoming Empress of India. When she published her Highland diaries and sent him a copy, he paid her the most famous compliment she ever received:*

'We authors, Ma'am . . .' Monypenny and Buckle, *Life of Disraeli*, V

THE TENNYSONS VISIT THE QUEEN 1862–3

The Poet Laureate's first visit to Osborne was occasioned by his poem 'In Memoriam', which struck a chord in the recently widowed queen; the description of Arthur Hallam (subject of the poem) reminded her of Albert, even to the blue eyes.

Tennyson was so moved by the interview that he could not give a very connected account of it afterwards. From what he reported to Emily [his wife], it seems that he was standing with his back to the fire when the Queen entered and that she came and stood about five paces from him with her arms crossed, very pale and like a little statue in her self-possession. She spoke in a quiet, sweet, sad voice and looked very pretty, with a stately innocence about her, different from other women. She said, 'I am like your Mariana now.'

Dear and Honoured Lady: The Correspondence between Queen Victoria and Alfred Tennyson, ed. Hope Dyson and Charles Tennyson (1969)

'Mariana and the Moated Grange' was another Tennysonian lament, with the refrain, 'I would that I were dead.'

The interview closed with Alfred saying, 'We are all grieved for Your Majesty', and the Queen replying, 'The country has been kind to me and I am thankful.' At parting she asked him whether there was anything that she could do for him, to which he answered: 'Nothing, Madam, but shake my two boys by the hand. It may keep them loyal in the troublous times to come.' Ibid.

In the following year the Tennysons were invited to bring over their sons Lionel and Hallam; Hallam aged ten wrote a long account of the visit, during which they

met the princesses Louise and Beatrice and the princes Alfred and Leopold as well as the queen. Hallam first described the royal children's Swiss Cottage, Fort and garden ('a great many mare's tail about Osborne, asparagus and radishes') ending with Leopold's paper boats, which the prince set fire to on the sea, before Hallam reached Queen Victoria herself:

The Queen is not stout. Her Majesty has a large mind and a small body to contain it therein . . . Her Majesty has a beautiful little nose and soft blue eyes . . .

Observations:—You must always say 'Mam' when in Her Majesty's presence. You must stand until the Queen asks you to sit down. Her Majesty does not *often* tell you to sit down.

FINIS

Ibid.

Foreign Travel

The queen chose the Pension Wallace in Lucerne for her first visit to Switzerland, perhaps attracted by its Scottish name.

She travelled as the Countess of Kent. Disraeli addressed her as 'our dear Peeress' and her daughter Louise signed herself 'Lady Louise Kent'. Her Master of the Household said to Ly. Ely:
'It won't do for you to have Marchioness of Ely on your baggage. You will be greater than her.'
'No dear, I shall put "Plain Lady Ely" on my boxes.'

Longford, *Victoria R.I.*, quoting a letter of Sir Henry Ponsonby to his wife, 4 August 1868

A VISIT TO THE SOUTH OF FRANCE, 1892

In the 1920s a local fisherman told Mr Laurens van der Post that he remembered as a boy watching with amazement a strange woman land on the rocky, off-shore islet where he and his family lived in caves. She shared their *bouillabaisse*, this unknown woman, out of the big pot . . . The old fisherman drank to her memory: her wonderful bearing, her sea-blue eyes, her stupendous purple bonnet. 'La Reine Victorie!'

Ibid. [*Told to the author by Sir Laurens van der Post.*]

Enter John Brown

Four years after Albert's death, the queen brought John Brown, a favourite ghillie of the royal couple, south to London, Windsor, and Osborne to be her 'permanent personal attendant' with the title, from 1872, of 'Esquire'. The queen's blatant favouritism combined with his own bluntness caused him to be envied and disliked by her court and family and even to be accused of having married her. But Sir Henry Ponsonby, her sagacious private secretary, recognized him for what he was: despite his faults, 'a first class servant'.

Brown first attracted the limelight through two carriage accidents:

Once the leading horse of the Queen's carriage fell and Brown saved the day by promptly sitting on its head. A few days later he rescued her when her drunken coachman drove her into a ditch at night, resulting for her in a black eye and a chronically crooked thumb. Longford, *Victoria R.I.*

JOHN BROWN'S LEGS

From the queen the Master of the Household once received a note of complaint:

It is, that my poor Brown has so much to do that it wd be a *gt* relief if—the Equerries cld receive a *hint* not to be *constantly* sending for him *at all hours* for trifling messages: he is often so *tired* from being so constantly on his legs that, he goes to bed with swollen feet and can't sleep for fatigue!

Ibid., quoting Royal Archives, 26 December 1866

The news of this royal complaint must have crossed the Atlantic, for an anonymous skit on Brown's daily duties was later published in New York, parodying the queen's style in her Leaves from the Journal of Our Life in the Highlands *and entitled 'John Brown's Legs':*

'We make it a point to have breakfast every morning of our lives . . . Brown pushed me (in a hand-carriage) up quite a hill and then ran me down again. He did this several times and we enjoyed it very much . . . He then put me in a boat on the lake and rocked me for about half an hour. It was very exhilarating.'

The parody ends with a message from New York that O'Donovan Rossa, the Fenian leader, is after Brown. The Legs scamper away from 'Windsoral' for ever. Ibid.

JOKES ABOUT JOHN BROWN, 1868

In Tinsley's Magazine *for October there appeared an article entitled 'Women in Britain' by 'An American'.*

Soon after my arrival in England, at a table where all the company were gentlemen by rank or position, there were constant references to and jokes about 'Mrs Brown' . . . Then came out all the stupid scandal about her Majesty's Highland servant . . . that the Queen was insane, and John Brown was her keeper; that the Queen was a spiritualist, and John Brown was her medium. *Tinsley's Magazine*, Oct. 1868

Henry Ponsonby described another rumour to his brother Arthur:

We have been rather surprised here by a statement that the Minister at Berne has complained of a libel on the Queen in the *Lausanne Gazette*—a foolish thing to do—and it has brought the matter into notoriety but no further steps are taken as he has apologised. We do not know what the libel is—and I believe the Queen is as ignorant as any of us, but I hope she will not hear it, as I believe it to be a statement that she has married John Brown, and the idea that it could be said she was marrying one of the servants would make her angry and wretched. Brown . . . has lately been raised to be personal attendant—that is, all messages come by him—as he is always dressed as a highlander he is conspicuous and so is talked of. Besides which he certainly is a favourite—but he is only a Servant and nothing more—and what I suppose began as a joke about his constant attendance has been perverted into a libel that the Queen has married him.

Ponsonby Letters

Punch *printed an imaginary Court Circular*

7 July 1866 Balmoral, Tuesday.

Mr John Brown walked on the Slopes [of Windsor].

He subsequently partook of a haggis. In the evening, Mr John Brown was pleased to listen to a bag-pipe.

Mr John Brown retired early.

Punch Magazine

BROWN STORIES

Sir Arthur Bigge, the queen's private secretary, later Lord Stamfordham, had been promised a day's fishing at Balmoral.

He was dismayed when Brown appeared in his room with a stern countenance. 'Ye'll no be going fishing,' Bigge was told. 'Her Majesty thinks its about time ye did some work.'

Kenneth Rose, *Kings, Queens and Courtiers* (1986)

One day when Vicky and her daughter Princess Charlotte of Prussia were visiting the queen, Brown entered the queen's room, where Charlotte was with her grandmother:

Queen: Say how de do to Brown, my dear.
Charlotte: How de do.
Queen: Now go and shake hands.
Charlotte: No, that I won't. Mama says I ought not to be too familiar with servants.

Ponsonby Letters

Ponsonby reported 'no end of a row with Vicky' about her children's upbringing. At Balmoral there was an argument as to how many courtiers could share the queen's pew at Crathie church without overcrowding her when she took the sacrament. Brown brought the discussion to an end by roaring:

'She had better have the place all to herself and have done with this hombogging.' *Ponsonby Letters*

BROWN'S SECOND SIGHT

The queen believed that her Highland servant was gifted with his native second sight, particularly after the spate of deaths in her family in 1861. Her mother had died in the spring. She wrote in her Journal in the autumn:

In speaking and lamenting over our leaving Balmoral, Brown said to me he hoped we should all be well through the winter & return safe, '& above all that you may have no deaths in the family'. Well—& then the last day—he spoke of having lost (twelve years ago) in six weeks time of typhus fever three grown up brothers & one grown up sister . . . Now not four weeks after we left this same fever has entered a royal house nearly allied to us & swept away two and nearly a third . . . Journal, Royal Archives

The queen was referring to the Portuguese royal family. Brown must have seemed even more prophetic when the Prince Consort died of fever a few weeks later. But there is not a shred of evidence that Brown was a medium or held seances in the Blue Room at Windsor where Albert died.

Brown's brusque manner, even to Victoria herself, gave her confidence in him.

His admonishments to keep still while he tucked in her rug or pinned her cape and his habit of addressing her as 'wumman' seemed the plainest guarantee of devotion. 'Hoots, then, wumman,' he was overheard shouting at her one day by a chance tourist, after pricking her chin, 'Can ye no hold yerr head up?' Henry Ponsonby referred to Brown almost affectionately as 'the child of nature'. Longford, *Victoria R.I.*

He was also a child of the whisky bottle.

One day a maid-of-honour was told by Brown that she could have two hours off as the Queen was going out.

'To tea, I suppose?'

'Well, no, she don't much like tea—we tak oot biscuits & sperruts.'

The Queen had indeed disliked tea ever since her girlhood. She once congratulated John Brown on the best cup she had ever drunk.

'Well, it should be, Ma'am, I put a grand nip o' whisky in it.' Ibid.

'MY BELOVED JOHN'

I found these words in an old Diary or Journal of mine. I was in great trouble about the Princess Royal who had lost her child in '66—& dear John said to me: 'I wish to take care of my dear good mistress till I die. You'll never have an honester servant' & I took & held his dear, kind hand & I said I hoped he might long be spared to comfort me & he answered, 'But we all *must* die'

Often my beloved John would say: 'You haven't a more devoted servant than Brown'—and oh! *how* I felt *that*! Often & often I told him no one loved him more than I did or had a better friend than me: & he answered 'Nor you—than me.' 'No one loves you more.'

Tom Cullen, *The Empress Brown: The True Story of a Victorian Scandal* (Boston, 1969). *Cullen quotes from a letter of Queen Victoria to John Brown's brother, after John's death.*

THE DEATH OF JOHN BROWN, 1883

Brown died of delirium tremens and erysipelas. His sympathetic treatment by the queen's new doctor, James Reid, greatly endeared Reid to her, 'utterly crushed' as she was.

More than twenty years after the death of John Brown, Reid once more became involved, albeit indirectly, with the Queen and her servant, this time over the contents of a black trunk which contained more than 300 letters written by Queen Victoria to Dr Profeit about John Brown, 'many of them most compromising' as Reid recorded in his diary. The contents of these letters will never be known. Their secret was contained in Reid's green memorandum book which was burnt by his son.

Michaela Reid, *Ask Sir James* (1987)

The letters could have been about Brown's proposed marriage to one of the queen's dressers, mentioned rather mysteriously in Sir Henry's letters to Lady Ponsonby (Victoria R.I.). *The queen probably put a stop to it, as she would gladly have stopped the marriages of all her favourite servants, including Reid himself, an action which must have seemed to Reid both selfish and compromising. It is also possible that she was trying to get the local Scottish doctor to help her cover up on Brown's numerous attacks of drunkenness. When another of the Brown family, Donald, died of drink while in her service, she had it hushed up, believing that stimulants were necessary to hard-working servants—though not to the upper classes!*

Disraeli's Magic Touch—With a Trowel

A combination of blown-up and unaffected fervour was the secret of Disraeli's phenomenal success with Queen Victoria. On the subject of his exaggerated devotion he once said to the poet Matthew Arnold:

You have heard me called a flatterer, and it is true. Everyone likes flattery and when you come to royalty, you should lay it on with a trowel.

W. F. Monypenny and G. E. Buckle, *Life of Disraeli* (1929)

BIRTH OF 'THE FAERY'

After the wedding of the Prince and Princess of Wales in 1863, to which Mr and Mrs Disraeli were invited—a social triumph—Disraeli was sent for by the queen. He described his audience:

Disraeli was summoned to wait for his audience in [the late] Prince Albert's special room.

'In less than five minutes from my entry, an opposite door opened, and the Queen appeared.

She was still in widow's mourning and seemed stouter than when I last saw her but this was perhaps only from her dress. I bowed deeply when she entered and raised my head with unusual slowness, that I might have a moment for recovery. Her countenance was grave but serene and kind, and she said in a most musical voice: "It is some time since we met" . . .

At last she asked after my wife, hoped she was well, and then, with a graceful bow, vanished.'

Already, one feels, Disraeli is investing the Queen with some strange magical quality. She *appears*. She *vanishes*. The myth of 'the Faery' is being born.

Robert Blake, *Disraeli* (1966)

DISRAELI AND HIS 'BIRD'

Of his first visit as prime minister to the queen at Osborne in 1874, Disraeli later told Lady Bradford that he half expected to be kissed:

'She was wreathed with smiles and, as she tattled, glided about the room like a bird.'

Monypenny and Buckle, *Disraeli*

A GIFT WITH THE PERSONAL TOUCH

A year later, in 1875, Disraeli waved his wand over the Suez Canal and, with the help of a loan from the Rothschilds bought it from the bankrupt Egyptian government, and 'presented' it (a majority of its shares) to the queen:

'It is just settled; you have it, Madam.'

Ibid.

And in the very next year, 1876, he passed in parliament the Royal Titles Bill, declaring Victoria 'Queen-Empress'.

DISRAELI AND THE QUEEN

As prime minister, Disraeli referred to his sovereign as 'the Faery' and compared his own attitude towards her with Gladstone's:

'Gladstone treats the Queen like a public department; I treat her like a woman.'

Blake, *Disraeli*

BEWARE OF THE QUEEN!

Sir Henry Ponsonby, Private Secretary to Queen Victoria, and Colonel Arthur Haigh, equerry to Prince Alfred, experienced the terror inspired by the queen.

Yesterday Haig and I went out towards the garden by a side door when we were suddenly nearly carried away by a stampede of royalties, headed by the Duke of Cambridge and brought up by [Prince] Leopold, going as fast as they could. We thought it was a mad bull. But they cried out: 'The Queen, the Queen,' and we all dashed into the house again and waited behind the door till the road was clear. When Haig and I were alone we laughed immensely. This is that 'one-ness' we hear of.

Arthur Ponsonby, *Henry Ponsonby* (1942)

Next moment the short sad figure hobbled into Sandringham House.

BRAVING THE QUEEN

There is a story of Queen Victoria and the Emperor William at a grand Windsor dinner for the family. It shows that one member of the family, at least, was not afraid of her.

On such solemn occasions as this, the Queen arranged the seating according to closeness of kinship . . . The Emperor of Germany, treated as a grandson, was relegated to the end of the table . . . At dessert, when one of the guests proposed a toast to the Queen of England, another to the Empress of India, and a third to some of the other grand titles of the mighty Sovereign—the Emperor raised his glass, like a child sent to stand in the corner, and just said, with a mischievous smile:

'To Grandmama!'

Joanna Richardson, *Portrait of a Bonaparte* (1987)

IMPRESSIONS OF A YOUNG GRANDDAUGHTER—SOPHIE OF PRUSSIA

'My dear Grandmama is very tiny—a very, very pretty little girl—and wears a veil like a bride.'

Darling Child: Private Correspondence of Queen Victoria and the Crown Princess of Prussia, 1871–78, ed. Fulford, Vicky to Queen Victoria, reporting her daughter's remark, 27 July 1874

DEATH OF DIZZY, LORD BEACONSFIELD, 1881

His characteristic ironical wit did not desert him. 'Take away that emblem of mortality,' he said when given an air cushion to lie on, and on March 31 when correcting his last speech for Hansard: 'I will not go down to posterity talking bad grammar.' It was suggested to him that he might like to be visited by the Queen. 'No it is better not. She would only ask me to take a message to Albert.' Disraeli was much in the Queen's thoughts. She sent him spring flowers and before departing for Osborne wrote on April 5 her last letter, with instructions that it should be read to him if he was not well enough to read it himself. Disraeli held it in his hand for a moment as if in deep thought. 'This letter ought to be read to me by Lord Barrington, a Privy Councillor,' he said—and it duly was. Blake, *Disraeli*

AN IMPERIAL DISASTER, 1885

Queen Victoria felt so deeply the disgrace of General Gordon's death at Khartoum at the hands of the Mahdi, before the relieving force arrived, that she took revenge on Mr Gladstone, her Liberal prime minister. She sent him a telegraphic rebuke en clair, *which was handed to him by the station master as he left the train on which he was travelling:*

These news from Khartoum are frightful, and to think that all this might have been prevented and many precious lives saved by earlier action is too frightful. Longford, *Victoria R.I.*

Gladstone thought of resigning.

GLADSTONE'S SECOND MINISTRY

Victoria was able to make life uncomfortable for Gladstone. She tried to avoid sending for him and was criticized by the Liberals. She replied:

The Queen does not the least care but rather wishes it should be known that she has the greatest possible disinclination to take this half-crazy & really in many ways ridiculous old man . . . Ibid.

Though aware of her duties as a constitutional monarch, Victoria's intense dislike of Gladstone's policies led her to overstep her royal rights. She was in the habit of keeping Lord Salisbury, the Opposition leader, privately informed of Gladstone's intentions. She could not, however, exclude him from power. When Sir William Harcourt, the Liberal statesman, assured the queen how anxious the new government were to please her, she refused to converse with him but 'merely bowed'. She described them as 'a motley crew' because after they were sworn in, instead of rising to kiss her hand, they crawled forward in a body on their hands and knees.

Ibid.

THE JUBILEE BONNET, 1887

The queen as usual since her widowhood refused to wear a crown and robes of state, even for the most solemn events during the Golden Jubilee celebrations. In desperation her children sent her much loved daughter-in-law, Alexandra princess of Wales to make her change her mind. Alix came out from the presence precipitately:

'I never was so snubbed!'

Ibid.

Lord Halifax begged the queen to realize that people wanted 'gilding for their money' and Lord Rosebery argued that the Empire should be ruled by a sceptre not a bonnet. Even the coachman of Victoria's cousins, the Cambridges, 'deplored' her driving to the Abbey 'with a bonnet on'.

But the humble bonnet meant more to Victoria than mere headgear. It was a symbol of both widowhood and motherhood. Well into the twentieth century Ben Cooper, gamekeeper at the Astors' stately home, Cliveden, still remembered Queen Victoria frequently driving over from Windsor and taking tea with the housekeeper on the terrace. She came in a one-horse carriage wearing a bonnet 'that couldn't have cost five shillings'.

An old lady who had joined in both the Golden and Diamond Jubilee celebrations recognized one of the 'Golden' bonnets reappearing ten years later.

VICTORIA'S ASCENSION

At the Golden Jubilee thirty thousand schoolchildren were entertained in Hyde Park.

As a huge balloon rose from the grass a small girl cried out, 'Look, there's Queen Victoria going up to Heaven!'

Ibid.

A GRANDCHILD'S JUBILEE VISIT, 1887

Princess Alice, countess of Athlone, recalled the Golden Jubilee of her grandmother Queen Victoria. She was talking to a British ambassador in Africa at the time of Queen Elizabeth II's Silver Jubilee.

She and her brother had been driven from their parents' home at Claremont Park, Esher, to Buckingham Palace, changing horses at Richmond. After lunch her grandmother had sent for her and placed her on her lap. Queen Victoria had very short legs, and her black silk dress was extremely slippery—'Grandmama's lap was like a landslide'—but she was well aware that it would never do for her to fall off until she was given permission to leave. In order to stay in place she had to keep pushing herself up with her feet—'it was like treading water'—and she was greatly relieved when she was finally allowed to get down.

David Scott, *Ambassador in Black and White* (1981)

THE QUEEN'S BAD NIGHTS, 1889

Like her uncle George IV, Queen Victoria rang the bell the moment she woke up in the night. Her chief dresser, Mrs Tuck, had a nervous breakdown. Dr Reid wrote about the problem to the queen's former physician, Sir William Jenner:

I gave the Queen last night no Dover's Powder, but a draught with 25 grains of Ammonium Bromide and 30 drops of Tincture of Henbane. After grumbling about her very bad night she said that perhaps after all she had more sleep than she thought, as, except once, she did not think she remained awake longer than five or six minutes at a time! Every time she wakes, even for a few minutes, she rings for her maids, who of course don't like it, and naturally call the night a 'bad' one. She has got into the habit of waking up at night, and I fear it may not be easy to break this habit.

Reid, *Ask Sir James*

Dr Reid tried to improve her digestion by substituting a milky cereal dish at night for over rich concoctions. She liked the cereal—but ate both!

THE GLASS EYE

In 1892 Prince Christian, husband of the queen's third daughter Helena, was accidentally shot in the eye by his brother-in-law Prince Arthur. The queen's doctors had great difficulty in persuading her that Christian must have the injured eye removed. The prince himself was to make an art of his injury.

He collected a great number of glass eyes and at dinner-parties would ask the footman to bring in the tray containing them. He would explain the history of each at great length, his favourite being a blood-shot eye which he wore when he had a cold.

John van der Kiste, *Queen Victoria's Children* (Gloucester, 1986)

GLADSTONE'S TRAUMATIC RESIGNATION

After a profoundly impressive career—Liberal prime minister four times—he resigned in 1894. The queen was deeply relieved, for whereas his admiring public had christened him the 'GOM' (Grand Old Man), she referred to him as 'the abominable old G. man', terrified that he and his hated Liberal policies would go on forever. Gladstone sourly described the conversation with HM at the resignation audience as 'neither here nor there'. At his last audience ever, conducted a year before his death, he was even more bitter:

To speak frankly, it seemed to me that the Queen's peculiar faculty . . . of conversation had disappeared. It was a faculty, not so much the free offspring of a rich and powerful mind (!), as the fruit of assiduous care with long practice and much opportunity. John Morely, *Life of Gladstone*, III (1903)

Gladstone's criticism of the queen's conversation was printed in his official biography. A far more savage attack on her attitude during his resignation audience appeared later. Here, HM's parting from him had seemed like his parting from a mule, after a tour of Sicily in 1831:

I had been on the back of the beast for many scores of hours. It had done me no wrong. It had rendered me much valuable service. But it was in vain to argue. There was the fact staring me in the face. I could not get up the smallest shred of feeling for the brute. I could neither love nor like it.

Philip Magnus, *Gladstone* (1954)

Too true, alas; except that Queen Victoria would have loved the mule, a dumb animal, far more than her eloquent prime minister.

The Queen's Indian Secretary

Abdul Karim the Munshi (secretary) 'occupied the same position as John Brown', according to Sir Henry Ponsonby, 'in the Queen's affections.' The court disliked him for putting on airs and also because they suspected him of being the ignorant accomplice of dangerous Indians. On 27 April 1897, while Ponsonby's son, Sir Frederick, was with the queen in the south of France, he wrote explosively to Sir Henry Babington-Smith, private secretary to the Viceroy of India:

My dear Babs . . . the following is *strictly confidential*, so please treat it like the confessional. We have been having a good deal of trouble about the

Munshi here, and although we have tried our best, we cannot get the Queen to realise how very dangerous it is for Her to allow this man to see every confidential paper relating to India, & in fact to all state affairs. The Queen insists on bringing the Munshi forward as much as she can, & if it were not for our protests, I don't know where she would stop. Fortunately he happens to be a thoroughly stupid & uneducated man, & his one idea in life seems to be to do nothing & eat as much as he can . . . I don't know whether you remember a man of the name of Ruffudin Ahmed, who tried to stand for Parliament. Well he supplies the brains that are deficient in the Munshi, & being a very clever man, he tries to extract all he can out of the Munshi, & that, I think, is where the real danger comes in. The Munshi is even allowed to read the Viceroy's letters, & any letters of importance that come from India. Things have now come to such a pass that the police have been consulted & have furnished some rather interesting details about both the Munshi & Ruffudin. But it is no use, for the Queen says that it is 'race prejudice' & that we are all jealous of the poor Munshi (!)

India Office Library

There had been a row three years before, according to Ponsonby, when the Munshi sent the Viceroy a Christmas card, which the Viceroy did not answer; 'HM wished to know why.'

THE QUEEN'S DEFENCE

To make out that the poor good Munshi is so *low* is really *outrageous* & in a country like England quite out of place . . . She has known 2 Archbishops who were sons respectively of a Butcher & a Grocer . . . Abdul's father saw good and honourable service as a Doctor [in fact a prison apothecary], and he feels cut to the heart to be thus spoken of. It probably comes from some low jealous Indians or Anglo-Indians . . . The Queen is so sorry for the poor Munshi's sensitive feelings.

Ponsonby, *Henry Ponsonby*

'MUNSHIMANIA', 1897

The Household turned more and more against the Munshi as the queen got older, believing that he tyrannized over his fellow Indian servants, at least the Hindus among them, he being a Muslim, and even frightened the queen herself; yet her 'Munshimania' made her stand up for him. In 1897, in spite of his having contracted venereal disease, Victoria determined to take him with her to France. Lady-in-waiting Harriet Phipps was chosen to inform the queen that she would have to choose between the Munshi and the Household.

The Queen, in a rage, swept all the paraphernalia from her crowded desk onto the floor. This was an impasse from which there seemed to be no way out. The Queen would not be dictated to by the Household and the Household would not associate with the Munshi. Reid, *Ask Sir James*

In the end the Munshi travelled in the 'ordinary train' both there and back from Cimiez, but the rows continued all that year and after, the queen veering from tears to fury.

DIAMOND JUBILEE, 1897

Amid scenes of great rejoicing the House of Commons presented a loyal address to the queen at Buckingham Palace.

I came in with the Queen's suite into an empty ballroom, and everything was most dignified . . . and she sat down while all of us courtiers grouped ourselves around her. The doors at the other end of the ballroom were then opened and in came the House of Commons like a crowd being let onto the ground after a football match. There seemed to be no order, and the Speaker, Prime Minister, and Leader of the Opposition were lost in the struggling mass of MPs. This dishevelled mass of humanity came at the Queen, and instinctively the men of the Household felt that they were called upon to do or die. We moved out, formed a protecting screen, and stemmed the tide while the Lord Chamberlain and Lord Steward [members of the Household] tried to find the Speaker, etc. Meanwhile the Queen was thoroughly put out at the mismanagement of the function and did not hesitate to let the two great officers of State know what she thought.

Frederick Ponsonby, *Recollections of Three Reigns* (1951)

HOW VILLAGE RADICALS SPOILED THE JUBILEE

It was like this—Queen Victoria was going to have her Jubilee, I can't remember which one it was, and the squire called us all together for a meeting and said, 'We must celebrate her Majesty's Jubilee as loyal subjects, we shall have a fête and we shall plant some trees on the common.' Now we radicals didn't bear the old lady any ill will, nothing like that, but we didn't like being told what we were going to do so we made our plans. We made them carefully over several months. When the day came there were marquees erected, there were teas, there were side shows, there were sports and there were some holes dug on the common ready to receive the trees and, at the appropriate time in the afternoon, the Lord Lieutenant, who'd come to open proceedings, went with some estate workers to the holes to plant the trees and he was going to make a little speech. When he got there one of us radicals was sitting in each hole and the squire said to the first one, 'Get out of that hole my man so his Lordship can plant the tree.' He never said anything, he wasn't rude or anything like that, he just sat in the hole and because it's common land nobody could move him and at each hole there was the same thing—a radical sitting quite quietly, we were not swearing sort of people or anything like that so we never said anything; and that's the way we spoiled Queen Victoria's Jubilee.'

Tony Harman, author of *Seventy Summers*, BBC 1986, talking to an old man in Buckinghamshire, unpublished

BLAST SIR JAMES

At the age of eighty Queen Victoria was amazed and furious to learn that her indispensable doctor, Reid, intended to marry her youngest maid-of-honour, the Hon. Susan Baring. She let off steam to her daughter Vicky, revealing a rare trace of snobbery.

I must tell you of a marriage (wh. annoys me vy. much) wh. will surprise you gtly. Sir J.Reid !!! and my late M. of H. Susan Baring! It is incredible. How she cld. accept him I cannot understand! If I had been younger I wld. have let him go rather—but at my age it wld. be hazardous and disagreeable and so he remains living in my House wherever we are!! And she quite consents to it. But it is too tiresome and I can't conceal my annoyance. I have never said a word to her yet. It is a gt. mésalliance for her. But he has money of his own.

Reid, *Ask Sir James*

The queen managed to conceal her annoyance from Sir James and eventually forgave the couple; but the story that he won her over is not true. It was to one of the queen's ladies that he said 'he certainly would not do it again', not to the queen herself.

THE AGA KHAN KNIGHTED, 1898

I kissed the hand which she held out to me. She remarked that the Duke of Connaught was a close friend of my family and myself. She had an odd accent, a mixture of Scotch and German—the German factor in which was perfectly explicable by the fact that she was brought up in the company of her mother, a German princess, and a German governess, Baroness Lehzen. She also had the German conversational trick of interjecting 'so'—pronounced 'tzo'—frequently into her remarks. I was knighted by the Queen at this meeting but she observed that, since I was a prince myself and the descendant of many kings, she would not ask me to kneel, or to receive the accolade and the touch of the sword upon my shoulder, but she would simply hand the order to me. I was greatly touched by her consideration and courtesy.

The Memoirs of the Aga Khan (1954)

THE QUEEN'S LAUGHTER

The Journals are full of the phrase 'I was very much amused' but there is no documented record of her saying 'We are not amused.' The following anecdotes illustrate her keen sense of the absurd.

Another trait in the Queen's character was her cheerfulness; in fact, when she started to laugh she found it difficult to stop, and her laugh was no *company* laugh, but thoroughly hearty. Mr Gibson, RA [a sculptor] tells a story that when Her Majesty was sitting to him, he asked permission to measure her mouth. 'Oh, certainly,' replied the Queen, 'if I can only keep it still and not laugh.' The proposal was so unexpected and droll that it was some time before the Queen could compose herself; directly she closed her mouth she burst out laughing again. Tooley, *Personal Life of Victoria*

One moonlight night she was leaning from her window at Windsor Castle, and was softly addressed by a sentimental sentry below. It was with the most full blooded laughter that she related how 'he took me for a housemaid'.

John Gore, *George V* (1941), quoting a letter from Queen Victoria to her grandson Prince George

The queen amused and amusing:

At one dinner-party the Queen described how her mother the Duchess of Kent had once carried a fork out from the dining-room mistaking it for her fan. At another dinner-party she related with roars of laughter how shocked her Master of the Household had been by the dreadful design for the Ashanti medals: 'Roman soldiers with nothing—nothing at all—but helmets on!'

Ponsonby Letters

A prayer by wee Dr Macgregor in Crathie church at Balmoral for the government:

'That the Almighty would send down His wisdom on the Queen's Meenisters, who sorely needed it.' Her Majesty turned purple in the face trying not to laugh. Ibid.

Lord Dufferin amused the queen.

He told her one day at dinner about an American visitor who had said to his English hostess, 'How old are you? How long have you been married? I should like to see your nuptial bed.' The Queen burst out laughing but raised her napkin to protect Princess Beatrice and the maids-of-honour who sat on the other side of the table. Ibid.

The queen joked about legs.

Queen Victoria once imprudently enquired from a male person of her court, on which part of the body were the rheumatic pains which had invalided one of her maids-of-honour, and since she had asked, he was obliged to tell her that they were in her legs. She replied, no doubt humorously, that when she came to the throne young ladies . . . 'did not use to have legs.' E. F. Benson, *As We Were: A Victorian Peep-show* (1930)

It was possible also to laugh at as well as with the queen.

Queen Victoria had . . . instructed a décolletée granddaughter before going into dinner: 'A little rose in front, dear child, because of the footmen.' Rose, *Kings, Queens and Courtiers*

Nevertheless Victoria was as likely to be amused as not amused during her later years, and particularly by ludicrous mishaps, as when a duchess dropped her bustle, a horsehair object resembling a large sausage which was retrieved from the floor by a footman and handed back to her.

'I believe it belongs to your Grace.'
Her Grace, denying it, ordered its removal, only to be told later that her maid had identified it as hers.

'The Queen burst into fits of laughter.'

Marie Mallet, *Life with Queen Victoria*, ed. Victor Mallet (1968)

The Hon. Reginald Brett (later Lord Esher) and his wife Nellie dined for the first time at Windsor Castle in November 1897.

When Nellie rose from her curtsy, her dress gave a loud crack like a pistol shot, much to the Sovereign's amusement.

James Lees-Milne, *The Enigmatic Edwardian: Life of Reginald Brett, Viscount Esher* (1986)

'MOMENTS OF TROUBLE'

The Queen was at Osborne, and she went out for her customary drive with Lady Errol, who was then in waiting. These dear, elderly ladies, swathed in crêpe, drove in an open carriage, called a sociable. The Queen was very silent, and Loelia (Lady Errol) thought it time to make a little conversation. So she said, 'Oh, Your Majesty, think of when we shall see our dear ones again in Heaven!'

'Yes,' said the Queen.

'We will all meet in Abraham's bosom,' said Loelia.

'I will *not* meet Abraham,' said the Queen.

An entry in Queen Victoria's diary for this day runs:
'Dear Loelia, not at all consolatory in moments of trouble!'

Princess Marie-Louise, *My Memories of Six Reigns*

'MY PETTICOATS'

One afternoon, I was sitting in my room when I received an SOS from Her Majesty's page telling me that the Queen wished me to go to her at once. I leapt out to the corridor and found her half sitting and half lying in a little passage. 'My dear, I have had a terrible accident.'

'Good heavens, what?' I said.

Apparently the horses had shied and nearly upset the carriage and, in Grandmama's words, 'Dear Frankie Clark (who succeeded John Brown) lifted me out of the carriage and, would you believe it, all my petticoats came undone!'

Ibid.

'MY TROOPS'

Everyone knows that the so-called Battle of Colenso in the South African War was far from being a brilliant success, but here is a story regarding it. My sister was sitting with the Queen, when the latter, turning to her, said: 'Thora, go and tell Sir Arthur Bigge (afterwards Lord Stamfordham) to clear the line as I wish to telegraph to the troops.'

Thora went to Lord Stamfordham with this message, and on her return said: 'Grandmama, Sir Arthur says it is only customary for the Sovereign to telephone to the troops if they win a victory, and this is not a victory.'

The Queen replied, 'And since when have I not been proud of my troops whether in success or defeat? Clear the line.'

Ibid.

SAYINGS OF QUEEN VICTORIA

On the Black Country: 'It is like another world. In the midst of so much wealth, there seems to be nothing but ruin.'

On Florence Nightingale: 'I wish we had her at the War Office.'

On an alcoholic she had visited in hospital: 'I *know* he has *now* taken the pledge, and I will not give him up.'

On Albert's death: 'He *would* die; he seemed not to care to live.'

On the unecumenical clergy at King Leopold's funeral: 'Nasty "Beggars" as Brown would say.'

On a proposed new royal portrait by von Angeli: 'It better be done before I get too hideous to behold.'

On Gladstone's Irish Land Bill: 'She *cannot* and will not be Queen of a *democratic monarchy*.'

On Black Week in the Boer War: 'Please understand that there is no one depressed in this house; we are not interested in the possibilities of defeat; they do not exist.'

On the secret of her own eternal youth: 'Beecham's Pills'.

Queen Victoria's Death-Bed, 1901

We have more than one eyewitness account, though only Sir James Reid her doctor for twenty years was present almost the whole time, from the moment when she began to fail until the end—and after. The Royalties, as he records in his diary, moved in and out, so that different eyewitnesses did not always see them in the same positions around the dying queen's bed. The Duke of Argyll, husband of Princess Louise, produced the most famous simile for the queen's last hours, which he handed on to Sir Frederick Ponsonby. She went down:

like a great three-decker ship. Ponsonby, *Recollections of Three Reigns*

SIR JAMES REID'S ACCOUNT

21 January: . . . In the evening I took the Prince of Wales to see the Queen and to speak to her. After the Prince of Wales left the Queen, Mrs Tuck [the chief dresser] and I went to her bedside, and HM took my hand and repeatedly kissed it. She evidently in her semi-conscious state did not realise the Prince had gone, and thought it was *his* hand she was kissing. Mrs Tuck, realising this, asked her if she still wanted the Prince of Wales, and she said 'yes'. The Prince returned to her bedside and spoke to her and she said to him 'Kiss my face'. Reid, *Ask Sir James*

Susan Reid, who was staying in a nearby cottage, had decided to meet her husband outside Osborne House so as not to get entangled with the Royalties at such a time. Reid sent her a note on the morning of 21 January, using their nickname for the queen.

Bipps was very bad last night, and we thought she was going to 'bat': but she has rallied and is rather better again, but is almost unconscious. Come up to the bicycle house about *12.45* and I'll come to you. Ibid.

Tuesday 22 January: About 9.30 a.m., when I had gone to my room for a short time to wash and change my clothes, and had asked Powell [Sir Richard Douglas Powell, another of the three doctors in attendance] to go and take my place, he rushed up to my room, and asked me to hurry back as she looked like dying. All the family were summoned, and the Bishop of Winchester said prayers for the dying while I kept plying her with oxygen. The Princesses Christian, Louise, and Beatrice kept telling her who was beside her (the Queen for long too blind to see), mentioning each other's names and those of all the rest of the family present, but omitting the name of the Kaiser who was standing at her bedside. I whispered to the Prince of Wales, 'Wouldn't it be well to tell her that her grandson the Emperor is here too.' The Prince turned and said to me 'No it would excite her too much', so it was not done. Ibid.

The Kaiser's arrogance towards his English relations, especially his uncle the Prince of Wales, had made him so unpopular that at one time the queen had remarked that if he insisted on adopting 'imperial' airs in private as well as in public, 'he better not come here'. But William never lost his devotion to his grandmother, nor she her love for her first grandchild.

Reid continued:

Later she rallied, and Powell and Barlow [Sir Thomas Barlow, the third doctor in attendance] said her vitality was phenomenal. I sent out all the family to let her rest. She began to take food again, talked better and got clearer in her head, and I could not help admiring her clarity.

In the forenoon I went to tell the Kaiser I meant to take him to see the Queen when none of the family was there. He was very grateful and said, 'Did you notice this morning that everybody's name in the room was mentioned to her except mine.' I replied '*yes* and that is one reason why *I* specially wish to take you there' . . . I went to the Prince of Wales to report about the Queen, and said I would like to take the Kaiser to see her. He replied, 'Certainly, and tell him the Prince of Wales wished it.' I took the Kaiser to see her, and sent all the maids out and took him up to the bedside, and said, 'Your Majesty, your grandson the Emperor is here; he has come to see you as you are so ill,' and she smiled and understood. I went out and left him with her five minutes alone. She said to me afterwards, 'the Emperor is very kind.'

She was again getting weaker, and at about 1.45 she got very bad again, and at 3 we summoned the family once more, who stayed in and about the room . . . I asked the family to go out again about 4 . . . At this time we sent the 4 p.m. bulletin: 'The Queen is sinking.' I returned to her room after five minutes absence, and did not leave till she died at 6.30 p.m. The family returned soon after me, and kept going in and out, but the Kaiser remained the whole time standing on the opposite side to me, as did [Alexandra] the Princess of Wales. . . . A few minutes before she died her eyes turned fixedly to the right and gazed on the picture of Christ in the

'Entombment of Christ' over the fireplace. Her pulse kept beating well till the end when she died with my arm round her. I gently removed it, let her down on the pillow, and kissed her hand before I got up . . . I had for the last hour been kneeling at her right side . . . helped by the Kaiser who knelt on the opposite side of the bed supporting her. The Prince of Wales was sitting behind me at the end, and Princess Louise kneeling on my right. All the rest were round about, Nurse Soal sitting on the bed at the top, and Mrs Tuck standing beside her. Reid, *Ask Sir James*

It is astonishing to note that Reid had never seen the queen in her bed until six days before she died; nor did he realize till after her death that she had suffered from prolapse and ventral hernia. Such inhibition seems extraordinary until one remembers that it was virtually still the nineteenth century and that Lady Flora Hastings when dying of cancer had agreed to be examined 'under her clothes' only when she wished to have her honour vindicated, since the tumour had been mistaken for pregnancy.

SIR THOMAS BARLOW'S ACCOUNT

He wrote to his brother on 23 January:

Yesterday's experience was too full of the pathetic human side to think of it as a great historical event.

There was the poor Queen not suffering much as far as we could judge but sinking by slow degrees, her splendid constitution showing itself to the last . . .

Her face was quite beautiful in its way: she had but little pain; her expression was for the most part calm. There was a simple dignity like that of an old Roman.

. . . The Princess of Wales [Alexandra] was on one side holding her hand and Princess Louise on the other holding the other. The Queen had a pathetic little symptom of her illness. Whenever she was raised and often when lying down she had a feeling of insecurity and liked her hand to be held. Often her devoted dressers held her hands but at the last it was the two princesses.

The end approached: It was at 6.20 p.m., the daylight was passing and a few candles and a lamp were lit. The day might have been in September—the cedars and the ilexes were in splendid condition and the grass so green. There wasn't a grace of winter gloom. There was scarcely a sound but that of a fountain in the garden close by . . .

The Queen lay on a simple, narrow mahogany bedstead with a quite small dark chocolate coloured canopy at the head. [Reid had moved her from her double bed to make the nursing easier.]

. . . When Reid took me to her on the Monday and told her Dr Barlow had come, she gave me her most charming smile, like that of a child.

She was so helpful to us in everything and though she said little, she showed *sense* to the last, for now and then she made us understand that she wished to be left quiet without any talk.

There were fine touches sometimes. Once when her dressers had been doing something for her, she looked up and said with infinite tenderness, 'My poor girls', as though she felt sorry for the rather trying work they had to do . . .

The most interesting of the bystanders was the Prince of Wales who sat and knelt to the right of the bed and the Emperor, who stood near the Queen's left shoulder . . . But the Emperor's was the figure that to us was the most striking personality in the room next to the Queen.

There he stood with his eyes immovably fixed on his grandmother, apparently with no thought but of her. When asked to speak he said he had come to tell her about the Empress Frederick [his mother Vicky who was dying of cancer of the spine], that she was a little better, she was taking drives again, that she sent her love, and then quietly, he took his place of watching again—no selfconsciousness or posing there but simple dignity and intense devotion.

But in the earlier part of the day, when the family had been summoned he had showed himself so ready and deft in putting in a pillow here and there and when some of the others said 'more air' he was away to the window to lift it himself if I had not forestalled him. Longford, *Victoria R.I.*

THE COFFIN AND ITS MEMORABILIA

Queen Victoria left many written 'Instructions' about her death and burial, including two to Sir James Reid her doctor in the 1870s and one for Mrs Tuck the dresser, who gave it to Reid on 23 January. The first, in her own hand, stated:

In case of the Queen's death she wishes that her faithful and devoted personal attendant (and true friend) Brown should be in the room and near at hand, and that he should watch over her earthly remains and place it in the coffin, with Löhlein [Prince Albert's valet] or, failing, one who may be most generally in personal attendance on her. This her Physicians are to explain in case of necessity to those of her children who may be there.

Reid, *Ask Sir James*

The second, in Sir William Jenner's handwriting, noted the queen's 'command' that no strange consultants should be brought in to see her. (Even Sir James had never seen her in any bed but her death-bed, and she did not intend to have strange men approaching her with stethoscopes.)

Reid described Mrs Tuck's private instructions from the queen in his diary for 24 January:

I had a talk with Mrs Tuck who, the night before, had read me the Queen's instructions about what the Queen had ordered her to put in the coffin, some of which none of the family were to see, and as she could not carry out Her Majesty's wishes without my help, she asked me to cooperate. Later I helped her and the nurse to put a satin dressing gown and garter ribbon and star etc: on the Queen. We cut off her hair to be put into lockets, and rearranged the flowers.

25 January: At 9.30 I went to the Queen's room, and arranged with Mrs Tuck and Miss Stewart to put on the floor of the Queen's coffin, over the layer of charcoal 1½ inches thick, the various things (dressing gown of Prince Consort, a cloak of his own embroidered by Princess Alice, the Prince Consort's plaster hand, numerous photographs etc:) which Her Majesty had left instructions with them to put in. Reid, *Ask Sir James*

Lady Reid, editor of Reid's diary, added that there were 'rings, chains, bracelets, lockets, photographs, shawls, handkerchiefs, casts of hands etc.' all the souvenirs from her life—early, middle, and late.

Reid continued:

Over these was laid the quilted cushion made to fit the shape of the coffin, so that it looked as if nothing had been put in. Ibid.

Reid then organized the lifting of the queen into her coffin: he and Mrs Tuck took the head, two other dressers the feet,

the King [Edward VII] and Duke of Connaught, the Emperor and Prince Arthur the straps (laid under the body) on the left, over the coffin, and Woodford, Scott and Spenser [the undertaker and his men] on the bed on the right side of the body. Then all the Royalties went out, leaving Mrs Tuck, Misses Stewart and Ticking with me, and they rearranged the Queen's dressing gown, the veil and lace. Then I packed the sides with bags of charcoal in muslin and put in the Queen's left hand the photo of Brown and his hair in a case (according to her private instructions) which I wrapped in tissue paper, and covered with Queen Alexandra's flowers. After all was done I asked . . . the Royal ladies there to come and have a last look. Ibid.

They would not have been pleased to know that the memorabilia of John Brown were hidden under Queen Alexandra's flowers, which is no doubt why Queen Victoria kept the fact secret. But to wear the picture on her left wrist would have been characteristic of her. After all, her left hand had rested on Brown's strong arm, especially just before he died, when she was suffering from rheumatism after a fall. She may even have told Brown that this picture was to be there when she arranged for him to put her in the coffin. After the premature death of her 'devoted personal attendant (and true friend)' she would have seen no reason to change her mind.

Edward VII

1901–1910

'Good old Teddy', as the racing crowds delighted in calling Queen Victoria's eldest son, was vastly popular but in the opposite way to his revered mother. Where she had become a myth, he was intensely human, with the virtues and vanities of that status. As greedy as Henry VIII for lovely women and noble repasts, he created a buoyant court that had some of the glamour if not the culture of a sun-king's, especially in the light of Queen Alexandra's beauty. He helped to sway the national interest away from Berlin and toward Paris. His long wait for the throne did not prevent him from giving the name 'Edwardian' to the age. A picture of genial élitism and pre-war solid comfort was evoked by his nine-year reign. His rather unsophisticated idea of fun was the practical joke and his nickname was 'Tum-Tum'.

GOSSIP ABOUT PRINCE BERTIE AGED SIX, JANUARY 1848

Lady Beauvale, having reported to Charles Greville that seven-year-old Vicky the Princess Royal was 'very strong in body and mind', related what was said of the little brother, all of which Greville reported:

The Prince of Wales is weaker, more timid, and the Queen says he is a stupid boy; but the hereditary and unfailing antipathy of our Sovereigns to their Heirs Apparent seems thus early to be taking root, and the Queen does not much like the child. He seems too to have an incipient propensity to that sort of romancing which distinguished his Uncle, George IVth. The child told Lady Beauvale that during their cruise he was very nearly thrown overboard and was proceeding to tell her how, when the Queen overheard him, sent him off with a flea in his ear, and told her it was totally untrue. *The Greville Memoirs*

ROYAL SUCCESSION DISCLOSED, 1852

At the same age as his mother, the prince was told about his destiny by his tutor Mr Gibbs. Instead of reacting with a prompt 'I will be good', he turned to the queen for an explanation of what was still to him a mystery.

12 February 1852: Walked out with Bertie . . . He generally lets out to me, when he walks with me, something or other that is occupying his mind. This time it was how *I* came to the throne . . . He said that he had always believed Vicky would succeed, but now he knew that in default of *him*,

Affie, little Arthur & 'another brother, if perhaps we have one', would come before Vicky. I explained to him the different successions . . . He took it all in, very naturally. Longford, *Victoria R.I.*, quoting the queen's Journal

Hitherto Bertie had lived in a formidable matriarchy. Henceforth he would have to adjust his former sense of inferiority in the light of this newly discovered source of self-importance.

VISIT TO NAPOLEON III'S COURT, 1856

At the age of fourteen, the prince was carried away by the attention paid to him and asked the Empress Eugénie if he might stay on in Paris after his parents had returned home. She replied that they would not be able to do without their children in England. Bertie retorted:

'Not do without us! Don't fancy that! They don't want us, and there are six more of us at home!' *The Greville Memoirs*

THE PRINCE'S BALL IN CANADA, 1860

Never has the Prince seemed more manly nor in better spirits. He talked away to his partner . . . He whispered soft nothings to the ladies as he passed them in the dance, directed them how to go right, & shook his finger at those who mixed the figures . . . In short was the life of the party. During the evening though he and the Duke of Newcastle enquired for a pretty American lady Miss B. of Natchez, whom they met at Niagara Falls and with whom the Prince wished to dance. His Royal Highness looks as if he might have a very susceptible nature, and has already yielded to several twinges in the region of his midriff. *New York Herald*, 19 September 1860

NO TO BLONDIN

From Montreal he went on to Ottawa, where he laid the cornerstone of the Federal Parliament building and rode a timber shoot down the Ottawa River; then on, past Kingston, to Toronto and across Lake Ontario to the Niagara Falls, where he saw Charles Blondin, the French acrobat, walk across the Falls on a tightrope, pushing a man in front of him in a wheelbarrow. Blondin offered to put the Prince into the wheelbarrow for the return journey across the tightrope to the United States. The Prince accepted the offer, but was naturally prevented from going. So Blondin went back by himself, this time on stilts, leaving the Prince to travel on to Hamilton. Christopher Hibbert, *Edward VII* (1976)

TIME FOR TALK

At Philadelphia, which he thought the 'prettiest town' he had seen in America, he went to the opera—where the audience stood up to sing 'God Save the Queen'—and he visited the big, modern penitentiary, where he met a former judge, Vandersmith, who was serving a sentence for forgery. He asked him if he would like to talk. 'Talk away, Prince,' Vandersmith replied breezily. 'There's time enough. I'm here for twenty years!'

Hibbert, *Edward VII*

THE NELLIE CLIFTON AFFAIR, 1861

The prince's first sexual adventure took place at the Curragh in Ireland where he was stationed with his regiment. His fellow officers played a trick on their nineteen-year-old colleague by smuggling Nellie Clifton, a young actress, into his bed after a party. The consequences were out of all proportion to the prince's 'fall' and demonstrated the changed morality of the monarchy since his grandfather's day. His horrified father wrote to him:

If you were to try and deny it [that he was father of a possible child], she can drag you into a Court of Law to force you to own it & there with you [the Prince of Wales] in the witness box, she will be able to give before a greedy Multitude disgusting details of your profligacy for the sake of convincing the Jury, yourself crossexamined by a railing indecent attorney and hooted and yelled at by a Lawless Mob!! horrible prospect, which this person has in her power, any day to realise! And to break your poor parents' hearts!

Philip Magnus, *King Edward the Seventh* (1964)

In a letter to Vicky the queen expressed her revulsion, and the 'shudders' were to increase when a few weeks later the Prince Consort died. There were times when Victoria attributed his death to acute anxiety over Bertie.

'Oh! that boy—much as I pity him I never can or shall look at him without a shudder as you can imagine.'

Ibid.

THE PRINCE OF WALES'S PROPOSAL TO PRINCESS ALEXANDRA OF DENMARK, 1862

Both Victoria and Albert thought it was a case for Bertie of 'marry or burn', and they therefore encouraged Vicky to make the preliminary approaches. Bertie described his success in a long letter to his mother.

The prince began by saying that he was grateful for being accepted by Alix though he still felt as if he were in a dream. He went on to describe how he first received permission from Prince and Princess Christian of Denmark (Alix's parents) to propose, telling them how much he loved her. There followed a ride round the town in carriages and a visit to the zoo.

We then arranged that I should propose to her (today) out walking.

The night before he had sat between mother and daughter;

though rather shy we conversed a good deal together, and I fell in increasing love toward her every moment.

They reached the Palace of Laeken (home of his great-uncle Leopold King of the Belgians, who was also in the plot) at twelve noon; Uncle Leopold spoke kindly to them all, suggesting they should all go into the garden. Princess Christian and her other daughter Marie walked in front, Bertie and Alexandra some distance behind. Prince Philip of Brabant (Leopold's son)

took charge of the rest . . . After a few commonplace remarks Alexandra said that you had given her the white heather. I said I hoped it would bring her good luck. I asked her how she liked her own country and if she would some day come to England and how long she would remain. She said she hoped some time. I then said I hoped she would always remain there, and then offered my hand and my heart. She immediately said *yes*. But I told her not to answer too quickly but to consider over it. She said she had long ago. I then asked her if she liked me. She said *yes*. I then kissed her hand and she kissed me.

Letter from the Prince of Wales to Queen Victoria, annexed to her Journal, Royal Archives

THE WEDDING, 10 MARCH 1863

Alexandra's beauty was a magnet to all eyes, though her husband's bearing also attracted some praise:

The Prince of Wales, plump and nervous, but radiant in Garter robes and a gold collar over a general's uniform, 'looked', Lord Clarendon recorded, 'very like a gentleman amd more *considerable* than he is wont to do.'

Magnus, *Edward the Seventh*

Apart from the bridal pair, Prince William of Prussia, Vicky's eldest son and Bertie's nephew, attracted most attention. He was already at loggerheads with his little 'Aunt Beatrice' [Queen Victoria's youngest daughter] whose muff he had thrown out of the carriage window while driving around Windsor; now, having hurled the cairngorm from his dirk across the floor of St George's chapel, he set upon his two young uncles who were trying to keep him in order:

You will be amused to hear that the little royal boys were drolly insensible to the privilege of being at the wedding on the great 10th. Little Prince Waldemar [Vicky's younger son] resisted being taken from the donkey somebody had given him to play with in the corridor at Windsor, & when his Mamma insisted, he said he was too ill to go, with a very bad cold. Little P. [Prince] William of Prussia was committed to the charge of his young uncles during the service. The Queen enquired afterwards whether he had been good—to wh Ps Arthur & Leopold were obliged to answer 'No–o–o!' very solemnly. 'What was the matter?' . . . 'He bit us all the time!'

Harriet Martineau, Letter to a friend, Vera Wheatley (1957)

THE RETURN JOURNEY

On the shockingly overcrowded train which took the guests back to London Disraeli had to sit upon his wife's knee, whilst the Duchess of Westminster, wearing half a million pounds' worth of jewels, pushed her way into a third-class carriage with Lady Palmerston. Even so, she was luckier than Count Lavradio, who had his diamond star torn off and stolen by roughs. Georgina Battiscombe, *Queen Alexandra* (1969)

THE WALES'S COURT

Lady Waterford is not to be placed about the Princess of Wales. That precise little stick Lady Macclesfield is to be appointed—in short the Queen wishes the new Court to be as dull & stupid as her own . . . the Courts will be below the calibre & character of good society & will have nothing to recommend them but virtue & respectability. Too much of this will make such virtue tiresome to the Prince.

The Stanleys of Alderley, ed. Nancy Mitford (1939), quoting Lord Stanley's letter to Lady Stanley, 18 November 1862.

HUNTING IN LONDON

The Prince amused Londoners, but provoked some tart newspaper criticism, by a great day's run with the Royal Buckhounds on 2 March, 1868, when a carted deer was chased from Harrow through Wormwood Scrubs to the Goods Yard at Paddington Station. It was killed there before the astonished eyes of the railway guards and porters; and the Prince and his friends rode merrily through Hyde Park and down Constitution Hill to Marlborough House. Magnus, *Edward the Seventh*

HUNTING IN EGYPT, 1869

On 6 February the Prince's party started up the Nile in six blue and gold steamers, each towing a barge filled with luxuries and necessities including four riding horses, and a milk-white donkey for the Princess; 3,000 bottles of champagne and 4,000 of claret; four French chefs and a laundry . . .

Famous monuments and ruins were explored, and the Prince, who killed his first crocodile—a female, nine feet long, containing eighty eggs—with an expanding bullet on 28 February, failed to shoot a hyena, but killed quantities of cormorants, cranes, doves, flamingoes, hawk-owls, heron, hoopoes, mallards, merlins and spoonbills.

Ibid.

SHOOTING AN ELEPHANT, 1875

The Prince often told the story of the first elephant which he thought that he had killed. He hacked its tail off, while Charles Beresford climbed on to its rump and started to dance a hornpipe; but the beast rose unexpectedly to its feet and tottered off into the jungle before anyone could grab a rifle and fire. The Prince wrote fully to his sons, promising to bring a baby elephant to Sandringham for them to ride, and he complained of the jungle leeches 'which are very bad, and climb up your legs and bight [sic] you'.

Ibid.

TIGER HUNTING

'Do you go up trees?' asked the Prince of Wales, who—being stout—had doubtless recent and rueful memories of being pushed and pulled up trees in this most exciting and aristocratic of all varieties of big-game shooting.

'No,' said my father, whose girth, though considerable, was not as great as his guest's, 'I am too fat for tree work. I can't climb up. I stand and shoot.'

The Memoirs of the Aga Khan

PRACTICAL JOKES

At a ball which the Prince gave at Gunton on 10 January, 1870, it is on record that his friend, Christopher Sykes, became so drunk that he collapsed and had to be put to bed, and that his hosts retaliated by ordering that a dead seagull should be laid beside him; the joke answered so well that a live trussed rabbit was substituted on the following night.

Magnus, *Edward the Seventh*

THE KING AND THE DUCHESS

Edward liked teasing the old Duchess of Cambridge's favourite, Tosti. Princess Augusta of Hesse-Cassel, widow of the Duke of Cambridge and grandmother of Queen Mary, lived to be ninety-two. In old age she fell for the Italian singer Tosti.

Every day she gave him a present. She even gave him precious family heirlooms. Tosti had to pretend to accept them so as not to vex her, but he regularly took them to the Duke of Cambridge [her son]. The situation of the old Duchess's favourite musician naturally excited jealousy and jokes. One day the Prince of Wales, who always entered rooms unannounced, found Tosti at the invalid's bedside. He was leaning over her to hear what she was saying to him in her tremulous voice. 'Aha!' said His Royal Highness, 'David Rizzio!' [The man who murdered Lord Darnley and married his wife Mary Queen of Scots.] Tosti made no reply, but he promised himself that if another unpleasant comment was made to him, he would answer . . . A few days later, the Prince of Wales passed Tosti in a corridor; he gave a low bow, drew aside, and said: 'After you, Uncle!' (Hadn't people spread a rumour that the aged Duchess had offered to marry the young maestro to attach him to her for life?) At this fresh insult from the irreverent nephew, Tosti could no longer contain himself.

Richardson, *Portrait of a Bonaparte*

He complained to the Duke of Cambridge: 'If I am to be ridiculed, I shall not come again.' Immediately the jokes ceased and Tosti visited the duchess every day until she died.

Princess Marie-Louise, niece of Edward VII, heard this story of him in Paris, at a party when he was still Prince of Wales.

My uncle, who always had a discriminating eye for a pretty face, started what might be called a very definite flirtation with one of the young and charming guests. At the end of the evening he murmured that he would like to continue this acquaintance, and could he possibly come up and see her when all the other guests had left. She was slightly surprised and, I presume, very flattered, and agreed. She told HRH that she would place a rose outside her door, in order that he should be able to identify it. The guests left, and HRH found his way upstairs, saw the rose and discreetly knocked on the door. A thin voice said, '*Entrez*', which he did. But instead of his fair lady it was the kitchen maid who sat up in bed!

Princess Marie-Louise, *My Memories of Six Reigns*

THE MORDAUNT CASE, 1870

Sir Charles Mordaunt brought a divorce suit against his wife, as a result of which the Prince of Wales was subpoenaed. Twelve letters from the prince to Lady Mordaunt (who had confessed to having 'done wrong' with the prince and others, but who by this time was in a mental home) were read out in court. So patently innocuous were they as greatly to disappoint the scandal-mongers. Nevertheless the Lord Chancellor commented that

'it was as bad as a revolution for Wales'.

And Sir Henry Ponsonby reported that

London was black with the smoke of burnt confidential letters.

Ponsonby Letters

ROYAL UNPOPULARITY

When, referring to the extravagant number of officials at court, Dilke said in a speech at Manchester that one of them was a court undertaker, a man in his crowded audience shouted out that it was a pity there was not more work for him to do.

Hibbert, *Edward VII*

The Prince's Illness, 1871

The prince's serious attack of typhoid swept the Mordaunt case out of the public mind, but the royal family, who gathered at his bedside at Sandringham, suffered from being at too close quarters.

The house was crammed so full that Princess Louise and Princess Beatrice were obliged to sleep in one bed. The Queen was 'charming, so tender and quiet' and the Princes Arthur and Leopold, '*very* nice, so amiable and anxious about their brother'. On the whole, however, the presence of 'this extraordinary family' was a hindrance rather than a help—'it is quite impossible to keep a house quiet as long as it is swarming with people and really the way in which they all squabble and wrangle and abuse each other destroys one's peace'.

Battiscombe, *Queen Alexandra*, quoting Lady Macclesfield, lady-in-waiting to Princess Alexandra, Royal Archives

The Princess had need of all the comfort she could get. At one point, when the doctors told her that her presence excited their patient too much, she crawled into his room on her hands and knees.

Ibid.

Even the prince's delirium caused embarrassment for he imagined he had come to the throne and was planning radical changes of personnel at court.

PRAYING FOR THE PRINCE IN CHURCH

Sunday, 10 December 1871, Hereford. We do not know whether the Prince of Wales is alive or dead. Contradictory telegrams have been flying about, and we did not know whether to mention the Prince's name in the Litany or not. Mr Venables read prayers, and when he came to the petition in the Litany for the Royal Family he made a solemn pause and in a low voice prayed 'that it may please Thee (if he still survive) to bless Albert Edward Prince of Wales . . . Before afternoon service a form of prayer for the Prince came down by telegraph from the Archbishop of Canterbury, the first prayer that I have ever heard of as coming by telegraph.

Kilvert's Diary, ed. W. Plomer (1971)

THE SUSPENSE WAS CELEBRATED IN VERSE

Across the wire the electric message came:
'He is no better, he is much the same.'

Thought to be by the Poet Laureate, Alfred Austin—at his very best.

THE KINDLY PRINCESS

Disraeli was created Lord Beaconsfield in old age but his wit as a courtier was as sharp yet honied as ever.

Once, sitting at dinner by the Princess of Wales, he was trying to eat a hard dinner-roll. The knife slipped and cut his finger, which the Princess, with her natural grace, instantly wrapped up in her handkerchief. The old gentleman gave a dramatic groan, and exclaimed, 'When I asked for bread they gave me a stone; but I had a Princess to bind my wounds.'

G. W. E. Russell, *Collections and Recollections* (1898)

THE CLEVELAND STREET SCANDAL

The Prince was shocked and amazed, in October, 1889, when the superintendent of his stables, Lord Arthur Somerset, of the Blues, whom he always called 'Podge' and treated as an intimate friend, was discovered in a homosexual brothel which the police raided in Cleveland Street, off the Tottenham Court Road. A rumbling scandal was caused when some of the facts became known a month later; but the Prince, who said at first, 'I won't believe it, any more than I should if they accused the Archbishop of Canterbury', argued that any man addicted to such a filthy vice must be regarded as an 'unfortunate lunatic'. He expressed to Lord Salisbury his satisfaction that Lord Arthur had been allowed to flee the country, and asked that, if he should 'ever dare to show his face in England again', he should be allowed to visit his parents quietly in the country 'without fear of being apprehended on this awful charge'.

Magnus, *Edward the Seventh*

The Baccarat Case, 1891

Baccarat had become the prince's regular entertainment after his physique put an end to the dancing years. He carried his own counters engraved with the Prince of Wales's feathers. While a guest at Tranby Croft for the St Leger where baccarat was played, the prince heard that Sir William Gordon-Cumming, a fellow guest, had been observed by five players to be cheating. While denying the charge, Sir William promised never to play cards again provided that the affair was kept secret. Of course a secret known to so many was leaked and Sir William retaliated by bringing a civil action against his original accusers, which resulted in discredit for both sides in the eyes of the public.

The trial of Gordon-Cumming's Action was delayed until 1 June, and a gnawing anxiety affected the Prince of Wales's health. 'The whole thing', he informed Prince George [his son], 'has caused me the most serious annoyance and vexation, and that is one of the reasons why I thought it best not to go abroad—not knowing what might turn up.' The Princess [Alexandra] wrote that 'Papa' was 'quite ill' from worry, and that he had been dragged into the affair 'through his good nature . . . and made to suffer for trying to save . . . this worthless creature' and 'vile *snob*', Cumming, whom she had always detested and who had 'behaved too abominably'.

The case was tried by the Lord Chief Justice, Lord Coleridge, on 1–9 June 1891; and it seemed to the mass of the nation as though the Prince of Wales were on trial. Subpoenaed as a witness for the prosecution, he was in court every day except the last, and the Solicitor-General, Sir Edward Clarke, who represented Cumming, and who professed belief until his dying day in his client's innocence, went out of his way to be offensive to the Prince of Wales. He suggested in court that previous instances had been known of men who were willing 'to sacrifice themselves to support a tottering throne or prop a falling dynasty', and he claimed that Sir William was being victimised to save the honour of a Prince who encouraged habitually an illegal game; . . . Lord Coleridge summed up in favour of the defendants, and the jury returned a verdict against Gordon-Cumming after an absence of less than a quarter of an hour. The baronet was dismissed from the Army, expelled from all his clubs and socially annihilated; and the Prince of Wales wrote to Prince George: 'Thank God!—the Army and Society are now well rid of such a damned blackguard. The crowning point of his infamy is that he, this morning, married an American young lady, Miss Garner (sister to Mme de Breteuil), with money!' . . .

It would be difficult to exaggerate the momentary unpopularity of the Prince of Wales; and Queen Victoria informed the Empress Frederick that 'the Monarchy almost is in danger if he is lowered and despised'. She explained that 'it is not this special case . . . but the light which has been thrown on his habits which alarms and shocks people so much, for the example is so bad'.

Ibid.

PRESS REACTIONS TO THE BACCARAT SCANDAL

A German comic paper produced a cartoon showing the great door into Windsor Castle, surmounted by the Prince of Wales's feathers and the motto 'Ich Deal' . . . Stead, in his *Review of Reviews*, applied the test of the 'Prayer Gauge'. He calculated with ruthless arithmetic how many times in the various churches of the United Kingdom prayer had been offered during the last fifty years on behalf of the Prince of Wales since the day of his birth . . . and drew the conclusion that the baccarat scandal had been the only answer vouchsafed from on high to these millions of petitions . . . *The Times* published a leader at the end of the trial, which, in conclusion, expressed regret that the Prince, as well as Sir William Gordon-Cumming, had not signed a declaration that he, too, would never play cards again. Benson, *As We Were*

THE BERESFORD CASE, 1891

The prince, infatuated with the social siren Frances ('Daisy') Brooke, afterwards Lady Warwick, was called upon to help her retrieve a compromising letter she had sent to a former lover, Lord Charles Beresford. Beresford's injured wife, having accidentally got hold of the letter, deposited it with George Lewis, Society's leading solicitor, as a safeguard against a possible renewal of 'Daisy's' intrigue with 'Charlie'.

Quite properly, George Lewis refused to part with his client's property, even at the request of the Prince of Wales, but he nevertheless allowed the Prince to read the letter. The hot-tempered Irishman, 'Charlie' Beresford, was not unnaturally furious at this piece of royal interference; he called his one-time friend a coward and a blackguard to his face and was only just restrained from striking him. Battiscombe, *Queen Alexandra*

Beresford decried royal interference but the prince ostracized the couple.

In July 1891 Lord Charles took the only revenge which was open to him by threatening to publish an account of the whole sorry affair, 'publicity being evidently our only remedy'. Ibid.

By then Lady Charles had become the prince's mistress. The princess was never to prove as tolerant of Daisy as she had been of Mrs Langtry.

Lady Charles's sister, Mrs Gerald Paget, had written . . . 'a defamatory pamphlet', telling the whole sorry story from her sister's point of view. Although apparently only three copies were made, these three circulated freely, even in America. Far too many people 'in Society' now knew the story; it was even said that the duchess of Manchester had read the pamphlet aloud to her guests at a dinner-party. Ibid.

ROYAL SLUMMING

The prince's attendance at the Housing Commission sessions reflected at least a genuine interest.

The expedition to St Pancras and other London slums had been undertaken at the suggestion of Lord Carrington, a fellow-member of a Royal Commission on the Housing of the Working Classes. He, Carrington and the Chief Medical Officer of Health in the Local Government Board, all of them dressed in workmen's clothes, had left Carrington's house in a four-wheeler escorted by a police cab. The Prince had wandered about the narrow streets, dismayed and sickened by the appalling poverty, squalor and misery to which he was introduced, the background to so many thousands of Londoners' lives. He found a shivering, half-starved woman with three ragged, torpid children lying on a heap of rags in a room bereft of furniture. Asked by her landlord where her fourth child was, she replied, 'I don't know. It went down into the court some days ago and I haven't seen it since.' Distressed by her plight, the Prince took a handful of gold coins from his pocket and would have handed them over to her had not Carrington and the doctor warned him that such a display of wealth might lead to his being attacked by the woman's neighbours.

On their way back to Marlborough House, they were joined by one of the doctor's subordinate medical officers. Not recognizing the Prince, and supposing him to be some rich man out for a morning's slumming, and evidently irritated by his reflective silence and aloof demeanour, he slapped him on the back with some such familiar jocularity as 'What do you think of that, old Buck!' The Prince 'kept his temper and behaved very well'.

Hibbert, *Edward VII*

DEATH OF PRINCE EDDY, 1892

The loss of their eldest son from pneumonia was a grievous blow to the Wales parents, even though they knew that his weak character would have rendered him unfit for the throne.

For years the hat which Prince Eddy had been wearing when he went out shooting for the last time, and which he had waved to his mother as, glancing back, he had caught sight of her at a window, was kept hanging on a hook in her bedroom. And for years, too, his own room was kept exactly as it had been when he was alive to use it, his tube of toothpaste being preserved as he had left it, the soap in the washbasin being replaced when it mouldered, a Union Jack draped over the bed, and his uniforms displayed behind the glass door of a wardrobe.

'Gladly would I have given my life for his,' the Prince [George] told his mother, 'as I put no value on mine . . . Such a tragedy has never before occurred in the annals of our family.' Hibbert, *Edward VII*

PRINCESS ALEXANDRA AND THE LAST MISTRESS: MRS KEPPEL

A tiny anecdote illustrates her tolerant, faintly teasing attitude. One day she chanced to look out of the window at Sandringham just as her husband and his mistress were returning from a drive in an open carriage. The Princess herself never lost her graceful slimness but Alice Keppel, her junior by twenty-five years, had already grown very stout, whilst the Prince of Wales had long merited his disrespectful nickname of 'Tum-Tum'. The sight of these two plump persons sitting solemnly side by side was too much for her equanimity; calling to her lady-in-waiting to come and view the joke with her, she dissolved into fits of laughter.

Battiscombe, *Queen Alexandra*

THE PRINCE IN PARIS

[Count Deym] told his neighbours that the Prince was much too familiar in Paris with La Goulue, a famous dancer at the Moulin Rouge, and that on a recent occasion, when La Goulue had greeted his appearance with a ringing shout of 'Ullo Wales!', he had merely chuckled and ordered that all dancers and members of the orchestra should be supplied immediately with champagne. Magnus, *Edward the Seventh*

A SOCIALIST ASSESSMENT OF THE HEIR TO THE THRONE

6 February 1897—A great gathering last night in Queen's Hall—nine hundred LCC scholars receiving their certificates from the Prince of Wales. Sat close to HRH and watched him with curiosity. In his performance of the ceremony, from his incoming to his outgoing, he acted like a well-oiled automaton, saying exactly the words he was expected to say, noticing the right persons on the platform, maintaining his own dignity while setting others at ease, and otherwise acting with perfectly polished discretion. But observing him closely you could see that underneath the Royal automaton there lay the child and the animal, a simple kindly unmoral temperament which makes him a good fellow . . . But one sighs to think that this unutterably commonplace person should set the tone to London Society. There is something comic in the great British nation with its infinite variety of talents, having this undistinguished and limited-minded German bourgeois to be its social sovereign. A sovereign of real distinction who would take over as his peculiar province the direction of the *voluntary side of social life* . . . what might he not do to further our civilization by creating a real aristocracy of character and intellect. As it is, we have our social leader proposing in this morning's papers as a fit commemoration of his august mother's longest reign, the freeing of the hospitals from debt, the sort of proposal one would expect from the rank and file of 'scripture readers' or a committee of village grocers intent on goodwill on earth and saving the rates.

The Diary of Beatrice Webb, II, ed. N. and J. MacKenzie (1983)

The prince was persuaded to give a dinner to celebrate the new Dictionary of National Biography. *The occasion was not to his taste.*

It is said that on looking round the table his eye fell on Canon Ainger, who had written the entries on Charles and Mary Lamb. 'Who is the little parson?' he asked. 'Why is he here? He is not a writer.' It was explained to him that Ainger was 'a very great authority on Lamb'. At this the Prince put down his knife and fork, crying out in bewilderment, 'On *lamb!*'

Hibbert, *Edward VII*

THE PERILS OF FAMILIARITY

The Prince would cheerfully indulge a regrettable pleasure in practical jokes. According to Mrs Hwfa Williams, sister-in-law of the Prince's friend, Colonel Owen Williams, he would place the hand of the blind Duke of Mecklenburg on the arm of the enormously fat Helen Henneker, observing, 'Now, don't you think Helen has a lovely little waist?' And he would be delighted by the subsequent roar of laughter—'in which no one joined more heartily than Helen'. Similarly, he would pour a glass of brandy over Christopher Sykes's head or down his neck or, while smoking a cigar, he would tell Sykes to gaze into his eyes to see the smoke coming out of them and then stab Sykes's hand with the burning end. Shouts of laughter would also greet this often-repeated trick as the grave and snobbish Sykes responded in his complaisantly lugubrious, inimitably long-suffering way, 'As your Royal Highness pleases.'

Yet the idea of anyone pouring a glass of brandy over the Prince's head was unthinkable. Nor must anyone ever refer to him slightingly. A guest at Sandringham, a friend of the Duchess of Marlborough, who went so far as to call him 'My good man' was sharply asked to remember that he was not her 'good man'. And once in the green-room of the Comédie Française, while in conversation with Sarah Bernhardt and the comedian, Frederick Febvre, the Prince was approached by a man who asked him what he thought of the play. The Prince turned his hooded, blue-grey eyes on the interloper and replied, 'I don't think I spoke to you.'

When a newcomer to his circle mistook the nature of its atmosphere for a tolerance of familiarity and called across the billiard table after a bad shot, 'Pull yourself together, Wales!' he was curtly and coldly informed that his carriage was at the door. Similarly, when another of his guests, Sir Frederick Johnstone, was behaving obstreperously late at night in the billiard room at Sandringham and the Prince felt obliged to admonish him with a gentle reproachful, 'Freddy, Freddy, you're very drunk!', Johnstone's reply—made as he pointed to the Prince's stomach, rolled his *r*'s in imitation of his host's way of speaking and addressed him by a nickname not to be used in his presence—'Tum-Tum, you're *verrry* fat!' induced the Prince to turn sharply away and to instruct an equerry that Sir Frederick's bags were to be packed before breakfast. Hibbert, *Edward VII*

ASSASSINATION ATTEMPT, 1900

The anti-British feeling that resulted from the Boer War inspired the only attempt upon the prince's life.

This took place at the Gare du Nord, Brussels, at 5.30 in the afternoon, on 4 April. A Belgian youth, Jean Baptiste Sipido [aged 15] jumped on the footboard of the royal carriage as it steamed out of the station and fired several shots through the open window of the Prince's compartment. One bullet lodged in the back of the seat between the Prince and Princess. The occupants of the carriage . . . remained imperturbably calm, except for Alix's lap dog, which shivered with fright. HRH described his would-be assassin as '*un pauvre fou*', and observed how fortunate it was that anarchists were such poor shots: it was almost inconceivable to miss at a range of six feet. Giles St Aubyn, *Edward VII* (1979)

ALEXANDRA VISITS A PATIENT IN HOSPITAL

She was told that he had been wounded in the leg and that he had just realised that his knee would be permanently stiff and useless. Immediately the Queen was at his bedside: 'My dear, dear, man, I hear you have a stiff leg; so have I. Now just watch what I can do with it'—and lifting up her skirt she swept her lame leg clear over the top of his bedside table.

Battiscombe, *Queen Alexandra*

Her lameness was caused by a severe illness and rheumatism after her first pregnancy.

CORONATION POSTPONED, 1902

Edward VII acceded to the throne in January 1901. An emergency operation for appendicitis caused the postponement of his crowning.

Queen Alexandra remained in the room whilst the anaesthetic was administered and as King Edward began to throw his arms about and to grow black in the face she struggled to hold him down, crying out in great alarm 'Why don't you begin?' To his horror Treves [the surgeon] realised that she intended to be present throughout the operation. His reaction was eloquent of the changes that the last sixty years have seen both in surgical procedure and in etiquette. 'I was anxious to prepare for the operation', he wrote afterwards, 'but did not like to take off my coat, tuck my sleeves, and put on an apron while the Queen was present.' However, when told she must leave, Queen Alexandra went without further ado, joining her son and her two daughters next door in her bedroom. There forty minutes later, Treves came in to tell her that the operation had proved a complete success. Ibid.

ACCOUNTS OF THE CORONATION, 1902

Sir Charles Oman the historian related that

The Archbishop of Canterbury was an octogenarian . . . and had a giddy fit while doing homage; but some of the younger bishops got him on his feet and helped him away. He was better that the Dean, who actually rolled right over, luckily while he was still reaching for the chalice. It was said that the King, in true Sir Philip Sidney vein, refused a cup of soup prepared for him and insisted on it being given to the Archbishop of Canterbury. Oman was struck by both their Majesties' appearance. 'Her long neck was swathed with [a diamond collar to emphasize its slenderness] from chin to shoulder. Her hair was rather red-brown today . . . He looked like a man thoroughly relieved to get the business over.

Carola Oman, *An Oxford Childhood* (1976)

The account of the coronation by Louisa Antrim, a lady-in-waiting to Queen Alexandra:

The princesses were seated in the chancel, looking extremely well with all their crowns on the edge of the boxes in front of them. Just over them was the 'loose-box'—& well named it was—to me the one discordant note in the Abbey—for to see the row of lady friends in full magnificence did rather put my teeth on edge—La Favorita [Mrs Keppel] of course in the best place, Mrs Ronny Greville, Lady Sarah Wilson, Feo Stuart, Mrs Arthur Paget & that ilk . . . *Louisa, Lady-in-Waiting*, ed. E. Longford (1979)

THE KING INCOGNITO

The king and Lord Haldane in Marienbad:

Once while both were staying at Marienbad for a cure the King took Haldane for a motor trip into the country. They stopped at a little roadside inn with a ricketty wooden table in front of it. 'Here I will stand treat,' the King said. He ordered coffee for two and then said, 'Now I am going to pay. I shall take care to give only a small tip to the woman who serves the coffee, in case she suspects who I am.' The woman of course knew at a glance and was presumably disappointed by the meagre gratuity.

Lees-Milne, *The Enigmatic Edwardian*

CLOTHES AND THE KING

Pointing once to Haldane, who arrived in a shabby soft hat at a garden party, King Edward exclaimed to the ladies who surrounded him. 'See my War Minister approach in the hat which he inherited from Goethe!'; but he appreciated that some men were incorrigibly careless, like the Duke of Devonshire, or incorrigibly perverse, like Lord Rosebery. To Rosebery, who often offended and who came once to an evening party at Buckingham Palace wearing trousers instead of knee-breeches, King Edward remarked: 'I presume that you have come in the suite of the American Ambassador!' . . . He told Frederick Ponsonby, who had proposed to accompany him in a tail coat to a picture exhibition before luncheon: 'I thought everyone must know that a *short* jacket is always worn with a silk hat at a private view in the morning.' Ibid.

He even knew what the answer was when the Russian Ambassador asked him if it would be proper for him to attend race-meetings while in mourning: 'To Newmarket, yes, because it means a bowler hat, but not to the Derby because of the top hat.'

He selected his own clothes with the nicest care, and earnestly discussed with his tailor the exact manner in which he thought the cut of the evening dress waistcoat could be improved or the precise reduction that ought to be made in the length of the back of a tail coat. Austen Chamberlain, accompanying the King on a cruise as Minister in Attendance, was 'very much amused' to overhear an instruction issued to a Swiss valet as the yacht approached the Scottish coast: '*Un costume un peu écossais demain.*'

The King's taste in clothes was generally conservative: he attempted to prevent the demise of the frock-coat and to revive the fashion of wearing knee-breeches with evening dress. He refused to wear a Panama hat and derided those who did; he continued to wear a silk hat while riding in Rotten Row long after this was considered old-fashioned. Yet he made several new fashions respectable. His adoption of a short, dark blue jacket with silk facings, worn with a black bow tie and black trousers, on the voyage out to India led to the general acceptance of the dinner jacket.

The King's adoption of the loose, waist-banded Norfolk jacket made this type of jacket popular all over England; while photographs of him wearing a felt hat with a rakishly curved brim brought back from Homburg, or a green, plumed Tyrolean hat from Marienbad, led to thousands of others being sold at home. He found it more comfortable—then decided it looked elegant—to leave the bottom button of his waistcoat undone, and soon no gentleman ever did that button up.

Sometimes he went too far. The sight of the King on a German railway station in a green cap, pink tie, white gloves and brown overcoat induced the *Tailor and Cutter* to express the fervent hope 'that his Majesty [had] not brought this outfit home'. Other observers were driven to complain about the tightness of his coats, and the excessive size of his tie-pins, as well as the ungainly figure he cut in those foreign uniforms which he loved to wear even when their short coats, as those of the Portuguese cavalry, 'showed an immense expanse of breeches', or when their huge, shaggy greatcoats, as those of the Russian dragoons, made him look 'like a giant polar bear'.

Hibbert, *Edward VII*

THE KING'S GAMBLING

But in his gambling, the Prince was very much of his time and social group. He would bet on anything. He placed £200 on a tennis match at Ascot played by his friend, Lord Suffield, encouraging him with shouts of 'Play up Charlie . . . you'll lose all my money,' and he backed him in a wrestling match against Lord Charles Beresford, losing both wagers. He was usually unlucky in his bets, unlucky too at the cardtable. Through his friend, Arthur Paget . . . he placed bets widely, displaying his punctiliousness in paying his losses as well as his weakness in arithmetic in a letter to Paget in September 1890. 'I owe you £650 for Doncaster, £50 for Sandown and £24 for whist, and I owe R. Moncrieffe, £329. Please also send £25 to O. Williams and £25 to R. Sassoon, *but do not pay them at Tattersalls*. I enclose a cheque for £1,003 which I hope you will find correct.

W. S. Adams, *Edwardian Portraits* (1956)

AN OPERATIC SCENE AT BERLIN, 1909

At a gala performance of the Kaiser's ballet, *Sardanapal*, King Edward fell asleep and woke up during the last scene when the Babylonian monarch lights his own funeral pyre. Imagining the theatre to be on fire the King called loudly for the fire engines. Longford, *Louisa, Lady-in-Waiting*

Like his mother the king was superstitious.

'The third day at Friedrichshof the King sent for a list of those who had dined the previous nights, and to his horror found we had been thirteen each night. He seemed much upset by this, but later told me it was all right. Feeling I was getting out of my depth I said, 'Why?' And he explained that Princess Frederick Charles of Hesse was *enceinte.*'

Ponsonby, *Recollections of Three Reigns*

Lady Brougham was a hostess famous for her mastery of the unexpected adjective.

I was told by a friend that when King Edward VII went to stay at Brougham, he arrived in a mood that rendered him difficult to please. Plainly something had gone wrong. At dinner, the King was still silent, so Lady Brougham began to talk, asking,

'Did you notice, sir, the soap in Your Majesty's bathroom?'

'No!'

'I thought you might, sir . . . It has such an amorous lather!'

After that the King's geniality returned.

Osbert Sitwell, *Left Hand Right Hand* (1948)

QUEEN ALEXANDRA'S TASTE

When Alexandra inherited Balmoral in 1901 she got rid of some of the tartans and thistles on the drawing-room carpets and curtains—Queen Victoria's 'tartanitis', as it was called. She wrote to her daughter-in-law Queen Mary when she in turn became mistress of Balmoral in 1910:

I wonder whether you have made any alterations in your rooms upstairs—as I confess dear Grand-Mama's [Victoria's] taste in wallpapers was rather sad and very doubtful!! that washed out pink moiré paper in the sitting-room is *sickly* and the one in the bedroom *appalling* but I never liked to touch anything of hers so left it all exactly as she had it.

Battiscombe, *Queen Alexandra*

THE QUEEN'S NOTORIOUS UNPUNCTUALITY

She had been unpunctual as a child, so it was not entirely a form of revenge on her husband for his own misdemeanours, though it caused him great irritation. One day when she had kept him waiting half an hour and he was afraid of being late for an engagement, she said in her gay, casual way to a courtier:

'Keep him waiting; it will do him good!'

Mabel, Countess of Airlie, Thatched with Gold, ed. J. Ellis (1962)

There was another case soon after the king's accession when it was arranged for the queen to help her husband receive various deputations. The king and the deputations waited from noon till one o'clock:

The King sat in the Equerries' room drumming on the table and looking out of the window with the face of a Christian martyr. Finally at 1.50 p.m. the Queen came down looking lovely and quite unconcerned. All she said was, 'Am I late?' The King swallowed and walked gravely out of the room.

Ponsonby, *Recollections of Three Reigns*

ALEXANDRA'S LAST MALAPROPISM

No one who heard it was quite sure whether it was due to deafness or a deliberate leg-pull.

'Did you know, Ma'am, that His Majesty has a new car?'

'A new cow?'

'No, Ma'am, a new *car*.'

'Yes, yes, I hear you. I understand, the old one has calved.'

Battiscombe, *Queen Alexandra*

PRINCESS 'DARLING'

Alexandra's biographer, Georgina Battiscombe, has pointed out that she was the true soul-mate of James Barrie's 'Mrs Darling' in his Peter Pan, *known as 'Motherdear' to her boys and girls.*

Here is more than a little of *Peter Pan*, the embarrassing whimsy, the undoubted charm, the understanding of children, the curious horror of growing up.'

Ibid.

One day in the 1880s, Princess 'Darling' tucked up her guest, Mrs Gladstone, in bed.

SAYING OF EDWARD PRINCE OF WALES

On waiting for the throne: 'I don't mind praying to the eternal Father, but I must be the only man in the country afflicted with an eternal mother.'

SAYINGS OF KING EDWARD

On art: 'I do not know much about art, but I think I know something about ar-r-rangement.'

To his grandchildren who came to see him dressed for the Coronation: 'Am I not a funny-looking old man?'

SAYINGS OF QUEEN ALEXANDRA

In reply to a long harangue by the Kaiser: 'Willy, dear . . . I am afraid I have not heard a single word you were saying.'

On wearing a real orange in her hair at her silver wedding: 'I am now no bud but the ripened fruit.'

Death of Edward VII

MRS KEPPEL AT THE KING'S BEDSIDE

Regy [Lord Esher] was extremely indignant over the account which Mrs Keppel put about implying that she had been summoned to the King's deathbed by the Queen who thereupon fell upon her neck and wept with her. 'Mrs. Keppel has lied about the whole affair ever since, and describes quite falsely, her reception by the Queen,' he wrote, on learning from Knollys later what actually had happened.

The Queen did *not* kiss her, or say that the Royal Family would 'look after her'. The Queen shook hands, and said something to the effect, 'I am sure you always had a good influence over him', and walked to the window. The nurses remained close to the King, who did not recognize Mrs K. and kept falling forward in his chair. Then she left the room with Princess Victoria almost shrieking, and before the pages and footmen in the passage, kept on repeating, 'I never did any harm, there was nothing wrong between us,' and then, 'What is to become of me?' Princess Victoria tried to quiet her, but she then fell into a wild fit of hysterics, and had to be carried into Freddy's [Sir F. Ponsonby's] room, where she remained for some hours; altogether it was a painful and rather theatrical exhibition, and ought never to have happened. It never would only she sent to the Queen an old letter of the King's written in 1901, in which he said that if he was dying, he felt sure those about him would allow her to come to him. This was written in a moment of weak emotion when he was recovering from appendicitis.

The criticism sounds harsh for whatever her faults may have been—and one of them was rapacity over money—Mrs Keppel's affection for the King was deep and genuine. And there is no doubt that her advent as *mâitresse en titre* did much to humanize her lover. As Millie [Lady] Sutherland observed to Regy at the beginning of the reign, 'the King is a funny man—a child, such a much pleasanter child since he changed mistresses.'

Lees-Milne, *The Enigmatic Edwardian*

GRIEVING

Grief at King Edward's death took many forms, some of them bizarre. One hostess of the late monarch threaded black ribbons through her daughter's underclothes; another tied a large black bow of crepe round a tree which he had planted in her garden five years before. A grocer in Jermyn Street saluted the passing of a dedicated trencherman by filling his window with black Bradenham hams. Sir Arthur Herbert, who as British Minister to Norway had allowed dancing in his Legation on what turned out to be the evening of the King's death, was reprimanded by the Foreign Office and retired prematurely in the following year.

Kenneth Rose, *King George V* (1984)

MOURNING

Everybody has gone into black for the King's death, and some enthusiasts talk of going on mourning for a year. It is all very absurd, considering what the poor King was, but the papers are crammed with his praises as if he had been a saint of God. All the week since his death has been one of storms and tempests . . . and last night one of the great beech trees was thrown down in the park. I saw it lying uprooted on my way to the station this morning, a symbol of the dead King, quite rotten at the root, but one half of it clothed with its spring green.

W. S. Blunt, *My Diaries* (1920)

QUEEN ALEXANDRA'S VIGIL

Queen Alexandra received Pom McDonnell, of the Office of Works on the twelfth. With tears in her eyes, she talked about the suddenness of the King's illness, her complete ignorance of its serious nature until she reached Calais, and 'the providential instinct which warned her to return in spite of all the arrangements having been made to remain in Venice'. McDonnell said 'it was the finger of God which had beckoned her home!' She liked this and repeated it softly to herself twice. Then she took him into the King's bedroom where he lay in a grey military greatcoat. 'They want to take him away,' she said piteously, 'but I can't bear to part with him. Once they hide his face from me everything is gone for ever.'

Longford, *Louisa, Lady-in-Waiting*

THE HOUSE OF WINDSOR

The new family name was proclaimed in 1917 during the Great War against the Kaiser's Germany, because of the Teutonic associations of the Royal Family's previous name—though whether the name inherited from Prince Albert had been Saxe-Coburg or Wettin remained in doubt even by the College of Heralds. From then on Windsor Castle had its eponymous dynasty.

George V

1910–1936

Another 'sailor-king' could not come amiss. His popularity was based on firm grounds. A strong sense of constitutional duty made him partake as fully as possible in the experiences of the Great War. He was thrown from his horse and injured while visiting the front. It was this sense of duty which decided him against giving asylum to his Russian cousins, the Tsar and his family, in 1917 for fear of the political consequences. He turned his back on a courtier who commiserated with him on the country's first Labour government. A happy marriage, stately queen, and abundant children added to his success, which was not affected by an unhealthy red complexion (caused by early typhoid) and loud voice, both at first wrongly put down to drink.

MIDSHIPMAN ON *BRITANNIA*

The first-born son of Edward and Alexandra, Albert Victor (Eddy) died of pneumonia in 1892. As the second son, Georgie was destined to spend many years in the Navy, which accounted for his quarterdeck manner and delight in risqué jokes; also his practical ability and sense of duty. In some ways it was a hard school. He gave his own early recollections:

It never did me any good to be a Prince, I can tell you, and many was the time I wished I hadn't been. It was a pretty tough place and, so far from making any allowances for our disadvantages, the other boys made a point of taking it out of us on the grounds that they'd never be able to do it later on . . .

They used to make me go up and challenge the bigger boys—I was awfully small then—and I'd get a hiding time and again. But one day I was landed a blow on the nose that made my nose bleed badly. It was the best blow I ever took because the doctor forbade my fighting any more.

Then we had a sort of tuck-shop on land, up the steep hill; only we weren't allowed to bring any eatables back into the ship, and they used to search you as you came aboard. Well, the big chaps used to fag me to bring them back a whole lot of stuff—and I was always found out, and got into trouble in addition to having the stuff confiscated. And the worst of it was, it was always *my* money; they never paid me back—I suppose they thought there was plenty more where that came from, but in point of fact we were only given a shilling a week pocket-money, so it meant a lot to me, I can tell you.

John Gore, *King George the Fifth: A Personal Memoir* (1941)

VISIT TO SOUTHERN AFRICA AGED FIFTEEN

They were conducted by the Governor to visit Cetywayo the Zulu king. 'He has got a little farm for himself,' wrote Prince George on February 26; 'we gave him each our photographs and he gave us his. He himself is eighteen stone and his wives 16 & 17 stone; there are four of them, they are very fine women, all over six feet.'

Harold Nicolson, *King George the Fifth: His Life and Reign* (1952)

VISIT OF NICKY THE TSAREVICH

The Duke of Windsor recalled later how his father, then Prince George, had been created Duke of York and that he bore an uncanny resemblance to his cousin.

When, as Czarevitch, Cousin Nicky came to London in 1893 for my father's wedding, my father was mistaken for him by a well intentioned diplomat who asked if he had come over especially for the Duke of York's wedding. My father loved to relate the confusion of the embarrassed envoy when he replied, 'I am the Duke of York, and I suppose I should attend my own wedding.'

Edward Duke of Windsor, *A King's Story* (1951)

THE BIGAMY LIBEL 1893–1910

On 25 April 1893, at the end of a Mediterranean holiday, he [Prince George] wrote to his father's private secretary from the British Embassy in Rome: 'The story of my being already married to an American is really very amusing. Cust has heard the same thing from England only he heard that my wife lived at Plymouth, why there I wonder?' On 3 May, the very day of his betrothal, the *Star* newspaper in London published a more circumstantial account: that the Duke of York had lately contracted a secret marriage in Malta with the daughter of a British naval officer.

At first he took it lightly. 'I say, May,' he told his fiancée one day, 'we can't get married after all. I hear I have got a wife and three children.' . . . Towards the end of the year [1910], however, there came an opportunity to destroy the lie once and for all. A republican paper called the *Liberator*, published in Paris but sent free to every British Member of Parliament, printed an article by E. F. Mylius entitled 'Sanctified Bigamy'. It asserted that in 1890 the future King George V had contracted a lawful marriage in Malta with a daughter of Admiral Sir Michael Culme-Seymour; that children had been born of the union; and that three years later the bridegroom, having by his brother's death found himself in direct line of succession to the throne, 'foully abandoned his true wife and entered into a sham and shameful marriage with a daughter of the Duke of Teck'. The article continued:

The Anglican Church, with its crew of emasculated, canting priests, presents little more resemblance to Christianity than if it were some idol-fetish of a tribe of South Sea Cannibals . . .

Our very Christian King and Defender of the Faith has a plurality of wives just like any Mohammedan Sultan, and they are sanctified by the Anglican Church.

The next issue of the *Liberator* returned to the theme: 'The *Daily News* of London tells us that the King plans to visit India with his wife. Would the newspaper kindly tell us which wife?' Rose, *King George V*

Since the chief witnesses to this lie were all alive the Crown instituted libel proceedings. Eventually Mylius was sentenced to twelve months' imprisonment and the king's honour was vindicated.

THE ART OF CHAFF

The Prince of Wales [the future George V], like his father, was a lifelong exponent of the art, practising it relentlessly on family and friends. During an inspection of the fleet he caused agonies of shame to the future Lord Mountbatten by loudly inquiring about a rag doll which the young midshipman had dearly loved in childhood. Archbishop Lang, on resuming his duties after an illness that cost him his hair and made him look twenty years older, was greeted by his sovereign 'with characteristic guffaws'. Ibid.

A LOOK AT YORK COTTAGE IN 1949

York Cottage, the favourite home of George V, prince and king, was at Sandringham.

I spend the morning visiting York Cottage the 'nest', the dairy, the gardens and the big house. There is nothing to differentiate the cottage from any of the villas at Surbiton. How right the Duke of Windsor was to say to me,

'Until you have seen York Cottage you will never understand my father.' It is almost incredible that the heir to so vast a heritage lived in this horrible little house. It is now partly an estate office and partly flats. But it is still untenanted in the upper floors and we went all over it. The King's and Queen's baths had lids that shut down so that when not in use they could be used as tables. His study was a monstrous little cold room with a north window shrouded by shrubberies, and the walls are covered in red cloth which he had been given while on a visit to Paris. It is cloth from which the trousers of the French private soldiers used to be made.

Harold Nicolson, *Diaries and Letters 1945–1962*, ed. Nigel Nicolson (1968)

AN ANARCHIST'S ATTACK, 1906

Prince George was an eyewitness of the attempt by Morales, the Spanish anarchist, to kill Queen Ena (Queen Victoria's granddaughter) and King Alfonso of Spain at their coronation.

Just before our carriage reached the Palace, we heard a loud report and thought it was the first gun of a salute. We soon learned however that when about 200 yards from the Palace in a narrow street, the Calle Mayor, close to the Italian Embassy, a bomb was thrown from an upper window at the King and Queen's carriage. It burst between the wheel horses and the front of the carriage, killing about 20 people and wounding about 50 or 60, mostly officers and soldiers. Thank God! Alfonso and Ena were not touched although covered with glass from the broken windows . . .

Of course the bomb was thrown by an anarchist, supposed to be a Spaniard and of course they let him escape. I believe the Spanish police and detectives are about the worst in the world. No precautions whatever had been taken, they are most happy go lucky people here. Naturally, on their return, both Alfonso and Ena broke down, no wonder after such an awful experience. Eventually we had lunch about 3. I proposed their healths, not easy after the emotions caused by this terrible affair.

Diary of George V, quoted in Rose, *King George V*

DEATH OF EDWARD VII AND ACCESSION OF GEORGE V

Next morning I was awakened by a cry from my brother Bertie. From the window of our room he cried, 'Look, the Royal Standard is at half-mast!' . . . That morning, while Bertie and I were dressing, Finch [valet] appeared with word that my father wished to see us both downstairs. My father's face was grey with fatigue, and he cried as he told us that Grandpapa was dead. I answered sadly that we had already seen the Royal Standard at half-mast. My father seemed not to hear as he went on to describe in exact detail the scene around the deathbed. Then he asked sharply, 'What did you say about the Standard?' 'It is flying at half-mast over the Palace,' I answered.

My father frowned and muttered, 'But that's all wrong,' and repeating as if to himself the old but pregnant saying, 'The King is dead. Long live the King!' he sent for his equerry and in a peremptory naval manner ordered that a mast be rigged at once on the roof of Marlborough House.

Windsor, *A King's Story*

THE FUNERAL PROCESSION OF EDWARD VII

Afterwards the Spanish ambassador complained that his sovereign, despite seniority of accession, was made to walk behind the Kaiser.

This put the Foreign Office and the Court in a fix. An apology would have been worse than useless, because high officials of Court and State are not expected to make mistakes of this sort. Finally the problem reached the King. He solved it diplomatically and ingeniously; the Kaiser, he said, was King Edward's nephew and his own first cousin, and for these reasons alone he had been given precedence, not as a reigning sovereign, but as a family mourner.

The Memoirs of the Aga Khan

CHANGING THE ROYAL NAME, 1917

General press hysteria about all things German could not be ignored by the royal family. In 1915 foreigners were expelled from the Order of the Garter and two years later the royal family changed its name. Reactions to the change ranged from enthusiasm and wit to derision and hostility.

Punch *trumpeted in a hideous cartoon: 'Long Live the House of Windsor!'* The Times *put the same thought into Latin: 'Stet Domus'.*

The demise of the name Battenberg was marked by the new Lord Milford Haven [the former Prince Louis of Battenberg] in characteristic fashion. He was staying at his elder son's house, 'Keavil', near Rosyth, when the change of title was officially approved. He wrote in his son's visitors' book, 'June 9th arrived Prince Hyde; June 19th departed Lord Jekyll.'

E. Longford, *The Royal House of Windsor* (1974)

On hearing of his enemy's dynastic transmutation, the Kaiser remarked that he was going to the theatre to see *The Merry Wives of Saxe-Coburg Gotha.*

Ibid.

The new Earl of Athlone, formerly Prince Alexander of Teck, was 'furious'. 'He thought that kind of camouflage stupid and petty,' said his wife Princess Alice.

Theo Aronson, *Crowns in Conflict* (1986)

The Bavarian nobleman Count Albrecht von Montgellas lost hope:

'The true royal tradition died on that day in 1917, when, for a mere war, King George V changed his name.' Geoffrey Bocca, *The Uneasy Heads* (1959)

RATIONING

At York Cottage during the Great War rationing was taken very seriously. There was nothing left for whoever came down last to breakfast as most people helped themselves too generously. One morning a courtier was kept late on the telephone and entered the dining-room after everyone else had sat down:

He found nothing to eat and immediately rang the bell and asked for a boiled egg. If he had ordered a dozen turkeys he could not have made a bigger stir. The King accused him of being a slave to his inside, of unpatriotic behaviour, and even went so far as to hint that we should lose the war on account of his gluttony. Ponsonby, *Recollections of Three Reigns*

STAMPS

When he was prince as when he was king, stamp-collecting was George's favourite hobby. One day Prince George was asked to look through an old lady's stamp album which had been valued at £50:

The Prince of Wales glanced through the album and at once saw it was a very valuable collection. In it he saw a 2½*d.* Bahama stamp for which he had been looking for years. He told Derek Keppel to reply that the collection was a valuable one and that she should send it to Puttick & Simpson the auctioneers in London and pay £10 in advertising the sale. The old lady carefully followed his advice and the collection was put into a good sale. Meanwhile, thc Prince of Wales gave instructions to his curator of stamps to buy this particular Bahama stamp at any price. The whole collection fetched over £7,000 and the Bahama stamp alone realised as much as £1,400. A week later Sir Arthur Davidson, Equerry to King Edward, had occasion to telephone to the Prince of Wales about something, and having finished he added: 'I know how interested Your Royal Highness is in stamps. Did you happen to see in the newspapers that some damned fool had given as much as £1,400 for one stamp?' A quiet and restrained voice answered: 'I was that damned fool.' Ibid.

A British representative in the Middle East, hearing of a suspected case of smallpox in the local printing works, feared that the royal tongue might be contaminated; so he assiduously boiled his entire offering of 400 stamps in a saucepan before despatching them to London.

Rose, *Kings, Queens and Courtiers*

MASSACRE SHOOTING

On home ground, too, he would blaze away until he stood on a carpet of spent cartridges, each bearing a tiny red crown. Behind the King and his loaders, a detective clicked up each addition to the bag on a pocket instrument that could record a four-figure total. Even his entourage sometimes flinched. Lord Lincolnshire wrote: 'Seven guns in four days at Sandringham killed 10,000 head . . . The King would have been shocked had anyone questioned his love for the animal kingdom. A niece, walking with him at Windsor, noticed that when they came on a dead garden bird his eyes filled with tears.

Rose, *King George V*

REAL TENNIS

'We played tennis. The King, Derek Keppel, Harry Verney and me and a pretty rotten game we had. It appears that Wigram and Willy Cadogan were accustomed to send easy ones over to the King. So when we played the usual game the King sulked and refused to try after the first set, he told us we didn't understand the game and we ought to send easy ones. I was furious as pat ball is such rot. So I proceeded to exaggerate this and lopt slow easy ones in a babyish way over the net. This annoyed the King who saw how absurd it must look and we had an altercation at the net . . . After some heat the King said all right play any way you like. So I then proceeded to smash them at him and he sulked and wouldn't move. Then an awkward pause after Derek and I won two sets. Then I asked him to try my way of cutting down properly and proposed he and I should play Harry Verney and Derek and give them 15. I really knew we could give them 30. So we started and had a capital game, the King cutting them down beautifully, they never had a look in as we were much the best. HM made some really beautiful strokes and it was a different game so all ended happily but I mean to tell Wigram he must not kowtow to the King in this way.'

A middle-aged monarch seeking relaxation from the problems of Home Rule might have wished so insolent a tutor elsewhere. In Sir Charles Cust he already had one candid courtier; two could be depressing. The King nevertheless saw beyond Ponsonby's provocative qualities, promoted him to be Keeper of the Privy Purse and Treasurer, heaped him with honours. Created Lord Sysonby in June 1935, he died in the King's service.

Ponsonby, *Recollections of Three Reigns*, quoting the author's letter to his wife

THE KING'S FAVOURITE *OBJET D'ART*

There was one object which he cherished: a silver-gilt statuette of Lady Godiva. The reason was that the short-sighted Queen Olga of the

Hellenes had once been heard to murmur as she peered at it: 'Ah, dear Queen Victoria.' Ibid.

MUSIC

One day Williams [the Bandmaster of the Grenadiers] made an arrangement of *Elektra* by Richard Strauss for his men and after months of practice they gave it for the first time at the changing of the Guard at Buckingham Palace. A personal message for Williams arrived afterwards from George V.

The note was brief and ran, 'His Majesty does not know what the Band has just played, but it is *never* to be played again.'

Sitwell, *Left Hand Right Hand*

One summer morning Mr Anthony Eden, on his way to Geneva, had an audience with the King at Buckingham Palace. The King's private apartments were then under repair, and Mr Eden was received in the North-East corner room, hung with relics of the Royal Pavilion, and situated immediately above the band-stand in the forecourt. The King, on entering, apologised to Mr Eden for having to receive him in this unfamiliar drawing-room. 'It is all right, however,' His Majesty added, 'I have told the band not to play till I give the word.' The King then furnished Mr Eden with a *catalogue raisonné* of all the subjects which, at Geneva, he would have to discuss. At last he reached a conclusion, and Mr Eden, in the few minutes that remained, started to make some observations of his own. 'Just one second,' said the King, as he rang the small gold hand-bell at his side. A page appeared. 'Tell the bandmaster that he can start playing now . . . You were saying . . .?' Harold Nicolson, *King George the Fifth* (1952)

DRAMA

Lady Diana Cooper played the non-speaking, star part in The Miracle *before the king and queen.*

After a performance which he and Queen Mary attended, I was sent for, as is the custom. In the Royal Box the King said that he had enjoyed it and asked how I managed to keep so still and all the expected questions. But my laurels wilted when instead of: 'Wonderful that you can express so much with gesture only,' he said: 'Of course, you've got no words to learn or say, and that's half the battle.' Diana Cooper, *Autobiography* (1979)

PICTURES

The king and queen visited the National Gallery when Kenneth (later Lord) Clark was director. The king's hope was to persuade Clark to take charge of the royal pictures at Buckingham Palace and Windsor:

We then came to the real purpose of his visit. He stopped his routine progress, faced me and said 'Why won't you come and work for me?' I replied 'Because I wouldn't have time to do the job properly.' He snorted with benevolent rage, 'What is there to do?' 'Well, sir, the pictures need looking after.' 'There's nothing wrong with them.' 'And people write letters asking for information about them.' 'Don't answer 'em.' And then, with great emphasis 'I want you to take the job.' As he was accustomed to addressing reluctant Prime Ministers (if such exist), Viceroys and Governors-General, the force of his command could not be resisted.

K. Clark, *Another Part of the Wood* (1974)

THE QUEEN'S DOLLS HOUSE

Queen Mary came several times to Mansfield Street [Lutyens' house in London] to see how it was getting on, and once the King came with her. They asked to be left alone with it—'to play with it', Father said. The Queen's favourite item was the miniature stamp album donated by Stanley Gibbons. On one embarrassing occasion she got her ear-ring caught in the beard of the engineer who was showing her that the lift and lavatory plugs really worked. Father had no hesitation in telling her some of his jokes which did not seem to shock her. Two tiny pillow cases had been made with M G embroidered on one and G M on the other. Father explained to her that these initials stood for 'May George?' and 'George May'. He also drew for her a picture of the King and Queen in bed with the caption 'Lazy Majesties', and another drawing showing a surprised hen looking over her shoulder to see a tiny King George just hatched from the egg she had laid. The caption to this was 'Lays Majesty'. One drawing he did not show her was of the King sitting on his crown as on a chamber-pot.

Mary Lutyens, *Edwin Lutyens* (1980)

ROYAL UNPUNCTUALITY

A family legend says that my Mother read the entire three volumes of Motley's *Dutch Republic* while waiting meals for Queen Alexandra her mother-in-law.

Windsor, 'My Hanoverian Ancestors'

THE KING AND THE BUS

I once had an amusing conversation with King George V during the First World War. I had no car or carriage, and as taxis were very scarce and in any case rather expensive for me, I used to travel by bus. I said: 'George, do you object to my going by bus?' He looked at me gravely for a moment and then said: 'What would Grandmama have thought! But I think you are quite old enough to travel by bus. Do you strap-hang?'

Princess Marie-Louise, *My Memories of Six Reigns*

Once when Derek Keppel [a courtier] was seen entering the Palace in a bowler hat during the London Season, the King assailed him roughly: 'You scoundrel, what do you mean by coming in here in that rat-catcher fashion? You never see me dressing like that in London.'

'Well, Sir,' Keppel replied, 'you don't have to go about in buses.'

'*Buses!* Nonsense!' Gore, *King George the Fifth*

A FAVOURITE AFTER-DINNER STORY

The King, who loved to hear his favourite stories repeated, would again and again ask Lord Louis Mountbatten to describe the visit of his sister, Crown Princess (later Queen) Louise of Sweden to Uppsala Cathedral. The Archbishop, determined to show off his knowledge of English, approached a chest of drawers in the sacristy with the startling announcement: 'I will now open these trousers and reveal some even more precious treasures to Your Royal Highness.' Rose, *Kings, Queens and Courtiers*

ROYAL PROPRIETY

The king and queen stayed with Lord Derby in 1921.

The royal party were spared the impropriety of watching a nature film in which a duck laid an egg. The egg episode was cut out. Ibid.

Queen Mary on the war-path:

At Ascot one year she ordered a sporting peeress to be ejected from the Royal Enclosure for wearing a sailor's cap with the legend in gold lettering, 'HMS Good Ship Venus'. Ibid.

ANALOGIES FOR QUEEN MARY

Her appearance was formidable, her manner—well, it was like talking to St Paul's Cathedral.

Chips: Diaries of Sir Henry Channon, ed. Robert Rhodes James (1967)

The novelist E. M. Forster, portly but myopic, bowed to the cake at [Lord Harewood's] wedding reception, thinking it was Queen Mary.

Rose, *Kings, Queens and Courtiers*

KEEPING IT IN THE FAMILY

It was typical of her [Queen Mary's] practicality and devotion to the family that she left her great collection of antiques and jewellery to the Queen [Elizabeth II] with nothing to any individual, although of course she expected things to be shared out a bit, as they were. I remember a courtier's comment: 'Poor Cynthia Colville [her longest-serving lady-in-waiting]! Thirty years of devotion and not even a toque to show for it!'

Princess Marie-Louise, *My Memories of Six Reigns*

SAYING OF QUEEN MARY

On being shown a haystack when evacuated to Badminton in 1939: 'So that's what hay looks like'.

THE ROYAL PHILISTINE

[Thomas Beecham the conductor] told me that George V went to the opera once a year—always to *La Bohème*. Once Beecham asked him if it was his favourite. 'Yes,' said the King. 'That's most interesting, Sir. I'd be most interested to know why.' 'Because it's much the shortest', said His Majesty.

The Lyttelton Hart-Davis Letters, Correspondence of George Lyttelton and Rupert Hart-Davis, II, ed. Rupert Hart-Davis (1979)

MUCH FORGIVEN TO GEORGE V

George Lyttelton wrote to Rupert Hart-Davis:

I shall not read abour dear Queen Mary, who (oddly?) does not interest me, any more than her second-rate son (eldest) did. She and the bearded saint, as Roger Fulford called him, must have been very indifferent parents. But much may be forgiven him for (a) when asked what film he would like to see when convalescing, announcing 'Anything except that damned Mouse' and (b) when the footman, bringing in the early morning royal tea, tripped and fell with his load and heard from the pillow 'That's right; break up the whole bloody place'. The old autocratic touch.

Ibid., IV (1982)

'THAT DAMN CHILD . . .'

After King George's dangerous illness in 1931 he was terrified of catching a cold. The Harewoods, including their elder son, his grandson George, came to see him. George Harewood wrote:

I had started to get hay fever and at the end of April, as we went to say goodbye to him in his sitting room after breakfast, I started to sneeze, either from the pollinating grass or sheer nerves, and no amount of assurance that I had hay fever could stop the shouts of 'Get that damn child away from me', which made a rather strong impression on an awakening imagination.

The Tongs and the Bones: The Memoirs of Lord Harewood (1981)

KING GEORGE AND CHARLOTTE

His love for his parrot Charlotte was a legend, its privileged existence occasionally a trial to his Household. Charlotte travelled with him whenever it was possible. At Sandringham every morning, punctual to the second, the King came in to breakfast with Charlotte on his finger, and she ranged over the breakfast-table, messing things up to her heart's content.

Gore, *King George the Fifth*

Kenneth Rose adds in his book King George V *that if she actually made a mess, the king slid the mustard pot over it.*

GEORGE V AND HIS SONS

I [Monckton] was sitting with [King George VI] in his room at Balmoral Castle when he told me that this was the only room in which he had real discussions with his father. He added how difficult it was to have serious talks with him. He said that his father had always treated his brothers as if they were all the same whereas in fact they were totally different in character. He then said: 'It was very difficult for David. My father was so inclined to go for him. I always thought that it was a pity that he found fault with him over unimportant things—like what he wore. This only put David's back up. But it was a pity that he did the things which he knew would annoy my father. The result was that they did not discuss the important things quietly. 'I think' he added, 'that is why David did not tell him before he died that he meant to marry.'

Lord Birkenhead, *Walter Monckton* (1969)

—AND HIS GRANDDAUGHTER, 1927

From this time forward references to 'sweet little Lilibet' grow more and more frequent. He loved to play with her the games of childhood. Archbishop Lang recalled an occasion when Princess Elizabeth was the groom and the King played the part of horse. The archbishop saw . . . the King-Emperor shuffling on hands and knees along the floor, while the little Princess led him by his beard.

Gore, *King George the Fifth*

—AND HIS SISTER PRINCESS VICTORIA

The death of his deeply loved sister Victoria in December 1935 broke him up. 'How I shall miss her,' he wrote, '& our daily talks on the telephone.' One of these daily talks had begun with the Princess ringing up her brother and saying affectionately, 'Is that you, you old fool?' The operator interrupted: 'Beg pardon, your Royal Highness, His Majesty is not yet on the line.'

Longford, *The Royal House of Windsor*

SAYINGS OF KING GEORGE

On looking bored or cross at public functions: 'We sailors never smile on duty.'

On H. G. Wells's remark about 'an alien and uninspiring court': 'I may be uninspiring, but I'll be damned if I'm an alien.'

To a lady whose hairpin fell into the soup: 'Did you come here expecting to eat winkles?'

To a very small child who said she was Ann Peace Arabella Mackintosh of Mackintosh: 'Ah, I'm just plain George.'

On the queen's table-talk: 'There you go again, May—always furniture, furniture, furniture!'

On the painter Turner: 'I tell you what, Turner was mad. My grandmother always said so.'

On the first Labour government: 'My grandfather would have hated it; my father could hardly have tolerated it; but I march with the times.'

On authors: 'People who write books ought to be shut up.'

On recovering from his illness: 'Yes I'm pretty well again—but not well enough to walk with the queen round the British Industries Fair.'

A SAYING FROM THE COURT CIRCLE

A man was once presented to George V.
The King: 'I think we have met before.'
Man: 'I don't think so Sir.'
Afterwards the king commented:
'A very nice fellow but he'd never make a good courtier.'

VERSIONS OF THE KING'S DYING WORDS

It has been alleged that when he again lay gravely ill seven years later, one of his doctors sought to soothe a restless patient with a whispered, 'Cheer up, Your Majesty, you will soon be at Bognor again.' To this the King is said to have replied, 'Bugger Bognor', and instantly expired. The tale carries a certain plausibility. The King was always emphatic in his language, not least when being fussed by his medical advisers. There is, however, a happier variant of the legend which rests on the authority of Sir Owen Morshead, the King's Librarian. As the time of the King's departure from Bognor drew near, a deputation of leading citizens came to

Craigwell to ask that their salubrious town should henceforth be known as Bognor Regis. They were received by Stamfordham [private secretary], who, having heard their petition, invited them to wait while he consulted the King in another room. The sovereign responded with the celebrated obscenity, which Stamfordham translated for the benefit of the delegation. His Majesty, they were told, would be graciously pleased to grant their request.

Rose, *Kings, Queens and Courtiers*

11 a.m. 20 January: He murmured something about the Empire, and I replied that 'all is well, Sir, with the Empire'.

F. Watson, *Dawson of Penn* (1950), quoting a memorandum of Lord Wigram, the king's private secretary

12 noon Council meeting: For 10 minutes or more the King was unable to sign with either hand. Lord Dawson kneeling at his side proposed first the right hand and then the left. HM then remarked to Lord Dawson 'You don't want me to sign with both hands?'

F. Watson, 'The Death of George V', *History Today* (December 1986)

THE SUPPRESSED STORY OF THE DEATH OF GEORGE V, 20 JANUARY 1936

'The life of the King is moving peacefully to its close.'

Thus the famous bulletin drawn up on a menu card at dinner by Lord Dawson of Penn, the royal physician. Though 'beautifully' written, as John Gore the king's biographer pointed out, it did not tell the whole story. For the peaceful close to the king's life was to be accelerated by the doctor himself. The full story was told in 1986 by Francis Watson, Lord Dawson's biographer, thirty-six years after the official biography. In his Sandringham notebook, Dawson described how he resorted to euthanasia, not to spare his patient pain—the king was already in a coma—but for the sake of the assembled family—and the morning papers. The announcement would appear more 'appropriately' in their columns than in the 'evening journals'.

At about 11 o'clock [at night] it was evident that the last stage might endure for many hours, unknown to the Patient but little comporting with that dignity and serenity which he so richly merited and which demanded a brief final scene. Hours of waiting just for the mechanical end when all that is really life has departed only exhausts the onlookers and keeps them so strained that they cannot avail themselves of the solace of thought, communion or prayer. I therefore decided to determine the end and injected (myself) morphia gr. 3/4 and shortly afterwards cocaine gr. 1 into the distended jugular vein: 'myself' because it was obvious that Sister B.

[Catherine Black, the King's devoted nurse since his first grave illness] was disturbed by this procedure. In about an hour—breathing quieter—appearance more placid—physical struggle gone.

Then the Queen and family returned and stood round the bedside—the Queen dignified and controlled—others with tears, gentle but not noisy. Intervals between respirations lengthened, and life passed so quietly and gently that it was difficult to determine the actual moment. Ibid.

Nevertheless the time was given as five minutes before midnight.

Postscript by the most recent biographer of George V:

Margot Asquith, the widow of George V's first prime minister and a woman addicted to outrageous flights of fancy, used to say in old age: 'The King told me he would never have died if it had not been for that fool Dawson of Penn.'

The unveiling of the deathbed secrets of January 20, 1936, invests her with awesome insight. Kenneth Rose, *Sunday Telegraph*, 1986

UNVEILING GEORGE V'S STATUE, 1947

The ceremony itself was over in twenty minutes, but then followed that interminable pause whilst the Royalties greeted each other, interkissed and chatted. It is only in England that a crowd of several thousands can stand happily in the rain and watch one family gossip.

Chips

Edward VIII

1936

A 'Year of Three Kings' had come round again, the last one being 1483. Edward Prince of Wales, called David in the family, eldest son of George V and Queen Mary, seemed all set to be a welcome contrast to his somewhat archaic father. Outgoing, up-to-date, wildly popular on both sides of the Atlantic, Prince Charming lacked nothing but the legendary Princess. Towards the end of his eleven months' reign, however, the majority of the public were ready for a return to Georgian 'stern duty' in place of Edwardian charm.

There had been no abdication since that of Richard II. In 1399 the change of rulers had involved a total rejection of the monarch's style. In 1936 it involved the rejection only of a twice-divorced woman as queen. Perhaps for that reason the effect on the monarchy was minimal—paperback romance not hardback history.

CHILDHOOD

Queen Mary's maternal instincts were fairly rudimentary. She saw her baby son David only twice a day and these visits were ruined by a psychotic nurse. Years later her son remembered:

Before carrying me into the drawing-room, this dreadful 'Nanny' would pinch and twist my arm—why, no one knew, unless it was to demonstrate, according to some perverse reasoning, that her power over me was greater than that of my parents. The sobbing and bawling this treatment invariably evoked understandably puzzled, worried, and finally annoyed them. It would result in my being peremptorily removed from the room before further embarrassment was inflicted upon them and the other witnesses of this pathetic scene. Eventually, my mother realised what was wrong, and the nurse was dismissed. Windsor, *A King's Story*

But visits to his grandfather, Edward VII, were sunlit memories.

Prince David was so little afraid of him, in fact, that he was even capable, on one occasion at least, of interrupting his conversation at table. He was reprimanded, of course, and sat in silence until given permission to speak. 'It's too late now, grandpapa,' Prince David said unconcernedly. 'It was a caterpillar on your lettuce but you've eaten it.' Hibbert, *Edward VII*

A TASTE OF THE NAVY—

Within a day or two of their return several sixth, or senior, termers decided that Cadet Prince Edward would look much better with his fair hair dyed red. So one evening, before 'quarters' (evening parade), I was cornered by my betters and made to stand at attention while one of them poured a bottle of red ink over my head. The ink dropped down my neck, ruining one of the few white shirts that I possessed; a moment later the bugle sounded for quarters, and the sixth termers dashed away to fall in their ranks, leaving me in a terrifying dilemma for which nothing that I had ever learnt under Mr Hansell [his tutor] seemed to supply a solution.

Windsor, *A King's Story*

—AND OF THE ARMY

I had a very small distinction of my own in 1917: I was the first Prince of Wales to step upon the battlefield of Crécy since the Black Prince (also an Edward) fought there in 1346. Windsor, 'My Hanoverian Ancestors'

THE FIRST TRUE LOVE

The soldier-prince, on leave in March 1918, was attending a party in Belgrave Square when the air-raid warning sounded. He and the other guests trooped down to the cellar, where he met two strayed revellers who had taken shelter in the house and been conducted to the cellar by Mrs Kerr-Smiley, the hostess. The uninvited guests were Mrs Freda Dudley-Ward and her escort, Buster Dominguez.

It was at this moment that in the semi-darkness a young man appeared at Mrs Dudley-Ward's side and started an animated conversation with her. He asked her where she lived and she replied for the moment at her mother-in-law's house in London, and asked him in return where he lived, and he said in London, too, and sometimes at Windsor. When the air raid was over Mrs Dudley-Ward and her escort tried to leave, but Mrs Kerr-Smiley came over to her and invited her and her escort to come upstairs and join the party.

'His Royal Highness is so anxious that you should do so,' she said.

So Mrs Dudley-Ward went upstairs and danced with the Prince of Wales until the early hours of the morning, when he took her home, Buster Dominguez having at some time disappeared for ever into the night. The next day the Prince of Wales called on Mrs Dudley-Ward and, after a Cinderella-like sequence in which he established which Mrs Dudley-Ward it was he wished to see, there began a relationship which was to last for sixteen years.

The coincidence of their having met in an air raid is matched by one equally improbable. The Mrs Kerr-Smiley in whose house this meeting occurred was the sister of that same Ernest Simpson who would figure so largely in the life of Edward VIII. Frances Donaldson, *Edward VIII* (1974)

THE ROLE OF PRINCE

Edward once asked Sir Frederick Ponsonby, a courtier, how he, Edward, was shaping as Prince of Wales:

'If I may say so, Sir, I think there is risk in making yourself too accessible,' he answered unhesitatingly.

'What do you mean?' I asked.

'The Monarchy must always retain an element of mystery. A Prince should not show himself too much. The Monarchy must remain on a pedestal.'

I maintained otherwise, arguing that because of the social changes brought about by the war, one of the most important tasks of the Prince of Wales was to help bring the institution nearer the people.

'If you bring it down to the people,' Fritz Ponsonby said coldly, 'it will lose its mystery and influence.'

'I do not agree,' I said. 'Times are changing.'

He replied severely, 'I am older than you are, Sir; I have been with your father, your grandfather, and your great-grandmother. They all understood. You are quite mistaken.' Windsor, *A King's Story*

THE TOUCHING MANIA

The unofficial diary kept by my staff . . . recorded in Melbourne: 'Confetti is appearing in great and unpleasant quantities, and the touching mania has started, only owing to the hearty disposition of the Australians the touches are more like blows and HRH and the Admiral arrived half blinded and black and blue.'

The 'touching mania', one of the most remarkable phenomena connected with my travels, took the form of a mass impulse to prod some part of the Prince of Wales. Whenever I entered a crowd, it closed around me like an octopus. I can still hear the shrill, excited cry, 'I touched him!' If I were out of reach, then a blow on my head with a folded newspaper appeared to satisfy the impulse. Ibid.

EXAMPLE OF THE PRINCE SUDDENLY TURNING 'ROYAL'

On several occasions the Prince of Wales went up to Oxford for a day or two in after years. Once on entering the Junior Common Room at Magdalen, he bade everybody be seated, telling them he was a Magdalen man and did not wish to be treated ceremoniously but as a member of the

College. The next time he went into the JCR nobody stirred, and he asked sharply if that was the way to treat the heir to the Throne.

Compton Mackenzie, *Windsor Tapestry* (1938)

'A GREAT BOY' IN THE UNITED STATES

The prince asked that Will Rogers, the comedian, should be invited to a party given by the Piping Rock Club. Afterwards Rogers told interviewers:

Why yes he's a great boy, no kidding. And at that dinner! . . . You know at a dinner like that where there's a great man present the people always watch him. And whether your stuff goes over or not depends a good deal on how he takes it . . .

New York Times, 7 September 1924

He was sitting on my right and he'd think up gags for me to spill. And laugh—Gosh. He'd just double up. And every time the Prince would laugh, everyone in the house would laugh . . . The Prince is a good kid. Too bad I can't afford to carry a guy like that around with me. I'd have a swell act if I did.

New York Herald Tribune, 7 September 1924

THE PRINCE AND RELIGION

His confidante for the evening was the famous Socialist Mrs Beatrice Webb, wife of Sidney Webb, Lord Passfield. (*Beatrice did not take the title.*) *Sidney was an agnostic though Beatrice had religious leanings.*

An informal dinner at York House—my first introduction into the Fort circle. The Prince, having devoted himself at dinner to the young Countess (Minto) and the middle-aged Duchess (Abercorn), settled down afterwards by the aged Baroness (Passfield) and opened out into an oddly intimate talk about his religious difficulties.

'What do you really believe, Mrs Webb?' he asked in an agitated tone. (I was there as Lady Passfield.) He is a neurotic and takes too much alcohol for health of body or mind. If I were his mother or grandmother I should be very nervous about his future. He clearly dislikes having to go to the Anglican Church, but whether he has leanings to Catholicism or is becoming an unbeliever there was not time to explore . . . I felt sorry for the man; his expression was unhappy—there was a horrid dissipated look as if he had no settled home either for his intellect or his emotions. In his study there were two pictures of the Queen, one over the mantelpiece and the other on his desk, but no symbol of the King. On one side of the wall hung a huge map of the world; on another side there were shelves filled with expensively bound library editions, obviously never read—there were no books obviously in general use. Like all those royal suites of apartments, there was no homeliness or privacy—the rooms and their trappings were all designed for company and not for home life.

But it was the unhappiness of the Prince's expression, the uneasy restlessness of his manner, the odd combination of unbelief and hankering after sacerdotal religion, the reactionary prejudice about India and the morbid curiosity about Russia revealed in his talk that interested me. The Anglican Church, whose services he said he 'had to attend', he clearly resented. He must be a problem to the conventional courtiers who surround him! Will he stay put in his present role of the most popular heir-apparent in British history? As I talked to him he seemed like a hero of one of Shaw's plays; he was certainly very unconventional in his conversation with a perfect stranger. Was it the Dauphin in *St Joan* or King Magnus in *The Apple Cart* that ran in my head? Not so mean as the first, not so accomplished as the second of GBS's incarnations of kingship!

The Diary of Beatrice Webb, IV, July 1930

The First Meeting with Wallis Simpson?

With the aid of Wallis's letters to her Aunt Bessie, it is now at last possible to establish the exact date of her first meeting with the Prince of Wales. Writing in the 1950s, the Duchess of Windsor placed the event in November 1930; the Duke thought it might have been in the autumn of 1931. Both their memories were at fault. The fateful encounter took place at Burrough Court, Lady Furness's country house at Melton Mowbray in Leicestershire, on Saturday, 10 January 1931.

Wallis & Edward: Letters 1931–1937, ed. Michael Bloch (1986)

WALLIS'S FIRST LETTER TO HER AUNT BESSIE AFTER THE MEETING

Tuesday, Jan. 13th. I never finished the letter of Thursday, January 8th. On Friday I got up and spent the entire day on hair and nails etc as Saturday we were going to Melton Mowbray to stay with Lady Furness [Thelma, the Prince's American mistress] . . . and the Prince of Wales was also to be a guest. In spite of cold [Wallis had a feverish cold but her husband Ernest insisted on their going] we took the 3:20 train . . . arrived at 6.30 and the Prince & Thelma Furness came about 7.30 . . . you can imagine what a treat it was to meet the Prince in such an informal way. There was no dinner party Sat night but Sunday she had 10 for dinner, Prince George [Edward's brother] returning for that . . . It was quite an experience and as I've had my mind made up to meet him ever since I've been here I feel relieved. I never expected however to accomplish it in such an informal way and Prince George as well. *Wallis & Edward*, Wallis to Aunt Bessie

EDWARD'S ACCOUNT OF THE FIRST MEETING

It was one of those week-ends for which our winters are justly infamous—cold, damp, foggy. Mrs Simpson did not ride and obviously had no interest in horses, hounds, or hunting in general. She was also plainly in misery from a bad cold in the head. Since a Prince is by custom expected to take the lead in conversing with strangers . . . I was prompted to observe that she must miss central heating, of which there was a lamentable lack in my country and an abundance in hers . . . a verbal chasm opened under my feet. Mrs Simpson did not miss the great boon that her country had conferred upon the world. On the contrary, she liked our cold houses. A mocking look came into her eyes. 'I'm sorry, Sir,' she said, 'but you have disappointed me.'

'In what way?'

'Every American woman who comes to your country is always asked the same question. I had hoped for something more original from the Prince of Wales.' Windsor, *A King's Story*

Within two years, Wallis had achieved 'something more original', a close friendship with the prince. But she was still assuring Aunt Bessie that she did not intend to let down either her friend Thelma or her husband Ernest:

Sunday, Feb. 18th [1934] . . . PS It's all gossip about the Prince. I'm not in the habit of taking my girlfriends' beaux. We are around together a lot and of course people are going to say it. I think I do amuse him. I'm the comedy relief and we like to dance together—but I always have Ernest hanging around my neck so all is safe.

Four days later:

Dearest Aunt B: I understand from PW [Prince of Wales] that Thelma is sailing between 15th and 25th of March . . . *Wallis & Edward*

Just before Thelma Furness sailed for America she wrote in her memoirs that Wallis suddenly said to her:

'Oh, Thelma, the little man is going to be so lonely.'

Double Exposure, a Twin Autobiography of Gloria Vanderbilt and Thelma Lady Furness (1959)

When Thelma returned Wallis had taken over the 'little man'.

LORD BEAVERBROOK'S FIRST MEETING WITH MRS SIMPSON

I was greatly interested by the way the other women greeted her. There were about six women who were present at the dinner or who came in afterwards. All but one of them greeted Mrs Simpson with a kiss. She received it with appropriate dignity, but in no case did she return it.

Lord Beaverbrook, *The Abdication of King Edward VIII* (1966)

MRS BELLOC LOWNDES'S FIRST MEETING

Several of my fellow-guests asked me what I thought of her. I said what had struck me most were her perfect clothes and that I had been surprised, considering that she dressed so simply, to see that she wore such a mass of dressmakers' jewels. At that they all screamed with laughter, explaining that all the jewels were real, that the then Prince of Wales had given her fifty thousand pounds' worth at Christmas, following it up with sixty thousand pounds' worth of jewels a week later at the New Year. They explained that his latest gift was a marvellous necklace which he had bought from a Paris jeweller. *Diaries and Letters of Marie Belloc-Lowndes* (1971)

BEING CLEVER

Monday, Nov 5th 1934 . . . Don't listen to such ridiculous gossip. E and myself are far from being divorced and have had a long talk about PW and myself and also one with the latter and everything will go on just the same as before, namely the 3 of us being the best of friends which will probably prove upsetting to the world as they would love to see my home broken up I suppose. I shall try and be clever enough to keep them both. E is away.

Wallis & Edward, Wallis to Aunt Bessie

But Wallis's cleverness was no match for the prince's growing infatuation. While taking part in a naval review in 1935, he was suddenly moved to write to her from on board ship in the middle of the night, 'WE' (Wallis Edward) being their kind of logo.

Tuesday [23 July], one o'clock a.m. HMS Faulknor

Wallis—A boy is holding a girl so very tight in his arms tonight. He will miss her more tomorrow because he will have been away from her some hours longer and cannot see her till Wed-y night. A girl knows that not anybody or anything can separate WE—not even the stars—and that WE belong to each other for ever. WE love [twice underlined] each other more than life so God bless WE. Your [twice underlined]

David

Ibid., Edward to Wallis

THE PRINCE AT 'THE FORT'

Prince Edward entertained at his weekend folly, Fort Belvedere. In 1935 the Duff Coopers were guests and Lady Diana wrote a descriptive letter to a friend:

This stationary is disappointingly humble—not so the conditions. I am in a pink bedroom, pink-sheeted, pink venetian-blinded, pink-soaped, white-telephoned and pink-and-white maided . . .

We arrived after midnight (perhaps as chaperones). Jabber and beer and bed was the order. I did not leave the 'cabin's seclusion' until 1 o'clock, having been told that no one else did. HRH was dressed in plus-twenties with vivid azure socks. Wallis admirably correct and chic.

The Prince changed into a Donald tartan dress-kilt with an immense white leather purse in front, and played the pipes round the table after dinner, having first fetched his bonnet. We 'reeled' to bed at 2 a.m. The host drinks least.

Cooper, *Autobiography*

WALLIS AT KING GEORGE V'S SILVER JUBILEE BALL, 1935

After the King and Queen had made their entrance and taken their seats on the dais at the end of the room, the dancing began. As David and I danced past, I thought I felt the King's eyes rest searchingly on me. Something in his look made me feel that all this graciousness and pageantry were but the glittering tip of an iceberg that extended down into unseen depths I could never plumb, depths filled with an icy menace for such as me.

The Duchess of Windsor, *The Heart Has its Reasons* (1956)

THE NIGHT THAT GEORGE V DIED

We paced the pavement for ten minutes as the anxious crowd thickened . . . The new bulletin was not encouraging—'increased cardiac of the heart' . . . My heart goes out to the Prince of Wales tonight, as he will mind so terribly being King. His loneliness, his seclusion, his isolation will be almost more than his highly strung and imaginative nature can bear. Never has a man been so in love . . . How will they re-arrange their lives, these people?

Chips

OMINOUS

George V's coffin travelled from King's Cross station to Westminster.

The Royal Crown had been taken from its glass case in the Tower and secured to the lid of the coffin over the folds of the Royal Standard. The jolting of the heavy gun-carriage must have caused the Maltese cross on the top of the crown—set with a square sapphire, eight medium-sized diamonds and one hundred and ninety-two smaller diamonds—to work loose. At the very moment the small procession turned into the gates of Palace Yard, the cross rolled off and fell on the road. Two Members of Parliament—Walter Elliot and Robert Boothby—stood on the pavement watching the procession. As a company sergeant-major, bringing up the rear of the two files of Grenadier Guardsmen flanking the carriage, bent down and in a swift movement picked up the cross and dropped it into his pocket, they heard the King's voice say: 'Christ! what will happen next?' 'A fitting motto', Walter Elliot remarked to his companion, 'for the coming reign.'

Donaldson, *Edward VIII*

KING EDWARD VIII'S PROFILE ON STAMPS

Successive issues of English coins traditionally show the successive monarchs facing left or right, alternately. It was George IV's turn to face right, and mine to face right; nevertheless I insisted on facing left, my left profile seeming somewhat more photogenic. As things turned out it appeared only on postage stamps; there were no coins of my reign.

Windsor, 'My Hanoverian Ancestors'

MRS SIMPSON AT A PARTY

12 February 1936 [shortly after George V's death] . . . to tea with the Brownlows [he was Edward's personal lord-in-waiting]. There we found assembled the 'new Court', Mrs Simpson very charming and gay and vivacious. She said she had not worn black stockings since she gave up the Can-Can.

Chips

KING EDWARD VIII'S CONCEPTS OF LITERATURE

I recall Lady Desborough telling me that on three separate occasions, at intervals of some years, on which, as Prince of Wales, King Edward VIII had gone to stay with her, she had been surprised, cumulatively surprised, when, at dinner on the first evening of each visit, he had treated her to the same opening sentences, obviously prepared beforehand and carefully calculated to interest, and to ingratiate himself with, his hostess whom he knew to be a discriminating amateur of literature. Like a refrain, these words came back to haunt her, nor, she assured me, could they have been due to a misplaced sense of humour on the part of the Prince: he was too serious in his enthusiasm.

'Lady Desborough, I know you're a bookish sort of person. At the moment, I'm reading *such* an interesting novel. I think it would appeal to you: it's called *Dracula*!'

Osbert Sitwell, *Rat Week: An Essay on the Abdication* (1986)

THE KING'S CONCERN WITH MONEY

One day King Edward sent for his head house-maid and asked her what happened to the guests' soap after they had left the house. She replied that it was taken to the servants' quarters and finished there. The King instructed her in future to bring it to his rooms for his own use.

There is an odd little pendant to this story. One evening, more than twenty years later, the Duke of Windsor confided in amused tones to a guest at dinner in his house in Paris that his wife had a strange little foible. All soap from the guests' rooms was gathered up, he said, and taken to her room where she used it up herself. Donaldson, *Edward VIII*

WALLIS AT ROYAL LODGE

I had seen the Duchess of York [wife of the king's brother Bertie] before on several occasions at the Fort and at York House. Her justly famous charm was highly evident. I was also aware of the beauty of her complexion and the almost startling blueness of her eyes. Our conversation, I remember, was largely a discussion of the merits of the garden at the Fort, and that of Royal Lodge. We returned to the house for tea, which was served in the drawing-room. In a few moments the two little Princesses joined us . . . They were both so blonde, so beautifully mannered, so brightly scrubbed, that they might have stepped straight from the pages of a picture book . . . David and his sister-in-law carried on the conversation with his brother throwing in only an occasional word. It was a pleasant hour; but I left with a distinct impression that while the Duke of York was sold on the American station wagon, the Duchess was not sold on David's other American interest. Wallis Windsor, *The Heart Has its Reasons*

A PHOTOGRAPH OF THE KING CARRYING AN UMBRELLA

Did you see that newspaper photograph of His Majesty walking from the Palace in the rain?' asked Wallis's neighbour, an MP, at dinner.

Wallis had of course seen it and was about to remark that she had thought it natural and amusing when her neighbour exclaimed with a visible shudder, 'That umbrella! Since you know the King, won't you ask him to be more careful in the future as to how he is photographed?' His undoubted disapproval took her aback. After all, what could be the harm in the King's using an umbrella . . . However, the Member seemed so perturbed that she suppressed a temptation to make light of the matter. Instead she countered by suggesting that it would be presumptuous of her, an American, to advise the King of England upon a point of behaviour. The man seemed not to hear. 'The Monarchy must remain aloof and above the commonplace. We can't have the King doing this kind of thing. He has the Daimler.'

Windsor, *A King's Story*

A MUSICAL PARTY FOR THE KING

It did not go well from the start but the real misfortune came after dinner, when Sibyl [Lady Colefax] had persuaded Arthur Rubinstein to play the piano. He announced that he would play the Barcarolle, meaning, of course, the work of Chopin. As the piece proceeded the King looked bewildered, then irritated, and finally said 'That isn't the one we like': he was thinking of the popular intermezzo in *The Tales of Hoffmann*. Rubinstein then played a Chopin Prelude. The King looked even more irritated, and at the end rose to leave. It was 10.15. Consternation. Sibyl on the verge of tears . . . Arthur Rubinstein, seeing that he was no longer needed, returned to the dining-room, and philosophically consumed the whisky and soda that he had denied himself before. We accompanied him, filled with anger and humiliation. By this time the King had reached the front door, but by good fortune Mr Churchill was arriving, which delayed the royal departure. At this moment there came from the drawing room the strains of 'Mad Dogs and Englishmen': Nöel [Coward], like the kind man he was, had put his artistic scruples in his pocket in order to save his friend's evening. And it *was* saved. The King returned to the drawing room and stayed, I believe, to a late hour. We went out with Arthur Rubinstein, found a taxi and took him back to his hotel.

Clark, *Another Part of the Wood*

CRUISE OF *THE NAHLIN*, 17 JUNE 1936

Lady Diana Cooper was at the party:

No sooner was the yacht sighted than the whole village turned out—a million children and gay folk smiling and cheering. Half of them didn't know which the King was and must have been surprised when they were told. He had no hat (the child's hair gleaming), *espadrilles*: the same little shorts and a tiny blue-and-white singlet bought in one of their own villages.

The King walks a little ahead talking to the Consul or Mayor, and we follow adoring it. He waves his hand half-saluting. He is utterly himself and unselfconscious. That I think is the reason why he does some things (that he likes) superlatively well. He does not *act*. In the middle of the procession he stopped for a good two minutes to tie up his shoe. There was a knot and it took time. We were all left staring at his behind. You or I would have risen above the lace, wouldn't we, until the procession was over? But it did not occur to him to wait, and so the people said: 'Isn't he human! Isn't he natural! He stopped to do up his shoe like any of us!'

Cooper, *Autobiography*

WALLIS'S ATTEMPT TO END IT

Thursday [14 October]
My dear

This is really more than you or I bargained for—this being haunted by the press. Do you feel you still want me to go ahead as I feel it will hurt your popularity in the country. Last night I heard so much from the Hunters that made me shiver—and I am very upset and ill to-day from talking until 4. It nearly ended in a row as naturally it wasn't pleasant things I heard of the way the man in the street regards me. I hear you have been hissed in the cinema, that a man in a white tie refused to get up in the theater when they played God save the King and that in one place they added and Mrs Simpson. Really David darling if I hurt you to this extent isn't it best for me to steal quietly away. Today Ernest called up to say he was deluged with cables from the US press and also that it had been broadcast in America last night . . . We can never stop America but I hope we can get small announcements after it is over from Beaverbrook which will be your Friday's job should we decide to go ahead. I can't help but feel you will have trouble in the House of Commons etc and may be forced to go. I can't put you in that position. Also I'm terrified that this judge here will lose his nerve [over her divorce from Ernest]—and then what? I am sorry to bother you my darling—but I feel like an animal in a trap and these two buzzards [the Hunters, her best friends] working me up over the way you are losing your popularity—through me. Do please say what you think best for all concerned when you call me after reading this. Together I suppose we are strong enough to face this mean world—but separated I feel eanum and scared for you, your safety etc. Also the Hunters say I might easily have a brick thrown at my car. Hold me tight please David.

Wallis & Edward, Wallis to Edward, 14 October 1936

'EANUM?'

Edward and Wallis had a simple lovers' language but no one knows what they meant by the favourite word 'Eanum'. Perhaps 'lonely'.

Hello my sweetheart. How are you? Missing a boy I hope. Here is the card for Kitty Brownlow's flowers. Hurry here please as a boy is longing to see a girl and will be all set and waiting at five thirty. Eanum?

David says more and more. Ibid.

The king discussed his coronation with Dr Lang, Archbishop of Canterbury, who told his biographer about the interview:

I noted at the time—and the facts seem strangely significant now—that he summoned his brother [Bertie] to be present [during interview with Lang], and when . . . I gave him a book of the Service as used at his father's Coronation he gave it to his brother, saying, 'I think *you* had better follow this.' I wonder whether even then he had in the back of his mind some thought that the Coronation might not be his, but his brother's.

J. G. Lockhart, *Cosmo Gordon Lang* (New York, 1940)

INNOVATION AT BALMORAL

My contribution to the traditional grandeur of Balmoral was the introduction of the three-decker toasted sandwich as a late supper item, after the movies. This proved so popular that it created a minor crisis in the kitchen through the heavy demand for repeat orders. I am sure that this innovation, so patently mine, hardly endeared the new reign to the household staff.

Wallis Windsor, *The Heart Has its Reasons*

The king's famous speech to the unemployed in South Wales—'Something must be done'—lowered his reputation afterwards as much as it had raised it at the time.

'These works brought all these people here. Something must be done to find them work.' And next day 'You may be sure that all I can do for you I will.' All he could do for them was to abdicate three weeks later.

Donaldson, *Edward VIII*

WALLIS'S INFLUENCE

5 November 1936: The King's attention to Wallis was very touching. He worships her, and she seems tactful and just right with him, always prefacing her gentle rebukes with 'Oh, Sir . . .' She confessed . . . that she always kicks him under the table hard when to stop and gently when to go on. Sometimes she is too far away and then it is difficult.

Chips

TROUBLE IN PARLIAMENT

10 November 1936: During questions, someone asked, innocuously, about the coming coronation. McGovern [Independent Labour] jumped up and shouted, 'Why bother, in view of the gambling at Lloyd's that there will not be one?' There were roars of 'Shame! Shame!' and he called out, 'Yes . . . Mrs Simpson.' This was the first time her name has been used in the House of Commons, although the smoking room and lobbies have long buzzed with it.

Ibid.

INCIDENTS THAT CAUSED THE CRISIS

11 November 1936: . . . the situation is extremely serious and the country is indignant; it does seem foolish that the monarchy, the oldest institution in the world after the Papacy, should crash, as it may, over dear Wallis. Yet why should we forsake our Sovereign? He has been foolish, indeed almost brazen. The Mediterranean [*Nahlin*] cruise was a Press disaster, the visit to Balmoral was a calamity, after the King chucked opening the Aberdeen Infirmary [on the ground that he was still in mourning for his father] and then openly appeared at Ballater railway station on the same day, to welcome Wallis to the Highlands. Aberdeen will never forgive him. The Simpson divorce has caused all this talk, and the American newspapers have had a Roman holiday. The headline in one referring to the Ipswich divorce, ran, 'The King's Moll. Reno'd in Wolsey's Home Town.' A pleasanter tale is of Wallis taking a taxi on her now famous journey to Scotland. 'King's Cross,' she is reported to have said. 'I'm sorry, lady,' answered the driver.

Ibid.

The last anecdote was also told of Mrs Keppel and Edward VII.

BALDWIN'S INTERVIEW WITH THE KING ON THE ABDICATION

7 December 1936: Oliver Baldwin (the Prime Minister's son) came to see me this morning. He told me that his father and the King walked round and round the garden at Fort Belvedere discussing the business, and then returned to the library having agreed that HM must abdicate. Stanley Baldwin was feeling exhausted. He asked for a whisky-and-soda. The bell was rung: the footman came: the drink was produced. S.B. raised his glass and said (rather foolishly to my mind), 'Well, Sir, whatever happens, my Mrs and I wish your happiness from the depths of our souls.' At which the King burst into floods of tears. Then S.B. himself began to cry. What a strange conversation-piece, those two blubbering together on a sofa!

Nicolson, *Diaries*

Baldwin had explained that the king's choice was a clear-cut either/or: either the crown, or marriage with a twice-divorced woman.

THE KING MEETS MR LINCOLN ELSWORTH, THE AMERICAN POLAR EXPLORER

He had just returned from his flight across the Antarctic, and I was surprised to hear from him, in the course of a description of that region's peculiarities, that it was wholly uninhabited.

'Not even Eskimos?' I asked.

'Not one at all, Sir,' he answered with the authority of an expert.

'Then, Mr Elsworth,' I said, 'if there are no people there, there are no politics.'

He looked at me startled. 'I am not sure, Sir, that I quite understand.'

'Ah!' I went on. 'To think of a whole continent with no Prime Minister, no Archbishop, No Chancellor of the Exchequer—not even a King. It must be paradise.' Windsor, *A King's Story*

INTERVIEW WITH CHURCHILL DURING THE CRISIS

'HM appeared to me to be under a very great strain and very near breaking point,' wrote Winston Churchill who dined at the Fort on Friday and Saturday, 4 and 5 December. 'He had two marked and prolonged blackouts, in which he completely lost the thread of his conversation.'

K. Middlemas and J. Barnes, *Stanley Baldwin* (1969)

THE KING AND HIS BROTHER

Walter Monckton, the king's private secretary, on the last dinner party before his abdication.

This dinner party was, I think, his *tour de force*. In that quiet panelled room he sat at the head of the table with his boyish face and smile, with a good fresh colour while the rest of us were pale as sheets, rippling over with bright conversation . . . On Mr Baldwin's right was the Duke of York, and I was next to him, and as the dinner went on the Duke turned to me and said: 'Look at him. We simply cannot let him go.' Donaldson, *Edward VIII*

THE FAREWELL TO THE FAMILY

It took place at Royal Lodge, Windsor, and there the ex-King said goodbye to his family. His mother, Queen Mary, ever magnificent, was mute and immovable and very royal . . . At last he left, and bowing over his brother's hand, the brother whom he had made King, he said, 'God bless you, Sir. I hope you will be happier than your predecessor', and disappeared into the night, leaving the Royal Family speechless. *Chips*

THE PARLIAMENTARY ACT OF ABDICATION

11 December 1936: 'The King is gone, Long live the King.' We woke in the reign of Edward VIII and went to bed in that of George VI. Honor [Lady Honor Channon, Chips's wife] and I were at the House of Commons by eleven o'clock . . . When the Bill came it was passed into law with the minimum of time . . . Then the Royal Commission was sent for, and the Lords Onslow, Denman and one other, filed out of the Chamber, and returned in full robes and wigs. Black Rod was sent to summon the Speaker, who, followed by his Commons, appeared at the bar. The Clerk read the Royal Commission. The three Lords bowed, and doffed their hats. The Bill was read. The King was still King Edward. The Clerk bowed, 'Le Roi le veult' [It is the King's will] and Edward, the beautiful boy King with his gaiety and honesty, his American accent and nervous twitching, his flair and glamour, was part of history. It was 1.52.

We went sadly home, and in the street we heard a woman selling newspapers saying, 'The Church held a pistol to his head.' In the evening we dined at the Stanleys' cheerless, characterless house, and at ten o'clock turned on the wireless to hear 'His Royal Highness Prince Edward' speak his farewell words in his unmistakable slightly Long Island voice. It was a manly, sincere farewell . . . There was a stillness in the Stanley's room. I wept, and I murmured a prayer for he [*sic*] who had once been King Edward VIII.

Then we played bridge.

Chips

THE KING JUST BEFORE THE BROADCAST

Account by Sir Eric Mieville, one of George VI's secretaries:

The end was amazing—everyone in a terrible sad state except the Chief Conspirator, who was honestly quite unmoved—except, I believe, for the brief period when he said Good-Bye to his Mother and Brothers. His last act prior to broadcasting his message and then leaving the country was to sit in his bedroom with a whisky and soda having his toe-nails seen to.

Kenneth Rose, *Daily Telegraph*, 7 December 1986

THE ABDICATION BROADCAST

Walter Monckton was an eyewitness when the king renounced the crown for 'the woman I love'.

The King ran through the draft broadcast rapidly in the last five minutes before he was due to begin. He also tried his voice on the microphone and was told that everything was in order. At 10 o'clock Sir John Reith [Director-General of the BBC] came in and stood over the King, who sat before the microphone, and announced 'His Royal Highness Prince Edward' and left the room. The King began, I thought, a little anxiously, but with the sentences his confidence grew, and the strength of his voice, and the final sentence 'God Save The King' was almost a shout. When it was over the King stood up and, putting his arm on my shoulder, said: 'Walter, it is a far better thing I go to.'

Donaldson, *Edward VIII*, quoting Monckton Papers

NO HERO TO HIS VALET

Windsor turned from the door and rang for his valet, Crisp. 'How's the packing coming along? . . . Good. Collect what you need for yourself. You'll be going with me.'

Crisp said, 'Sorry, Sir, but I'm not. I'm staying in England. I shall be leaving your service after you leave the Fort.' The only explanation Crisp ever offered was a gruff 'He gave up *his* job, I gave up mine.'

J. Bryan III and Charles J. V. Murphy, *The Windsor Story* (1979)

'OUR HAPPINESS'

After the Abdication was over Queen Mary told more than one person that to all her appeals he had answered: 'All that matters is our happiness,' and repeated this over and over again. Donaldson, *Edward VIII*

While waiting for Wallis's divorce decree to be made absolute, the Windsors had to live apart, she at Chateau Lou Viei in Cannes, he in Schloss Enzesfeld in Austria, lent by a member of the Rothschild family. Brownlow was the duke's former lord-in-waiting.

In mid-December, Brownlow had made a detour to Enzesfeld on his way home from Cannes, bringing messages from Wallis and an assurance that all was well with her. They told him at the Schloss that the Duke was resting. The bedroom was ajar, and in the wintry afternoon light Brownlow saw Windsor asleep on a bedspread strewn with photographs of his adored. He was smiling beatifically and clutching a small yellow pillow that had belonged to *her*. 'It was quite frightening,' Brownlow said.

The Windsor Story

LORD LOUIS MOUNTBATTEN'S VISIT TO THE DUKE

A day or so later, Mountbatten found an opportunity to tell the Duke, 'I have special permission from Bertie to offer myself as your best man.'

'Thank you but no,' the Duke answered. 'I want a proper royal wedding, with my two younger brothers as supporters.' There is no best man in royal weddings, only personages of rank, called 'supporters'.

Mountbatten knew that members of the Royal Family had agreed not to associate themselves publicly with the wedding, so he changed the subject: 'What are you going to wear?'

'Why, my uniform as Colonel of the Welsh Guards.'

'You can't do that, David! The Welsh Guards are getting a new colonel.' The monarch is automatically Colonel-in-Chief of each of the Guards regiments—the title passes with the Crown; but the *colonelcy* is by appointment, and always goes to a member of the Royal Family or a distinguished officer.

Windsor bit his lip. 'You're right. I've lost that too.'

Mountbatten remembered the remark because this was the only time during his visit that he heard the Duke say anything suggesting regret for what he had left. *The Windsor Story*

THE DUKE'S TRYING EXPERIENCE

Tonight he was told at dinner that HM wanted to talk on the phone to him. He said he couldn't take the call but asked it to be put through at 10 p.m. The answer to this was that HM said *he would talk at 6.45 p.m. tomorrow* as he was *too busy to talk any other time*. It was pathetic to see HRH's face. He couldn't believe it! He's been so used to having everything done as he wishes. I'm afraid he's going to have many more shocks like this.

Donaldson, *Edward VIII*, quoting Major Edward ('Fruity') Metcalfe to his wife Lady Alexandra, 2 January 1937

This experience had been matched by his brother Bertie's during the abdication crisis, when Edward was always too busy to see him.

THE WEDDING: PREPARATIONS

Two never to be forgotten scenes. On Wednesday & Thursday morning a figure in a dressing-gown with tousled hair sitting on the floor going through the mail helped by Mr Carter, his old clerk. The second, even more memorable perhaps than the ceremony today, was the rehearsal before dinner last night. A small pale green room with an alcove in one corner. The organist, Dupres, from Paris trying out the music in the room next door. Fruity . . . with HRH stands on the right of the alcove, Wallis on Herman's arm comes in—under the tutelage of Jardine, a large-nosed red-faced little man [the clergyman], they go over the service—HRH's jaw working the whole time exactly the same as I saw the King's all through the Coronation.

THE CEREMONY

It could be nothing but pitiable & tragic to see a King of England of only 6 months ago, an idolised King, married under those circumstances, & yet pathetic as it was, his manner was so simple and dignified & he was so sure of himself in his happiness that it gave something to the sad little service which it is hard to describe. He had tears running down his face when he came into the salon after the ceremony. She also could not have done it better. We shook hands with them in the salon. I realised I should have kissed her but I just couldn't, in fact I was bad the whole of yesterday . . . If she occasionally showed a glimmer of softness, took his arm, looked at him as though she loved him one would warm towards her, but her attitude is so correct. The effect is of a woman unmoved by the infatuated love of a younger man. Let's hope that she lets up in private with him otherwise it must be grim. Ibid., quoting Lady Alexandra Metcalfe, 3 June 1937

THE WINDSORS' VISIT TO GERMANY, 1937

To those who welcomed him with 'Heil Hitler' the duke

responded with what the reporter to the *New York Times* described as a modified Nazi salute—something between the real thing and a wave. On two occasions, however, he was reported as giving the full Hitler salute—the first time at a training school in Pomcrania when a guard of honour from the Death's Head Division of the Hitler Elite Guards was drawn up for his inspection, the second time for Hitler himself. Ibid.

'Her Royal Highness'

It was decided that Mrs Simpson, upon her marriage to the Duke of Windsor, should become a royal duchess but not a royal highness; a decision that did not please the duke. The duke and duchess of Windsor at Somerset Maugham's Villa Mauresque, Cap Ferrat, 5 August 1938:

When they arrived Willy [Maugham] and his daughter went into the hall. We stood sheepishly in the drawing-room. In they came. She, I must say, looks very well for her age. She has done her hair in a different way. It is smoothed over her brow and falls down the back of her neck in ringlets. It gives her a placid and less strained look. Her voice has also changed. It now mingles the accents of Virginia with that of a Duchess in one of Pinero's plays. He entered with his swinging naval gait, plucking at his bow tie. He had on a tussore dinner-jacket. He was in very high spirits. Cocktails were brought and we stood around the fireplace. There was a pause. 'I am sorry we were a little late', said the Duke, 'but Her Royal Highness couldn't drag herself away.' He had said it. The three words fell into the circle like three stones into a pool. Her (gasp) Royal (shudder) Highness (and not one eye dared to meet another). Nicolson, *Diaries*

THE DUKE'S EXPLANATION

The duke in 1956 gave his theory of the banned 'HRH'.

The letter was obviously written by Sir John Snake [Simon], probably with the help of someone in the Palace secretariat and God knows who else. My brother just took a piece of paper that was handed to him and copied it. It was not an idea he'd have thought of himself. Even less was the language his own—legalistic, no loopholes. I never blamed him, I've always given him the benefit of the doubt. Without question, other influences were working on him—somebody close to him, perhaps, others possibly of Ministerial rank, and I dare say the Archbishop of Canterbury. Yes, the Primate almost certainly had a hand in it. Windsor, *A King's Story*

ANOTHER EXPLANATION

A common friend of both the Royal Family and the Windsors was once asked by Wallis in the South of France: 'Why do the King and Queen treat us as they do?' Her friend replied: 'The public would not allow them to behave in any other way.' She said, 'Yes, I quite understand.' Shortly afterwards the same friend, finding himself alone on the Scottish moors with King George VI, decided to tell him the story. The King simply said, 'Yes, that's right.' E. Longford, *The Queen Mother* (1981)

'BANISHED' TO BERMUDA

The duke's war work was to be Governor of the Bahamas. The duchess gave her view of it while on board ship, sailing from Portugal:

Naturally we loathe the job but it was the only way out of a difficult situation—as we did not want to return to England except under our own conditions.

Michael Bloch, *The Duke of Windsor's War* (1982) quoting a letter of Wallis to Aunt Bessie

The Windsors' essential condition for returning to England was the never-to-be-granted 'HRH' for the duchess.

A VISIT TO DETROIT, 1941

As the Duke stood hatless listening to the national anthem on the platform at Detroit Station . . . a middle-aged man stepped from the crowd and came up to him. 'Your Royal Highness,' he said, 'I just wanted to welcome you; God bless you.' The Duke looked intently at him and asked: 'Where have I met you?' 'In Winnipeg in 1919,' said the man. 'You were a kiltie.' 'A Cameron Highlander, Sir.' Ibid.

The duke's memory remained 'royal'.

The Windsors in France

In 1958 James Pope-Hennessy visited the Windsors at their house in France.

The Duke of Windsor is, on first sight, much less small than I have been led to believe; he is not at all a manikin, but a well-proportioned human being. Just then his hair was blown out in tufts on either side of his head, and he was looking crumple-faced and wild . . . The hair is nicotine-coloured; but when he emerges from his shower and his valet's hands he looks very silken and natty and well-arranged; he has his father's eyes, and some, I fancy, of his mannerisms. He was drinking milk, for what the Duchess calls 'that lil' old ulcer'.

A Lonely Business: A Self-Portrait of James Pope-Hennessy ed. Peter Quennell (1981)

The author added that the duke was

exceedingly intelligent, original, liberal-minded . . . also one of the most considerate men I have ever met of his generation.

THE DUCHESS

She is, to look at, phenomenal. She is flat and angular, and could have been designed for a medieval playing-card. The shoulders are small and high; the head, very large, almost monumental; the expression is either anticipatory (signalling to one, 'I know it is going to be loads of fun, don't yew?') or appreciative—the great giglamp smile, the wide, wide open eyes, which are so very large and pale and veined, the painted lips and the cannibal teeth. There is one further facial contortion reserved for speaking of the Queen Mother, which is very unpleasant to behold, and seemed to *me* akin to frenzy . . . I only got this one completely on the last evening.

Ibid.

The author added that she was 'wildly good-natured and friendly.'

THE DUKE'S DOCUMENTS

He unlocked a white tin filing cabinet in a spare bedroom and they sat together on the green chintz bed looking at its contents.

There's a lot of valuable stuff here, you know,' he said. 'Unlike the Duchess I am very well-documented. But I keep them all under years, not under people. Let's take a look now at 1936.' He seized one of the two 1936 files and showed me various letters—until we reached one from Queen Mary, begging him not to broadcast. 'Surely you might spare yourself this strain and emotion' etc. A look of real disgust crossed over his face. 'She even tried that! Well, I ask you . . . If I hadn't even done that . . .'

We began to talk about the Abdication. 'People can say what they like for it or against it, I don't care; but one thing is certain: *I acted in good faith*. And I was treated bloody shabbily.' *A Lonely Business*

THE CALYPSO

A Panamanian folk legend by 'Blind Blake' was discussed by Pope-Hennessy and the duchess.

There was a record player outside the drawing-room door. The Duchess went off to change a record, walking with difficulty in her sheath of orange satin. I went with her. 'Have you heard "Love, Love, Love"?'

'Yes, I mean no. I'm not sure.'

'The record about *us*?'

'Oh no.'

'Well, I'm going to put it on for you. The Dook hasn't heard it. I only heard it two nights ago after dinner in Paris. They put it on as a kind of surprise, and it certainly sur-prised me, I can tell you.'

She then put on the record, a calypso: 'It was love, love, love and love alone that caused King Edward to leave his throne'—'That lady from Baltimore' etc.

'I'm going to call our lawyers Monday about it. *I* think it's libellous.'

We played it twice, the Duke jigging vaguely to it. 'I don't quite see where the libel would reside, Duchess,' I volunteered. 'Shouldn't you ignore it? You can't *now* say it *wasn't* love, so to speak.'

'You're right there. But I think it just so undignified. *And so offensive to the Monarchy*,' with dark emphasis.

'But you must expect to pass into folklore, Duchess. I don't see what you can do about it.'

'I jest think it is un-dig-ni-fied,' with a squaring of the angular shoulders and a slight, stiff flounce. 'I'm going to call our lawyers all the same.' Ibid.

THE OLD COMPLAINT

'I dined alone with David,' Mountbatten recorded in February 1970. 'Wallis was away "resting" so David and I had a delightful evening entirely to ourselves.' But the old sores had never healed. Repeatedly the Duke would revert to his old complaint that the Duchess had never been created a Royal Highness. 'I explained that it was his own mother's opposition, then followed by his sister's and sister-in-law's . . . which really made it impossible, and I advised him to give up the struggle.' The Duke could never give up.

P. Ziegler, *Mountbatten* (1985), quoting Mountbatten's Diaries, 2 March 1970

HIS WIFE'S VOICE

Sometimes she would call him from a distance—from the garden or from another part of the house. Then he would leave whatever he was doing and go to her, hurrying eagerly. You could hear his voice calling to her from afar: 'Coming, darling,' 'Yes, sweetheart!' I have seen him in the middle of a haircut in his dressing-room get up and run to his wife, leaving his astonished hairdresser agape. Dina Wells Hood, *Working for the Windsors* (1967)

HER KIND OF PERFECTIONISM

'Her day was quite extraordinary,' says Laura, Duchess of Marborough. 'She would book five or six fittings with couturiers. Her life's work was shopping . . . I went to look at the flowers at the funeral. It was tragic. They were all from dressmakers, jewellers, Dior, Van Cleef, Alexandre [her coiffeur]. Those people were her life.' S. Menkes, *The Windsor Style* (1987)

She had three separate face-lifts to keep up with her clothes.

THE MONARCHY FROM THE INSIDE

The Duke of Windsor was interviewed by the New York Daily News *in 1966:*

Being a Monarch, whether man or woman in these egalitarian times, can surely be one of the most confining, the most frustrating, and over the duller stretches, the least stimulating jobs open to an educated, independent-minded person. Even a saint would on occasion find himself driven to exasperation by the taboos which invisibly and silently envelop a constitutional monarchy. This is not meant in disrespect. It is only the way it looked to me from the inside. Birkenhead, *Walter Monckton*

SAYING OF THE DUCHESS OF WINDSOR

On art: 'My approach to art, whether modern or traditional, is decorative. When I look at a picture I never see it by itself, I see it as part of a room.'

EDWARD'S SAYINGS

On his first tail coat: 'I shall look an ass.'

On his Investiture: 'What would my Navy friends say if they saw me in this preposterous rig?'

On his father's best pieces of advice: 'Never to refuse an invitation to take the weight off my feet and to seize every opportunity I could to relieve myself.'

On risking his life at the front: 'What does it matter if I am killed? I have four brothers.'

On watching his Proclamation at St James's Palace: 'The thought came to me that I'd like to see myself proclaimed King.'

On having the clocks altered at Sandringham (they had been kept half an hour fast by his father and grandfather) while George V was dying: 'I'll fix those bloody clocks.'

On cruising publicly with Mrs Simpson: 'Discretion is a quality which, though useful, I have never particularly admired.'

On travelling without permits to the Spanish border in 1940: Je suis le Prince de Galles. Laissez-moi passer, s'il vous plait.' *The French veterans recognized him and let him through.*

On buying swimming trunks in Bermuda: 'It's I who wear the shorts in this family, you know.'

THE DUKE'S LAST WORDS

Sydney Johnson, his valet, was with him at the end.

'Just before His Royal Highness died, I heard him say four words, Mama, Mama, Mama, Mama.' Suzy Menkes, *The Windsor Style* (1987)

These words may be compared with his letter to Wallis while awaiting his mother's death in 1953:

... the bulletins from Marlborough House proclaim the old lady's condition to be slightly improved! Ice in the place of blood in the veins must be a fine preservative.

Quoted in Michael Bloch, *The Secret File of the Duke of Windsor* (1988)

George VI

1936–1952

By contrast with his brother Edward, nothing except athletics came easy to King George VI: public speaking, mastering state papers, winning 'the woman he loved'—for them all he had to work hard and success came as a result of indomitable will rather than facility. His pursuit of the Lady Elizabeth Bowes-Lyon was long and arduous. He was a one-woman man and had chosen the most popular debutante of her day. A stammer rendered public speaking at first torture, always a strain both for him and for his listeners. He had never seen a political document until the abdication of his brother pitched him into kingship. Two years later the country was at war. This was a challenge to which the king, who had been the 'Sailor-Prince' in the Great War, rose naturally with courage and resolution. His biographer noted that in other circumstances he would have made a good doctor, such was his intelligent interest in his own illnesses. As things turned out he made a good king.

BERTIE IN THE NURSERY

Prince Albert the nurse frankly ignored to a degree which amounted virtually to neglect. So completely did she disregard his wants and comforts that he was frequently given his afternoon bottle while driving in a C-sprung victoria, a process not dissimilar from a rough Channel crossing—and with corresponding results. It is not surprising that the baby developed chronic stomach trouble, which may well have laid the foundation for the gastric complaint from which he was later to suffer so acutely.

John Wheeler-Bennett, *King George the Sixth* (1958)

PRINCE BERTIE'S HAT-TRICK

Sport was one of the areas in which he excelled. He was a very good cricketer, the best in his family, and this prowess had given him one unchallengeable claim to fame: as a youth, he had once bowled out King Edward VII and the future monarchs George V and Edward VIII one after the other. This kingly hat-trick stands unrivalled in the records of English cricket.

Denis Judd, *King George the Sixth*

THE PRINCE'S ACCOUNT OF THE BATTLE OF JUTLAND, 1915

Though the prince had been bottom of his class at Osborne, being tongue-tied through his stammer, and later had developed a gastric ulcer, he insisted on seeing active service at the age of twenty. He wrote:

At the commencement I was sitting on the top of A turret and had a very good view of the proceedings. I was up there during a lull, when a German ship started firing at us, and one salvo 'straddled' us. We at once returned the fire. I was distinctly startled and jumped down a hole in the top of the turret like a shot rabbit!! I didn't try the experience again. The ship was in a fine state on the main deck. Inches of water sluicing about to prevent fires from getting a hold on the deck. Most of the cabins were also flooded . . .

My impressions were very different to what I expected. I saw visions of the masts going over the side and funnels hurtling through the air etc. In reality none of these things happened and we are still quite sound as before . . . It was certainly a great experience to have been through and it shows that we are at war and that the Germans can fight if they like.

Wheeler-Bennett, *King George*

Bertie and Elizabeth

FIRST MEETING WITH ELIZABETH

He was ten and she was five.

They had first met in 1905 at a children's party in Montague House, when, it is said, she gave him the crystallised cherries off her sugar cake; but they did not meet again until the summer of 1920 at a small dance given by Lord Farquhar.

Ibid.

At this dance he is said to have seized her metaphorically from his equerry's arms with the words, 'That's a lovely girl you've been dancing with. Who is she?' . . . He was to recognise 20 May 1920 as a red-letter day in his life.

Longford, *The Queen Mother* (1981)

MEMORIES OF A PARTY AT GLAMIS CASTLE

The then Duke of York, afterwards King, used to come into my bedroom in the evening, and we would talk of the Glamis monster and the admittedly sinister atmosphere in the castle, and of the other ghosts . . . One rainy afternoon, we were sitting about and I pretended that I could read cards, and I told Elizabeth Lyon's fortune, and predicted a great and glamorous royal future. She laughed, for it was obvious that the Duke of

York was much in love with her. As Queen she has several times reminded me of it. I remember the pipers playing in the candlelit dining-room, and the whole castle heavy with atmosphere, sinister, lugubrious, in spite of the gay young party. *Chips*

THE ENGAGEMENT

His first proposal to her was two years before she accepted him on 13 January 1923. He sent a telegram to his parents, King George and Queen Mary, announcing the great news:

All right. Bertie. Wheeler-Bennett, *King George*

The duke had had rivals during his determined courtship. Many years later one of them said:

I was madly in love with her. Everything at Glamis was beautiful, perfect. Being there was like living in a Van Dyck picture. Time, and the gossiping, junketing world, stood still. Nothing happened. Nothing, except that the seventeen-stone Leveson-Gower [married to Elizabeth's sister Rose] was once thrown out of bed by a ghost. But the magic gripped us all. I fell *madly* in love. They all did.

Her charm was indescribable, an indefinable atmosphere. She was also very kind and compassionate. And she could be very funny—which was rare in those circles. She was a wag. Longford, *The Queen Mother*

PRINCE ALBERT AND HIS FATHER

'Dickie' Mountbatten, the prince's cousin, had planned to take a popular girl called Audrey James to the Oxford and Cambridge rugger match in a party that included Prince Henry (Harry). At the last moment King George V decided to attend.

'Harry is so young, he can't stand up to his father like I can,' commiserated Prince Albert with all the hauteur of an elder brother. 'Of course he ought to have told him he had a party of his own. He doesn't understand, like you and me, the trouble it is to get these girls to do anything, otherwise he wouldn't have let the King spoil it all.' Ziegler, *Mountbatten*

ELIZABETH AND HER FATHER

As a small child she ran out of pocket-money and sent her father a telegram:

S.O.S. L.S.D. R.S.V.P. ELIZABETH. D. Duff, *George and Elizabeth* (1983)

GEORGE VI ON GEORGE V

Sir George Sitwell was noted for his irascibility.

The Queen, with her interest in people, had evidently been entertained by what friends had told her concerning my father [Sir George Sitwell] and my relations with him, and steered the conversation round to that subject. I gave in my turn a rather restrained account of him, though I did not seek to disguise his essential characteristics. When I had finished, the Duke said suddenly, 'He sounds just like my father! Won't listen to a word you say. Always knows better. There's no doing anything with them, when they're like that!'

Sitwell, *Rat Week*

THE LANGUAGE OF RHODODENDRONS

Bertie became a keen gardener after his father installed him and his family at Royal Lodge, Windsor. He wrote a thank-you letter after a strenuous visit to Lochinch Castle in Scotland, commenting on the state of his feet:

Having had time to examine my feet, Denudatum (naked) & Detersile (clean), I am glad to find that they are neither Hypoglaucum (blue beneath) Hyponepidotum (scaly) nor Hypopheum (grey) but merely Russatum (reddened).

Wheeler-Bennett, *King George*

THE FIRST MEETING OF THE DUKE WITH LIONEL LOGUE

The Australian speech therapist gave the duke of York confidence and a programme of breathing exercises, to relieve the agonies of his stammer. Logue wrote:

He entered my consulting room at three o'clock in the afternoon, a slim, quiet man, with tired eyes and all the outward symptoms of the man upon whom habitual speech defect had begun to set the sign. When he left at five o'clock, you could see that there was hope once more in his heart.

Ibid.

Two years later the duke was still having consultations with Logue.

He gave the Duke tongue-twisters to practise on, and both men laughed over the patient's variable success with 'Let's go gathering healthy heather with the gay brigade of grand dragoons.' Or with 'She sifted seven thick-stalked thistles through a strong, thick sieve.'

Judd, *George the Sixth*

Harold Nicolson failed to recognize the duchess at a party.

20 February 1936: . . . when I got in, there was a dear little woman in black sitting on the sofa, and she said to me, 'We have not met since Berlin.' I sat down beside her and chattered away all friendly, thinking meanwhile, 'Berlin? Berlin? How odd. Obviously she is English, yet I do not remember her at all. Yet there is something about her which is vaguely familiar.' While thus thinking, another woman came in and curtsied low to her and I realised it was the Duchess of York. Did I show by the tremor of an eyelid that I had not recognised her from the first? I did not. I steered my conversation onward in the same course as before but with different sails: the dear old jib of comradeship was lowered and very gently the spinnaker of 'Yes, Ma'am' was hoisted in its place. I do not believe that she can have noticed the transition. She is charm personified. Nicolson, *Diaries*

THE ABDICATION, 1936

The new king expressed his sense of shock to his cousin Dickie (*Lord Louis*) *Mountbatten.*

'Dickie, this is absolutely terrible. I never wanted this to happen; I'm quite unprepared for it. David has been trained for this all his life. I've never even seen a State paper. I'm only a Naval Officer, it's the only thing I know about.' And Lord Louis was able to give him consolation. 'This is a very curious coincidence. My father once told me that, when the Duke of Clarence died, your father came to him and said almost the same things that you have said to me now, and my father answered: 'George, you're wrong. There is no more fitting preparation for a king, than to have been trained in the Navy.' Wheeler-Bennett, *King George*

A DINNER-PARTY AT BUCKINGHAM PALACE, MARCH 1937

The dining-table is one mass of gold candelabra and scarlet tulips. Behind us the whole of the Windsor plate is massed in tiers. The dinner has been unwisely selected since we have soup, fish, quail, ham, chicken, ice and savoury. The wine, on the other hand, is excellent and the port superb. When we have finished our savoury the King rises and we all resume our procession back to the drawing-rooms. On reaching the door of the fourth drawing-room the equerries tell us to drop our ladies and to proceed onwards to a drawing-room beyond where the men sit down for coffee and cigars. The King occupies that interval in talking to Baldwin and Lloyd George, and I occupy it in discussing with David Cecil the reasons why we've been asked. He says, 'I know why I have been asked. I have been asked as a young member of the British aristocracy.' I say that I have been asked as a rising politician, and I regret to observe that David is not as convinced by this explanation as I might have wished.

We then pass on into the Picture Gallery, where we are joined by the women and by the King and Queen . . . The Queen then goes the rounds. She wears upon her face a faint smile indicative of how she would have liked her dinner-party were it not for the fact that she was Queen of England. Nothing could exceed the charm or dignity which she displays, and I cannot help feeling what a mess poor Mrs Simpson would have made of such an occasion. It demonstrated to us more than anything else how wholly impossible that marriage would have been. The Queen teases me very charmingly about my pink face and my pink views.

Thereafter the Queen drops us a deep curtsy which is answered by all the ladies present. We then go away . . . Nicolson, *Diaries*

A RAT

Osbert Sitwell wrote a poem and an essay called 'Rat Week' in which he castigated those members of Society who had surrounded the Windsors and were the first to scuttle when the duke abdicated. Lady Diana Cooper knew the poem and is said to have introduced herself to the new king with the words:

'I'm afraid I'm a Rat, Sir.' Sitwell, *Rat Week*

THE CHANGE IN THE REIGN

18 April 1937: Diana and Duff Cooper have returned from Windsor where they 'dined and slept'. Diana said it was all very different from the atmosphere at the Fort and the late regime. 'That was an operetta, this is an institution.' *Chips*

Nine years later Sandringham was neither an operetta nor an institution but a home—or so said Lady Airlie, Queen Mary's lady-in-waiting.

I thought—regretfully at first—how much the atmosphere had changed, but then I realized that this was inevitable for a new generation had grown up since I had last seen it. In the entrance hall there now stood a baize-covered table on which jig-saw puzzles were set out. The younger members of the party—the Princesses, Lady Mary Cambridge . . . and several young Guardsmen—congregated around them from morning till night. The radio, worked by Princess Elizabeth, blared incessantly.

Before the end of the week I revised my impressions. There was no denying that the new atmosphere of Sandringham was very much more friendly than in the old days, more like that of any home. One senses far more the setting of ordinary family life in this generation than in the last. It was the way in which the King said, 'You must ask Mummy,' when his daughters wanted to do something—just as any father would do.

Mabel, Countess of Airlie, Thatched with Gold

THE SHADOW OF THE ABDICATION

Any reminder of Edward VIII seemed to trigger a relapse to George VI's childhood traumas, and in later years Lord Plunket used to say he could always tell when the ex-King was due to call on one of his London stop-overs because of the sudden chill in the atmosphere—though nothing was said—and the way in which George VI's wife would 'drive out' of Buckingham Palace. It was the dark side of the family life the King worked so hard to keep sunny for his wife and daughters.

Robert Lacey, *Majesty* (1977)

THE SHADOW OF THE DUCHESS OF WINDSOR

A dialogue between George VI and David Lloyd George, former prime minister:

'She [the duchess] would never dare to come back here,' said HM.

'There you are wrong', replied David.

'She would have no friends,' said HM.

D. did not agree.

'But not you or me?' said the King anxiously.

Diaries of Frances Stevenson (Lady Lloyd George), ed. A. J. P. Taylor (1971)

EXTRACTS FROM KING GEORGE VI'S DIARY ON HIS CORONATION, 1937

Elizabeth's procession started first but a halt was soon called, as it was discovered that one of the Presbyterian chaplains had fainted & there was no place to which he could be taken. He was removed however after some delay & the procession proceeded & arrived in position . . .

I had two Bishops, Durham, & Bath & Wells, one on either side to support me & to hold the form of Service for me to follow. When this great moment came neither Bishop could find the words, so the Archbishop held his book down for me to read, but horror of horrors his thumb covered the words of the Oath.

My Lord Great Chamberlain was supposed to dress me but I found his hand fumbled & shook so I had to fix the belt of the sword myself . . . As I turned after leaving the Coronation Chair I was brought up all standing, owing to one of the Bishops treading on my robe. I had to tell him to get off it pretty sharply as I nearly fell down.

Wheeler-Bennett, *King George*

LEAVING THE ABBEY AFTER THE CORONATION CEREMONY

12 May 1937: . . . By now there was much general chaff and when one of the Gold Staff Officers lost his sense of humour, and called 'I say, a Baron has got out before the Viscounts', there was a roar of laughter.

Chips

THE KING MEETS A KING

A few years before the war King Levinsky [an American boxer] came to Britain to fight . . . He trained at Windsor and was introduced to George VI. 'Hey, majesty,' he said, 'so you're George de Sixth and I'm Levinsky de Foist!' Frank Keating, *Sports Guardian*, 17 October 1987

VISIT OF THE KING AND QUEEN TO PARIS, 1938

The idea was to cement the friendship with France, Britain's ally in the event of war.

We saw the King and Queen from a window, coming down the Champs Elysées with roofs, windows and pavements roaring exultantly, the Queen, a radiant Winterhalter, guarded by too many security measures. The Minister who was responsible for their safety told me that their fears and safeguards were such as to put a plain-clothes policeman in every window on the route and to have hefty citizens lean in a ring against the suspect trees lest they should fall on the procession.

Each night's flourish outdid the last. At the Opera we leant over the balustrade to see the Royal couple, shining with stars and diadem and the Légion d'Honneur proudly worn, walk up the marble stairs preceded by *les chandeliers*—two valets bearing twenty-branched candelabra of tall white candles. This custom seems to have died, for in 1957 no candles lit Queen Elizabeth to her Royal Box.

The Elysée and the Quai d'Orsay outshone each other in splendour and *divertissement*. Malmaison, decked doubly with roses, received the Queen. It was here that I talked to two crying old ladies who begged for my place on the Royal path. 'Vous la voyez toujours,' one said. 'Si seulement nous avions un roi,' said the other. Monarchy dies slow in many French hearts. A cook-general at a friend's house, serving a *blanquette de veau*, had said to me when the King was acclaimed earlier in the year: 'Enfin, nous avons un roi!' Cooper, *Autobiography*

THE VOYAGE TO CANADA, 1939

Visits to Canada and the United States were planned, again with a view to solidarity in the event of war. The royal couple were regarded as Britain's best ambassadors.

For three & a half days we only moved a few miles. The fog was so thick, that it was like a white cloud round the ship, and the fog horn blew incessantly. Its melancholy blasts were echoed by the icebergs like the twang of a piece of wire. Incredibly eerie, and really very alarming . . . We very nearly hit a berg the day before yesterday, and the poor Captain was nearly demented because some kind cheerful people kept on reminding him that it was about here that the Titanic was struck, & *just* about the same date!

Wheeler-Bennett, *King George*, quoting a letter of Queen Elizabeth to Queen Mary

VISIT OF THE KING AND QUEEN TO THE UNITED STATES, 1939

A garden party was laid on in the British Embassy, Washington, and John Wheeler-Bennett, the British historian, was whistled up 'as a sort of extra-equerry'.

It was the hottest day I have ever known and I suffered gravely in my morning dress. I was attached to Queen Elizabeth's party and all went famously, as we stopped here and there for momentary introductions. The Queen was superb . . . She was so utterly unlike anything they had expected, queenly but human, regal but sympathetic. She was a revelation.

But then we reached the marquee. In that oppressive heat the orders about refreshments [nothing to be touched till the Sovereigns arrived] were set at naught. We had just reached the entrance when an unfortunate man came out carrying two glasses of iced tea. Before him he beheld the Queen of England and, like a peccant schoolboy caught at the jam-jar, his nerve forsook him. 'My God, the Queen,' he ejaculated, and dropped both glasses. The Queen laughed delightedly and passed on her triumphant way.

J. Wheeler-Bennett, *Special Relationships* (1975)

VISIT TO THE PRESIDENT

At Hyde Park, the Roosevelt family estate, the President greeted them with a tray of cocktails. 'My mother', he told them, 'thinks you should have a cup of tea. She doesn't approve of cocktails.' 'Neither does mine,' the King said, and took one.

Rose, *Kings, Queens and Courtiers*

War was declared against Hitler's Germany on 3 September 1939. In March 1940 the king inspected the Dover Patrol and watched troops of the British Expeditionary Force entraining for leave.

A sergeant of the BEF, coming ashore at a south-east coast port yesterday on leave, hurried to the barrier to reach the waiting train. He thrust his papers into the hands of a man in naval uniform standing by the ticket collectors.

Suddenly he gave a gasp of surprise, straightened to attention and saluted. The ticket collector was the King.

The King tore the sergeant's ticket from his book of passes and handed the book back with a smile, and the sergeant hurried on.

The King was on the quayside to watch nearly a thousand men come ashore and few, in their eagerness to continue their journey, recognised him.

Daily Mirror, 15 March 1940

WARTIME PORTRAITS AT WINDSOR

I do not know who gave drawing lessons to the Princesses, unless Gerald Kelly did. The King frankly admitted to having only small knowledge of art, confessing once to my brother that when he saw a name under a portrait he was not always sure whether it was that of the artist or of the sitter. Kelly had established himself as a permanent guest at Windsor Castle, where he spent a comfortable war painting the state portraits of the King and Queen. He spun the task out, year after year, till it was finally rumoured that, like Penelope, he undid each night what he had added during the day. Even a caricature of him in a Christmas pantomine organised by the two Princesses—as 'Kerald Jelly, the immovable guest'—failed to dislodge him.

Wilfrid Blunt, *Slow on the Feather: Further Autobiography* (1986)

THE QUEEN'S RESOLUTION

I told the Queen today that I got home-sick, and she said, 'But that is right. That is personal patriotism. That is what keeps us going. I should die if I had to leave.' She also told me that she is being instructed every morning how to fire a revolver. I expressed surprise. 'Yes', she said, 'I shall not go down like the others.'

Nicolson, *Diaries*

SELF-DEFENCE AT THE PALACE

A German parachute invasion was on the cards.

One day King Haakon of Norway asked King George to demonstrate an anti-parachute alert in the Palace garden. King George pressed a button and—nothing happened. The outside warning to the lodge had not been received, and so with British phlegm the lodge-keeper told those who should have reported at action stations to take no notice of the alarm.

Meanwhile the range of Buckingham Palace was alive with pops and bangs. Lord Halifax, who had been given permission to use the Palace garden as a short cut to his office, was puzzled the first time he heard the shots and enquired their cause. 'Her Majesty's target practice,' was the reply. As his path ran nearby, he decided it might prove a short cut to the next world and chose another route to work.

Longford, *The Queen Mother*

NO SCUTTLE

Probably only the united resolve of King and Prime Minister prevent a wholesale scuttle of Court, Government and Parliament into the country. Such plans had been made by the previous government. George VI and Churchill tore them up. A. J. P. Taylor, *Sunday Express*, 22 September 1957

AN INTRUDER

German parachutists were not the only potential invaders whom Queen Elizabeth had to fend off. One day a half-crazed deserter whose family had all been killed in a raid found his way into the Queen's bedroom, threw himself at her feet and seized her by the ankles. It was like some scene from the middle ages. 'For a moment my heart stood absolutely still,' remembered the Queen, then, 'Tell me about it,' she said quietly, realising that if she screamed he might attack her. He poured out his sad tale as she moved step by step towards the bell. Longford, *The Queen Mother*

CHARITY BEGAN AT HOME

An elderly couple living near Royal Lodge were bombed out. A friend visited them expecting to find them depressed:

But I found them in wonderful spirits, the wife telling me she was wearing one of the Queen's dresses, with handbag, and her husband one of the King's suits. Ibid.

THE QUEEN IN SHEFFIELD, JANUARY 1941

I dined with Billy Harlech, the Regional Commissioner [Lord Harlech; North-Eastern Regional Commission for Civil Defence]. He had been spending the day with the Queen visiting Sheffield. He says that when the car stops, the Queen nips out into the snow and goes straight into the middle of the crowd and starts talking to them. For a moment or two they just gaze and gape in astonishment. But then they all start talking at once. 'Hi! Your Majesty! Look here!' She has that quality of making everybody feel that they and they alone are being spoken to. It is, I think, because she has very large eyes which she opens very wide and turns straight upon one.

Nicolson, *Diaries*

THE KING'S VISIT TO NORTH AFRICA, 1942

As he walked out on the verandah of his villa, first one man, then another, recognised him.

And as if called by one voice, the thousands of men, most of them semi-nude, many of them still dripping with water, raced up the beach like a human wave.

Then as if the wave had suddenly frozen, they stood silently below the verandah, a solid mass of tanned and dripping men.

There was one of those strange silences one sometimes gets among a huge crowd.

A voice started 'God Save the King'. In a moment the National Anthem was taken up everywhere . . .

As the last notes of the Anthem died out, the King suddenly turned, stepped down from the verandah. He stood there, surrounded by hundreds of men, talking to them, asking them about their experiences.

Then the men broke into song again, this time with 'For he's a Jolly Good Fellow'. Ann Morrow, *The Queen Mother* (1985)

THE QUEEN IN LANCASHIRE

On one visit . . . a council put on a spectacularly lavish meal which she sat through growing more and more embarrassed. Finally she turned to the mayor and said: 'You know, at Buckingham Palace we're very careful to observe the rationing regulations.' He said to the Queen: 'Oh, well then Your Majesty, you'll be glad of a proper do.' Ibid.

NO JOY-RIDES ON D DAY 1944

Less than a week before the date scheduled for D Day (5 June, though it was postponed at the last moment to the 6th owing to the execrable weather), both the king and the prime minister had decided independently to 'go in' with the invading ships. The argument that followed lasted for five days, until D Day, in fact, was only three days away.

Day 1. On Tuesday, 30 May the king recorded the result of his usual luncheon-audience with Winston Churchill:

I asked W. where he would be on D Day . . . & he told me glibly he hoped to see the initial attack from one of the bombarding ships . . . I was not surprised & when I suggested I should go as well (the idea has been in my mind for some time) he reacted well . . . W. cannot say no if he goes himself, & I don't want to have to tell him he cannot. So? I told E. [Queen Elizabeth] about the idea & she was wonderful as always & encouraged me to do it.

Day 2. 31 May. After sleeping on it, George VI very reluctantly agreed with his agitated private secretary, Sir Alan Lascelles, that the dual plan was not on. He wrote to Winston:

Our presence, I feel, would be an embarrassment to those responsible.

Day 3. 1 June. The king, Churchill and Lascelles met Admiral Ramsay in the Map Room at 3.15, where Ramsay was astounded and horrified to learn for the first time that the king, as well as the prime minister, had intended to go. King George reported the interview in his diary:

I said I very much deprecated the idea of his [the PM's] going as a passenger for a 'joy ride'. He said he had flown to USA, Middle East, Moscow & Teheran & had crossed the Atlantic by sea already & that this was nothing. I said that he had to pay those visits on duty for the future strategy of the war . . . When I left I could see Ramsay was a bit shaken & wished to stop the PM going. I saw Gen. Ismay later who was very upset.

Wheeler-Bennett, *Special Relationships*

Churchill's own account of the scene in the Map Room began with Admiral Ramsay and himself agreeing that the king must not go.

The King said that if it was not right for him to go neither was it right for me. I replied I was going as Minister of Defence in the exercise of my duty. Sir Alan Lascelles, who the King remarked was 'wearing a very long face', said that 'His Majesty's anxieties would be increased if he heard his PM was at the bottom of the English Channel.' I replied that that was all arranged for, and that moreover I considered the risk negligible. Sir Alan said that he had always understood that no Minister of the Crown could leave the country without the Sovereign's permission. I answered that this did not apply as I should be in one of HM's ships. Lascelles said the ship would be well outside territorial waters. The King then returned to Buckingham Palace.

Winston S. Churchill, *The Second World War*, V (1952)

Day 4. 2 June. The king wrote again:

My dear Winston: I want to make one more appeal to you not to go to sea on D Day. Please consider my position . . .

Day 5. 3 June. After denying any rights of the Cabinet to stop him, Churchill caved in to the king. Neither went. But Churchill's obstinate view, expressed eight years later, was that those who had to take grave and terrible decisions might need the refreshment of adventure.

Ibid.

THE KING AND MR ATTLEE

Clement Attlee was the Labour leader in the wartime coalition government. The king and 'Clem' were both shy but firm characters who respected each other. Before the war Attlee was leader of the opposition and described in his terse way a social meeting with the king:

I was discussing methods of pipe construction with the King when we met at dinner the other day. He had a most ingenious one which scraped the bowl every time so that it always kept the same size.

Kenneth Harris, *Attlee* (1982)

After the unexpected landslide Labour victory of 1945 Attlee was summoned to the Palace by the king on 27 July, to form a government. Attlee's wife drove him to the Palace and waited outside in her car during the audience. Attlee reported:

The King pulled my leg a bit. He told me I looked more surprised by the result than he felt.

Ibid.

'Victory in Europe' was celebrated on 8 May 1945. On the 17th the king and queen visited Parliament.

The King and Queen attended a ceremony in the Royal Gallery. It was almost wholly domestic. The Lord Chancellor made a short speech; the Speaker made a speech; the King read a long speech. He has a really beautiful voice and it is to be regretted that his stammer makes it almost intolerably painful to listen to him. It is as if one read a fine piece of prose written on a typewriter the keys of which stick from time to time and mar the beauty of the whole. It makes him stress the wrong word. 'My Lords and Members . . . *of* the House of Commons.' Then they walked down the aisle which separated the Lords from the Commons; very slowly they walked, bowing to right and left. The Queen has a truly miraculous faculty of making each individual feel that it is him whom she has greeted and to him that was devoted that lovely smile. She has a true genius for her job. But we listened in silence to the King's speech: a silence which seemed ungrateful for all the excellent work that he has done. But Winston, with his sense of occasion, rose at the end and waved his top hat aloft and called for three cheers. All our pent-up energies responded with three yells such as I should have thought impossible to emanate from so many elderly throats.

Nicolson, *Diaries*

THE ROYAL FAMILY ON TOUR IN SOUTH AFRICA 1947

The Royal Family were invited to tour South Africa in 1947. Princess Elizabeth was engaged to Prince Philip of Greece, but George VI persuaded his daughter to postpone the announcement until the family 'firm' of four had returned to England. The king's health was deterioriating.

Their favourite standing joke on the tour was the question asked of each other whenever they were in gala dress: 'Is this a special occasion?' It appears that during the King's tour of Canada before the war, he once noticed that a local mayor was not wearing a mayoral chain. The King, planning to present him with one, asked him whether or not he had a chain.

'Oh, yes, Sir,' answered the mayor. 'I have.'

'But I notice you are not wearing it,' said the King.

'Oh,' explained the mayor, 'but I only wear it on special occasions.'

Theo Aronson, *Royal Ambassadors* (Cape Town, 1976)

The Royal Family travelled through South Africa in the famous White Train. The king was moved by the warmth of their welcome but shocked by the hostility of the Nationalist party towards the Anglophile Jan Smuts.

On the royal dining-table in the White Train was spread a tablecloth printed with the South African motto: '*Ex Unitate Vires*'—In Unity is Strength. The first time the King saw it he exclaimed, 'Not much unity here!' for when the King bestowed the Order of Merit upon Jan Smuts at Cape Town, many of the Nationalists stayed away. The King was shocked. To insult Jan Smuts! Hero of the South African 'volk' in the Boer War, Smuts was now a great Imperial statesman. After one particularly virulent example in South Africa of dissension rather than unity, the King burst out characteristically, 'I'd like to shoot them all!' to which the Queen replied in her voice of gentle remonstrance, half-smiling, 'But Bertie, you can't shoot *everybody*'—as though he could at least shoot *some*.

Longford, *The Queen Mother*

THE TOUR OF RHODESIA (ZIMBABWE), 1947

In Salisbury [George VI] opened Parliament as King of Rhodesia, and near Bulawayo the party walked up the granite hill slope to the grave of Cecil Rhodes until the Queen found she could not undertake another step in her high-heeled cutaway shoes, and Princess Elizabeth had to lend her mother her own sandals and continue the climb in her stockinged feet. 'It was so like Mummy to set out in those shoes,' said the Princess. It was so like Queen Elizabeth . . . always to wish to look her formal best.

Helen Cathcart, *The Queen Mother Herself* (1979)

THE QUEEN AT A CONCERT

Osbert Sitwell gave this anecdote as an example of how well she could manage people. He and his friend Malcolm Bullock, MP were to meet her in the hall before a concert, sponsored by the BBC and its Director-General Sir John Reith.

. . . Malcolm came up, after a talk at the door with Reith, and said to the

Queen, 'I don't know what to do, Ma'am. Sir John Reith insists that I should lead the way into the hall, and I can't make him understand that it's wrong and that I can't do it!'

'Never mind, Malcolm,' Her Majesty said, 'Do what he tells you, and leave it to me.'

So, as we were about to go upstairs, with Malcolm in front of the Queen, and Sir John Reith watching with a now benevolent air, Her Majesty stopped Malcolm, and said in a voice the tones of which were clear and pleading, 'Malcolm, do you mind if I go in first? I do so much want to go in first, *just for once*.'

Sitwell, *Rat Week*

Sitwell hated Reith and described him as the tyrant of the BBC, 'an overbearing Scotch giant of Scotch principles and an obstinate nature'.

REX WHISTLER AND OSBERT SITWELL STAYING AT BALMORAL

In the evening there was a gillies' ball, and it was after one o'clock when we retired.

I said good night to Rex upstairs, but about half an hour later he rushed into my room, looking distracted.

'Osbert! there's someone in the room beyond me!'

'Well, what of it?'

'It's the King! I can recognise his voice, and he's talking to himself!'

'Well, what did he say?' I asked, very practically.

'He said: "I've never been so *tired* in my life—it's all these bloody guests!" '

Ibid.

REX WHISTLER ON THE VISIT TO BALMORAL

Now he was describing the visit in five letters written in his bedroom at Balmoral. After dinner on the first night, with the thrill of the pipers 'swaggering round the table (only three times, thank God!)', there was the Gillies' Ball where the King and Queen, having opened it, 'hopped and skipped and capered in the wildest way the *entire* time we were there'. He thought it must be like Elizabethan revelry; no pompous dignity, and no one taking particular notice of them among the 'roars of laughter, in a sea of whirling arms and legs'. Taking part himself in the simpler reels, he was lost, but it did not matter.

Laurence Whistler, *The Laughter and the Urn; the Life of Rex Whistler* (1985)

Vita Sackville-West, wife of Harold Nicolson and joint creator with him of Sissinghurst garden, was made a Companion of Honour by the king in 1948 for her services to literature and culture. Knole was the Sackvilles' historic home.

She received the CH from King George VI on 12 February.

He had asked her about Knole. She said that it had gone to the National Trust. He raised his hands in despair. 'Everything is going nowadays. Before long, I shall also have to go.' Nicolson, *Diaries*

The king conferred a knighthood on his surgeon in 1949. The operation was a right lumbar sympathectomy.

On the occasion of Professor Learmonth's final examination His Majesty asked him, at its conclusion, to give him his bath-robe and slippers; then, pushing forward a stool and picking up a sword which he had hitherto concealed, he said: 'You used a knife on me, now I'm going to use one on you', and bidding him kneel, bestowed upon him the accolade of knighthood. Wheeler-Bennett, *King George*

The king died in 1952.

SAYINGS OF KING GEORGE VI

On running the 'Duke of York's, Camp': 'I'll do it, provided that there's no damned red carpet about it.'

On the Princess Royal's horsiness: 'My sister was a horse until she came out.'

On wartime unity: Voice from the crowd, 'Thank God for a good King!' George VI, 'Thank God for a good people.'

On hearing that the dashing but allegedly accident-prone Mountbatten was to command an aircraft-carrier: 'Well, that's the end of the *Illustrious*!'

On the artist John Piper's sketches of Windsor Castle, with their typically brooding backgrounds: 'Why is it, Mr Piper, that it always seems to be raining when you do a sketch of Windsor? You've been very unlucky in the weather.'

On the 'Skylon', a famous gimmicky exhibit at the Festival of Britain: 'Like the British economy, it has no visible means of support.'

ROYAL THANKS

In the year that she was widowed Queen Elizabeth was sent a copy of Dame Edith Sitwell's literary anthology, Book of Flowers, *and thanked her:*

I started to read it, sitting by the river, and it was a day when one felt engulfed by great black clouds of unhappiness and misery, and I found a sort of peace stealing round my heart as I read such lovely poems and heavenly words.

I found a hope in George Herbert's poem, 'Who could have thought my shrivel'd heart, could have recovered greennesse? It was gone quite underground' and I thought how small and selfish is sorrow. But it bangs one about until one is senseless, and I can never thank you enough for giving me such a delicious book wherein I found so much beauty and hope, quite suddenly one day by the river. Victoria Glendinning, *Edith Sitwell* (1981)

THE QUEEN AT A RACING CATASTROPHE, 1956

At luncheon yesterday I sat between Michael Adeane and the young Duke of Devonshire. They had both been standing with the royal party at Aintree when Devon Loch [the Queen's horse] collapsed. They said it was a really horrible sight. The public and the people in the enclosure took it for granted that the horse had won and turned towards the royal box and made a demonstration, yelling and waving their hats. The someone shouted out that there had been an accident, and the ovation stopped suddenly as if a light had been switched off. There was a complete hush. The Princess Royal panted, 'It can't be true. It can't be true!' The Queen Mother never turned a hair. 'I must go down', she said, 'and comfort those poor people.' So down she went, dried the jockey's tears, patted Peter Cazalet [the trainer] on the shoulder and insisted on seeing the stable-lads who were also in tears. Nicolson, *Diaries*

ROYAL PERFORMANCE

The Queen Mother conferred a knighthood on Peter Hall, director of the National Theatre, on 1 November 1977.

'The Queen Mum officiated. It is remarkable that a lady of 76 or 77 can stand in the same spot for an hour and three quarters and, apparently without being prompted and without a crib, remember a little something significant to say to over a hundred people in the right order.

Peter Hall's Diaries, ed. J. Goodwin (1983)

SAYINGS OF QUEEN ELIZABETH

To the Captain of their touring ship who asked if she had realized the danger of their boiler-room fire: 'Yes, indeed. Every hour someone said there was nothing to worry about, so I knew there was real trouble.'

On the Blitz: 'The destruction is so awful, the people so wonderful, they deserve a better world.'

On the impossibility of Royal evacuation during the blitz: 'The children won't leave without me; I won't leave without the King; and the King will never leave.'

On the bombing of Buckingham Palace: 'I'm glad we have been bombed; I feel I can look the East End in the face.'

To a Boer who said he could never quite forgive the British for having conquered his country: 'I understand that perfectly. We feel very much the same in Scotland.'

Elizabeth II

1952–

Succeeding to the crown at the same age—twenty-five—as her distant predecessor Elizabeth I, the young queen could not but inspire hopes of a new 'Elizabethan Age'. This was never possible. There was no Shakespeare, no Drake, but a nation battling with the aftermath of war and fall of empire. The Queen herself was not scholarly like the first Elizabeth but practical, sensible, extremely able, and genuinely caring. Instead of dallying with political suitors she married young and for love, founding a royal family of three boys and a girl. Her consort, Prince Philip Duke of Edinburgh, is intellectually gifted and has handed on to their son and heir a capacity for original thought. Parents, children, and royal cousins alike have shown an active sense of duty and responsibility never before shared by so many members of a British royal house at the same time. Born on 26 April 1926, the Queen has passed the age of sixty without the possibility of abdication in favour of the heir ever being discussed at the Palace, despite the media's inordinate interest in the subject. Her reign of thirty-seven years has seen democratic changes in the Court system, many of them introduced by the Queen and Prince Philip.

PRINCESS ELIZABETH IN 1928

She perched on a little chair between the King and me, and the King gave her biscuits to eat and to feed his little dog with, the King chortling with little jokes with her—she just struggling with a few words, 'Grandpa' and 'Granny' and to everyone's amusement has just achieved addressing the very grand-looking Countess of Airlie as 'Airlie'. After a game of bricks on the floor with the young equerry Lord Claud Hamilton, she was fetched by her nurse, and made a perfectly sweet little curtsy to the King and Queen and then to the company as she departed. Rose, *George V*

Later that year she accompanied her grandparents to Balmoral, where Winston Churchill was a fellow guest. 'There is no one here at all', he wrote to his wife, 'except the family, the household and Princess Elizabeth—aged 2. The latter is a character. She has an air of authority and reflectiveness astonishing in an infant.' Even in the nursery she was no stranger to royal duties. Sir Owen Morshead liked to recall a morning at Windsor Castle, when the officer commanding the guard strode across to where a pram stood, containing Princess Elizabeth: 'Permission to march off, please, Ma'am.' There was an inclination of a small bonneted head and a wave of a tiny paw. Ibid.

Princess Elizabeth announced the birth of her sister to Lady Cynthia Asquith:

'I've got a baby sister, Margaret Rose, and I'm going to call her Bud.' 'Why Bud?' 'Well, she's not a real Rose yet, is she? She's only a bud.'

Cynthia Asquith, *Illustrated Magazine*, 4 April 1963

THE TWO PRINCESSES AND THE ABDICATION

The royal assent was given to the Act of Abdication at 1.52 p.m. on Thursday, December 10. That afternoon Princess Elizabeth was puzzled by the bustle outside her front door; crowds collecting, noisy cheering crowds. Who were they? What was it all about? At last she went down and asked a footman. From him she learnt for the first time that uncle David had abdicated and Papa was King. Up the stairs she flew to tell Margaret the news.

'Does that mean that you will have to be the next Queen?' asked the younger sister.

'Yes, some day,' replied Lilibet.

'Poor you,' said Margaret. Elizabeth Longford, *Elizabeth R.* (1983)

Princess Elizabeth wrote an account of the morning of the coronation for her parents:

To Mummy and Papa. In Memory of Their Coronation, from Lilibet by Herself.

At 5 o'clock in the morning I was woken up by the band of the Royal Marines striking up just outside my window. I leapt out of bed and so did Bobo [Margaret MacDonald, the Princesses' nursemaid and later the Queen's dresser]. We put on dressing-gowns and shoes and Bobo made me put on an eiderdown as it was so cold and we crouched in the window looking on to a cold, misty morning. There were already some people in the stands and all the time people were coming to them in a stream with occasional pauses in between. Every now and then we were hopping in and out of bed looking at the bands and the soldiers. Royal Archives

LESSONS

Thousands of school children, girls and boys, had learnt their history from Warner and Marten's *History of England*, but this was not the same thing as being instructed by the master himself. Marten kept lumps of sugar in his pocket, as though his first ever girl-pupil might turn out to be a pony. He munched them himself, though, between bites at his handkerchief. He never looked directly at the Princess but occasionally addressed her in the way he addressed the Eton boys, as 'Gentlemen'. Longford, *Elizabeth R.*

THE QUEEN'S RIDING

Mountbatten got particular pleasure from coaching the Queen in her riding. 'I put Lilibet through her paces on Surprise on the lawn,' he noted in his diary, 'standing in the middle as I made her do dressage round me. She seemed to enjoy it as much as I did.' Whether she did or did not, she recognised his skill as a teacher and his eye for detail. Try as he might, however, he could never persuade her of the merits of riding side-saddle, a practice which he felt to be as elegant as it was functional, which she disliked, refusing to have recourse to it except at the ceremony of Trooping the Colour. Philip Ziegler, *Mountbatten* (1981)

Even this side-saddle riding ended in 1987 and the Queen rode alone in a small open carriage for Trooping the Colour.

Incidents from Prince Philip's Childhood

He was Elizabeth's third cousin and nephew of Lord Mountbatten.

Among the rare authentic stories of Philip's childhood is the account of how an insensitive grown-up arrived one day on the beach with toys for all but an invalid child, assumed to be ruled out for playthings. Philip, who was five, went into the house and collected all his personal treasures and presented them to her, the latest acquisition on top. It could have been showing-off. It was more probably an early glimpse of character. One of his equerries recently came out with something on this: 'What people don't realise is that he's immensely kind. No one has a bigger heart, or takes greater pains to conceal it.'

There's one other reliable tale of those times. It was at Berck Plage that Philip, having observed nomadic salesmen of oriental works on the beach, dragged out a couple of his hostess's carpets and tried setting up in business on his own. They were repossessed before he made a sale.

Basil Boothroyd, *Philip: An Informal Biography* (1981)

ACTING IN THE SCHOOL PLAY AT GORDONSTOUN

Philip's self-deprecation concealed phenomenal success at school, where he was head boy and won many trophies.

In *Macbeth* he only got the part of Donalbain (two-lines-and-a-spit) and gave as the reason why he got that, 'There was nobody else who could be trusted to enter on horseback and not fall off.' Ibid.

THE FIRST MEETING WITH PRINCESS ELIZABETH

George VI, in nostalgic mood, took his family to visit Dartmouth Naval College in July 1939. Philip was there, aged eighteen, Elizabeth thirteen. On the last evening Philip dined with them on board and then, with a flotilla of small craft, saw them off:

Half of Dartmouth College followed her out in all sorts of craft, from launches to sailing dinghies and rowing boats. Then all but one gave up. And who would you guess was in it, alone and still rowing, as 'Elizabeth watched him follow it through an enormous pair of binoculars'? Right. He was 'extremely attracted to his pretty little third cousin who looked at him with adoring blue eyes'. Through enormous binoculars. Ibid.

THE FIRST AIR-RAID WARNING

The two princesses were at Windsor during the blitz with their nurse 'Alah' and nursery governess 'Crawfie'.

We were all ensconced in the Castle, with Alah in the nursery, at the time the first bombs fell on Windsor. About two nights after we were settled in, the alarm bell went . . .

At the sound of the alarm bell I went at once to the shelter. There was no sign of the children and no sign of Alah, and everyone was in a state of fuss. Sir Hill Child [Master of the Household] came and said, 'This is impossible. They simply must come.'

I ran all the way to the nurseries, where I could hear a great deal of commotion going on. I shouted 'Alah!' . . .

Alah was always very careful. Her cap had to be put on, and her white uniform.

Lilibet called, 'We're dressing, Crawfie. We must dress.'

I said, 'Nonsense! You are not to dress. Put a coat over your night clothes, at once.'

They finally came to the shelter. By this time Sir Hill Child was a nervous wreck. He stood rather in awe of Alah, but he said, 'You must understand that the Princesses must come down at once. They must come down whatever they are wearing.'

The shelter was in one of the dungeons, not a particularly inviting place anyway. Marion Crawford, *The Little Princesses* (1950)

Princess Margaret says that in fact it was not a dungeon but a basement in the Brunswick Tower where they lived.

The atmosphere was gloomy, and there were beetles. The walls had been reinforced, and beds put up, but that first night for some reason nothing was ready . . .

The little girls [Lilibet was fourteen] were very good. They took it all most calmly. Margaret fell asleep on my knee. Lilibet lay down and read a book . . .

It was two in the morning before the all-clear sounded . . . Sir Hill Child bowed ceremoniously to Lilibet. 'You may go to bed, ma'am,' he said.

Ibid.

Princess Margaret later elaborated on the castle's defences:

They dug trenches and put up some rather feeble barbed wire, and the feeble barbed wire of course wouldn't have kept anybody out but it kept us in . . . Princess Margaret speaking on *Desert Island Discs*, BBC Radio 4, 3 April 1981

GLEE SINGING AT WINDSOR CASTLE DURING THE WAR

On one occasion, some last-minute military duty having prevented two of the three basses from being present, I found myself called in as a stop-gap. Miss Barham, vastest and most splendid of the Eton Dames, presided—sitting in the centre of the front row between the two princesses—and I was just behind Princess Margaret who, as I had been told, usually managed to enliven an otherwise rather solemn affair. At the end, Dr Harris [organist at St George's Chapel] always invited Princess Elizabeth to choose a favourite song, which this time was one entitled (so far as I can remember) 'Oh! that I were but a little tiny bird'. Princess Margaret did not fail us. Turning to the officer seated on the other side of her she said in a very audible whisper, 'I think Miss Barham would have to be jet-propelled.' Blunt, *Slow on the Feather*

THE PRINCESSES' VE DAY, 8 MAY 1945

The king was persuaded to let his daughters join the rejoicing throng outside the Palace gates. 'Poor darlings,' he wrote in his diary, 'they have not had any fun yet.' The account of their evening was given by Madame de Bellaigue, the Princesses' French teacher.

On V-E Day the Princesses were allowed, chaperoned by Major Phillips, Crawfie and myself and accompanied by young officers, to mix with the crowd.

The King drew the line about Picadilly Circus, which was to be avoided. I shall never forget running wildly down St James's Street, with a puffing Major of the Grenadiers, to keep pace with the Princesses.

When we reached the Palace they shouted like the other people, 'We want the King.' 'We want the Queen.' On the whole we were not recognised. However, a Dutch serviceman, who attached himself to the end of our file of arm-in-arm people (the Princesses being in the centre of the file) realised who the Princesses were. He withdrew discreetly and just said, 'It was a great honour. I shall never forget this evening.'

All our group got back to the Palace through a garden gate. The Queen was anxiously waiting for us. Her Majesty provided us with sandwiches she made herself. Longford, *Elizabeth R.*

Many years after the event, Elizabeth, as Queen, added a touch of her own to the memories of V-E Day. She met Hammond Innes at a literary party:

Innes said that for some extraordinary reason they started talking about police helmets. 'How would you know about them, Ma'am?' she was asked, and replied, 'Of course I do, I knocked one off on VE Day.'

Ann Morrow, *The Queen* (1983)

IN THE AUXILIARY TERRITORIAL SERVICE

Princess Elizabeth joined the ATS in 1945.

The Queen recently held a private reunion at Buckingham Palace with the girls—now women nearing 60—who had been her comrades during the war when she enlisted in the ATS and trained as transport driver. They remembered her as 2nd Subaltern No 230873 Windsor, a 19-year-old who had never driven before and, in the early days, stalled her heavy transport on Windsor Hill.

'What d'yer think you're doing?' demanded a policeman another time, not recognising her in the driving seat of a truck she had stranded broadside. 'I couldn't tell him because I didn't know!' the Queen laughed, in telling the story against herself. *Majesty Magazine* (1986)

PRINCESS MARGARET OFFENDED BY THE PRESS, AGED FIFTEEN

I was with some of my fellow Sea Rangers in a boat on the lake at Frogmore. And *what* do you think appeared in the newspapers? They said I had pulled the bung from the bottom of the boat! That made me frightfully cross. I was part of a *team* and very proud of it, I might tell you. I would never have dreamt of doing something so irresponsible.

Christopher Warwick, *Princess Margaret* (1983)

Marriage

WHEN DID PRINCE PHILIP FIRST THINK ABOUT MARRIAGE?

Boothroyd asked him this question and he replied:

I suppose one thing led to another. I suppose I began to think about it seriously, oh, let me think now, when I got back in 'forty-six and went to Balmoral. It was probably then that we, that it became, you know, that we began to think about it seriously, and even talk about it. And then there was their excursion to South Africa, and it was sort of fixed up when they came back. That's really what happened.' Boothroyd, *Philip*

The date set for the wedding of Princess Elizabeth and Prince Philip at Westminster Abbey was 20 November 1947.

A 'CONGRATULATORY' LETTER TO THE PRINCESS

One of the many [letters] was an attack on the whole Royal Family ending unexpectedly with the words, 'Best wishes to you from the lawful Queen of England, commonly known as Mrs E. M. Ottewell'. Ibid.

A CONTROVERSIAL WEDDING PRESENT

The Duke of Windsor recalled an 'intimate gift' once received by Queen Victoria from an African chieftain: she had sent him a present of a Court uniform and he replied with a loin-cloth. Perhaps Queen Mary knew this story and told it to her eldest son. If so, it may have come into her mind again when she saw what she took to be a distasteful offering displayed among Lilibet's wedding presents.

Gandhi wove the thread for a crocheted tray-cloth. Unfortunately Queen Mary, who was not usually straightlaced, mistook the tray-cloth for a loin-cloth. 'What a horrible thing', she said, and Philip had to drown the hisses of 'indelicate' in loud testimonies to Gandhi's greatness. 'Queen Mary moved on in silence,' noted Lady Airlie her lady-in-waiting. Next time round the display-tables, Princess Margaret tactfully hid the object from Granny's view.

Great people of the past were recalled in two other gifts, both of food and no doubt thought appropriate in those days of rationing, though less practical than the 500 cases of tinned pineapple from the Government of Queensland: a piece of condensed soup from the stores of HMS *Victory* at the Battle of Trafalgar and some chocolate sent by Queen Victoria in 1900 to her troops in South Africa. Longford, *Elizabeth R.*

VISIT TO PARIS, 1948

The princess was pregnant, the Press merciless. Sir John Colville was the princess's private secretary.

One evening was to be arranged as a private one. We went to a most select three-star restaurant; the French had been turned out, so we found a table, just a party of us all alone in this vast restaurant. Prince Philip spotted a round hole in a table just opposite us, through which the lens of a camera was poking. He was naturally in a frightful rage. We went on to a night club, again the French all turned out. One of the most appalling evenings I have ever spent. Everybody dressed up to the nines—nobody in either place—except the lens.

Ibid., quoting Colville

BIRTH OF PRINCE CHARLES, 14 NOVEMBER 1948

Princess Elizabeth wrote of her baby's hands to a friend:

Fine, with long fingers—quite unlike mine and certainly unlike his father's. It will be interesting to see what they become.

Anthony Holden, *Charles Prince of Wales* (1979)

George Bernard Shaw's secretary, Miss Blanche Patch, once recounted some of his eccentric views on the Royal Family:

He had a singular comment upon the birth of Prince Charles, remarking to me that perhaps it was a pity that Princess Elizabeth had had an heir. I asked him why. It would probably be better, said he, if Margaret were allowed to come to the throne because the second child of a reigning monarch often made a better sovereign than the elder; and he indicated both George the Fifth and George the Sixth to support the theory.

Dear Mr Shaw, ed. Vivian Elliot (1987)

A VISIT TO GREECE, 1950

Prince Philip took his wife to see his homeland a few months before the birth of Princess Anne. She sailed in the more comfortable HMS Surprise, *the C-in-C's Despatch Vessel, while Philip escorted her as commander of* Magpie.

Signals of great gaiety passed between the ships. Some were in clear. *Surprise* to *Magpie*, 'Princess full of beans': *Magpie* to *Surprise*, 'Is that the best you can give her for breakfast?'

Boothroyd, *Philip*

THE ACCESSION OF QUEEN ELIZABETH II, 22 JANUARY 1952

The princess and Prince Philip were on tour in Kenya when they heard that the king had died suddenly in his sleep.

6 February: She became Queen while perched in a tree in Africa, watching the rhinoceros come down to the pool to drink. Nicolson, *Diaries*

The Queen's cousin and lady-in-waiting, Lady Pamela Mountbatten, afterwards described how the Queen had reached her 'perch' in the wild fig tree into which was built the Treetops Hotel.

At first all looked tranquil. Colobus monkeys swinging in the Cape chestnut trees whose purple flowers were reflected in the still pool. But in fact the Princess's party were entering the oval clearing—200 yards by 100 yards—an hour after forty-seven elephants including five cows with calves and three trumpeting bulls, had crashed out of the forest. A white pillowcase was fluttering on the roof of Treetops to warn the party of danger. When they saw a huge old cow elephant with two calves standing besides Treetops flapping her ears, a whispered consultation took place. There was fifty yards of open ground to cover in front, and behind them a narrow forest path, with safety ladders hanging from some of the trees. Should they advance or retreat? They were ten minutes away from their cars. The cow had not scented them and there were three guns in their party including Prince Philip's. 'Go ahead!' he said.

It was decided to divide the party in two . . . The Princess would run less risk by going steadily forward than going back. Steadily and silently she went, not treading on a twig; up the Treetops ladder into the thirty-foot tree, from which she would descend as Queen.

Before sunset they had a hilarious time watching the elephants blowing dust over some pigeons and themselves, with plenty of trumpeting. Treetops had been decorated with white bunting as if in their honour, some baboons having reached through a window and stolen the toilet rolls out of the little house, and draped the branches of the fig tree.

Longford, *Elizabeth R.*

The Coronation, 2 June 1953

TELEVISION AT THE CORONATION?

When the Cabinet papers of 1952 were opened under the 'thirty year' rule on 1 January 1983, it was revealed that there had been an unsuspected controversy over coronation television. After a first inspection of the papers, journalists reported on 2 January . . . 'The Queen wanted to ban TV at Coronation' (*Sunday Telegraph*); 'When the Queen said No but the nation said Yes' (*Sunday Express*).

A week passed, and this hasty—and false—impression was corrected by Sir John Colville, who had been the Prime Minister's representative on the

Coronation Joint Committee. To his dismay, he had heard a chorus of weighty voices, including the Earl Marshal, Archbishop of Canterbury, Churchill and the Cabinet urging the Queen to spare herself the strain, heat and glare of the TV cameras by refusing to be televised.

To their surprise the Queen received this message coldly. In fact she favoured television, believing that all her subjects should have the chance to see her coronation. Churchill at once agreed. 'After all it was the Queen who was to be crowned and not the Cabinet!' A second Cabinet meeting was called to reverse their earlier decision. 'Thus it was', wrote Colville, 'that the new 26-year-old Sovereign personally routed the Earl Marshal, the Archbishop, Sir Winston Churchill and the Cabinet, all of whom submitted . . . with astonishment, but with a good grace.' No one had guessed that this reserved young woman would be the first to propel her reign into the television age. Yet her motivation was clear. Nothing must stand between her crowning and the nation's right to participate. Ibid.

THE ABBEY CARPET

The correct length of pile for the carpet at Westminster Abbey was insisted on by the Queen. (At King George VI's coronation the pile had been so deep that some of the old peers could hardly drag their robes across it; while at Edward VII's the heels of Queen Alexandra and her ladies became embedded.) Even so the pile was laid the wrong way, so that the metal fringe of Elizabeth II's golden mantel caught in it and clawed her back when she tried to move forward. She had to signal the Archbishop of Canterbury,

'Get me started!' Ibid.

THE HOLY OIL FOR THE CORONATION

The oil made for Edward VIII and used by George VI had been bombed and the firm who made it up had gone out of business.

It took some time to track down an elderly relative of the firm who had kept, for sentiment's sake, a few ounces of the original base to the compound, and into the breach stepped J. D. Jamieson, a Bond Street chemist. He made up a fresh batch of oil to a formula almost identical to that employed for Charles I, and the whole nation applauded not only his expertise but also his sacrifice when it was learnt that in order to improve his sense of smell, he gave up smoking for a whole month before starting work. Lacey, *Majesty*

Robert Graves the poet after an audience with the newly crowned Queen:

'The holy oil has taken for that girl. It worked for her all right.'

Alastair Forbes, 'After the Royal Wedding', *Books & Bookmen* (October 1981)

CRAWFIE'S DOWNFALL

Marion Crawford, former governess of the princesses, much to the royal family's annoyance turned to journalism, after publishing The Little Princesses *in 1950.*

In 1955 the magazine columns proved the undoing of the ex-governess, for the magazine [*Woman's Own*] went to press some time before the event on which Miss Crawford chose to peg her copy, and, not content to write background pieces, she purported to be a reporter on the spot. 'The bearing and dignity of the Queen at the Trooping of the Colour ceremony at the Horse Guards Parade last week', she wrote in *Woman's Own* dated 16 June 1955, 'caused admiration among the spectators . . .'

Yet unfortunately the Trooping of the Colour in 1955 was cancelled because of a rail strike, and Royal Ascot was postponed. So Crawfie's sparkling picture of the green turf, white rails and open carriages spanking down the course—'Ascot this year had an enthusiasm about it never seen there before'—created a sensation she did not intend. She concluded her career as a writer more rapidly than that as a governess. Lacey, *Majesty*

LORD ALTRINCHAM AND THE QUEEN'S IMAGE

In August 1957 Lord Altrincham's National and English Review *was devoted to the modern monarchy. Altrincham (John Grigg) attacked the Queen's entourage for being 'almost without exception the "tweedy" sort'. They put speeches into her mouth:*

The personality conveyed by the utterances which are put into her mouth is that of a priggish schoolgirl, captain of the hockey team, a prefect and a recent candidate for confirmation . . . [Her style of speaking is] a pain in the neck . . . [Yet when she is older] the Queen's reputation will depend, far more than it does now, upon her personality . . . She will have to say things which people can remember and do things on her own initiative which will make people sit up and take notice.

The reaction was either violent or aloof:

Lord Strathmore: 'Young Altrincham is a bounder. He should be shot.'
Duke of Argyll: 'I would like to see the man hanged, drawn and quartered.'
Lord Scarborough: 'I am not interested in Lord Altrincham's views.'

Duke of Beaufort (Master of the Horse) replied to the Press that he spent most of his time hunting in Gloucestershire and his only influence on the Palace affected 'a certain class of horse belonging to the Queen'.

Robert Lacey wrote:

When Lord Altrincham was slapped in front of television cameras by a sixty-four-year-old representative of the League of Empire Loyalists, the story seemed blessed with eternal life. *The Observer*, for whom Altrincham had written some articles on progressive Toryism, disowned its former contributor, and so did 'the elected representatives of the rate payers of this ancient town of Altrincham' . . . [They] wished 'completely to dissociate' from [young Altrincham]. 'No town has a greater sense of loyalty to the Crown than the Borough of Altrincham.'

The *Daily Mail* found to its horror that a majority of its readers aged between sixteen and thirty-four agreed with Lord Altrincham and that all age groups felt that the Court circle around Elizabeth II should be widened . . .

Today Lord Altrincham, who became John Grigg in 1963 . . . blames the 1957 storm on the 'Shintoistic atmosphere of the post-Coronation period . . . There was a tendency—quite alien to our national tradition—to regard as high treason any criticism of the monarch however loyal and constructive its intent' (*Sunday Times*, 1972). Ibid.

THE EFFECT OF ALTRINCHAM ON CHRISTMAS BROADCASTS

'We listened to the Queen on the wireless, and in spite of the long wash of the Caribbean seas she came across quite clear and with a vigour unknown in pre-Altrincham days.' Nicolson, *Diaries*

ROYAL ENTERTAINMENT

The Queen has always enjoyed Commonwealth parties, especially when Churchill's impish humour played upon them.

After a dinner at Buckingham Palace, Churchill engaged the Muslim Prime Minister of Pakistan in conversation.

'Will you have a whisky and soda, Mr Prime Minister?' 'No, thank you!' 'What's that?' 'No, thank you!' 'What, why?' 'I'm a teetotaller, Mr Prime Minister.' 'What's that?' 'I'm a teetotaller.' 'A teetotaller. Christ! I mean God! I mean Allah!'

R. Menzies, *Afternoon Light: Some Memories of Men and Events* (1967)

Menzies went on to relate that there was at once a general rush to tell the story to the Queen. Menzies thought he had got there first but the Queen interrupted him.

'You're too late; Tommy Lascelles has told me about it and Tommy says that as the footman, in his astonishment, dropped the tray and caught it before it reached the carpet, without spilling a drop, he ought to be put into the English cricket team, where the slip-fielding needs improving.'

Menzies, *Afternoon Light*

AN INVESTITURE

At an Investiture, the writer, Anthony Powell, was behind a man who was asked by the Queen, 'What do you do?' 'I kill mosquitoes,' he replied, and the Queen said with relief, 'Oh, good.' Morrow, *The Queen*

'GET ON YOUR TOES' AT THE PALACE

It did not need film cameras to show that if the royal family are with people who are themselves interesting, they respond. Once, escaping for a minute from a cluster of worthies, the Queen called the attention of the former German Ambassador, the silver-haired, courtly Dr Hans Reute, to a barefooted Franciscan monk in a group nearby. 'I am always fascinated by their toes, aren't you?' said the Queen, and anyone who had not heard her might have thought from her serious expression that she was talking about the mark against the pound in the Stock Exchange that day. Ibid.

THE QUEEN'S HUMOROUS STYLE

Two women in headscarves and those green quilted jackets known as 'Husky's' were about to leave a teashop in Norfolk when they were stopped by another customer at a nearby table, who craned forward, her elbow almost in the cream, to say to one of them, 'Excuse me, but you do look awfully like the Queen.' 'How very reassuring,' the Queen smiled, exiting with her cake for the drive back to Sandringham. Ibid.

It is sad that the Queen is not able to show in public this highly developed sense of fun . . . Once, when she was being taken round an artificial insemination unit by the urbane chairman of the Milk Marketing Board, Sir Richard Trehane, they came to an object which prompted the Queen to ask, 'What is that?' Trehane replied, 'It is a vagina, Ma'am,' and the Queen looked up at him without a flicker of the eyelids, 'Ask a silly question . . .!' Ibid.

'HO-HO'

The Queen enjoys political gossip. On an evening when Sir Harold Wilson, Labour prime minister, turned up for his weekly audience at the Palace, he said:

'Have you seen the latest editions of the *Evening Standard* today, Ma'am?' The evening papers were full of a story about Giscard d'Estaing driving all over France with some ladies of doubtful virtue in his car. Sir Harold told the Queen about this as she always loved stories about the French: 'Ho-ho, Mr Wilson!' she said with relish.

Mr Wilson then had to go to France and Prime Ministers always have to ask the Queen's permission to leave the country. 'It is my humble duty to beg Your Majesty's leave to go to Paris, Ma'am.' The Queen chuckled and was obviously very amused, 'thinking I'd be taking Giscard's car all over Paris, etc.' He went to France, had a private dinner with the French President; reported to the Foreign Office officials travelling with him that he did not feel too well, and had an early night.

As always, with that slightly mischievous interest in political gossip, the Queen could hardly wait to ask the Prime Minister the following week how he had got on. 'Well, how did it go?' she asked immediately, and Mr Wilson, gravely sitting on the other side of the fireplace, reported in detail about the meeting and the points raised until the Queen leant forward and, with a solemn look, inquired, 'Ho-ho, Mr Wilson?'

'Ma'am,' he replied with a saintly expression. 'There was no ho-ho; everyone went to bed early.'

Ibid.

THE SILVER WEDDING, 1972

The Queen has acquired the confidence to lighten her official speeches with dry wit. At the Guildhall banquet she began:

I think everyone will concede that today, of all occasions, I should begin my speech with 'My Husband and I'. She then went on to tell the story of a bishop who, when asked what he thought of sin, replied that he was against it. If anyone asked her the same question about marriage and family life, she would say 'I am for it.'

Lacey, *Majesty*

THE QUEEN NOT UNAWARE OF RESENTMENTS

Once a woman was splashed by the royal vehicle with mud in a lane near Sandringham.

Queen: 'I quite agree with you, madam.'
Prince Philip: 'Hmm? What did she say, darling?'
Queen: 'She said, "Bastards!" '

Ibid.

SCENES FROM THE QUEEN'S LUNCHEONS AT BUCKINGHAM PALACE

In order to meet more people with a variety of backgrounds and interests, the Queen began an experiment during the first decade of her reign of giving small dinners and luncheons for about a dozen people. They have proved most successful. On one of the occasions she described her first televised Christmas broadcast:

First the electricians drilled holes through the walls at Sandringham to get the cables in, and incidentally let in so much icy air that she trembled even more from cold than fright. 'The family looked absolutely horrified at Christmas luncheon and thought I was going to break down with nerves.' The theme of her message—'I welcome you to the peace of my own home'—was beginning to look somewhat ironical. Next the make-up girls dabbed spots of bright yellow paint on her forehead, cheekbones, nose and chin, to take off the 'shine'; while the parting in her hair, which looked on the screen like a long white road, had to be toned down too. At least this part of the proceedings made her laugh. *Eyewitness account*

FAMOUS PEOPLE AT THE QUEEN'S TABLE

Cecil Day-Lewis, Poet Laureate, 1968–72, was at a royal luncheon.

He put his feet on what he thought was a well-placed footstool. It turned out to be a recumbent corgi. Rose, *Kings, Queens and Courtiers*

To the great conductor Barbirolli she said,

'Tell me, Sir John, you have been in the public eye for many years. You must have received some adverse criticism from time to time. How do you react to it?'

'I do nothing about it, ma'am. I made up my mind long ago not even to notice it. It has no effect whatever.'

Her Majesty looked thoughtful. 'I wonder if that can really be possible?'

If Sir John had offered her an unattainable ideal others gave her moments of pure relaxation. Hugh Scanlon of the Trades Union Congress was to see a piece of his roast potato fly off his plate on to the carpet. He hoped that Her Majesty had not noticed—until one of the corgies approached the morsel, sniffed it, turned up its nose and stalked away.

'It's not your day, Mr Scanlon, is it?' said the Queen.

Asparagus was another source of fun. A guest sitting immediately on her left realised that HM would be served first and himself last. He was eager to see how she would deal with the stout, buttery, home-grown stems. After he was served, the Queen turned to him with a sweet smile: 'Now, it's my turn to see you make a pig of yourself.' Longford, *Elizabeth R.*

A RECEPTION FOR THE MEDIA

11 February 1975: To Buckingham Palace for the Queen's reception for the media, at least I suppose that's where we were. Newspaper editors; television controllers; journalists and commentators; Heath looking like a tanned waxwork; Wilson; Macmillan a revered side show, an undoubted star; a few actors (Guinness, Ustinov, Finney); and all the chaps like me—

John Tooley, George Christie, Trevor Nunn. And Morecambe and Wise.

It was two and a half hours of tramping round the great reception rooms, eating bits of Lyons paté, drinking over-sweet warm white wine, everyone looking at everyone else, and that atmosphere of jocular ruthlessness which characterises the Establishment on its nights out. Wonderful paintings, of course, and I was shown the bullet that killed Nelson.

As we were presented, the Queen asked me when the National Theatre would open. I said I didn't know. The Duke asked me when the National Theatre would open. I said I didn't know. The Queen Mother asked me when the National Theatre would open. I said I didn't know. The Prince of Wales asked me when the National Theatre would open. I said I didn't know. At least they all knew I was running the National Theatre.

Home by 2 a.m. with very aching feet. Who'd be a courtier?

Peter Hall's Diaries

ROYAL PERFORMANCE AT THE NATIONAL THEATRE

25 October 1976: *Campiello* went very badly indeed and I was deeply ashamed. The presence of royalty nearly always ossifies the public in a theatre, but this particular play meant nothing to this posh audience. The actors were like men struggling through a nightmare. And the special fanfare version of God Save the Queen, which we commissioned from Howarth Davies, and which sounded well when it was rehearsed, sounded horrible, with many many mistakes by the Household Cavalry trumpeters.

The one undoubted success of the entire opening ceremony was Larry who, before *Campiello* started, made an elegant, though over-written, speech. The audience gave him a standing ovation. So they should have done. But it was difficult for a play to follow that.

HM did her job magnificently. She didn't eat and she didn't drink. She chatted well everywhere, was extremely gracious to everybody, and worked with a will. The atmosphere was warm and friendly and it was a party. But the play was dreadful. Ibid.

THE SILVER JUBILEE, 1977

Wednesday 20 July: . . . Off to Buckingham Palace for the Queen's Jubilee party. About 800 people there, the public rooms all open . . . The Royal Family wandered through speaking to whoever took their fancy. Breakfast was served at 1 a.m. I came out of the press of people bearing three plates of sausages and scrambled eggs, a cigar between my teeth, and nearly knocked the Duke of Edinburgh over. He greeted me cheerfully, and a red-coated flunkey removed the cigar from my mouth so that I could speak. Ibid.

A NEW MASTER OF THE HOUSEHOLD: 'ALL THAT BLISTERS'

Making his debut, he took a great deal of trouble to ensure that the food should remain hot—even on gold plate. To his chagrin, it was tepid. Afterwards he apologised to the Queen. 'Don't worry,' she replied. 'People come here not for hot food but to eat off gold plate.'

K. Rose, 'Albany', *Sunday Telegraph*, 20 April 1986

PARALYSING EFFECT OF ROYALTY

An eminent man of letters who was also a radical visited the Palace. He accepted the invitation 'in a spirit of mingled curiosity and ribaldry'. The mood survived until the Queen appeared and her guests were presented. 'Suddenly I felt physically ill,' he said. 'My legs felt weak, my head swam and my mind went totally blank. "So you're writing about such-and-such, Mr—" said the Queen. I had no idea what I was writing about, or even if I was writing a book at all. All I could think of to say was, "What a pretty brooch you're wearing, ma'am." So far as I can recall she was not wearing a brooch at all. Presumably she was used to such imbecility. Anyway she paid no attention to my babbling and in a minute or two I found that I was talking sense again.'

Philip Ziegler, *Crown and People* (1978)

A ROYAL GARDEN PARTY, NEW STYLE, 1979

Thursday, 24 July: . . . I always enjoy this—the centre courtyard and the tweaked curtains in the upper floors, the glimpse of that lovely low-ceilinged oval drawing room, the sudden burst of green grass, brass bands, striped tents, multi-coloured hats. Spot my old friend John W. (he's clearly on the same rota as me)—carrying, as usual, his plastic shopping bag into which he pops the occasional cup-cake to take back to his village children . . . 'straight from the Queen's tea table', he tells them . . . Tea in the Royal Tent, standing in its own circus ring guarded by Yeomen of the Guard. A red ticket, a permissive wave from a gloved hand, a table on which stand twenty toppers, upturned as if waiting to be filled up from a teapot, royal footmen, helpful ushers, instructions. Royalty in close proximity always charges the air and causes behaviour to go into a different gear. Preoccupations with falling crumbs, top heavy teaspoons, the tendency of high heels to sink inexorably and anchor-like into the turf. Beyond the ropes, the guests sit on chairs or stand gazing with frank curiosity at the Queen and us downing eclairs. We behave under such scrutiny like extras in the background of a Drury Lane musical . . . feigned conversation interest . . . tiny forced laughs . . . exaggerated courtesies . . .

Hugh Casson, *Diary* (1981)

THE QUEEN AND THE (LABOUR) PRIVY COUNCIL, 1964

We drove to the Palace and there stood about until we entered a great drawing-room. At the other end was this little woman with a beautiful waist, and she had to stand with her hand on the table for forty minutes while we went through this rigmarole. We were uneasy, she was uneasy. Then at the end informality broke out and she said, 'You all moved backwards very nicely,' and we all laughed. And then she pressed a bell and we all left her. Richard Crossman, *Diaries of a Cabinet Minister* (1975)

CROSSMAN'S GAFFE, 1966

One summer's evening Crossman was dining in the country with the Queen's close friend Lord Porchester. Oblivious of the fact that he was going to meet the Queen in a fortnight's time, he denounced her 'snobbish' and 'dreary' court. When he arrived at the Palace, the Queen, he said, showed herself 'very clever' in immediately clearing the air by subtly referring to his gaffe:

'Ah, Lord Porchester was telling me about you.' Ibid.

GAME AND SET TO THE QUEEN, 1967

Crossman, as Lord President of the Council, was expected to attend the ceremonies involved in the Opening of Parliament by the Queen and to wear morning dress. Not possessing such a garment and in any case being against dressing up, he wrote, unwisely, to the old Duke of Norfolk asking to be excused. The Duke replied icily that only Her Majesty could excuse him. With the temperature still falling, Crossman found himself calling on Sir Michael (later Lord) Adeane in his mournful 'little office'. Adeane said,

'You mucked things up terribly by writing to the Duke of Norfolk . . . I can clear it now if you really do not want to go.'

Then he sent a fast one that the Lord President failed to return:

'Of course the Queen has as strong a feeling of dislike of public ceremonies as you do. I don't disguise from you the fact that it will certainly occur to her to ask herself why you should be excused when she has to go, since you are both officials.' Ibid.

THE QUEEN AND MARGARET THATCHER

It is said that Mrs Thatcher felt embarrassed at a public ceremony because her frock closely resembled that of the Queen. Afterwards, Downing Street discreetly asked the Palace whether there was any way in which the Prime Minister could know of the Queen's choice on such occasions. The reply was reassuring and dismissive: 'Do not worry. The Queen does not notice what other people are wearing.' Rose, *Kings, Queens and Courtiers*

THE QUEEN AND HER PRIME MINISTERS

James Callaghan, a former prime minister, described his weekly audiences.

I used to be at the Palace for 1 to 1½ hours, never less than 1 hour, unless both had dinner engagements—evening audiences always—no drink—that was the rule apparently—all [her prime ministers] treated the same. But each thinks he is treated in a much more friendly way than the one before! Though I'm sure that's not true. The Queen is more even-handed. What one gets is friendliness but not friendship.

Longford, *Elizabeth R.*

The weekly meetings between the Queen and Mrs Thatcher—both of the same age—are dreaded by at least one of them . . . [the Queen is] more matter of fact . . . [the Prime Minister] more like a Queen.

Anthony Sampson, *The Changing Anatomy of Britain* (1982)

THE QUEEN ON HER FAVOURITE PRIME MINISTER

'Winston, of course, because it was always such fun.'

THE QUEEN'S COURAGE, 1961

Harold Macmillan, the prime minister, approved of the royal resolution to visit Ghana, despite the risk of a bomb being aimed at Nkrumah and killing the Queen as well:

The Queen has been absolutely determined all through. She is grateful for MPs' and Press concern about her safety, but she is impatient of the attitude towards her to treat her as a *woman*, and a film star or mascot. She has indeed 'the heart and stomach of a man'.

Harold Macmillan, *Pointing the Way* (1972)

THE QUEEN 'UNDER FIRE'

Saturday, June 13, 1981, was the date chosen by an unhappy youth of seventeen to arm himself with six blank cartridges in a replica pistol and perform a sad little act of self-advertisement—firing them at the Queen to frighten her during her official Birthday Parade known as Trooping the Colour . . .

The Queen might have been caught at a disadvantage, for she was riding side-saddle. An expert equestrian, she normally rides astride. But since the earliest days she has always worn a long navy blue riding-skirt with her military tunic for this ceremonial occasion, and had spent three weeks beforehand practising in the Royal Mews. She would not have chosen a side-saddle on which to control a startled, prancing horse. Yet her control was complete.

Two days later it happened to be the Garter ceremony at Windsor Castle. Every post was bringing in sackfuls of congratulatory letters to the Queen, and her guests at Windsor added their voices to the chorus of praise for her calm and skilful management of the black mare, Burmese. 'It wasn't the shots that frightened her,' said the Queen in defence of her nineteen-year-old charger, 'but the cavalry!' Two officers of the Household Cavalry had quite correctly spurred their horses forward to take up their positions on each side of their Sovereign, so that Burmese began to prance; it was at the sight of these unexpected companions, not lack of courage in the firing-line.

Longford, *Elizabeth R.*

THE INTRUDER IN THE QUEEN'S BEDROOM, FRIDAY, 9 JULY 1982

The *Daily Express* scooped a sensation: 'INTRUDER AT THE QUEEN'S BEDSIDE—*She kept him talking for 10 minutes . . . Then a footman came to her aid.*'

That afternoon the Home Secretary confirmed the news to a shocked House of Commons. A man had entered Her Majesty's bedroom . . . and was under arrest . . .

At 6.45 a.m. on the Friday, Michael Fagan, aged thirty-five and under the delusion at times that his father was Rudolf Hess, was seen climbing the Palace railings by an off-duty constable who telephoned the Palace police. They made a cursory outside inspection while Fagan was making an inside inspection of the door leading from the Queen's Stamp Room into the Palace's interior. It was locked, so he left by the unlocked window through which he had entered and began looking for another way into the Palace. The published report did not include the remark of a sergeant in the Palace's police control room, where the Stamp Room alarm had rung twice: 'There's that bloody bell again!'

After shinning up a drainpipe and removing his sandals and socks, Fagan padded along a corridor leading to the private apartments, picking up and smashing a glass ashtray on the way. Domestic troubles at home involving his wife, four children, two stepchildren and parents had suddenly made him think of slashing his wrists with the jagged edges in front of the Queen. The whole escapade, said a relative in court afterwards, was in no way directed against the Queen, who he admired, 'It was a cry for help.'

The 'cry for help' might well have come from the Queen, when she was awakened at 7.18 a.m. to see that her bedroom curtains were being drawn, not by the usual maid bringing her tea and papers at 7.45, but by a bare-footed young man in jeans and tee-shirt who sat down on her bed and dripped blood from his right thumb on to the bed clothes.

The visitor launched into his unhappy family affairs, the Queen responding with unhurried interest . . . while wondering how she could get help without frightening him. First she pressed her night alarm button connected with the police control room. It was not working. Then she rang her bedside bell into the corridor. No answer. The maid was cleaning in an adjacent room with the door shut and the footman on duty was exercising the posse of eleven corgis. The only armed guard in the corridor had gone off duty according to schedule at 6.0 a.m.

After the Queen had twice telephoned the Palace operator to send police to her bedroom—no one made haste, HM's voice sounding so 'calm'—she was immensely relieved when her visitor asked for a cigarette. This gave the Queen her chance.

'You see I have none in this room,' she said. 'I will have some fetched for you,' and she went out into the corridor and found the maid.

'Bloody hell, ma'am,' exclaimed the horrified Yorkshire woman. 'He oughtn't to be in there.' At that moment the footman returned with the eleven corgis. He and the maid got Fagan into a pantry, where the footman plied him with cigarettes while the Queen kept off the indignant corgis; for Fagan was showing signs of panic. At last the police arrived—eight minutes after the Queen's first call—and Fagan was led away. Ibid.

THE QUEEN AS A SITTER

Norman Hepple the portrait painter described his sessions with the Queen.

She is an extremely amusing person. Once I was talking about the portraits that had been painted of her and saying that she had not been very lucky with her likenesses. She said, 'One day when I was driving out of the Palace the car stopped just outside the gate and an old lady came up and peered in at me and said'—and she mimicked a cockney voice beautifully—'She ain't very like her pictures is she.' It amused me that she put it that way round instead of complaining that 'my portraits are not really like me'. . . . I was painting her standing in very heavy robes, which is enormously tiring. After she had stood for an hour I said to her, 'Ma'am, wouldn't you like to sit down?' She said, 'No, I am used to standing. I have been standing all my life.' No one else would have put up with it.

Christopher Hibbert, *The Court of St James* (1979)

THE QUEEN'S CARIBBEAN TOUR, 1985

She has three basic expressions . . . a dour glare verging on a scowl, delight, and lively interest . . . It is the last she uses at garden parties. Some people take it for a genuine desire to learn all there is to know on the subject in hand. Thirty-odd years of experience have taught her how to cope. She simply walks away—then turns and flashes a brilliant smile from a few feet away. Short of yelling, 'Come back, I haven't finished yet,' there is nothing they can do and no way they can feel hurt . . .

At dinner on *Britannia* custom insists that nobody continues eating after HM has stopped. This used to leave slow eaters hungry and sometimes resentful . . . These days she pushes her last few peas around the plate until everyone else is finished.

Simon Hoggart, 'Caribbean Queen', in *The Queen Observed*, ed. Trevor Grove (1986)

VISIT TO MOROCCO 1986: COMING TO THE BOIL IN A TEA-TENT

Everything was hopelessly behind schedule—no food, no king.

Elizabeth II pointed irritably at the programme. Mopping her brow, she looked helplessly towards the Atlas Mountains, where a dramatic line of a thousand horsemen had waited patiently since dawn. She went over to the photographers: 'Keep your cameras trained; you may see the biggest walk-out of all time.'

Robert Fellowes, the Queen's Assistant Private Secretary, who always looks worried even when things are going well, studied the programme as the Queen tapped her foot. It seemed indecent to watch all the stress signals. Boiling with indignation, the men from the Palace were impotent. And still the King was reluctant to abandon his cool caravan. He did dart in and out once or twice, and the Queen, always correct, stood up each time he appeared; on one of these forays, this left her in an uncharacteristic pose, standing feet apart with her thumbs in the belt of her dress as she watched his disappearing back. The Moroccans explained that, of course. the King was supervising the food as he wanted everything to be perfect for the Queen. The Queen made it clear at this point that she wanted to leave. It was 3.40 pm; she was hungry, tired and hot. This posed a delicate problem. Not even the Queen of England could get away without the King's blessing. A Moroccan chauffeur would not drive even an English monarch into Marrakesh against the King's wishes. Even a Morrocan pilot could not suddenly take off from Marrakesh and get clearance for an unscheduled flight to London, and how much more difficult would this be for a British Caledonian crew?

So, the Queen stayed put, but asked her Secretary to let Hassan know in his caravan that she did not want to miss her appointment in Marrakesh with the ninety-six-year-old Field-Marshal Sir Claude Auchinleck. The King waved aside such notions. At 4.00 pm the Queen was offered some tea from a copper tray by a bowing servant and was presented with four Arab horses. Morrow, *The Queen*

It emerged afterwards that the Queen's ordeal had been caused by the requirements of security (a 'flexible' schedule), not intentional rudeness.

PRINCE PHILIP AT CAMBRIDGE

Specialist audiences do not welcome intrusion by the inspired amateur on their preserves. After Prince Philip had visited a Cambridge college as Chancellor of the University, a senior academic observed: 'He was wonderful with the kitchen staff, quite good with the undergraduates, lamentable with the dons.' Rose, *Kings, Queens and Courtiers*

PRINCE PHILIP IN BRAZIL

The bantering tone sometimes hurts unintentionally. He asked a Brazilian admiral if he had won his dazzling display of medals on the artificial lake of his country's capital, Brazilia. 'Yes, Sir,' the victim replied, 'not by marriage.' Ibid.

THE QUEEN AND THE DUKE IN ROUGH WEATHER

Crosland, as foreign secretary, and his wife Susan, were with the Queen on the royal yacht.

When we foregathered in the drawing-room before lunch complexions were better than the evening before. 'I have *never* seen so many grey and grim faces round a dinner table,' said the Queen. She paused. 'Philip was not at all well.' She paused. 'I'm glad to say.' She giggled. I'd forgotten her Consort is an Admiral of the Fleet. Susan Crosland, *Tony Crosland* (1982)

PRINCE ANDREW

7 August 1959: The Queen is to have another baby in January or February. What a sentimental hold the monarchy has over the middle classes! all the solicitors, actors and publishers at the Garrick were beaming as if they had acquired some personal benefit. Nicolson, *Diaries*

THE ATTACK ON PRINCESS ANNE, 1974

A madman tried to kidnap her from her car in the Mall and fired six shots, wounding her bodyguard and others. 'If the man had succeeded in abducting Anne, she'd have given him the hell of a time while in captivity,' remarked her father proudly. Philip Ziegler, *Elizabeth's Britain* (1986)

THE HIPPOMANE

Princess Anne has two loves: children (she is President of the Save the Children Fund) and horses. Her 'hippomania' has brought distinction to herself and her country in winning the Individual European Three-Day Event (1971) and Combined Championship (Hickstead 1973). She was a member of the British team at the Olympic Games in Montreal (1976):

'When I appear in public,' she said, 'people expect me to neigh, grind my teeth, paw the ground and swish my tail.' They were not far wrong. At a dinner party, the story went, she talked to one of her neighbours about horses throughout the entire meal, utterly ignoring the other. At last she turned: 'Could I have the sugar please?' The slighted young man placed two lumps on his palm and held them out to the hippomane.

Rose, *Kings, Queens and Courtiers*

She herself told this story at a Royal Academy dinner. She said people often handed the sugar to her on the palm of their hand.

PRINCE CHARLES AT THIRTY, 1978

The chores, frustrations and sheer tedium of his life were beginning to outweigh its considerable perks. Shortly before his birthday, a Qantas air hostess had settled into conversation with him and, instead of trotting out the usual star-struck platitudes, had said: 'God, what a rotten, boring job you've got!' When he told the story to the Callaghan Cabinet, at a private dinner in his honour at Chequers, they all laughed politely. 'But no,' wailed the Prince in desperation, 'you don't understand what I mean. She was right!' Anthony Holden, *Their Royal Highnesses* (1981)

THE TRAINING OF A FUTURE MONARCH

The Queen has always been aware of a difficult problem: the danger of what she calls 'the Edward VII situation'—a state of remoteness from the centre of government imposed on the Prince of Wales. To avert this danger, Prince Charles has seen and discussed state papers and sat in on some Cabinet meetings.

Although all this is better than nothing, the Prince remains very conscious of the limitations of his position, of the fact that there is really no way in which he can ever become fully involved in the machinery of government. Nothing that he might say could ever alter the course of events. A future Prince of Wales, he muses, should somehow become more closely concerned, and at an earlier age, with the monarch's day-to-day work. 'I hope that somebody reminds me about this in about twenty years' time,' he adds with a characteristically wry smile.

Theo Aronson, *Royal Family: Years of Transition* (1983)

A ROYAL WEDDING

The marriage of the Prince of Wales to Lady Diana Spencer took place on 29 July 1981.

Two astrologers, one American and one British, had advised that the marriage should not after all take place.

Much more in the spirit of things was the suggestion of Mr George Foulkes, MP for South Ayrshire, who urged that Lady Diana should support British industry on her wedding day by wearing jeans up the aisle of St Paul's; . . . Mr Foulkes later conceded that there were four denim-making factories in his constituency. Holden, *Their Royal Highnesses*

FACTS FACED AT THE WEDDING

It was a scene which ten years ago would have been unthinkable for the Queen in her role as Head of the Church of England: a divorced man and his ex-wife [Earl Spencer and Mrs Shand Kydd] sitting side by side, and then walking arm in arm after the Archbishop of Canterbury to sign the register. Moreover Princess Margaret's divorced husband, Lord Snowdon, was sitting in Row A with his second wife beside him. But when the statistics indicate that one in three marriages ends in divorce, even the Royal Family has had to change its attitude.

Penny Junor, *Diana Princess of Wales* (1982)

MASTER OF THE QUEEN'S MORALITY

Lyon, King of Arms, who vetted guests to the Queen's Scottish palace, was at one time Sir Thomas Innes.

He frowned upon divorce. When a peccant Scottish nobleman pleaded that he should not be excluded from the Queen's presence as he had been remarried in church, Lyon's reply was magisterial: 'That may well admit him to the Kingdom of Heaven but it will noo get him through the gates of the Palace of Holyroodhouse.' Rose, *Kings, Queens and Courtiers*

COSTS

The fact that St Paul's was nearly three times further from Buckingham Palace than the Abbey would involve extra security arrangements and transport which would more than double the cost. At a time when nearly three million people were unemployed and the British economy was at an all-time low, the Queen and her councillors were understandably nervous.

Prince Charles was apparently urged 'to think again, if for no other reason than that we are worried that we will not have enough soldiers to line the route properly', to which Charles replied caustically, 'Well, stand them further apart.' Junor, *Diana Princess of Wales*

Sayings

QUEEN ELIZABETH II

On the royal limousines waiting for her at the airport after her fathers death: 'Oh, they've sent those hearses.'

On the rigours of the Coronation: 'I'll be all right. I'm as strong as a horse.'

On being Queen of Canada: 'I am Queen of *all* Canadians, not just of one or two ancestral strains.'

On not wearing a crown at the hundredth birthday party of the musician Sir Robert Mayer: 'I thought it was Sir Robert's night, not mine.'

PRINCE PHILIP

On his children: 'I've always tried to help them master at least one thing because as soon as a child feels self-confidence in one area, it spills over into all the others.'

On the vanished monarchies of Europe: 'Most of the monarchies in Europe were really destroyed by their greatest and most ardent supporters.'

On the journalists perched on the Rock of Gibraltar: 'Which are the monkeys?'

On the future of the Monarchy: 'If . . . people feel it has no further part to play, then for goodness' sake let's end the thing on amicable terms without having a row about it.'

On the Press: 'You must sometimes stretch out your neck, but not actually give them the axe.'

On fidelity in marriage: 'Only a moral imperative can persuade husbands and wives to be faithful to each other.'

PRINCE CHARLES

On being prayed for in church on Sundays at prep school: 'I wish they prayed for the other boys too.'

On becoming aware of his inheritance: 'I didn't wake up in my pram one day and say "Yippee . . ." you know. But I think it just dawns on you, slowly, that people are interested in one . . .'

On learning his royal trade: 'I learnt the way a monkey learns—by watching its parents.'

On his grandmother, the Queen Mother: 'I can only admit from the very start that I am hopelessly biased and completely partisan.'

On his marriage: 'Diana will certainly help keep me young.'

On being present at the birth of his elder son Prince William: 'It is rather a grown up thing, I found. Rather a shock to my system.'

On his envy of Prince Andrew in the Falklands: '. . . I never had that chance to test myself. It's terribly important to see how you react, to be tested.'

On the rejected architectural plan for the addition to the National Gallery: 'Its like a monstrous carbuncle on the face of a beloved friend.'

To a group of journalists known as Charles-watchers: 'A happy new year to you, but a particularly nasty one to your editors.'

On his travels around Britain and abroad: 'I work bloody hard right now and will continue to.'

On the opinion of the world about him: 'If people think me square, then I am happy to be thought square.'

On marriage: 'Whatever your place in life, you are forming a partnership which you hope will last fifty years.'

PRINCESS DIANA

To her flatmates: 'Please telephone me. I'm going to need you.'

On being asked by a TV interviewer whether she was in love: 'Of course.'

PRINCESS ANNE

On sailing: 'It gives me an utterly detached sensation that I have only otherwise experienced on a galloping horse . . . testing your skill against Nature, your ideals and the person you would like to be.'

When asked on television why she was 'unco-operative with journalists': 'I don't do stunts. I don't go for them anyway. Why should I do it to please their editors?'

On her children and the royal 'system': 'I doubt if the next generation will be involved at all.'

PRINCESS MARGARET

On friendship: 'My friends are old friends.'

On the Press: 'I've been misreported and misrepresented since the age of seventeen and I gave up long ago reading about myself.'

On her family: 'My children are not royal. They just happen to have the Queen for an aunt.'

'THE FROG PRINCE'

Queen Elizabeth II celebrated her sixtieth birthday on 26 April 1986. The Observer *produced a symposium on the Queen, in which Katherine Whitehorn wrote the first essay, entitled 'Queen of Hearts', pointing out that she was certainly the object of every tourist's interest. This poses a paradox:*

When asked why Americans should be so besotted with royalty after all their efforts to get rid of George III, one historian explained it neatly: 'It's the fairy stories that keep it going,' he said. 'Whoever heard of a girl kissing a frog and it turning into a handsome senator?' *The Queen Observed*

Epilogue

THIS has been the story of many dynasties and one royal line. That line goes back a thousand years, yet it has shown infinite variety rather than recognizable family traits. Indeed it seems to cover the whole human spectrum, though in heightened or exaggerated form because of the royal ambience. The Plantagenets may have been strapping while the Windsors have had a gift for looking younger than their years, but far more remarkable are the differences between these kings and queens, over fifty in number. There are murderers and martyrs; soldiers and sailors; scholars, versifiers, bards; scallywags, imbeciles, rakes; lawgivers, law-breakers, saints. They have come from all parts of the United Kingdom, Scotland, Wales, and England north and south, and from many European countries—France, Spain, Germany. And in a sense they have been representative of their people, becoming in their own persons more civilized and less savage as the centuries pass but also less starkly glittering and definitely more ordinary to write about. The hushed abdication broadcast from Windsor Castle has replaced the crunch of the axe on Tower Green. Even anthologists can have no regrets.

Acknowledgements

EXTRACTS from the Royal Archives are reproduced by kind permission of The Librarian (Oliver Everett), Windsor Castle.

The editor and publisher gratefully acknowledge permission to include copyright material in this volume as follows:

from *Adam of Eynsham, Magna Vita Sancti Hugonis*, The Life of St Hugh of Lincoln, 2 vols., ed. by D. L. Douie and D. H. Farmer (1961; corrected reprint with new preface 1985). Reprinted by permission of Oxford University Press.

W. S. Adams: from *Edwardian Portraits* (Secker & Warburg, 1956).

from *Alfred the Great: Asser's Life of King Alfred and Other Contemporary Sources*, trans. by Simon Keynes and Michael Lapidge (Penguin Classics, 1983), © Simon Keynes and Michael Lapidge, 1983. Reprinted by permission of Penguin Books Ltd.

from *The Anglica Historia of Polydore Vergil*, ed. and trans. by Denys Hay (Third Camden Series, 1950). Reprinted by permission of Boydell & Brewer Ltd.

from *The Anglo-Saxon Chronicle*, ed. & trans. by G. N. Garmonsway (Everyman's Library, 1953). Reprinted by permission of J. M. Dent & Sons Ltd.

Janet Arnold: from *Lost From Her Majesties Back* (The Costume Society, 1980). Used with permission.

Theo Aronson: from *Royal Family: Years of Transition* (1983). Reprinted by permission of John Murray (Publishers) Ltd.; and from *The Royal Ambassadors* (1976). Reprinted by permission of David Philip Publisher (Pty) Ltd.

Maurice Ashley: from *Charles II* (1971). Reprinted by permission of Weidenfeld & Nicolson Ltd; and from *James the Second* (Dent, 1971). Reprinted by permission of the publisher.

Stanley Ayling: from *George III* (1972). Reprinted by permission of Collins Publishers.

Geoffrey le Baker, *see Chronicles of the Age of Chivalry*.

Richard Barber: from *King Arthur: Hero and Legend* (1986). Reprinted by permission of Boydell & Brewer Ltd.

F. Barlow: from *Thomas À Becket* (1986). Reprinted by permission of Weidenfeld & Nicolson Ltd.

Georgina Battiscombe: from *Queen Alexandra* (1969). Reprinted by permission of Constable Publishers and A. M. Heath & Co. Ltd.

Lord Beaverbrook: from *The Abdication of King Edward VIII* (1966). Reprinted by permission of Hamish Hamilton Ltd.

from *Bede: A History of the English Church and People*, trans. by Leo Sherley-Price and revised by R. E. Latham (Penguin Classics, 1955, 1968). © Leo Sherley-Price, 1955, 1968. Reprinted by permission of Penguin Books Ltd.

Michael Bennett: from *Lambert Simnel and The Battle of Stoke* (1981). Reprinted by permission of Alan Sutton Publishing.

E. F. Benson: from *As We Were* (1985). Reprinted by permission of the estate of the author and The Hogarth Press.

Lord Birkenhead: from *Walter Monckton* (1969). Reprinted by permission of Weidenfeld & Nicolson Ltd.

John Blair: 'The Anglo-Saxon Period' in *The Oxford Illustrated History of Britain*, ed. by Kenneth O. Morgan (1984). Reprinted by permission of Oxford University Press.

Robert Blake: from *Disraeli* (Eyre & Spottiswoode, 1966). Reprinted by permission of Associated Book Publishers Ltd.

Olivia Bland: from *The Royal Way of Death* (1986). Reprinted by permission of Constable Publishers.

Michael Bloch: from *The Duke of Windsor's War*. © Michael Bloch, 1982. Reprinted by permission of Curtis Brown, London.

Wilfrid Blunt: from *Slow on the Feather: Further Autobiography* (1986). Reprinted by permission of Michael Russell (Publishing) Ltd.

Basil Boothroyd: from *Philip: An Informal Biography* (1981). Reprinted by permission of Longman Group UK Ltd.

from *Boswell's Life of Johnson*, ed. by R. W. Chapman (3rd edn., 1970). Reprinted by permission of Oxford University Press.

Flora Brennan: from *Puckler's Progress*, ed. & trans. by Flora Brennan (1987). Reprinted by permission of Collins Publishers.

John Brooke: from *King George the Third* (1972). Reprinted by permission of Constable Publishers.

from *The Brut or The Chronicles of England*, II, ed. by F. W. D. Brie (EETS, 1908). Reprinted by permission of the Early English Text Society.

Bryan and Murphy: from *The Windsor Story* (1979). Reprinted by permission of Grafton Books—a Division of the Collins Publishing Group.

from *Byron's Letters & Journals*, ed. by L. A. Marchand (1973). Reprinted by permission of John Murray (Publishers) Ltd.

Henry Cary, Viscount Falkland: quoted in H. F. Hutchison, *Edward II: The Pliant King* (Eyre & Spottiswoode, 1971).

Hugh Casson: from Hugh Casson: *Diary* (1981). Reprinted by permission of Macmillan Administration Ltd.

Helen Cathcart: from *The Queen Mother Herself* (1979). Reprinted by permission of W. H. Allen & Co.

George Cavendish: from *The Life and Death of Cardinal Wolsey*, ed. by R. S. Sylvester (EETS, 1959). Reprinted by permission of the Early English Text Society.

F. Chamberlin: from *The Sayings of Elizabeth* (The Bodley Head, 1923).

H. W. Chapman: from *The Last Tudor King* (1958). Reprinted by permission of Jonathan Cape Ltd. (Aitken & Stone).

from *Chips: Diaries of Sir Henry Channon*, ed. by Robert Rhodes James (1967). Reprinted by permission of Weidenfeld & Nicolson Ltd.

from *The Chronicle of Richard of Devizes of the Time of King Richard the First*, ed. by J. T. Appleby (1963). Reprinted by permission of Oxford University Press.

from *The Chronicle of Walter of Guisborough*, ed. H. Rothwell (Camden Society, lxxxix, 1957). Reprinted by permission of Boydell & Brewer Ltd.

from *Chronicles of the Age of Chivalry*, ed. by Dr Elizabeth Hallam (1987). Reprinted by permission of Phoebe Phillips Editions.

Winston Churchill: from *The Second World War*, Volume V: *Closing the Ring* (1952). © Houghton Mifflin Company, 1951. © renewed 1979 by Lady Sarah Audley and The Honourable Lady Soames. Reprinted by permission of Cassell & Houghton Mifflin Company.

Kenneth Clark: from *Another Part of the Wood* (1974). © Kenneth Clark, 1974.

Reprinted by permission of John Murray (Publishers) Ltd., and Harper & Row Publishers, Inc.

John Clarke: from *George II*. Reprinted by permission of Weidenfeld & Nicolson Ltd.

Mary Clive: from *This Sun of York—A Biography of Edward IV* (1973). Reprinted by permission of Macmillan Administration Ltd.

Diana Cooper: from *Autobiography* (1979). Reprinted by permission of Michael Russell (Publishing) Ltd.

from *The Correspondence of George, Prince of Wales*, iii, ed. by A. Aspinall (Cassell, 1965).

Marion Crawford: from *The Little Princesses* (Cassell, 1950).

from *The Croker Papers 1808–1857*, ed. by Bernard Pool (B. T. Batsford Ltd., 1967).

Susan Crosland: from *Tony Crosland* (1982). Reprinted by permission of Jonathan Cape Ltd. on behalf of the author.

Richard Crossman: from *Diaries of a Cabinet Minister*, i (1975). © R. H. S. Crossman, 1975. Reprinted by permission of Hamish Hamilton.

from *The Crowland Chronicle Continuations 1459–1486*, ed. by Nicholas Pronay and John Cox. © 1986 Nicholas Pronay and John Cox. Reprinted by permission of Richard III & Yorkist History Trust.

Tom Cullen: from *The Empress Brown*. © Tom Cullen, 1969. Reprinted by permission of Houghton Mifflin Company.

Gila Curtis: from *The Life and Times of Queen Anne* (1972). Reprinted by permission of Weidenfeld & Nicolson Ltd.

G. P. Cuttino and Thomas W. Lyman: from 'Where is Edward II?', *Speculum*, 53 (1978). Reprinted by permission of The Medieval Academy of America.

from *Dear and Honoured Lady: The Correspondence between Queen Victoria and Alfred Tennyson*, ed. by Hope Dyson and Charles Tennyson (1969). Reprinted by permission of Macmillan Administration Ltd.

from *Dearest Child: Letters Between Queen Victoria and the Princess Royal*, ed. by Roger Fulford (Evans Bros., 1964). Reprinted by permission of Unwin Hyman Ltd.

from *The Diaries of Frances Stevenson*, ed. by A. J .P. Taylor (Century Hutchinson, 1971). Reprinted by permission of David Higham Associates Ltd.

from *The Diary of Beatrice Webb*, ed. by N. and J. Mackenzie (Virago & LSE, 1985). Reprinted by permission of Virago Press Ltd.

from *The Diary of John Evelyn*, ed. by John Bowle (1983). Reprinted by permission of Oxford University Press.

from *The Diary of Samuel Pepys*, vols. i, vi, vii, ix, edited by R. Latham and W. Matthews. © The Master, Fellows, and Scholars of Magdalene College, Cambridge, Robert Latham, and Mrs William Matthews. Reprinted by permission of Unwin Hyman and the University of California Press.

Frances Donaldson: from *Edward VIII* (1974). Reprinted by permission of Weidenfeld & Nicolson Ltd.

D. C. Douglas: from *William the Conqueror* (Eyre & Spottiswoode, 1964).

from *Eadmer's History of Recent Events in England*, ed. by G. Bosanquet (Cresset Press, 1964).

from *The Ecclesiastical History of Orderic Vitalis*, vol. ii, ed. & trans. by Marjorie Chibnall (1968). Reprinted by permission of Oxford University Press.

B. Emerson: from *The Black Prince* (Weidenfeld & Nicolson, 1976).

Carolly Erickson: from *The First Elizabeth* (1983). © Carolly Erickson, 1983. Reprinted by permission of Summit Books, a division of Simon & Schuster, inc.

ACKNOWLEDGEMENTS

Antonia Fraser: from *King Charles II* (1979), and from *Mary Queen of Scots* (1969). Reprinted by permission of Weidenfeld & Nicolson Ltd.

from *Froissart's Chronicles*, trans. John Joliffe (Collins, 1968). Reprinted by permission of A. D. Peters & Co. Ltd.

Roger Fulford: from *Your Dear Letter* (1971). Reprinted by permission of Unwin Hyman Ltd.

from *Gesta Henrici Quinti*, ed. by F. Taylor and J. S. Roskell (1975). Reprinted by permission of Oxford University Press.

from *Gesta Stephani*, ed. & trans. by K. R. Potter (Oxford Medieval Texts). © Oxford University Press, 1976. Reprinted by permission of Oxford University Press.

John Gillingham: from *Richard the Lionheart* (1978), and *The Wars of the Roses* (1981). Reprinted by permission of Weidenfeld & Nicolson Ltd.

from *The Girlhood of Queen Victoria*, ed. by Lord Esher. Reprinted by permission of John Murray (Publishers) Ltd.

Victoria Glendinning: from *Edith Sitwell: A Lion Among Unicorns*. © Victoria Glendinning, 1981. Reprinted by permission of Weidenfeld & Nicolson Ltd., and Alfred A. Knopf, Inc.

from John Gore: *King George V: A Personal Memoir* (1941). Reprinted by permission of John Murrray (Publishers) Ltd.

from *The Great Red Book of Bristol*, ed. by E. W. W. Veale, *Bristol Record Society*, xviii (1953). Used by permission.

Edward Gregg: from *Queen Anne* (1980). Reprinted by permission of Routledge & Kegan Paul Ltd.

John Guy: 'The Tudors' in *The Oxford Illustrated History of Britain*, ed. by Kenneth O. Morgan (1984). Reprinted by permission of Oxford University Press.

Peter Hall: from *Peter Hall's Diaries* ed. by John Goodwin. © Petard Productions Ltd., 1983. Reprinted by permission of Hamish Hamilton Ltd.

P. W. Hammond: letter to *The Times* and extract from *Richard III: Loyalty, Lordship and Law*, ed. by P. W. Hammond (1986). Reprinted by permission of the author.

from *Richard III: The Road to Bosworth Field* (1985), ed. by P. W. Hammond and A. F. Sutton. Reprinted by permission of Constable Publishers.

Lord Harewood: from *The Tongs and the Bones: The Memoirs of Lord Harewood* (1981). Reprinted by permission of Weidenfeld & Nicolson Ltd.

Tony Harman: from *Seventy Summers*. Reprinted by permission of BBC Enterprises Ltd.

Kenneth Harris: from *Attlee* (1982). Reprinted by permission of Weidenfeld & Nicolson Ltd., and W. W. Norton & Co., Inc.

John Harvey: from *The Black Prince and His Age* (1976). Reprinted by permission of B. T. Batsford Ltd.

M. A. Hemmings: from *England Under Henry III* (Longman, 1924). Used with permission.

Simon Hoggart: 'Caribbean Queen' from *The Queen Observed*, ed. by Trevor Grove (Pavilion Books, 1986). Reprinted by permission of the author.

Anthony Holden: from *Their Royal Highnesses* (1981). Reprinted by permission of Weidenfeld & Nicolson Ltd. and A. P. Watt Ltd., on behalf of Anthony Holden.

Henry of Huntingdon: from *History of the English*, newly trans. in *The Plantagenet Chronicles*, ed. by Dr Elizabeth Hallam (1986). Reprinted by permission of Phoebe Phillips Editions.

ACKNOWLEDGEMENTS

Christopher Hibbert: from *The Court of St. James* (1979). Reprinted by permission of Weidenfeld & Nicolson Ltd.; from *Edward VII* (Allen Lane, 1976). Reprinted by permission of David Higham Associates Ltd.; from *George IV* (Penguin Books, 1976). © Christopher Hibbert, 1972, 1973. Reprinted by permission of Penguin Books Ltd.

Thea Holme: from *Prinny's Daughter* (1976). Reprinted by permission of Hamish Hamilton Ltd.

Lucy Hutchinson: from *Memoirs of the Life of Colonel Hutchinson*, ed. by James Sutherland (1973). Reprinted by permission of Oxford University Press.

Paul Johnson: from *Elizabeth I* (Weidenfeld & Nicolson, 1974).

from *The Journals of Mrs Arbuthnot*, ed. by F. Bamford and the Duke of Wellington (Macmillan, 1950).

Denis Judd: from *King George the Sixth* (1982). Reprinted by permission of Michael Joseph Ltd., and David Higham Associates Ltd.

Penny Junor: from *Diana, Princess of Wales* (1982). © Penny Junor, 1982. Reprinted by permission of Doubleday, a division of Bantam, Doubleday, Dell Publishing Group, Inc., and Sidgwick & Jackson Ltd.

from *Kilvert's Diary*, ed. by William Plomer (1971). Reprinted by permission of Jonathan Cape Ltd., on behalf of Mrs Sheila Hooper.

from *A King's Story: The Memoirs of HRH The Duke of Windsor* (Cassell, 1951).

John van der Kiste: from *Queen Victoria's Children* (1986). Reprinted by permission of Alan Sutton Publishing Ltd.

Robert Lacey: from *Majesty*. © Robert Lacey, 1977. Reprinted by permission of Harcourt Brace Jovanovich and Curtis Brown Ltd., London.

James Lees-Milne: from *The Enigmatic Edwardian: Life of Reginald Brett, Viscount Esher* (1986). Reprinted by permission of Sidgwick & Jackson Ltd.

from *The Letters of Queen Victoria*, ed. by Lord Esher and A. C. Benson (1907). Reprinted by permission of John Murray (Publishers) Ltd.

from *The Life of King Edward the Confessor*, ed. by F. Barlow (1962). Reprinted by permission of Oxford University Press.

D. M. Loades: from *The Reign of Mary Tudor* (Ernest Benn, 1979). Reprinted by permission of the author.

Roger Lockyer: in *For Veronica Wedgwood*, ed. by Ollard and Tudor Craig (1986). Reprinted by permission of Collins Publishers.

Elizabeth Longford: from *Elizabeth R* (1983), and from *The Queen Mother* (1981). Reprinted by permission of Weidenfeld & Nicolson Ltd.; from *Louisa, Lady-in-Waiting* (1979). Reprinted by permission of Roxby & Lindsey Press.; from *Victoria R. I.* © Elizabeth Longford, 1964. Reprinted by permission of Curtis Brown Ltd., London and Weidenfeld & Nicolson Ltd.

from *Lord Eldon's Anecdote Book*, ed. by A. L. J. Lincoln and R. L. McEwen (Stevens & Sons, 1960). Reprinted by permission of Sweet & Maxwell Ltd.

Susan Lowndes: from *Diaries and Letters of Marie Belloc-Lowndes* (1971). Reprinted by permission of Chatto & Windus for the author.

Mary Lutyens: from *Edwin Lutyens* (1980). Reprinted by permission of John Murray (Publishers) Ltd.

from *The Lyttelton/Hart-Davis Letters*, vol. iv, ed. by Sir Rupert Hart-Davis. Reprinted by permission of John Murray Publishers Ltd.

Mabell Countess of Airlie: from *Thatched with Gold*, ed. by Jennifer Ellis (Century Hutchinson, 1962). Reprinted by permission of the Earl of Airlie.

ACKNOWLEDGEMENTS

Harold Macmillan: from *Pointing the Way* (1972). Reprinted by permission of Macmillan Administration Ltd.

Philip Magnus: from *King Edward the Seventh* (1964), and from *Gladstone* (1954). Reprinted by permission of John Murray (Publishers) Ltd.

Dominic Mancini: from *The Usurpation of Richard III*, ed. and trans. by C. A. J. Armstrong (1936). Reprinted by permission of Oxford University Press.

Walter Map: from *Courtier's Trifles*, ed. and trans. by M. R. James, rev. C. N. L. Brooke and R. A. B. Mynors (1983). © Sir Roger Mynors, 1983. Reprinted by permission of Oxford University Press.

Princess Marie-Louise: from *My Memories of Six Reigns* (Evans, 1956). Reprinted by permission of Unwin Hyman Ltd.

Richard Marius: from *Thomas More: A Biography* (1984). © Richard Marius, 1984. Reprinted by permission of J. M. Dent & Sons Ltd., and Alfred A. Knopf, Inc.

Joanot Martorell and Marti Joan de Galba: from *Tirant Lo Blanc*, trans. D. H. Rosenthal (1984). Reprinted by permission of Macmillan Administration Ltd., and Pantheon Books, a division of Random House, Inc.

D. Mathew: from *Lady Jane Grey: The Setting of a Reign* (Eyre Methuen, 1972).

Gervase Mathew: from *The Court of Richard II* (1968). Reprinted by permission of B. T. Batsford Ltd.

from *The Memoirs of the Aga Khan* (Cassell, 1954).

from *Memoirs of William Hickey*, ed. by Peter Quennell (1960). Reprinted by permission of Century Hutchinson Publishing Group Ltd.

Suzy Menkes: from *The Windsor Style* (1987). Reprinted by permission of Grafton Books, a division of the Collins Publishing Group.

R. Menzies: from *Afternoon Light: Some Memories of Men and Events* (Cassell, 1967).

Keith Middlemas: from *King George VI* (1969). Reprinted by permission of Weidenfeld & Nicolson Ltd.

Keith Middlemas and Dr John Barnes: from *Stanley Baldwin* (1969). Reprinted by permission of Weidenfeld & Nicolson Ltd.

A. A. Mitchell: 'Charles the First in Death' from *History Today*, (1966).

J. V. Mitchell: letter to the *Daily Telegraph*, 20 September 1987.

St Thomas More: from *The History of King Richard III and selections from the English and Latin Poems*, ed. by Richard S. Sylvester (1976). Reprinted by permission of Yale University Press.

Ann Morrow: from *The Queen Mother* (1985), and from *The Queen* (1983). Reprinted by permission of Grafton Books, a division of the Collins Publishing Group.

J. E. Neale: from *Queen Elizabeth* (1934). Reprinted by permission of Jonathan Cape Ltd. on behalf of the Estate of J. E. Neale.

Harold Nicolson: from *Diaries and Letters 1945–1962*, ed. by Nigel Nicolson (1968). Reprinted by permission of Collins Publishers; from *King George V: His Life and Reign* (1952). Reprinted by permission of Constable Publishers.

Carola Oman: from *An Oxford Childhood* (Hodder & Stoughton, 1976). Reprinted by permission of Curtis Brown.

S. Painter: from *William Marshal: Knight Errant, Baron & Regent of England* (1933). Reprinted by permission of Johns Hopkins University Press.

from *The Paston Letters and Papers of the Fifteenth Century*, ed. by Norman Davis, 2 vols. (1971–6). Reprinted by permission of Oxford University Press.

from *A Persian at the Court of King George 1809–1810*, ed. and trans. by Margaret Morris Cloake (Barrie & Jenkins, 1988).

ACKNOWLEDGEMENTS

J. H. Plumb: from *Sir Robert Walpole*, vol. i (Penguin Books, 1972). © J. H. Plumb, 1956. Reprinted by permission of Augustus M. Kelley, Publishers. Reprinted by permission of Penguin Books Ltd. (Augustus M. Kelley, Publishers, USA).

Arthur Ponsonby: from *Henry Ponsonby: His Life From His Letters* (1942). Reprinted by permission of Macmillan, London & Basingstoke.

Frederick Ponsonby: from *Recollections of Three Reigns* (Eyre & Spottiswoode, 1951); from *Sidelights on Queen Victoria* (1930). Reprinted by permission of Macmillan, London & Basingstoke.

James Pope-Hennessy: from *A Lonely Business: A Self-Portrait of James Pope-Hennessy*, ed. by Peter Quennell. The writings of James Pope-Hennessy, © John Pope-Hennessy, 1981. Selection and editorial matter, © Peter Quennell, 1981. Reprinted by permission of Curtis Brown, London.

M. Prestwich: 'The Art of Kingship: Edward I, 1272–1307', *History Today*, May 1985.

D. B. Quinn: in *A New History of Ireland*, vol. ii, ed. by Art Cosgrove (1987). Reprinted by permission of Oxford University Press.

Gwen Raverat: from *Period Piece: A Cambridge Childhood* (1952). Reprinted by permission of Faber & Faber Ltd.

Conyers Read: from *Mr Secretary Cecil and Queen Elizabeth*. Published 1955 by Alfred A. Knopf, Inc. All rights reserved. Reprinted by permission of Jonathan Cape Ltd., and Alfred A. Knopf, Inc.

Michaela Reid: from *Ask Sir James* (Hodder & Stoughton Ltd., 1987). The quotations from the Reid papers are by permission of Hodder & Stoughton Ltd., on behalf of Sir Alexander Reid, copyright holder.

from *Reminiscences of Captain Gronow*, ed. by John Raymond (1964). © The Bodley Head Ltd., 1964. All rights reserved. Reprinted by permission of The Bodley Head and Viking Penguin Inc.

Joanna Richardson: from *George IV: A Portrait* (1966). Reprinted by permission of Sidgwick & Jackson Ltd., and the author; and from *Portrait of a Bonaparte* (1987). Reprinted by permission of Quartet Books Ltd.

Jasper Ridley: from Henry VIII (1984). © Jasper Ridley, 1984, 1985. All rights reserved. Reprinted by permission of Constable Publishers and Viking Penguin, Inc.

Jasper Ridley: from *The Life and Times of Mary Tudor* (Weidenfeld & Nicolson, 1974).

William Rishanger, *see Chronicles of the Age of Chivalry*.

Kenneth Rose: from *George V* (1984), and from *Kings, Queens and Courtiers* (1986). Reprinted by permission of Weidenfeld & Nicolson Ltd.

S. Runciman: from *A History of the Crusades*, vol. iii (1954). Reprinted by permission of Cambridge University Press.

Giles St Aubyn: from *Edward II* (1979). Reprinted by permission of Collins Publishers.

L. F. Salzman: from *Edward I* (1968). Reprinted by permission of Constable Publishers.

John Scarisbrick: from *Henry VIII* (Eyre & Spottiswoode, 1968). Reprinted by permission of Associated Book Publishers Ltd.

David Scott: from *Ambassador in Black and White*. Reprinted by permission of Weidenfeld & Nicolson Ltd. (Peterborough Lit. Agency).

from *Selections from Clarendon's History of the Rebellion*, ed. by G. Huehns (1978). Reprinted by permission of Oxford University Press.

from *Six Town Chronicles*, ed. by R. Flentey (1911). Reprinted by permission of Oxford University Press.

ACKNOWLEDGEMENTS

Osbert Sitwell: from *Left Hand Right Hand*, vol. i (1945). Reprinted by permission of David Higham Associates Ltd.; from *Rat Week; An Essay on the Abdication* (1986). Reprinted by permission of Michael Joseph Ltd.

Lacey Baldwin Smith: from *The Masque of Royalty* (1971). Reprinted by permission of Jonathan Cape Ltd.

B. Stone: 'Models of Kingship: Arthur in Medieval Romance', *History Today* (Nov. 1987).

Publius Cornelius Tacitus: from *The Annals of Imperial Rome*, trans. by Michael Grant (Cassell, 1963).

from *The Times: Past, Present and Future* (1985). Reprinted by permission of Times Newspapers Ltd.

H. R. Trevor-Roper: from *Archbishop Laud* (1940). Reprinted by permission of Macmillan Administration Ltd.

from *Vita Edwardi Secundi*, ed. & trans. By N. Denholm-Young (1957). Reprinted by permission of Oxford University Press.

from *Wallis and Edward: Letters 1931–1937*, ed. by Michael Bloch (1986). Reprinted by permission of Weidenfeld & Nicolson Ltd.

Marina Warner: from *Queen Victoria's Sketchbook* (1979). Reprinted by permission of Macmillan Administration Ltd.

W. L. Warren: from *Henry II* (Methuen, 1973); and from *King John* (Eyre & Spottiswoode, 1961).

Christopher Warwick: from *Princess Margaret* (1983). Reprinted by permission of Weidenfeld & Nicolson Ltd.

Francis Watson: 'The Death of George V', *History Today* (Dec. 1986). Reprinted by permission of History Today Ltd.

C. V. Wedgwood: from *The Trial of Charles I* (1964). Reprinted by permission of Collins Publishers.

from *Wellington and His Friends*, ed. by the 7th Duke of Wellington. Reprinted by permission of Macmillan Administration Ltd.

Vera Wheatley: 'Letter to a Friend' from *Harriet Martineau* (1957). Reprinted by permission of Martin Secker & Warburg Ltd.

John Wheeler-Bennett: from *King George the Sixth* (1958), and from *Special Relationships* (1975). Reprinted by permission of Macmillan Administration Ltd.

Laurence Whistler: from *The Laughter and the Urn: The Life of Rex Whistler* (1985). Reprinted by permission of Weidenfeld & Nicolson Ltd.

William White: letter to *The Times* and extract from 'The Sons of Edward IV: A Re-examination of the evidence on their deaths and on the bones in Westminster Abbey' by W. J. White and P. W. Hammond in *Richard III, Loyalty, Lordship and Law*, ed. by P. W. Hammond (1986). Reprinted by permission of the author.

N. Williams: from *Elizabeth* (Weidenfeld & Nicolson, 1967).

The Duchess of Windsor: from *The Heart Has Its Reasons* (1956). © 1956 by the Duchess of Windsor. Reprinted by permission of Michael Joseph Ltd., and David McKay Co., a division of Random House, Inc.

Cecil Woodham-Smith: from *Queen Victoria: Her Life and Times* (1972). Reprinted by permission of Hamish Hamilton Ltd.

H. and B. van der Zee: from *William and Mary* (1973). Reprinted by permission of Macmillan Administration Ltd.

Philip Ziegler: from *Crown and People*. © Philip Ziegler 1978. Reprinted by permission of Curtis Brown Ltd.; from *Mountbatten* (1985), and from *King William IV* (1971).

ACKNOWLEDGEMENTS

Reprinted by permission of Collins Publishers, and from *Elizabeth's Britain* (Country Life Books, 1986).

While every effort has been made to secure permission, we may have failed in a few cases to trace the copyright holder. We apologize for any apparent negligence.

Index

INDEX

INDEX

INDEX

INDEX

INDEX